SCHAUM'S OUTLINE OF

BASIC ELECTRICITY
Second Edition

MILTON GUSSOW, M.S.

Principal Staff Engineer (Ret.)
Applied Physics Laboratory
The Johns Hopkins University

SCHAUM'S OUTLINE SERIES

New York   Chicago   San Francisco   Lisbon   London   Madrid
Mexico City   Milan   New Delhi   San Juan   Seoul
Singapore   Sydney   Toronto

To Libbie, Myra, Susan, Edward, Marc, Nicole, Sara, Laura, and Jeff

MILTON GUSSOW was a principal staff engineer at The Johns Hopkins University Applied Physics Laboratory. He received his B.S. degree with distinction from the U.S. Naval Academy, his B.S.E.E. from the U.S. Navy Postgraduate School, and his M.S. from the Massachusetts Institute of Technology. He has been an adjunct professor at The George Washington University, American University, and Johns Hopkins University, where he taught undergraduate and graduate courses in mathematics and electrical engineering. He was formerly Senior Vice President for Education at the McGraw-Hill Continuing Education Center, and he is the author of over seventy technical papers. He is a life member of IEEE and a full emeritus member of Sigma Xi, the Scientific Research Society. Gussow currently is an engineering consultant to the U.S. defense industry.

The McGraw·Hill Companies

Schaum's Outline of
BASIC ELECTRICITY

1 2 3 4 5 6 7 8 9 10 11 12 13 14 15 CUS/CUS 0 1 3 2 1 0 9 8 7 6

ISBN-13: 978-0-07-147498-6
ISBN-10: 0-07-147498-6

Preface

Basic Electricity, now in its second edition, is intended as a basic text or as a supplement to standard texts to cover the fundamentals of electricity and electric circuits. It may be used by beginning students in high schools, technical institutes, and colleges who have limited or no experience in electricity or electric circuits. It may also be used by other technology and engineering students as an introduction to basic electrical engineering.

The subject matter is divided into 22 chapters. Each chapter starts with pertinent definitions, principles, theorems, and equations with many illustrative examples to reinforce the student's learning. This is followed by sets of solved and supplementary problems, numbering over 1,150. Answers are provided with each supplementary problem.

Explanations and step-by-step solutions are deliberately detailed so that the text can stand alone. Thus it also can be used as a distance-study or reference book. Another use is as a refresher or remedial text for those students who have already had a course in basic electric circuits. Knowledge of basic algebra and trigonometry is assumed. Designed to provide a broad yet deep background in the nature of electricity and the operation and application of electric circuits, the text uses numerous and easy-to-follow solved problems accompanied by many circuit diagrams. Starting with the physics of electron current, the book describes and analyzes direct-current and alternating-current circuits, dc and ac generators and motors, single-phase as well as three-phase circuits, and transformers. To assure correlation to modern practice and design, illustrative problems are presented in terms of commonly used voltage and current ratings, covering circuits and equipments typical of those found in today's electrical systems.

There are several special features of this book. One is the prolific use of equation numbers for reference so that the reader will always know the source of each equation used. Another is the step-by-step procedure so that students can follow very easily the solution to a problem. Other features include simplified ways to solve problems on three-phase transformer windings, series and parallel resonance, and *RL* and *RC* circuit waveforms. A unique feature added in the second edition is explaining in simple terms and applying the fundamental tools of determinant solution techniques and complex numbers. With these new mathematical skills, the students will be able to analyze not only electric circuit problems more easily, but also more complex electric circuits. Further, these added skills will allow them to transition more smoothly to the study of advanced electric circuits.

I wish to thank Barbara Gilson of McGraw-Hill for her continuing efforts to get the second edition published. I also am indebted to the following individuals for their review and constructive criticism: Professor Robert S. Weissbach, Program Chair, Electrical Engineering Technology, Penn State Erie; Professor William T. Smith, Department of Electrical and Computer Engineering, University of Kentucky; Professor Robert Kahal, formerly Professor of Electrical Engineering, Syracuse University; and to Professor Howard S. Feldmesser of The Johns Hopkins University.

MILTON GUSSOW

Contents

Chapter 1

The Nature of Electricity

STRUCTURE OF THE ATOM

Matter is anything that has mass and occupies space. Matter is composed of very small particles called *atoms*. All matter can be classified into either one of two groups: *elements* or *compounds*. In an element, all the atoms are the same. Examples of elements are aluminum, copper, carbon, germanium, and silicon. A compound is a combination of elements. Water, for example, is a compound consisting of the elements hydrogen and oxygen. The smallest particle of any compound that retains the original characteristics of that compound is called a *molecule*.

Atoms are composed of subatomic particles of *electrons*, *protons*, and *neutrons* in various combinations. The electron is the fundamental negative (−) charge of electricity. Electrons revolve about the nucleus or center of the atom in paths of concentric "shells," or *orbits* (Fig. 1-1). The proton is the fundamental positive (+) charge of electricity. Protons are found in the nucleus. The number of protons within the nucleus of any particular atom specifies the atomic number of that atom. For example, the silicon atom has 14 protons in its nucleus so the atomic number of silicon is 14. The neutron, which is the fundamental neutral charge of electricity, is also found in the nucleus.

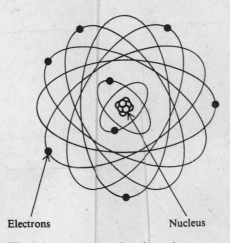

Electrons   Nucleus

Fig. 1-1   Electrons and nucleus of an atom

Atoms of different elements differ from one another in the number of electrons and protons they contain (Fig. 1-2). In its natural state, an atom of any element contains an equal number of electrons and protons. Since the negative (−) charge of each electron is equal in magnitude to the positive (+) charge of each proton, the two opposite charges cancel. An atom in this condition is electrically neutral, or in balance (Fig. 1-2).

Example 1.1   Describe the two simplest atoms.
The simplest atom is the hydrogen atom, which contains 1 proton in its nucleus balanced by 1 electron orbiting the nucleus (Fig. 1-2a). The next simplest atom is helium, which has 2 protons in its nucleus balanced by 2 electrons orbiting the nucleus (Fig. 1-2b).

A stable (neutral) atom has a certain amount of energy, which is equal to the sum of the energies of its electrons. Electrons, in turn, have different energies called *energy levels*. The energy level of an electron is proportional to its distance from the nucleus. Therefore, the energy levels of electrons in shells farther the nucleus are higher than those of electrons in shells nearer the nucleus. The electrons in the outerm are called *valence* electrons. When external energy such as heat, light, or electric energy is applied t

1

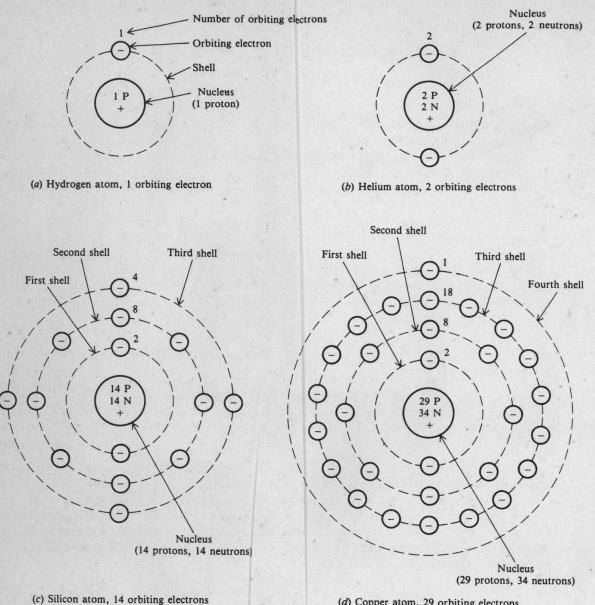

(a) Hydrogen atom, 1 orbiting electron

(b) Helium atom, 2 orbiting electrons

(c) Silicon atom, 14 orbiting electrons

(d) Copper atom, 29 orbiting electrons

Fig. 1-2   Atomic structure of four common elements

materials, the electrons gain energy. This may cause the electrons to move to a higher energy level. An atom in which this has occurred is said to be in an *excited state*. An atom in an excited state is *unstable*.

When an electron has moved to the outermost shell of its atom, it is least attracted by the positive charges of the protons within the nucleus of its atom. If enough energy is then applied to the atom, some of the outermost shell or valence electrons will leave the atom. These electrons are called *free* electrons. It is the movement of free electrons that provides electric current in a metal conductor.

Each shell of an atom can contain only a certain maximum number of electrons. This number is called the *quota* of a shell. The orbiting electrons are in successive shells designated K, L, M, N, O, P, and Q at increasing distances outward from the nucleus. Each shell has a maximum number of electrons for stability (Fig. 1-3). After the K shell has been filled with 2 electrons, the L shell can take up to 8 electrons. The maximum number

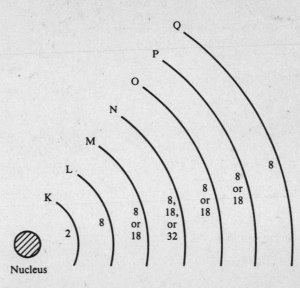

Fig. 1-3   Energy shells and the quota of electrons for each shell

of electrons in the remaining shells can be 8, 18, or 32 for different elements. The maximum for an outermost shell, though, is always 8.

Example 1.2   Structure the copper atom by identifying its energy shells (Fig. 1-2*d*).

In the copper atom there are 29 protons in the nucleus balanced by 29 orbiting electrons. The 29 electrons fill the K shell with 2 electrons and the L shell with 8 electrons. The remaining 19 electrons then fill the M shell with 18 electrons, and the net result is 1 electron in the outermost N shell.

If the quota is filled in the outermost shell of an atom, an element made up of such atoms is said to be *inert*. When the K shell is filled with 2 electrons, we have the inert gas helium (Fig. 1-2*b*). When the outer shell of an atom lacks its quota of electrons, it is capable of gaining or losing electrons. If an atom loses one or more electrons in its outer shell, the protons outnumber the electrons so that the atom carries a net positive electric charge. In this condition, the atom is called a *positive ion*. If an atom gains electrons, its net electric charge becomes negative. The atom then is called a *negative ion*. The process by which atoms either gain or lose electrons is called *ionization*.

Example 1.3   Describe what happens to the copper atom when it loses an electron from its outermost shell.

The copper atom becomes a positive ion with a net charge of +1.

THE ELECTRIC CHARGE

Since some atoms can lose electrons and other atoms can gain electrons, it is possible to cause a transfer of electrons from one object to another. When this takes place, the equal distribution of the positive and negative charges in each object no longer exists. Therefore, one object will contain an excess number of electrons and its charge must have a negative, or minus (−), electric polarity. The other object will contain an excess number of protons and its charge must have a positive, or plus (+), polarity.

When a pair of objects contains the same charge, that is, both positive (+) or both negative (−), the objects are said to have like charges. When a pair of bodies contains different charges, that is, one body is positive (+) while the other body is negative (−), they are said to have unlike or opposite charges. The law of electric charges may be stated as follows:

> Like charges repel each other; unlike charges attract each other.

If a negative (−) charge is placed next to another negative (−) charge, the charges will repel each other (Fig. 1-4a). If a positive (+) charge is placed next to a negative (−) charge, they will be drawn together (Fig. 1-4c).

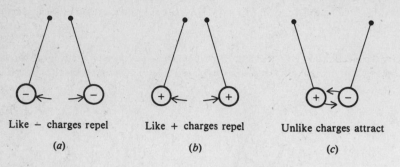

Like − charges repel   Like + charges repel   Unlike charges attract

        (a)   (b)   (c)

Fig. 1-4 Force between charges

THE COULOMB

The magnitude of electric charge a body possesses is determined by the number of electrons compared with the number of protons within the body. The symbol for the magnitude of the electric charge is Q, expressed in units of coulombs (C). A charge of one negative coulomb, $-Q$, means a body contains a charge of 6.25×10^{18} more electrons than protons.*

Example 1.4 What is the meaning of $+Q$?
A charge of one positive coulomb means a body contains a charge of 6.25×10^{18} more protons than electrons.

Example 1.5 A dielectric material has a negative charge of 12.5×10^{18} electrons. What is its charge in coulombs?
Since the number of electrons is double the charge of 1 C (1 C = 6.25×10^{18} electrons), $-Q = 2$ C.

THE ELECTROSTATIC FIELD

The fundamental characteristic of an electric charge is its ability to exert a force. This force is present within the electrostatic field surrounding every charged object. When two objects of opposite polarity are brought near each other, the electrostatic field is concentrated in the area between them (Fig. 1-5). The electric

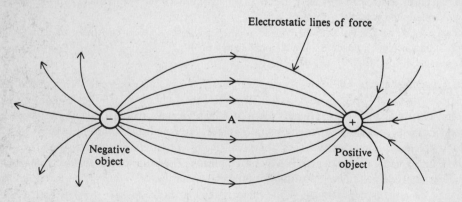

Electrostatic lines of force

Negative object   A   Positive object

Fig. 1-5 The electrostatic field between two charges of opposite polarity

*See page 16 (Chapter 2) for an explanation of how to use powers of 10.

field is indicated by lines of force drawn between the two objects. If an electron is released at point A in this field, it will be repelled by the negative charge and will be attracted to the positive one. Thus both charges will tend to move the electron in the direction of the lines of force between the two objects. The arrowheads in Fig. 1-5 indicate the direction of motion that would be taken by the electron if it were in different areas of the electrostatic field.

Example 1.6   Draw the electrostatic field that would exist between two negatively charged objects.
When two like charges are placed near each other, the lines of force repel each other as shown below.

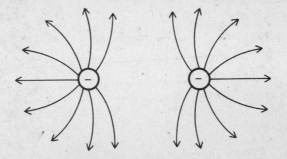

A charged object will retain its charge temporarily if there is no immediate transfer of electrons to or from it. In this condition, the charge is said to be *at rest*. Electricity at rest is called *static* electricity.

POTENTIAL DIFFERENCE

Due to the force of its electrostatic field, an electric charge has the ability to do the work of moving another charge by attraction or repulsion. The ability of a charge to do work is called its *potential*. When one charge is different from the other, there must be a difference in potential between them.

The sum of the differences of potential of all the charges in the electrostatic field is referred to as *electromotive force* (emf).

The basic unit of potential difference is the *volt* (V). The symbol for potential difference is V, indicating the ability to do the work of forcing electrons to move. Because the volt unit is used, potential difference is called *voltage*.

Example 1.7   What is the meaning of a battery voltage output of 6 V?
A voltage output of 6 V means that the potential difference between the two terminals of the battery is 6 V. Thus, voltage is fundamentally the potential difference between two points.

CURRENT

The movement or the flow of electrons is called *current*. To produce current, the electrons must be moved by a potential difference. Current is represented by the letter symbol I. The basic unit in which current is measured is the ampere (A). One ampere of current is defined as the movement of one coulomb past any point of a conductor during one second of time. Electricity can be termed as *electric current*.

Example 1.8   If a current of 2 A flows through a meter for 1 minute (min), how many coulombs pass through the meter?
1 A is 1 C per second (C/s). 2 A is 2 C/s. Since there are 60 s in 1 min, $60 \times 2\,C = 120\,C$ pass through the meter in 1 min.

The definition of current can be expressed as an equation:

$$I = \frac{Q}{T}$$

where I = current, A
    Q = charge, C
    T = time, s
or
$$Q = I \times T = IT \tag{1-2}$$

Charge differs from current in that Q is an accumulation of charge, while I measures the intensity of moving charges.

Example 1.9   Find the answer to Example 1.8 by using Eq. (*1-2*).
    Write down the known values:

$$I = 2\,\text{A} \quad T = 60\,\text{s}$$

Write down the unknown:

$$Q = ?$$

Use Eq. (*1-2*) to solve the unknown:

$$Q = I \times T$$

Substitute $I = 2$ A and $T = 60$ s:

$$Q = (2\,\text{A}) \times (60\,\text{s})$$

Solve for Q:

$$Q = 120\,\text{C} \qquad Ans.$$

CURRENT FLOW

In a conductor, such as copper wire, the free electrons are charges that can be forced to move with relative ease by a potential difference. If a potential difference is connected across two ends of a copper wire (Fig. 1-6),

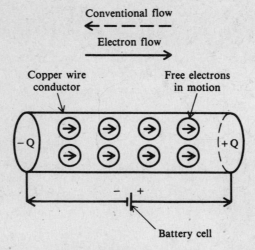

Potential difference = 1.5 V

Fig. 1-6   Potential difference across two ends of a wire conductor causes electric current

the applied voltage (1.5 V) forces the free electrons to move. This current is a drift of electrons from the point of negative charge, $-Q$, at one end of the wire, moving through the wire, and returning to the positive charge, $+Q$, at the other end. The direction of the electron drift is from the negative side of the battery, through the wire, and back to the positive side of the battery. The direction of electron flow is from a point of negative potential to a point of positive potential. The solid arrow (Fig. 1-6) indicates the direction of current in terms of electron flow. The direction of moving positive charges, opposite from electron flow, is considered the *conventional flow* of current and is indicated by the dashed arrow (Fig. 1-6). In basic electricity, circuits are usually analyzed in terms of conventional current because a positive potential is considered before a negative potential. Therefore, the direction of conventional current is the direction of positive charges in motion. Any circuit can be analyzed by either electron flow or conventional flow in the opposite direction. In this book, current is always considered as conventional flow.

SOURCES OF ELECTRICITY

Chemical Battery

A voltaic chemical cell is a combination of materials which are used for converting chemical energy into electric energy. A battery is formed when two or more cells are connected. A chemical reaction produces opposite charges on two dissimilar metals, which serve as the negative and positive terminals (Fig. 1-7). The metals are in contact with an electrolyte.

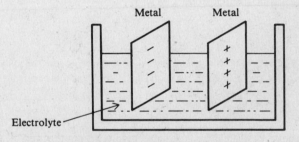

Fig. 1-7   Voltaic chemical cell

Generator

The generator is a machine in which electromagnetic inductance is used to produce a voltage by rotating coils of wire through a stationary magnetic field or by rotating a magnetic field through stationary coils of wire. Today, more than 95 percent of the world's energy is produced by generators.

Thermal Energy

The production of most electric energy begins with the formation of heat energy. Coal, oil, or natural gas can be burned to release large quantities of heat. Once heat energy is available, conversion to mechanical energy is the next step. Water is heated to produce steam, which is then used to turn the turbines that drive the electric generators. A *direct* conversion from heat energy to electric energy will increase efficiency and reduce thermal pollution of water resources and the atmosphere.

Magnetohydrodynamic (MHD) Conversion

In an MHD converter, gases are ionized by very high temperatures, approximately 3000 degrees Fahrenheit (3000°F), or 1650 degrees Celsius (1650°C). The hot gases pass through a strong magnetic field with current resulting. The exhausted gases are then moved back to the heat source to form a complete cycle (Fig. 1-8). MHD converters have no mechanical moving parts.

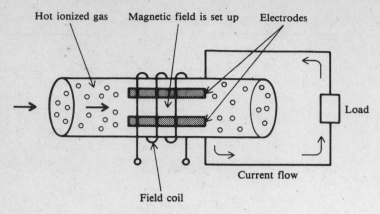

Fig. 1-8   Principles of MHD converter

Thermionic Emission

The thermionic energy converter is a device that consists of two electrodes in a vacuum. The emitter electrode is heated and produces free electrons. The collector electrode is maintained at a much lower temperature and receives the electrons released at the emitter.

Solar Cells

Solar cells convert light energy directly into electric energy. They consist of semiconductor material like silicon and are used in large arrays in spacecraft to recharge batteries. Solar cells are also used in home heating.

Piezoelectric Effect

Certain crystals, such as quartz and Rochelle salts, generate a voltage when they are vibrated mechanically. This action is known as the *piezoelectric effect*. One example is the crystal phonograph cartridge, which contains a Rochelle salt crystal to which a needle is fastened. As the needle moves in the grooves of a record, it swings from side to side. This mechanical motion is applied to the crystal, and a voltage is then generated.

Photoelectric Effect

Some materials, such as zinc, potassium, and cesium oxide, emit electrons when light strikes their surfaces. This action is known as the *photoelectric effect*. Common applications of photoelectricity are television camera tubes and photoelectric cells.

Thermocouples

If wires of two different metals, such as iron and copper, are welded together and the joint is heated, the difference in electron activity in the two metals produces an emf across the joint. Thermocouple junctions can be used to measure the amount of current because current acts to heat the junction.

DIRECT AND ALTERNATING CURRENTS AND VOLTAGES

Direct current (dc) is the current that moves through a conductor or circuit in one direction only (Fig. 1-9*a*). The reason for the unidirectional current is that voltage sources such as cells and batteries maintain the same

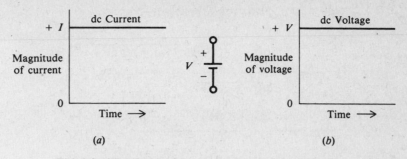

Fig. 1-9 Waveforms of a constant dc current and dc voltage

polarity of output voltage (Fig. 1-9b). The voltage supplied by these sources is called *direct-current voltage*, or simply *dc voltage*. A dc voltage source can change the amount of its output voltage, but if the same polarity is maintained, direct current will flow in one direction only.

Example 1.10   Assuming the polarity of the battery were reversed in Fig. 1-9b, draw the new curves of current and voltage.
      With polarity reversed, the current will now flow in the opposite direction. The curves would then appear as follows:

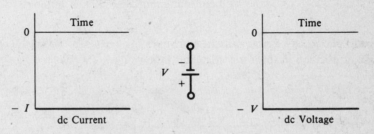

An alternating-current voltage (ac voltage) source periodically reverses or alternates in polarity (Fig. 1-10a). Therefore, the resulting alternating current also periodically reverses direction (Fig. 1-10b). In terms of conventional flow, the current flows from the positive terminal of the voltage source, through the circuit, and back to the negative terminal, but when the generator alternates in polarity, the current must reverse its direction. The ac power line used in most homes is a common example. The voltage and current direction go through many reversals each second in these systems.

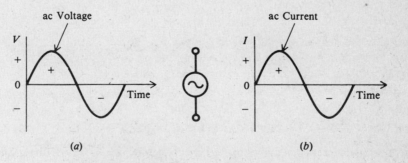

Fig. 1-10 Waveforms of ac voltage and ac current

Solved Problems

1.1 Match each term in column 1 to its closest meaning in column 2.

	Column 1		Column 2
1.	Electron	(*a*)	Positive charge
2.	Neutron	(*b*)	Same number of electrons and protons
3.	Compound	(*c*)	Electrons in first shell
4.	Neutral	(*d*)	Released electrons
5.	Valence electrons	(*e*)	Neutral charge
6.	Atomic number	(*f*)	Electrons in outermost shell
7.	Free electrons	(*g*)	Quota filled in outermost shell
8.	K shell	(*h*)	Number of electrons in nucleus
9.	Ion	(*i*)	Negative charge
10.	Inert	(*j*)	Quota of 2 electrons
		(*k*)	Combined elements
		(*l*)	Number of protons in nucleus
		(*m*)	Charged atom

Ans.   1. (*i*)   2. (*e*)   3. (*k*)   4. (*b*)   5. (*f*)   6. (*l*)   7. (*d*)   8. (*j*)   9. (*m*)   10. (*g*)

1.2 Show the atomic structure of the element aluminum with atomic number 13. What is its electron valence?

Because aluminum has 13 protons in the nucleus, it must have 13 orbiting electrons to be electrically neutral. Starting with the innermost shells (Fig. 1-3), we have

K shell	2 electrons
L shell	8 electrons
M shell	3 electrons
Total	13 electrons

The atomic structure for aluminum then is shown in Fig. 1-11. Its electron valence is -3 because it has 3 valence electrons.

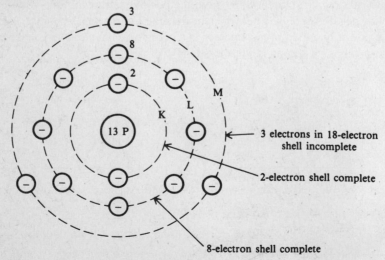

Fig. 1-11

1.3 In observing the maximum number of electrons in shells K, L, M, and N in Fig. 1-3, you will find that they are 2, 8, 18, and 32 electrons, respectively. Develop a formula that describes this relationship, where n is the shell number in sequential order outward from the nucleus.

The formula is $2n^2$ because the maximum number of electrons in the

K or first shell ($n = 1$) is $2(1^2) = 2(1) = 2$

L or second shell ($n = 2$) is $2(2^2) = 2(4) = 8$

M or third shell ($n = 3$) is $2(3^2) = 2(9) = 18$

N or fourth shell ($n = 4$) is $2(4^2) = 2(16) = 32$

This relationship is true for most elements.

1.4 What is the net charge of a body that contains 8 protons and 4 electrons?

The numerical value of the net charge is found by subtracting the number of one type of charge from the number of the other type. So a positive charge of 8 ($+8$) and a negative charge of 4 (-4) yields a positive charge of 4 ($+4$).

1.5 A charged insulator has deficiency of 50×10^{18} electrons. Find its charge in coulombs with polarity.

Since $1\,C = 6.25 \times 10^{18}$ electrons, $8\,C = 50 \times 10^{18}$ electrons. Deficiency of electrons means an excess of protons. So the insulator has a positive charge of $8\,C$, or $+Q = 8\,C$.

1.6 Write the word which most correctly completes each of the following statements:

(a) A rubber rod repels a second rubber rod, so both rods have _____ charges.

(b) Glass rubbed with silk attracts rubber rubbed with fur. If the rubber rod is negative, the glass rod must be _____.

(a) like (law of charges);   (b) positive (law of charges)

1.7 Find the current needed to charge a dielectric so that it will accumulate a charge of $20\,C$ after $4\,s$.

Known values: $Q = 20\,C$; $T = 4\,s$

Unknown: $I = ?$

Use Eq. (*1-1*) to find I:

$$I = \frac{Q}{T} = \frac{20\,C}{4\,s} = 5\,A \qquad Ans.$$

1.8 A current of $8\,A$ charges an insulator for $3\,s$. How much charge is accumulated?

Known values: $I = 8\,A$; $T = 3\,s$

Unknown: $Q = ?$

Use Eq. (*1-2*) to find Q:

$$Q = IT = (8\,A)(3\,s) = 24\,C \qquad Ans.$$

1.9 Write the word or words which most correctly complete each of the following statements.

(a) The ability of a charge to do work is its ___potential___.

(b) When one charge is different from the other, there is a _____ of _____.

(c) The unit of potential difference is the ___volt___.

(d) The sum of potential differences of all charges is called _____.

(e) The movement of charges produces _____.

(f) A greater amount of moving charges means a _____ value for the current.

(g) When the potential difference is zero, the value of current is _____.

(h) The rate of flow of charge is called _____.

(i) The direction of the conventional flow of current is from a point of _____ potential to a point of _____ potential.

(j) Electron flow is opposite in direction to _____ flow.

(k) Direct current (dc) has just _____ direction.

(l) A _____ is an example of a dc voltage source.

(m) An alternating current (ac) _____ its polarity.

Ans. (a) potential   (h) current
      (b) difference, potential (i) positive, negative
      (c) volt   (j) conventional
      (d) electromotive force   (k) one
      (e) current   (l) battery
      (f) higher   (m) reverses
      (g) zero

1.10   Match each device in column 1 to its closest principle in column 2.

	Column 1		Column 2
1.	Battery	(a)	Electromagnetic induction
2.	Generator	(b)	Free electrons
3.	TV camera tube	(c)	Ionized gases
4.	Vacuum tube	(d)	Chemical reaction
5.	Phonograph needle	(e)	Thermal energy
		(f)	Photoelectricity
		(g)	Mechanical motion

Ans. 1. (d)   2. (a)   3. (f)   4. (b)   5. (g)

Supplementary Problems

1.11   Match each term in column 1 to its closest meaning in column 2.

	Column 1		Column 2
1.	Proton	(a)	Negative charge
2.	Molecule	(b)	Quota of 8 electrons
3.	Quota	(c)	Excited state
4.	L shell	(d)	Maximum number of electrons in a shell
5.	Element	(e)	Atom negatively charged
6.	Unstable	(f)	Positive charge
7.	Shell	(g)	Mass and volume

	Column 1		Column 2
8.	Copper	(h)	Atomic number is 29
9.	Negative ion	(i)	Quota of 18 electrons
10.	Matter	(j)	Orbit
		(k)	Smallest particle having same characteristics
		(l)	Atomic number is 14
		(m)	All atoms the same

Ans.   1. (*f*)   2. (*k*)   3. (*d*)   4. (*b*)   5. (*m*)   6. (*c*)   7. (*j*)   8. (*h*)   9. (*e*)   10. (*g*)

1.12   Write the word or words which most correctly complete each of the following statements.

(*a*)   Electrons move about the nucleus of an atom in paths which are called _____.

(*b*)   The nucleus of an atom consists of particles called _____ and _____.

(*c*)   The number of protons in the nucleus of an atom is known as the _____ _____ of that atom.

(*d*)   When all the atoms within a substance are alike, the substance is called a chemical _____.

(*e*)   A _____ is the smallest particle of a compound which retains all the properties of that compound.

(*f*)   The energy _____ of an electron is determined by its distance from the nucleus of an atom.

(*g*)   If a neutral atom gains electrons, it becomes a _____ ion.

(*h*)   If a neutral atom loses electrons, it becomes a _____ ion.

(*i*)   Unlike charges _____ each other, while like charges _____ each other.

(*j*)   A charged object is surrounded by an _____ field.

    Ans.   (*a*)   shells *or* orbits   (*f*)   level
            (*b*)   protons, neutrons   (*g*)   negative
            (*c*)   atomic number   (*h*)   positive
            (*d*)   element   (*i*)   attract, repel
            (*e*)   molecule   (*j*)   electrostatic

1.13   Show the atomic structure of the element phosphorus, which has an atomic number of 15. What is its electron valence?   *Ans.*   See Fig. 1-12. Electron valence is −5.

1.14   Show the atomic structure of the element neon, which has an atomic number of 10. What is its electron valence?   *Ans.*   See Fig. 1-13. Electron valence is 0. Thus, neon is inert.

1.15   What is the net charge if 13 electrons are added to 12 protons?   *Ans.*   −1 electron

1.16   What becomes of the silicon atom when it loses all the orbiting electrons in its outermost shell? *Ans.*   It becomes a negative ion with a net charge of −4. See Fig. 1-2c.

1.17   A charged insulator has an excess of 25×10^{18} electrons. Find its charge in coulombs with polarity. *Ans.* $-Q = 4\,\text{C}$

1.18   A material with an excess of 25×10^{18} electrons loses 6.25×10^{18} electrons. The excess electrons are then made to flow past a given point in 2 s. Find the current produced by the resultant electron flow. *Ans.* $I = 1.5\,\text{A}$

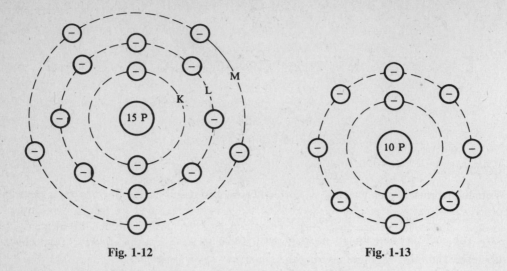

Fig. 1-12   **Fig. 1-13**

1.19   A charge of 10 C flows past a given point every 2 s. What is the current?
Ans. $I = 5\,\text{A}$

1.20   How much charge is accumulated when a current of 5 A charges an insulator for 5 s?
Ans. $Q = 25\,\text{C}$

1.21   Match each item in section 1 with its application in section 2.

Section 1				Section 2		
1.	Water	4.	Quartz	(a) Solar cell	(e)	Photoelectric cell
2.	Cesium oxide	5.	Carbon–zinc	(b) Generator	(f)	Turbine
3.	Silicon	6.	Iron–copper	(c) Battery	(g)	MHD converter
				(d) Crystal oscillator	(h)	Thermocouple

Ans.   1. (*f*)   2. (*e*)   3. (*a*)   4. (*d*)   5. (*c*)   6. (*h*)

1.22   Fill in the missing quantity:

	I, A	Q, C	T, s
(a)	?	10	2
(b)	5	?	4
(c)	?	9	2
(d)	7	?	3
(e)	2	6	?

Ans.

	I, A	Q, C	T, s
(a)	5	....	....
(b)	....	20	....
(c)	4.5	....	....
(d)	....	21	....
(e)	....	....	3

Chapter 2

Electrical Standards and Conventions

Units

INTRODUCTION

The international metric system of units of dimensions, commonly called SI, is used in electricity. The abbreviation SI stands for *système internationale.* The seven base units of SI are length, mass, time, electric current, thermodynamic temperature, light intensity, and amount of substance (Table 2-1). Formerly the MKS metric system was used, where M stands for meter (length), K for kilogram (mass), and S for seconds (time). The two supplementary units of SI are plane angle and solid angle (Table 2-2).

Table 2-1 Base Units of the International Metric System

Quantity	Base Unit	Symbol
Length	meter	m
Mass	kilogram	kg
Time	second	s
Electric current	ampere	A
Thermodynamic temperature	kelvin	K
Light intensity	candela	cd
Amount of substance	mole	mol

Table 2-2 Supplementary SI Units

Quantity	Unit	Symbol
Plane angle	radian	rad
Solid angle	steradian	sr

Other common units can be derived from the base and supplementary units. For example, the unit of charge is the coulomb, which is derived from the base units of second and ampere. Most of the units that are used in electricity are derived ones (Table 2-3).

METRIC PREFIXES

In the study of basic electricity, some electrical units are too small or too large to express conveniently. For example, in the case of resistance, we often use values in thousands or millions of ohms (Ω). The prefix *kilo* (denoted by the letter k) is a convenient way of expressing a thousand. Thus, instead of saying a resistor has a value of 10 000 Ω, we normally refer to it as a 10-kilohm (10-kΩ) resistor. In the case of current, we often use values in thousandths or millionths of an ampere. We use expressions such as milliamperes and microamperes. The prefix *milli* is a short way of saying a thousandth and *micro* is a short way of saying a millionth. Thus 0.012 A becomes 12 milliamperes (mA) and 0.000 005 A becomes 5 microamperes (μA). Table 2-4 lists the metric prefixes commonly used in electricity and their numerical equivalents.

Table 2-3 Derived SI Units

Quantity	Unit	Symbol
Energy	joule	J
Force	newton	N
Power	watt	W
Electric charge	coulomb	C
Electric potential	volt	V
Electric resistance	ohm	Ω
Electric conductance	siemens	S
Electric capacitance	farad	F
Electric inductance	henry	H
Frequency	hertz	Hz
Magnetic flux	weber	Wb
Magnetic flux density	tesla	T

Table 2-4 Metric Prefixes Used in Electricity

Prefix (Letter Symbol)	Value		Pronunciation
mega (M)	million	1 000 000	as in *megaphone*
kilo (k)	thousand	1 000	kill'oh
milli (m)	thousandth	0.001	as in *mili*tary
micro (μ)	millionth	0.000 001	as in *micro*phone
nano (n)	thousand-millionth	0.000 000 001	nan'oh
pico (p)	million-millionth	0.000 000 000 001	peek'oh

Example 2.1   A resistor has a value of 10 M stamped on its case. How many ohms of resistance does this resistor have? The letter M denotes mega, or million. Thus the resistor has a value of 10 megohms (MΩ) or 10 million ohms.

Example 2.2   A power station has a capacity of delivering 500 000 watts (W). What is the capacity in kilowatts (kW)? Refer to Table 2-4. *Kilo* stands for 1000. Thus, 500 000 W = 500 kW.

POWERS OF 10

We have seen that it is often necessary or desirable to convert one unit of measurement to another unit that may be larger or smaller. In the previous section, we did this by substituting a metric prefix for certain values. Another way would be to convert the number to a power of 10. Powers of 10 are often termed the "engineer's shorthand." Examples of expressing number as powers of 10 are shown in Table 2-5.

Table 2-5 Powers of 10

Number	Power of 10	Commonly Read As
0.000 001 =	10^{-6}	10 to the minus sixth
0.000 01 =	10^{-5}	10 to the minus fifth
0.000 1 =	10^{-4}	10 to the minus fourth
0.001 =	10^{-3}	10 to the minus third
0.01 =	10^{-2}	10 to the minus second
0.1 =	10^{-1}	10 to the minus one
1 =	10^{0}	10 to the zero
10 =	10^{1}	10 to the first
100 =	10^{2}	10 squared
1 000 =	10^{3}	10 cubed
10 000 =	10^{4}	10 to the fourth
100 000 =	10^{5}	10 to the fifth
1 000 000 =	10^{6}	10 to the sixth

Rule 1: To express numbers *larger* than 1 as a small number times a power of 10, move the decimal point to the *left* as many places as desired. Then multiply the number obtained by 10 to a power which is equal to the number of places moved.

Example 2.3

$3000 = 3,000.$ (Decimal point is moved three places to the left.)
$= 3 \times 10^3$ (Therefore the power, or exponent, is 3.)
$6500 = 65,00.$ (Decimal point is moved two places to the left.)
$= 65 \times 10^2$ (Therefore the exponent is 2.)
$880\,000 = 88,0000.$ (Decimal point is moved left four places.)
$= 88 \times 10^4$ (Therefore the exponent is 4.)
$42.56 = 4,2.56$ (Decimal point is moved left one place.)
$= 4.256 \times 10$ (Therefore the exponent is 1.)

Rule 2: To express numbers *less* than 1 as a whole number times a power of 10, move the decimal point to the *right* as many places as desired. Then multiply the number obtained by 10 to a *negative* power which is equal to the number of places moved.

Example 2.4

$0.006 = 0.006,$ (Decimal point is moved three places to the right.)
$= 6 \times 10^{-3}$ (Therefore the power, or exponent, is -3.)
$0.435 = 0.4,35$ (Decimal point is moved one place to the right.)
$= 4.35 \times 10^{-1}$ (Therefore the exponent is -1.)
$0.000\,92 = 0.000\,92,$ (Decimal point is moved right five places.)
$= 92 \times 10^{-5}$ (Therefore the exponent is -5.)
$0.578 = 0.57,8$ (Decimal point is moved right two places.)
$= 57.8 \times 10^{-2}$ (Therefore the exponent is -2.)

Rule 3: To convert a number expressed as a positive power of 10 to a decimal number, move the decimal point to the right as many places as the value of the exponent.

Example 2.5

$0.615 \times 10^3 = 0_{\circ}615_{\textstyle,}$ (The exponent is 3. Therefore move the decimal point three places to the right.)

            $= 615$

$0.615 \times 10^6 = 0_{\circ}615\,000_{\textstyle,}$ (Move the decimal point six places to the right.)

            $= 615{,}000$

$0.0049 \times 10^3 = 0_{\circ}004_{\textstyle,}9$ (Move decimal point right three places.)

            $= 4.9$

$84 \times 10^2 = 84_{\circ}00_{\textstyle,}$ (Move decimal point right two places.)

            $= 8400$

Rule 4: To convert a number expressed as a negative power of 10 to a decimal number, move the decimal point to the *left* as many places as the value of the exponent.

Example 2.6

$70 \times 10^{-3} = 0_{\textstyle,}070_{\circ}$ (The exponent is -3. Therefore move the decimal point three places to the left.)

            $= 0.07$

$82.4 \times 10^{-2} = 0_{\textstyle,}82_{\circ}4$ (Move decimal point left two places.)

            $= 0.824$

$60\,000 \times 10^{-6} = 0_{\textstyle,}060\,000_{\circ}$ (Move decimal point left six places.)

            $= 0.06$

$0.5 \times 10^{-3} = 0_{\textstyle,}000_{\circ}5$ (Move decimal point left three places.)

            $= 0.0005$

Rule 5: To multiply two or more numbers expressed as powers of 10, *multiply* the coefficients to obtain the new coefficient and *add* the exponents to obtain the new exponent of 10.

Example 2.7

$$10^2 \times 10^4 = 10^{2+4} = 10^6 \quad Ans.$$

$$10^{-1} \times 10^4 = 10^{-1+4} = 10^3 \quad Ans.$$

$$(40 \times 10^3)\,(25 \times 10^2) = 40 \times 25 \times 10^3 \times 10^2 \quad (40 \times 25 = 1000,\ 3+2 = 5)$$
$$= 1000 \times 10^5 \quad (\text{But } 1000 = 10^3)$$
$$= 10^3 \times 10^5$$
$$= 10^8 \quad Ans.$$

$$(2 \times 10^{-2})\,(50 \times 10^2) = 2 \times 50 \times 10^{-2} \times 10^2$$
$$= 100 \times 10^0 \quad (\text{But } 100 = 10^2)$$
$$= 10^2 \times 1 \quad (10^0 = 1)$$
$$= 10^2 \quad Ans.$$

$$(3 \times 10^{-4})\,(6 \times 10^6) = 3 \times 6 \times 10^{-4} \times 10^6$$
$$= 18 \times 10^2 \quad Ans.$$

Rule 6: To divide by powers of 10, use the formula

$$\frac{1}{10^n} = 1 \times 10^{-n}$$

We therefore can transfer any power of 10 from numerator to denominator, or vice versa, simply by changing the sign of the exponent.

Example 2.8

$$\frac{15}{10^{-1}} = 15 \times 10^1 = 150 \qquad \frac{1500}{10^4} = 1500 \times 10^{-4} = 0.15$$

$$\frac{15}{10^{-3}} = 15 \times 10^3 = 15\,000 \qquad \frac{0.25 \times 4}{10^{-2}} = 1.0 \times 10^2 = 100$$

The prefixes in Table 2-4 are expressed as powers of 10 in Table 2-6.

**Table 2-6 Metric Prefixes
Expressed as Powers of 10**

Metric Prefix	Power of 10
mega (M)	10^6
kilo (k)	10^3
milli (m)	10^{-3}
micro (μ)	10^{-6}
nano (n)	10^{-9}
pico (p)	10^{-12}

Example 2.9 Problem answers can be expressed in different but equivalent units. For example, $3\,000\,000\ \Omega$ is different but equivalent to $3\ M\Omega$.

(a) Express 2.1 V in millivolts (mV).

$$1\ V = 10^3\ mV$$

$$2.1\ V = 2.1 \times 10^3 = 2100\ mV \qquad Ans.$$

(b) Express 0.006 A in milliamperes (mA).

$$1\ A = 10^3\ mA$$

$$0.006\ A = 0.006 \times 10^3 = 6\ mA \qquad Ans.$$

(c) Change 356 mV to volts (V).

$$1\ mV = 10^{-3}\ V$$

$$356\ mV = 356 \times 10^{-3} = 0.356\ V \qquad Ans.$$

(d) Change $500\,000\ \Omega$ to megohms ($M\Omega$).

$$1\ \Omega = 10^{-6}\ M\Omega$$

$$500\,000\ \Omega \times 10^{-6} = 0.5\ M\Omega \qquad Ans.$$

(e)   Change 20 000 000 picofarads (pF) into farads (F).

$$1 \text{ pF} = 10^{-12} \text{ F}$$

$$20\,000\,000 \text{ pF} \times 10^{-12} = 0.000\,02 \text{ F} \qquad Ans.$$

SCIENTIFIC NOTATION

In scientific notation, the coefficient of the power of 10 is always expressed with one decimal place and the required power of 10. Several examples will make the procedure clear.

Example 2.10   Express the following numbers in scientific notation.

$$300\,000 = 3\underline{.000\,00}_\odot \times 10^5 \qquad \text{(Move decimal point left five places—power is 5 by Rule 1)}$$
$$= 3 \times 10^5$$

$$871 = 8\underline{.71}_\odot \times 10^2 \qquad \text{(Move decimal point left two places—power is 2 by Rule 1)}$$
$$= 8.71 \times 10^2$$

$$7425 = 7\underline{.425}_\odot \times 10^3 \qquad \text{(Move decimal point left three places—power is 3 by Rule 1)}$$
$$= 7.425 \times 10^3$$

$$0.001 = 0_\odot\underline{001}_\downarrow \times 10^{-3} \qquad \text{(Move decimal point right three places—power is } -3 \text{ by Rule 2)}$$
$$= 1 \times 10^{-3}$$

$$0.015 = 0_\odot\underline{01}_\downarrow 5 \times 10^{-2} \qquad \text{(Move decimal point right two places—power is } -2 \text{ by Rule 2)}$$
$$= 1.5 \times 10^{-2}$$

ROUNDING OFF NUMBERS

A number is rounded off by dropping one or more digits at its right. If the digit to be dropped is *less* than 5, we leave the digit as it is. For example, 4.1632, if rounded to four digits, would be 4.163; if rounded to three digits, 4.16. If the digit to be dropped is *greater* than 5, we increase the digit to its left by 1. For example, 7.3468, if rounded to four digits, would be 7.347; if rounded to three digits, 7.35. If the digit to be dropped is *exactly* 5 (that is, 5 followed by nothing but zeros), we increase the digit to its left by 1 if it is an odd number and we leave the digit to the left as it is if it is an even number. For example, 2.175, when rounded to three digits, becomes 2.18. The number 2.185 would also round to the same value, 2.18, if rounded to three digits.

Any digit that is needed to define the specific value is said to be *significant*. For example, a voltage of 115 V has three significant digits: 1, 1, and 5. When rounding off numbers, zero is not counted as a significant digit if it appears immediately after the decimal point and is followed by other significant digits. Such zeros must be retained and the count of significant digits must begin at the first significant digit beyond them. For example, 0.000 12 has two significant digits, 1 and 2, and the preceding zeros don't count. However, 18.0 has three significant digits; in this case zero is significant because it is not followed by other significant digits. In electricity, specific values are usually expressed in three significant digits.

Example 2.11   Round off the following numbers to three significant digits.

We look at the fourth significant digit to the right and observe whether this digit is less than 5, greater than 5, or equal to 5.

$$5.6428 = 5.64 \qquad 0.016\,95 = 0.0170$$
$$49.67 = 49.7 \qquad 2078 = 2080$$
$$305.42 = 305 \qquad 1.003 \times 10^{-3} = 1.00 \times 10^{-3}$$

$$782.51 = 783 \qquad 12.46 \times 10^5 = 12.5 \times 10^5$$
$$0.003\,842 = 0.003\,84 \qquad 1.865 \times 10^2 = 1.86 \times 10^2$$

Scientific notation is a convenient way to work problems in electricity. Often, we express a numerical answer in terms of a prefix rather than leave the answer in scientific notation.

Example 2.12   Express each of the following first in scientific notation and then with a prefix.

(a)   0.000 53 A to milliamperes (mA)

$$0.000\,53 \text{ A} = 5.3 \times 10^{-4} \text{ A}$$
$$= 0.53 \times 10^{-3} \text{ A}$$
$$= 0.53 \text{ mA}$$

(b)   2500 V to kilovolts (kV)

$$2500 \text{ V} = 2.5 \times 10^3 \text{ V}$$
$$= 2.5 \text{ kV}$$

(c)   0.000 000 1 F to microfarads (μF)

$$0.000\,000\,1 \text{ F} = 1 \times 10^{-7} \text{ F}$$
$$= 10 \times 10^{-6} \text{ F}$$
$$= 10 \,\mu\text{F}$$

Solved Problems

Express each of the following in the units indicated.

2.1   2 A to milliamperes

$$1 \text{ A} = 1000 \text{ mA} = 10^3 \text{ mA}$$

Multiply 2 by 1000 to get 2000 mA.   *Ans.*

or   Multiply 2 by 10^3 to get 2×10^3 mA, which is 2000 mA.

2.2   1327 mA to amperes

$$1 \text{ mA} = 0.001 \text{ A} = 10^{-3} \text{ A}$$

Multiply 1327 by 0.001 to get 1.327 A.   *Ans.*

or   Multiply 1327 by 10^{-3} to get 1327×10^{-3} A, which is 1.327 A.

2.3   8.2 kΩ to ohms

$$1 \text{ k}\Omega = 1000 \,\Omega = 10^3 \,\Omega$$

Multiply 8.2 by 1000 to get 8200 Ω.   *Ans.*

or   Multiply 8.2 by 10^3 to get 8.2×10^3 Ω which is 8200 Ω.

2.4   680 kΩ to megohms

Use two steps.

Step 1:   Convert to ohms.

Multiply 680 by 1000 to get 680 000 Ω.

Step 2:   Convert to megohms.

Multiply 680 000 Ω by 0.000 001 to get 0.68 MΩ.   *Ans.*

2.5   10 000 μF to farads

$$1\,\mu F = 0.000\,001\,F = 10^{-6}\,F$$

Multiply 10 000 by 0.000 001 to get 0.01 F.   *Ans.*

or   Multiply 10 000 by 10^{-6} to get $10\,000 \times 10^{-6}$, which is 0.01 F.

2.6   0.000 000 04 s to nanoseconds (ns)

$$1\,s = 1\,000\,000\,000\,ns = 10^9\,ns$$

Multiply 0.000 000 04 by 1 000 000 000 to get 40 ns.   *Ans.*

or   Multiply 0.000 000 04 by 10^9 to get $0.000\,000\,04 \times 10^9$ ns, which is equal to 4×10^1 or 40 ns.

Express the following numbers as decimal numbers.

2.7 0.75×10^3

Move decimal point right three places—Rule 3:

$$0.75 \times 10^3 = 0\underset{\smile}{750}_{,} = 750 \qquad \textit{Ans.}$$

2.8 0.75×10^{-3}

Move decimal point left three places—Rule 4:

$$0.75 \times 10^{-3} = 0_{,}\underset{\smile}{000}75 = 0.000\,75 \qquad \textit{Ans.}$$

2.9 $(2.1 \times 10^{-1})(4 \times 10^2)$

$$
\begin{aligned}
(2.1 \times 10^{-1})(4 \times 10^2) &= 2.1 \times 4 \times 10^{-1} \times 10^2 \\
&= 8.4 \times 10^1 \quad \text{(Rule 5)} \\
&= 84 \quad \text{(Rule 3)} \qquad \textit{Ans.}
\end{aligned}
$$

2.10   Express 4160 in scientific notation.

Move decimal point left three places—Rule 1:

$$4160 = 4\underset{\smile}{160}_{,} \times 10^3 = 4.160 \times 10^3 \qquad \textit{Ans.}$$

Express each quantity in the following problems in scientific notation and then perform the indicated arithmetic calculation.

2.11 0.072×1000

Express $0.072 = 7.2 \times 10^{-2}$ (Move decimal point right two places—Rule 2)

Then
$$0.072 \times 1000 = \left(7.2 \times 10^{-2}\right)\left(1 \times 10^3\right)$$
$$= (7.2 \times 1)\left(10^{-2} \times 10^3\right)$$
$$= 7.2 \times 10 \quad \text{(Rule 5)}$$
$$= 72 \quad \text{(Rule 3)} \quad Ans.$$

2.12 0.0045×100

Express $0.0045 = 4.5 \times 10^{-3}$ (Move decimal point right three places—Rule 2)

$100 = 1 \times 10^2$ (Move decimal point left two places—Rule 1)

Then
$$0.0045 \times 100 = \left(4.5 \times 10^{-3}\right)\left(1 \times 10^2\right)$$
$$= (4.5 \times 1)\left(10^{-3} \times 10^2\right)$$
$$= 4.5 \times 10^{-1} \quad \text{(Rule 5)}$$
$$= 0.45 \quad \text{(Rule 4)} \quad Ans.$$

2.13 $7500 \div 100$

Express $7500 = 7.5 \times 10^3$ (Move decimal point left three places—Rule 1)

$100 = 1 \times 10^2$ (Move decimal point left two places—Rule 1)

Then
$$\frac{7500}{100} = \frac{7.5 \times 10^3}{1 \times 10^2}$$
$$= 7.5\left(10^3 \times 10^{-2}\right) \quad \left(1/10^2 = 10^{-2}\right)$$
$$= 7.5 \times 10^1 \quad \text{(Rule 5)}$$
$$= 75 \quad \text{(Rule 3)} \quad Ans.$$

2.14 $\dfrac{4000}{2000}$

Express $4000 = 4 \times 10^3$ (Move decimal point left three places—Rule 1)

$2000 = 2 \times 10^3$ (Move decimal point left three places—Rule 1)

$$\frac{4000}{2000} = \frac{4 \times 10^3}{2 \times 10^3} = 2 \quad Ans.$$

Note that any factor divided by itself cancels out to 1. That is, $10^3/10^3 = 10^{3-3} = 10^0 = 1$.

2.15 $\dfrac{1000 \times 0.008}{0.002 \times 500}$

Express $1000 = 1 \times 10^3$ (Move decimal point left three places—Rule 1)

$0.008 = 8 \times 10^{-3}$ (Move decimal point right three places—Rule 2)

$0.002 = 2 \times 10^{-3}$ (Move decimal point right three places—Rule 2)

$500 = 5 \times 10^2$ (Move decimal point left two places—Rule 1)

Then
$$\frac{1000 \times 0.008}{0.002 \times 500} = \frac{\left(1 \times 10^3\right)\left(8 \times 10^{-3}\right)}{\left(2 \times 10^{-3}\right)\left(5 \times 10^2\right)}$$

$$= \frac{1 \times 8 \times 10^3 \times 10^{-3}}{2 \times 5 \times 10^{-3} \times 10^2}$$

$$= \frac{8 \times 10^0}{10 \times 10^{-1}} \quad \text{(Rule 5)}$$

$$= \frac{8 \times 1}{10^0} \quad \text{(Rule 5)}$$

$$= 8 \quad Ans.$$

2.16 $\dfrac{1}{4 \times 100\,000 \times 0.000\,05}$

Express
$$4 = 4$$
$$100\,000 = 1 \times 10^5 \quad \text{(Move decimal point left five places—Rule 1)}$$
$$0.000\,05 = 5 \times 10^{-5} \quad \text{(Move decimal point right five places—Rule 2)}$$

Then
$$\frac{1}{4 \times 100\,000 \times 0.000\,05} = \frac{1}{4\left(1 \times 10^5\right)\left(5 \times 10^{-5}\right)}$$

$$= \frac{1}{\left(4 \times 5\right)\left(10^5 \times 10^{-5}\right)}$$

$$= \frac{1}{20 \times 10^0} \quad \text{(Rule 5)}$$

$$= \frac{1}{2 \times 10 \times 1} \quad \text{(Rule 1)}$$

$$= \frac{10^{-1}}{2} \quad \text{(Rule 6, } 1/10 = 10^{-1}\text{)}$$

$$= 0.5 \times 10^{-1}$$

$$= 0.05 \quad \text{(Rule 4)} \quad Ans.$$

2.17   We might read 220 V on a certain type of voltmeter, but a precision instrument might show that voltage to be 220.4 V, and a series of precise measurements might show the voltage to be 220.47 V. How many significant digits does each measurement have?

220 V, three significant digits

220.4 V, four significant digits

220.47 V, five significant digits

If the accuracy of measurement required is five places, then the instrument must measure to at least five significant digits.

In Problems 2.18–2.20, perform the indicated operations. Round off the figures in the results, if necessary, and express answers to three significant digits as a number from 1 through 10 and the proper power of 10.

2.18 $\dfrac{0.256 \times 338 \times 10^{-9}}{865\,000}$

Express
$$0.256 = 2.56 \times 10^{-1}$$
$$338 = 3.38 \times 10^2$$
$$865\,000 = 8.65 \times 10^5$$

Then

$$\frac{0.256 \times 338 \times 10^{-9}}{865\,000} = \frac{\left(2.56 \times 10^{-1}\right)\left(3.38 \times 10^{2}\right)\left(10^{-9}\right)}{8.56 \times 10^{5}}$$

$$= \frac{2.56 \times 3.38}{8.65}\left(10^{-1} \times 10^{2} \times 10^{-9} \times 10^{-5}\right)$$

$$= (1.00)\left(10^{13}\right)$$

$$= 1.00 \times 10^{-13} \quad Ans.$$

2.19 $\dfrac{2800 \times 75.61}{0.000\,900\,5 \times 0.0834}$

Express

$$2800 = 2.8 \times 10^{3}$$

$$75.61 = 7.561 \times 10^{1}$$

$$0.000\,900\,5 = 9.005 \times 10^{-4}$$

$$0.0834 = 8.34 \times 10^{-2}$$

Then

$$\frac{2800 \times 75.61}{0.000\,900\,5 \times 0.0834} = \frac{\left(2.8 \times 10^{3}\right)\left(7.561 \times 10^{1}\right)}{\left(9.005 \times 10^{-4}\right)\left(8.34 \times 10^{-2}\right)}$$

$$= \frac{2.8 \times 7.561}{9.005 \times 8.34}\frac{10^{3} \times 10^{1}}{10^{-4} \times 10^{-2}} = \frac{21.17}{75.10}\frac{10^{4}}{10^{-6}}$$

$$= 0.2819 \times 10^{10}$$

$$= 2.819 \times 10^{-1} \times 10^{10}$$

$$= 2.82 \times 10^{9} \quad Ans.$$

2.20 $\dfrac{1}{6.28 \times 400 \times 10^{6} \times 25 \times 10^{-12}}$

Then

$$\frac{1}{6.28\left(4 \times 10^{2}\right)\left(10^{6}\right)\left(2.5 \times 10^{1}\right)\left(10^{-12}\right)} = \frac{1}{(6.28 \times 4 \times 2.5)\left(10^{2} \times 10^{6} \times 10^{1} \times 10^{-12}\right)}$$

$$= \frac{1}{62.80 \times 10^{-3}} = 0.0159 \times 10^{3}$$

$$= 1.59 \times 10^{-2} \times 10^{3}$$

$$= 1.59 \times 10^{1} = 15.9 \quad Ans.$$

Supplementary Problems

Express each of the following in the units indicated (use powers of 10 where applicable).

2.21 $5\,600\,000\,\Omega$ in megohms *Ans.* $5.6\,\text{M}\Omega$

2.22 $2.2\,\text{M}\Omega$ in ohms *Ans.* $2\,200\,000\,\Omega$ or $2.2 \times 10^{6}\,\Omega$

2.23 $0.330\,\text{M}\Omega$ in kilohms *Ans.* $330\,\text{k}\Omega$

2.24 $0.013\,\text{kV}$ in volts *Ans.* $13\,\text{V}$

2.25 $0.24\,\text{A}$ in milliamperes *Ans.* $240\,\text{mA}$

2.26 20 000 μA in amperes   *Ans.*   0.02 A

2.27 0.25 mA in microamperes   *Ans.*   250 μA

2.28 10 000 V in kilovolts   *Ans.*   10 kV

2.29 4 000 000 W in megawatts (MW)   *Ans.*   4 MW

2.30 5 000 kW in megawatts   *Ans.*   5 MW

2.31 200 ns in seconds   *Ans.*   0.000 000 2 s or 2×10^{-7} s

Express each of the following as decimal numbers.

2.32 0.006×10^2 *Ans.*   0.6

2.33 43.41×100 *Ans.*   4341

2.34 0.0053×10^3 *Ans.*   5.3

2.35 $400/10^3$ *Ans.*   0.4

2.36 3×10^{-2} *Ans.*   0.03

2.37 $100 000 \times 10^{-4}$ *Ans.*   10

2.38 $(0.5 \times 0.03)/10^{-2}$ *Ans.*   1.5

2.39 $(3.1 \times 10^{-1})(2 \times 10^{-2})$ *Ans.*   0.0062

2.40 $600/(5 \times 10^2)$ *Ans.*   1.2

Express each of the following in scientific notation, that is, as a number from 1 to 10 and the proper power of 10.

2.41 120 000   *Ans.* 1.2×10^5

2.42 0.006 45   *Ans.* 6.45×10^{-3}

2.43 2 300 000   *Ans.* 2.3×10^6

2.44 550×10^{-4} *Ans.* 5.5×10^{-2}

2.45 0.0008×10^3 *Ans.* 8×10^{-1}

Perform the indicated operations. Express the answer in scientific notation.

2.46 $\dfrac{200 \times 0.008}{0.02 \times 10^3}$ *Ans.* 8×10^{-1}

2.47 $\dfrac{1}{(4 \times 10^4)(0.5 \times 10^{-5})}$ *Ans.*   5

2.48 $\dfrac{(3.2 \times 10^2)(1.4 \times 10^{-1})}{(2 \times 10^{-3})(4 \times 10^2)}$ *Ans.* 5.6×10

2.49 $\dfrac{400\,000}{2 \times 10^7}$ *Ans.* 2×10^{-2}

2.50 $\dfrac{300(4 \times 10^{-5})(10^2)}{12 \times 10^2}$ *Ans.* 1×10^{-3}

Round off the following numbers to three significant digits.

2.51   3.824   *Ans.*   3.82

2.52   3.825   *Ans.*   3.82

2.53   3.826   *Ans.*   3.83

2.54   205.6   *Ans.*   206

2.55   0.004 152   *Ans.*   0.004 15

2.56   2096   *Ans.*   2100

2.57 7.803×10^2 *Ans.* 7.80×10^2

2.58 $0.001\,205 \times 10^{-3}$ *Ans.* $0.001\,20 \times 10^{-3}$

Perform the indicated operations. Round off the answers to three-place accuracy.

2.59 $\dfrac{(8.31 \times 10^0)(5.7 \times 10^3)}{(2.1 \times 10^{-1})(3.0 \times 10^6)}$ *Ans.* 7.52×10^{-2}

2.60 $\dfrac{(5 \times 10^2)(6 \times 10^4)(9 \times 10^{16})}{(7 \times 10^{-6})(5 \times 10^{10})(3 \times 10^{14})}$ *Ans.* 2.57×10^4

2.61 $\dfrac{170\,000(6910)(100\,000)}{9185(54\,000)}$ *Ans.* 2.37×10^5

2.62 $\dfrac{790(0.0014)(0.01)}{0.000\,006(500\,000)}$ *Ans.* 3.69×10^{-3}

Graphical Symbols and Electrical Diagrams

SCHEMATIC DIAGRAM

A simple electric circuit is shown in pictorial form in Fig. 2-1a. The same circuit is drawn in *schematic* form in Fig. 2-1b. The schematic diagram is a shorthand way to draw an electric circuit, and circuits usually are represented in this way. In addition to the connecting wires, three components are shown symbolically in Fig. 2-1b: the dry cell, the switch, and lamp. Note the positive (+) and the negative (−) markings in both pictorial and schematic representations of the dry cell. The schematic components represent the pictorial components in a simplified manner. A schematic diagram then is one that shows by means of graphic symbols the electrical connections and the functions of the different parts of a circuit.

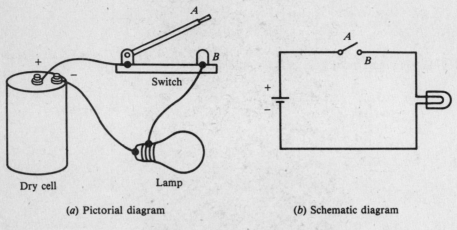

(a) Pictorial diagram   (b) Schematic diagram

Fig. 2-1 A simple lamp circuit

The standard graphic symbols for the commonly used electrical and electronic components are given in Fig. 2-2.

Examples of common letter symbols used to denote various circuit components are given in Table 2-7.

Table 2-7 Examples of Letter Symbols for Circuit Components

Part	Letter	Example
Resistor	R	R_3, 120 kΩ
Capacitor	C	C_5, 20 pF
Inductor	L	L_1, 25 mH
Rectifier (metallic or crystal)	CR	CR_2
Transformer	T	T_2
Transistor	Q	Q_5, 2N482 Detector
Jack	J	J_1

A schematic diagram of a two-transistor radio receiver is shown in Fig. 2-3. The circuit diagram in Fig. 2-3 shows the components in the order from left to right in which they are used to convert radio waves into sound waves. With the use of the diagram, it is then possible to trace the operation of the circuit from the incoming signal at the antenna to the output at the headphones. The components in a schematic diagram are identified by letter symbols such as R for resistors, C for capacitors, L for inductors, and Q for transistors (Table 2-7). Symbols are further identified by letter–number combinations such as R_1, R_2, and R_3 (sometimes written as $R1$, $R2$, $R3$) to prevent confusion when more than one type of component is used (Fig. 2-3). The letters B, C, and E near the transistor symbols indicate the base, collector, and emitter of the transistors (Fig. 2-3). The numerical values of components are often indicated directly in the schematic diagram, such as 220 kΩ for R_1 and 0.022 μF for C_2 (Fig. 2-3). When these values are not given in this way, they are stated in the parts list or the notes which accompany the diagram.

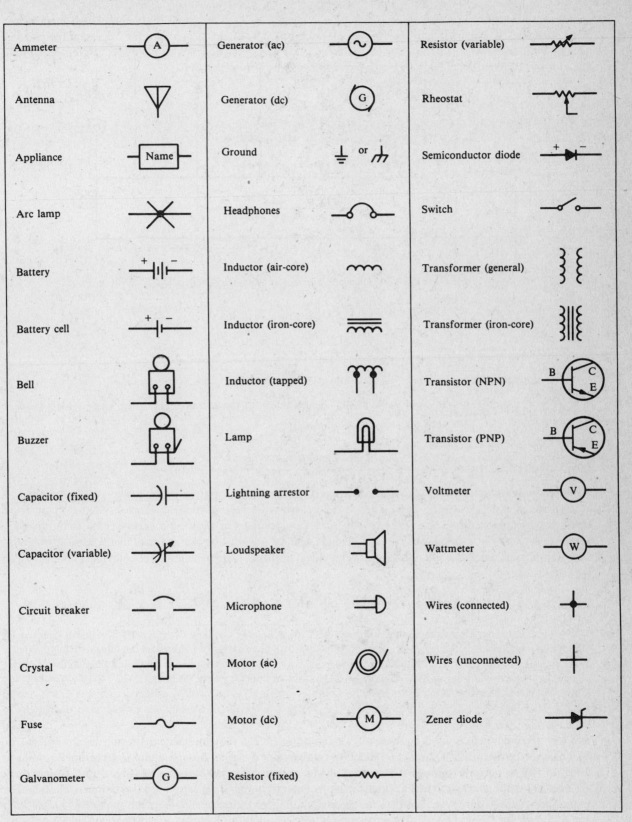

Fig. 2-2 Standard circuit symbols

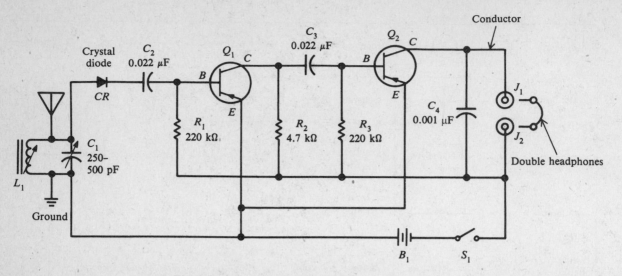

Fig. 2-3   Schematic diagram of a two-transistor radio

A schematic diagram does not show the physical location of the components or the wires which connect the components.

ONE-LINE DIAGRAM

A one-line, or single-line, diagram shows the component parts of a circuit by means of single lines and appropriate graphical symbols. The single lines represent the two or more conductors that are connected between the components in the actual circuit. The one-line diagram shows the necessary basic information about the sequence of a circuit but does not give the detailed information that is found in a schematic diagram. One-line diagrams are generally used to show complex electrical systems without the individual conductors to the various loads.

Example 2.13   Draw a one-line diagram showing the major equipment, switching devices, and connecting circuits of an electric power substation.

See Fig. 2-4. The single line running down represents three lines in this three-wire system. The power path can be traced from the shielded aluminum cable steel reinforced (ACSR) downward past a grounded lightning arrester, through a disconnect switch, fused disconnect switches, and a step-down transformer. It continues downward through an oil circuit breaker, disconnect switch, past another lightning arrester, and out through the ACSR line.

BLOCK DIAGRAM

The block diagram is used to show the relationship between the various component groups or stages in the operation of a circuit. It shows in block form the path of a signal through a circuit from input to output (Fig. 2-5). The blocks are drawn in the shape of squares or rectangles that are joined by single lines. Arrowheads placed at the terminal ends of the lines show the direction of the signal path from input to output as the diagram is read from left to right. As a general rule, the necessary information to describe the components or stages is placed within the block. On some block diagrams, devices such as antennas and loudspeakers are shown by standard symbols instead of by blocks (Fig. 2-5).

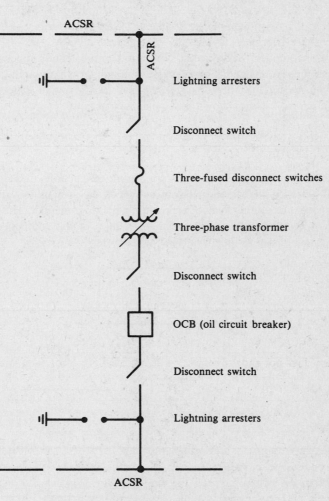

Fig. 2-4　A one-line diagram of a substation

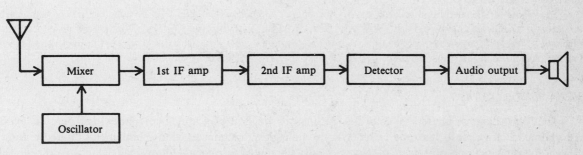

Fig. 2-5　Block diagram of a typical transistor radio receiver circuit

Because a block diagram shows the functions of the various stages of a circuit and uses single lines, it is a type of functional one-line diagram. The block diagram gives no information about specific components or wiring connections. Therefore, it is limited in use but does give a simple way of illustrating the overall features of a circuit. For this reason, block diagrams are frequently used by electricians, electronic technicians, and engineers as a first step in designing and laying out new equipment.

To show how easy it is to understand a circuit's operation by means of a block diagram, look at Fig. 2-5. The signal comes through the antenna and then progresses through the mixer circuit, through the intermediate-frequency (IF) amplifier stages and the detector stage, and finally to the output stage and speaker. The oscillator is in an auxiliary circuit, and so it is not shown in the main signal path.

WIRING DIAGRAM

The wiring, or connection, diagram is used to show wiring connections in a simple, easy-to-follow way. They are very commonly used with home appliances and electrical systems of automobiles (Fig. 2-6). The typical wiring diagram shows the components of a circuit in a pictorial manner. The components are identified by name. Such a diagram also often shows the relative location of the components within a given space. A color-coding scheme may be used to identify certain wires or leads (Fig. 2-6).

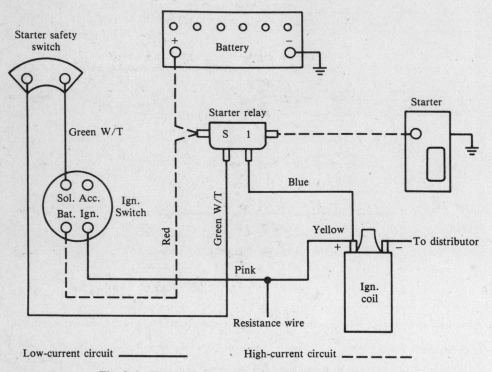

Fig. 2-6   Wiring diagram of an automobile starting circuit

ELECTRICAL PLAN

An integral part of any set of drawings for the construction of a building is the wiring plan or layout. Architects, electrical designers, and contractors use floor-plan diagrams to locate components of the building's electrical system, such as receptacle outlets, switches, lighting fixtures, and other wiring devices. These devices and wiring arrangements are represented by means of symbols (Fig. 2-7). The living-room plan (Fig. 2-7) shows two three-way switching arrangements. In one arrangement, a ceiling outlet is switched from two door locations. Similarly, two receptacle outlets on the north wall are switched from two separate locations. The connection between switch and ceiling outlet is drawn with a medium-weight solid line, indicating that the connection (conduit or cable) is to be concealed in the walls or ceiling above. The cross lines indicate the conductors in the conduit or cable. If cross lines are omitted, two wires are understood to be in the connection.

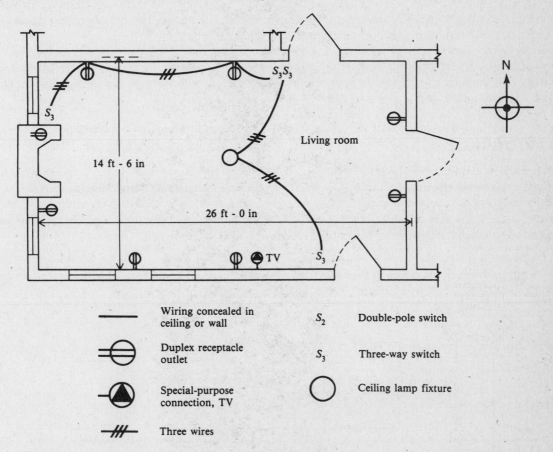

	Wiring concealed in ceiling or wall	S_2	Double-pole switch
	Duplex receptacle outlet	S_3	Three-way switch
	Special-purpose connection, TV		Ceiling lamp fixture
	Three wires		

Fig. 2-7 Floor plan of a room with fixtures, outlets, and switches. Standard wiring symbols are shown

Solved Problems

2.63 Write the word or words which most correctly complete the following statements.

(a) A picture showing the actual parts of a circuit and their connections is called a _____ diagram.

(b) On a schematic diagram, the components are represented by _____ .

(c) Schematic diagrams are often drawn with the input on the _____ and the output on the _____ .

(d) Example of letter symbols to identify components on a schematic diagram are _____ for diodes, _____ for capacitors, and _____ for inductors.

(e) A single-line diagram is also called a _____ diagram.

(f) Color coding is shown on a _____ diagram.

Ans. (a) pictorial; (b) symbols; (c) left, right; (d) CR, C, L; (e) one-line; (f) wiring or connection

2.64   In Fig. 2-3, identify the following symbols:

(a) C_1
(b) C_2, C_3, C_4 (g)
(c) CR (h)
(d) Q_1 and Q_2
(e) J_1 and J_2 (i) R_1, R_2, R_3
(f) S_1 (j) L_1

(k) B_1
(l)

(a)   Variable capacitor   (g)   Conductors crossing but not connected
(b)   Fixed capacitors   (h)   Conductor electrically connected
(c)   Crystal rectifier (CR) diode (i)   Fixed resistors
(d)   PNP transistors   (j)   Variable iron-core inductor (antenna coil)
(e)   Jacks   (k)   Battery
(f)   Single-pole single-throw switch   (l)   Antenna

2.65   Using the circuit symbols in Fig. 2-2, draw a schematic circuit containing an ac generator, a switch, an ammeter, a bell, and a buzzer. Label the diagram carefully.

See Fig. 2-8.

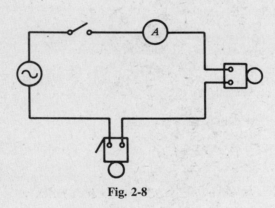

Fig. 2-8

2.66   Identify the symbols for transformers shown in Fig. 2-9.

Fig. 2-9

(a) Air-core transformer; (b) Iron-core transformer; (c) Iron-core variable transformer. The heavy lines between the coils represent an iron core. The iron core is simply a piece of iron around which the coils are wrapped. An arrow running through the coils of a transformer means the transformer is variable.

2.67   Since an electric circuit is a path for current to flow through, a break in this path would stop current flow. Switches are simply ways to break this path or to control the flow of current. Some more symbols for switches are shown in Fig. 2-10. To remember these switches, think of the number of poles as the number of wires coming to either side of the switch. The throws can be thought of as the number of *on* positions. Answer the following questions:

(*a*)   The number of wires coming to each side of a double-pole switch is _____ .

(*b*)   The number of *on* positions a DPST switch has is _____ .

(*c*)   The number of *on* positions a DPDT switch has is _____ .

(*a*) two;   (*b*) one;   (*c*) two

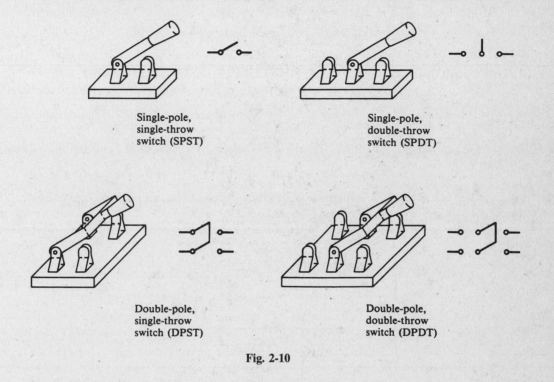

Single-pole,
single-throw
switch (SPST)

Single-pole,
double-throw
switch (SPDT)

Double-pole,
single-throw
switch (DPST)

Double-pole,
double-throw
switch (DPDT)

Fig. 2-10

2.68   The ground symbol is often used in schematic diagrams. Some components are grounded to the frame or chassis of the equipment in which they are located. Aircraft, automobiles, and TV sets are grounded in this fashion. A frame or other ground must be a good conductor. The ground (frame) is the return path for current back to the power source. For the circuit shown in Fig. 2-11, complete the circuit by putting in the ground symbols:

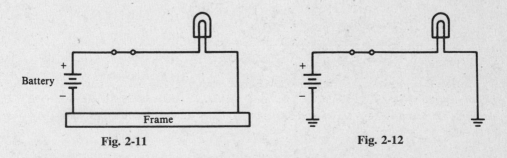

Battery

Frame

Fig. 2-11   **Fig. 2-12**

See Fig. 2-12. In this circuit, current flows from the plus terminal of the battery, through the switch to the lamp to ground, and then to the negative terminal of the battery. Since both the battery and the lamp are grounded to the frame, there is a complete circuit.

Supplementary Problems

2.69   Write the words which correctly complete the following statements.

(a)   The use of a schematic diagram makes it possible to trace the function of a circuit from _____ to _____ .

(b)   Examples of letter symbols to identify components on a schematic diagram are _____ for transformers, _____ for resistors, and _____ for transistors.

(c)   The _____ symbol is used to show that wires are electrically _____ at that point.

(d)   The _____ diagram is most often used to show complex electrical systems.

(e)   The _____ diagram is an easy way to show the relationships of various parts of a circuit.

(f)   Floor-plan diagrams are used with electrical _____ systems.

Ans.   (a) input, output;   (b) T, R, Q;   (c) dot, connected;   (d) one-line or single-line;   (e) block;
(f) wiring

2.70   Draw a schematic diagram showing a dc generator, a switch, a fuse, an arc lamp, and a resistor. Label the diagram carefully.   *Ans.*   See Fig. 2-13.

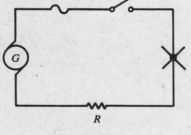

R

Fig. 2-13

2.71   A rectifier is a device that changes alternating current to direct current. Show a symbol for the rectifier.

Ans.

2.72   Match each symbol with its respective component.

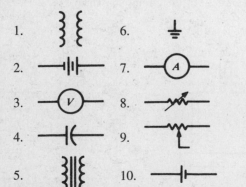

1.   6.
2.   7.
3.   8.
4.   9.
5.   10.

(a)   Voltmeter
(b)   Resistor
(c)   Cell
(d)   Ammeter
(e)   Battery
(f)   Rheostat
(g)   Iron-core transformer
(h)   Variable resistor
(i)   Ground
(j)   Air-core transformer
(k)   Capacitor
(l)   Lamp
(m)   Variable capacitor

Ans.   1. (*j*)   2. (*e*)   3. (*a*)   4. (*k*)   5. (*g*)   6. (*i*)   7. (*d*)   8. (*h*)   9. (*f*)   10. (*c*)

2.73   Place a ground symbol wherever one is needed in the diagram of Fig. 2-14.   *Ans.*   See Fig. 2-15

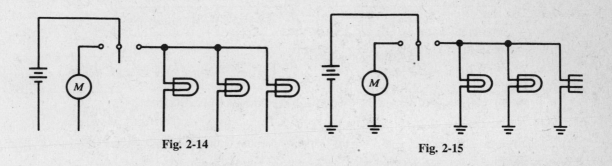

Fig. 2-14   Fig. 2-15

2.74   For the wire connections shown below, identify each as "connection" or "no connection."

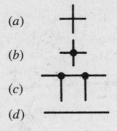

(*a*)

(*b*)

(*c*)

(*d*)

Ans.   (*a*) no connection;   (*b*) connection;   (*c*) connection;   (*d*) no connection

2.75   A fuse is a safety device which operates as a switch to turn a circuit off when the current exceeds a specific value. A circuit breaker performs the same protective function as a fuse, but unlike a fuse it can be reset. Show the symbols for a fuse and circuit breaker.

Ans.   Fuse: ⎓⌇⎓   Circuit breaker: ⎓⌒⎓

2.76   Match the name of each component with its respective symbol.

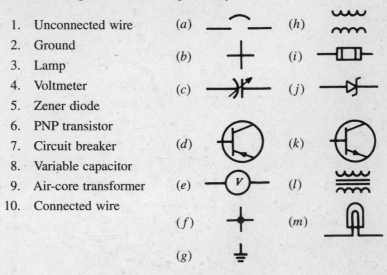

1.   Unconnected wire
2.   Ground
3.   Lamp
4.   Voltmeter
5.   Zener diode
6.   PNP transistor
7.   Circuit breaker
8.   Variable capacitor
9.   Air-core transformer
10.   Connected wire

Ans.   1. (*b*)   2. (*g*)   3. (*m*)   4. (*e*)   5. (*j*)   6. (*d*)   7. (*a*)   8. (*c*)   9. (*h*)   10. (*f*)

2.77 An elementary radio receiver consists of four major stages: an antenna, a tunable resonant circuit, a detector circuit, and headphones. Draw a block diagram of this elementary receiver.
Ans. See Fig. 2-16.

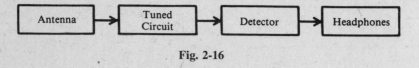

Fig. 2-16

2.78 Match each symbol with its meaning used in architectural floor-plan diagrams.

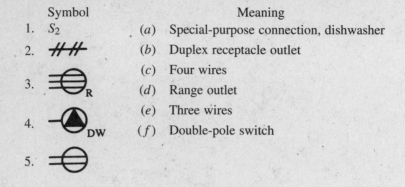

Symbol		Meaning
1. S_2	(a)	Special-purpose connection, dishwasher
2.	(b)	Duplex receptacle outlet
3.	(c)	Four wires
4.	(d)	Range outlet
5.	(e)	Three wires
	(f)	Double-pole switch

Ans.   1. (*f*)   2. (*c*)   3. (*d*)   4. (*a*)   5. (*b*)

Chapter 3

Ohm's Law and Power

THE ELECTRIC CIRCUIT

A practical electric circuit has at least four parts: (1) a source of electromotive force, (2) conductors, (3) a load, and (4) a means of control (Fig. 3-1). The emf is the battery, the conductors are wires that connect the various parts of the circuit and conduct the current, the resistor is the load, and the switch is the control device. The most common sources of emf are batteries and generators. Conductors are wires which offer low resistance to a current. The load resistor represents a device that uses electric energy, such as a lamp, bell, toaster, radio, or a motor. Control devices might be switches, variable resistances, fuses, circuit breakers, or relays.

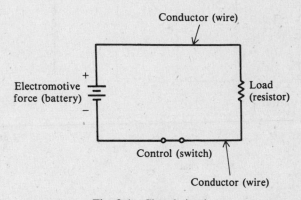

Fig. 3-1   Closed circuit

A complete or *closed circuit* (Fig. 3-1) is an unbroken path for current from the emf, through a load, and back to the source. A circuit is called *incomplete* or *open* (Fig. 3-2a) if a break in the circuit does not provide a complete path for current.

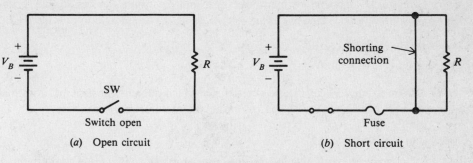

Fig. 3-2   Open and short circuits

To protect a circuit, a fuse is placed directly into the circuit (Fig. 3-2b). A fuse will open the circuit whenever a dangerously large current starts to flow. A fuse will permit currents smaller than the fuse value to flow but will melt and therefore break or open the circuit if a larger current flows. A dangerously large current will flow when a "short circuit" occurs. A short circuit is usually caused by an accidental connection between two points in a circuit which offers very little resistance (Fig. 3-2b).

A ground symbol is often used to show that a number of wires are connected to a common point in a circuit. For example, in Fig. 3-3a, conductors are shown making a complete circuit, while in Fig. 3-3b, the same circuit is shown with two ground symbols at $G1$ and $G2$. Since the ground symbol means that the two points are connected to a common point, electrically the two circuits (Fig. 3-3a and b) are exactly the same.

39

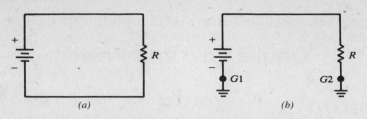

Fig. 3-3   Closed circuits *a* and *b* are the same

Example 3.1   Replace with ground symbols the return wire of the closed circuit (in Fig. 3-4*a*).
See Fig. 3-4*b*.

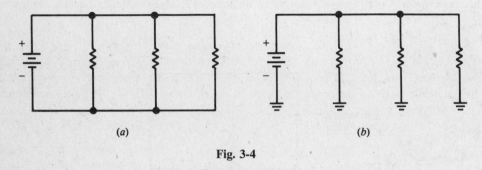

Fig. 3-4

RESISTANCE

Resistance is the opposition to current flow. To add resistance to a circuit, electrical components called resistors are used. A *resistor* is a device whose resistance to current flow is a known, specified value. Resistance is measured in ohms and is represented by the symbol *R* in equations. One ohm is defined as that amount of resistance that will limit the current in a conductor to one ampere when the voltage applied to the conductor is one volt.

Resistors are common components of many electrical and electronic devices. Some frequent uses for resistors are to establish the proper value of circuit voltage, to limit current, and to provide a load.

FIXED RESISTORS

A *fixed resistor* is one that has a single value of resistance which remains constant under normal conditions. The two main types of fixed resistors are carbon-composition and wire-wound resistors.

Carbon-Composition Resistors

The resistance element is primarily graphite or some other form of solid carbon carefully made to provide the desired resistance. These resistors generally are inexpensive and have resistance values that range from $0.1\,\Omega$ to $22\,\text{M}\Omega$.

Wire-Wound Resistors

The resistance element is usually nickel–chromium wire wound on a ceramic rod. The entire assembly is normally covered with a ceramic material or a special enamel. They have resistance values from $1\,\Omega$ to $100\,\text{k}\Omega$.

The actual resistance of a resistor may be greater or less than its rated or nominal value. The limit of actual resistance is called *tolerance*. Common tolerances of carbon-composition resistors are ±5, ±10, and ±20 percent. For example, a resistor having a rated resistance of $100\,\Omega$ and a tolerance of ±10 percent may have an actual resistance of any value between 90 and 110 Ω, that is, 10 Ω less or more than the rated value of 100. Wire-wound resistors usually have a tolerance of ±5 percent.

Resistors having high tolerances of ±20 percent can still be used in many electric circuits. The advantage of using a high-tolerance resistor in any circuit where it is permissible is that it is less expensive than a low-tolerance resistor.

The *power rating* of a resistor (sometimes called the "wattage" rating) indicates how much heat a resistor can dissipate, or throw off, before being damaged. If more heat is generated than can be dissipated, the resistor will be damaged. The power rating is specified in watts. Carbon-composition resistors have wattage ratings which range from 1/16 to 2 W, while wire-wound resistors have ratings from 3 W to hundreds of watts.

The physical size of a resistor is no indication of its resistance. A tiny resistor can have a very low or a very high resistance. The physical size, however, gives some indication of its power rating. For a given value of resistance, the physical size of a resistor increases as the power rating increases.

VARIABLE RESISTORS

Variable resistors are used to vary or change the amount of resistance in a circuit. Variable resistors are called *potentiometers* or *rheostats*. Potentiometers generally consist of carbon-composition resistance elements, while the resistance element in a rheostat is usually made of resistance wire. In both devices, a sliding arm makes contact with the stationary resistance element (Fig. 3-5).

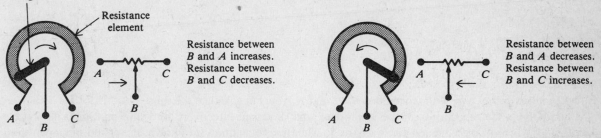

Fig. 3-5   When the sliding arm of a variable resistor is moved, the resistance between the center terminal and end terminals changes

As the sliding arm rotates, its point of contact on the resistance element changes, thus changing the resistance between the sliding arm terminal and the terminals of the stationary resistance (Fig. 3-5).

Rheostats are often used to control very high currents such as those found in motor and lamp loads (Fig. 3-6).

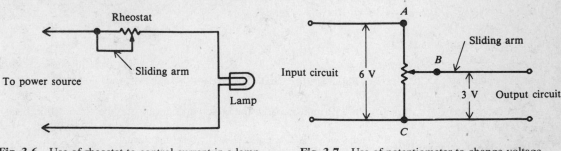

Fig. 3-6   Use of rheostat to control current in a lamp circuit

Fig. 3-7   Use of potentiometer to change voltage

Potentiometers can be used to vary the value of voltage applied to a circuit (Fig. 3-7). In this circuit, the input voltage is applied across the terminals *AC* of the stationary resistance. By varying the position of the sliding arm (terminal *B*), the voltage across terminals *BC* will change. As the sliding arm moves closer to terminal C, the voltage of the output circuit decreases. As the sliding arm moves closer to terminal A, the output voltage of the circuit increases. Potentiometers as control devices are found in amplifiers, radios; television sets, and electrical instruments. The rating of a variable resistor is the resistance of the entire stationary resistance element from one end terminal to the other.

OHM'S LAW

Ohm's law defines the relationship between current, voltage, and resistance. There are three ways to express Ohm's law mathematically.

1. The current in a circuit is equal to the voltage applied to the circuit divided by the resistance of the circuit:

$$I = \frac{V}{R} \qquad (3\text{-}1)$$

2. The resistance of a circuit is equal to the voltage applied to the circuit divided by the current in the circuit:

$$R = \frac{V}{I} \qquad (3\text{-}2)$$

3. The applied voltage to a circuit is equal to the product of the current and the resistance of the circuit:

$$V = I \times R = IR \qquad (3\text{-}3)$$

where I = current, A
        R = resistance, Ω
        V = voltage, V

If you know any two of the quantities V, I, and R, you can calculate the third.

The Ohm's law equations can be memorized and practiced effectively by using an Ohm's law circle (Fig. 3-8a). To find the equation for V, I, or R when two quantities are known, cover the unknown third quantity with your finger.

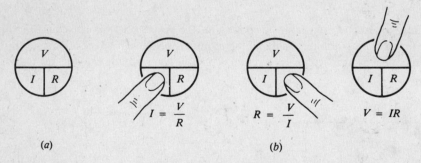

(a)   (b)

Fig. 3-8   The Ohm's law circle

The other two quantities in the circle will indicate how the covered quantity may be found (Fig. 3-8b).

Example 3.2   Find I when $V = 120$ V and $R = 30\ \Omega$. Use Eq. (3-1) to find the unknown I.

$$I = \frac{V}{R}$$
$$= \frac{120}{30} = 4\text{ A} \qquad Ans.$$

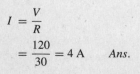
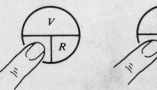

Example 3.3   Find R when $V = 220$ V and $I = 11$ A. Use Eq. *(3-2)* to find the unknown R.

$$R = \frac{V}{I}$$

$$= \frac{220}{11} = 20 \ \Omega \qquad Ans.$$

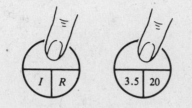

Example 3.4   Find V when $I = 3.5$ A and $R = 20 \ \Omega$. Use Eq. *(3-3)* to find the unknown V.

$$V = IR = 3.5(20) = 70 \text{ V} \qquad Ans.$$

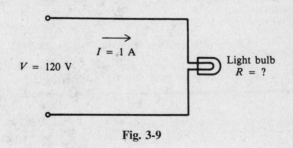

Example 3.5   An electric light bulb draws 1.0 A when operating on a 120-V dc circuit. What is the resistance of the bulb?

The first step in solving a circuit problem is to sketch a schematic diagram of the circuit itself, labeling each of the parts and showing the known values (Fig. 3-9).

$$I = 1 \text{ A}$$

$$V = 120 \text{ V}$$

Light bulb
$R = ?$

Fig. 3-9

Since I and V are known, we use Eq. *(3-2)* to solve for R.

$$R = \frac{V}{I} = \frac{120}{1} = 120 \ \Omega \qquad Ans.$$

ELECTRIC POWER

The electric power P used in any part of a circuit is equal to the current I in that part multiplied by the voltage V across that part of the circuit. Its formula is

$$P = VI \tag{3-4}$$

where P = power, W
V = voltage, V
I = current, A

Other forms for $P = VI$ are $I = P/V$ and $V = P/I$.

If we know the current I and the resistance R but not the voltage V, we can find the power P by using Ohm's law for voltage, so that substituting

$$V = IR \qquad (3\text{-}3)$$

into Eq. (*3-4*) we have

$$P = IR \times I = I^2 R \qquad (3\text{-}5)$$

In the same manner, if we know the voltage V and the resistance R but not the current I, we can find the power P by using Ohm's law for current, so that substituting

$$I = \frac{V}{R} \qquad (3\text{-}1)$$

into Eq. (*3-4*) we have

$$P = V\frac{V}{R} = \frac{V^2}{R} \qquad (3\text{-}6)$$

If you know any two of the quantities, you can calculate the third.

Example 3.6   The current through a 100-Ω resistor to be used in a circuit is 0.20 A. Find the power rating of the resistor.
Since I and R are known, use Eq. (*3-5*) to find P.

$$P = I^2 R = (0.20)^2(100) = 0.04(100) = 4\,\text{W} \qquad Ans.$$

To prevent a resistor from burning out, the power rating of any resistor used in a circuit should be twice the wattage calculated by the power equation. Hence, the resistor used in this circuit should have a power rating of 8 W.

Example 3.7   How many kilowatts of power are delivered to a circuit by a 240-V generator that supplies 20 A to the circuit?
Since V and I are given, use Eq. (*3-4*) to find P.

$$P = VI = 240(20) = 4800\,\text{W} = 4.8\,\text{kW} \qquad Ans.$$

Example 3.8   If the voltage across a 25 000-Ω resistor is 500 V, what is the power dissipated in the resistor?
Since R and V are known, use Eq. (*3-6*) to find P.

$$P = \frac{V^2}{R} = \frac{500^2}{25\,000} = \frac{250\,000}{25\,000} = 10\,\text{W} \qquad Ans.$$

HORSEPOWER

A motor is a device which converts electric power into the mechanical power of a rotating shaft. The electric power supplied to a motor is measured in watts or kilowatts; the mechanical power delivered by a motor is measured in horsepower (hp). One horsepower is equivalent to 746 W of electric power. The metric system will be used to express horsepower in watts. For most calculations, it is sufficiently accurate to consider 1 hp = 750 W or 1 hp = 3/4 kW. To convert between horsepower and kilowatt ratings, use the following equations.

$$hp = \frac{1000 \times kW}{750} = \frac{4}{3} \times kW \tag{3-7}$$

$$kW = \frac{750 \times hp}{1000} = \frac{3}{4} \times hp \tag{3-8}$$

Example 3.9   Change the following units of measurements: (*a*) 7.5 kW to horsepower, and (*b*) 3/4 hp to watts.

(*a*)   Use Eq. (*3-7*):

$$hp = \frac{4}{3} \times kW = \frac{4}{3}(7.5) = 10 \qquad Ans.$$

(*b*)   Use Eq. (*3-8*):

$$kW = \frac{3}{4} \times hp = \frac{3}{4} \times \frac{3}{4} = \frac{9}{16} = 0.563$$

$$1 \, kW = 1000 \, W$$

$$W = 1000(0.563) = 563 \qquad Ans.$$

ELECTRIC ENERGY

Energy and work are essentially the same and are expressed in identical units. Power is different, however, because it is the time rate of doing work. With the watt unit for power, one watt used during one second equals the work of one joule, or one watt is one joule per second. The joule (J) is a basic practical unit of work or energy (see Table 2-3).

The kilowatthour (kWh) is a unit commonly used for large amounts of electric energy or work. The amount of kilowatthours is calculated as the product of the power in kilowatts (kW) and the time in hours (h) during which the power is used.

$$kWh = kW \times h \tag{3-9}$$

Example 3.10   How much energy is delivered in 2 h by a generator supplying 10 kW?
Write Eq. (*3-9*) and substitute given values.

$$kWh = kW \times h = 10(2) = 20$$

$$\text{Energy delivered} = 20 \, kWh \qquad Ans.$$

Solved Problems

3.1   Write the word or words which most correctly complete the following statements.

(*a*)   The four basic parts of a complete circuit are the _____, _____, _____, and _____.

(*b*)   A fixed resistor is one which has a _____ resistance value.

(*c*)   In a carbon-film resistor, a film of _____ is deposited upon a ceramic core.

(*d*)   The _____ rating of a resistor indicates how much current the resistor can conduct before becoming _____.

(e) The physical size of a resistor has no relationship to its _____.

(f) The two most common types of variable resistors are called _____ and _____.

(g) The rated resistance of a variable resistor is the resistance between its _____ terminals.

(h) _____ are used as current-limiting devices.

(i) If the voltage applied to a circuit is doubled and the resistance remains the same, the current in the circuit will increase to _____ the original value.

(j) If the current through a conductor is doubled and the resistance is constant, the power consumed by the conductor will increase to _____ times the original amount.

Ans.   (a) voltage source, conductors, load, control device   (f) rheostats, potentiometers

        (b) specific or single   (g) end

        (c) carbon   (h) Rheostats

        (d) wattage or power, overheated or damaged   (i) twice $(I = V/R)$

        (e) resistance   (j) four $(P = I^2R)$

3.2 In Fig. 3-10, the resistor limits the current in the circuit to 5 A when connected to a 10-V battery. Find its resistance.

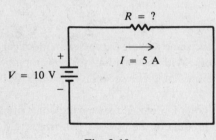

Fig. 3-10

Since I and V are known, solve for R by Ohm's law.

$$R = \frac{V}{I} \tag{3-2}$$

$$= \frac{10}{5} = 2\ \Omega \qquad Ans.$$

3.3 Figure 3-11 shows a doorbell circuit. The bell has a resistance of 8 Ω and requires a 1.5 A current to operate. Find the voltage required to ring the bell.

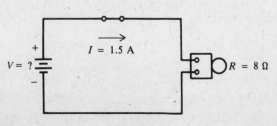

Fig. 3-11

Since R and I are known, solve for V by Ohm's law.

$$V = IR \qquad (3\text{-}3)$$

$$= 1.5(8) = 12\,\text{V} \qquad Ans.$$

3.4   What current will flow through a lamp when it has a resistance of 360 Ω and is connected to an ordinary house voltage of 115 V as shown in Fig. 3-12?

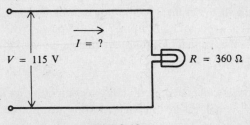

Fig. 3-12

Since R and V are given, calculate I by Ohm's law.

$$I = \frac{V}{R} \qquad (3\text{-}1)$$

$$= \frac{115}{360} = 0.319\,\text{A} \qquad Ans.$$

Values generally will be computed to three significant figures.

3.5   Find the current drawn by a 60-W incandescent lamp rated for 120-V operation. Also find the current drawn by a 150-W, 120-V lamp and a 300-W, 120-V lamp. As the wattage increases, what happens to the current?

P and V are known and we wish to find I. Solving for I in Eq. (3-4),

$$I = \frac{P}{V}$$

For the 60-W, 120-V lamp:

$$I = \frac{60}{120} = 0.5\,\text{A} \qquad Ans.$$

For the 150-W, 120-V lamp:

$$I = \frac{150}{120} = 1.25\,\text{A} \qquad Ans.$$

For the 300-W, 120-V lamp:

$$I = \frac{300}{120} = 2.5\,\text{A} \qquad Ans.$$

We see that if V remains unchanged, the greater the value of P, the greater will be the value of I. That is to say, higher wattages draw higher currents for the same voltage rating.

3.6 Find the power consumed by a fixed 25-Ω resistor for each of the following currents: 3 A, 6 A, and 1.5 A. What effect does a change in current have on the amount of power dissipated by a fixed resistor?

I and R are known and we wish to find P.

$$P = I^2 R \qquad\qquad (3\text{-}5)$$

at
$$\begin{aligned} 3\,\text{A}: \quad & P = 3^2(25) = 225\,\text{W} \qquad Ans. \\ 6\,\text{A}: \quad & P = 6^2(25) = 900\,\text{W} \qquad Ans. \\ 1.5\,\text{A}: \quad & P = (1.5)^2 = (25) = 56.2\,\text{W} \qquad Ans. \end{aligned}$$

If the current is doubled to 6 A from 3 A, the power will increase by 2^2, or 4, so 900 W $= 4 \times$ 225 W. If the current is halved to 1.5 A from 3 A, the power will decrease by $(1/2)^2$, or 1/4, so 56.2 W $= 1/4 \times$ 225 W. We see that if R does not change, power will change according to the square of the change in current.

3.7 The efficiency of a motor is calculated by dividing its output by its input. The output is measured in horsepower, while the input is measured in watts or kilowatts. Before the efficiency can be calculated, the output and the input must be expressed in the same units of measurement. Find the efficiency of a motor which receives 4 kW and delivers 4 hp.

Step 1. Express all measurements in the same units.

$$\text{Input} = 4\,\text{kW}$$

$$\text{Output} = \frac{3}{4} \times \text{hp} = \frac{3}{4}4 = 3\,\text{kW} \qquad\qquad (3\text{-}8)$$

Step 2. Find the efficiency by dividing output by input.

$$\text{Efficiency} = \frac{\text{output}}{\text{input}} = \frac{3\,\text{kW}}{4\,\text{kW}} = 0.75$$

Efficiency is not expressed in any units. To change the decimal efficiency into a percent efficiency, move the decimal point two places to the right and add a percent sign (%).

$$\text{Efficiency} = 0.75 = 75\% \qquad Ans.$$

3.8 The motor in a washing machine uses 1200 W. How much energy in kilowatthours is used in a week by a laundromat with eight washers if they are all in use 10 hours per day (h/day) for a 6-day week?

Change 1200 W to 1.2 kW.

For one motor:
$$\text{Energy} = 1.2\,\text{kW} \times \frac{10\,\text{h}}{\text{day}} \times 6\,\text{days} = 72\,\text{kWh}$$

For eight motors:
$$\text{Energy} = 8 \times 72\,\text{kWh} = 576\,\text{kWh} \qquad Ans.$$

3.9 A radio receiver draws 0.9 A at 110 V. If the set is used 3 h/day, how much energy does it consume in 7 days?

Find the power.

$$P = VI = 110(0.9) = 99\,\text{W} = 0.099\,\text{kW}$$

Then find the energy.

$$\text{Energy} = 0.099\,\text{kW} \times \frac{3\,\text{h}}{\text{day}} \times 7\,\text{days} = 2.08\,\text{kWh} \qquad Ans.$$

3.10 Electric utility companies establish their rates at a given number of cents per kilowatthour. Rates for electric energy in the United States depend upon the method by which the electricity is generated, the type and complexity of the transmission and distribution systems, the maintenance cost, and many other factors. By knowing the amount of energy you use (from the meter readings) and the cost per kilowatthour of energy in your area, you can calculate your own monthly electric bill.

One residence used 820 kWh of electric energy in one month. If the utility rate is 6 cents per kilowatthour, what was the owner's electrical bill for the month?

A convenient formula for calculating the total cost is:

$$\text{Total cost} = \text{kWh} \times \text{unit cost}$$
$$= 820 \times 6 \text{ cents} = 4920 \text{ cents} = \$49.20^* \qquad Ans.$$

---

*Here we are using an average unit cost. Most utility rates are stepped according to blocks of kilowatthour usage plus minimum charges, not to mention full adjustment charges and taxes.

Supplementary Problems

3.11 Write the word or words which most correctly complete the following statements.

(a) Common sources of energy used in electric circuits are _____ and _____.

(b) In a circuit, an incandescent lamp is treated as a _____ load.

(c) The resistance element of a wire-wound resistor is made of _____ wire.

(d) The amount by which the actual resistance of a resistor may vary from its rated value is called its _____.

(e) A large resistor of a given type has a higher _____ rating than a smaller resistor of the same type.

(f) A common resistor defect is an open or burned-out condition caused by excessive _____ through a resistor.

(g) The amount of resistance can be changed in a circuit by a _____ resistor.

(h) A variable resistor used to change the value of voltage applied to a circuit is the _____.

(i) If the resistance of a circuit is doubled and the current remains unchanged, the voltage will increase to _____ its original value.

(j) If a toaster rated at 1000 W is operated for 30 min, the energy used is _____ kWh.

Ans. (a) batteries, generators; (b) resistive; (c) nickel–chromium; (d) tolerance; (e) wattage or power; (f) current; (g) variable; (h) potentiometer; (i) twice ($V = IR$); (j) 0.5

3.12 Use Ohm's law to fill in the indicated quantity.

	V	I	R
(a)	?	2 A	3 Ω
(b)	120 V	?	2400 Ω
(c)	120 V	24 A	?

Ans.

	V	I	R
(a)	6 V	....	....
(b)	....	0.05 A	....
(c)	....	....	5 Ω

	V	I	R
(d)	?	8 mA	5 kΩ
(e)	60 V	?	12 kΩ
(f)	110 V	2 mA	?
(g)	?	2.5 A	6.4 Ω
(h)	2400 V	?	1 MΩ

Ans.

	V	I	R
(d)	40 V	....	....
(e)	....	5 mA	....
(f)	....	....	55 kΩ
(g)	16 V	....	....
(h)	....	2.4 mA	....

3.13 A circuit consists of a 6-V battery, a switch, and a lamp. When the switch is closed, 2 A flows in the circuit. What is the resistance of the lamp?   *Ans.*   3 Ω

3.14 Suppose you replace the lamp in Problem 3.13 with another one, requiring the same 6 V across it but drawing only 0.04 A. What is the resistance of the new lamp?   *Ans.*   150 Ω

3.15 A voltage of 20 V is measured across a 200-Ω resistor. What is the current flowing through the resistor? *Ans.*   0.10 A or 100 mA

3.16 If the resistance of the air gap in an automobile spark plug is 2500 Ω, what voltage is needed to force 0.20 A through it?   *Ans.*   500 V

3.17 The filament of a television tube has a resistance of 90 Ω. What voltage is required to produce the tube's rated current of 0.3 A?   *Ans.*   27 V

3.18 A 110-V line is protected with a 15-A fuse. Will the fuse "carry" a 6-Ω load?   *Ans.*   No

3.19 A sensitive dc meter takes 9 mA from a line when the voltage is 108 V. What is the resistance of the meter?   *Ans.*   12 kΩ

3.20 An automobile dashboard ammeter shows 10.8 A of current flowing when the headlights are lit. If the current is drawn from the 12-V storage battery, what is the resistance of the headlights? *Ans.*   1.11 Ω

3.21 A 160-Ω telegraph relay coil operates on a voltage of 6.4 V. Find the current drawn by the relay. *Ans.*   0.04 A

3.22 What is the power used by a soldering iron taking 3 A at 110 V?   *Ans.*   330 W

3.23 A 12-V battery is connected to a lamp that has a 10-Ω resistance. How much power is delivered to the load?   *Ans.*   14.4 W

3.24 An electric oven uses 35.5 A at 118 V. Find the wattage generated by the oven.   *Ans.*   4190 W

3.25 A 12-Ω resistor in a power supply circuit carries 0.5 A. How many watts of power are dissipated in the resistor? What must be the wattage rating of the resistor in order to dissipate this power safely as heat?   *Ans.*   3 W, 6 W

3.26 Find the power used by a 10-kΩ resistor drawing 0.01 A.   *Ans.*   1 W

3.27 Find the current through a 40-W lamp at 110 V.   *Ans.*   0.364 A

3.28 An electric dryer requires 360 W and draws 3.25 A. Find its operating voltage.   *Ans.*   111 V

3.29 Fill in the indicated quantity.

	hp	kW	W
(a)	$2\frac{1}{4}$	?	?
(b)	?	8.75	?
(c)	?	?	1000

Ans.

	hp	kW	W
(a)	....	1.69	1690
(b)	$11\frac{2}{3}$	....	8750
(c)	$1\frac{1}{3}$	1 kW	....

3.30 A motor delivers 2 hp and receives 1.8 kW of energy. Find its efficiency (see Problem 3.7).
Ans.   83.3 percent

3.31 A generator receives 7 hp and supplies 20 A at 220 V. Find the power supplied by the generator and its efficiency.   *Ans.*   4400 W, 83.8 percent

3.32 A 4-hp lathe motor runs 8 h/day. Find the electric energy in kilowatthours used in a day.
Ans.   24 kWh

3.33 How much power and energy is drawn from a 110-V line by a 22-Ω electric iron in 3 h?
Ans.   550 W, 1.65 kWh

3.34 What does it cost to operate a 5.5-kW electric range for $3\frac{1}{2}$ h at 3.8 cents per kilowatthour?
Ans.   73 cents

3.35 In a certain community, the average rate of electric energy is 4.5 cents per kilowatthour. Find the cost of operating a 200-W stereo receiver in this community for 12 h.   *Ans.*   11 cents

Chapter 4

Direct-Current Series Circuits

VOLTAGE, CURRENT, AND RESISTANCE IN SERIES CIRCUITS

A *series circuit* is a circuit in which there is only one path for current to flow along. In the series circuit (Fig. 4-1), the current I is the same in all parts of the circuit. This means that the current flowing through R_1 is the same as the current through R_2, is the same as the current through R_3, and is the same as the current supplied by the battery.

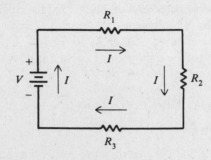

Fig. 4-1 A series circuit

When resistances are connected in series (Fig. 4-1), the total resistance in the circuit is equal to the sum of the resistances of all the parts of the circuit, or

$$R_T = R_1 + R_2 + R_3 \qquad (4\text{-}1)$$

where
$$R_T = \text{total resistance, } \Omega$$
$$R_1, R_2, \text{ and } R_3 = \text{resistance in series, } \Omega$$

Example 4.1 A series circuit has a 50-Ω, a 75-Ω, and a 100-Ω resistor in series (Fig. 4-2). Find the total resistance of the circuit.

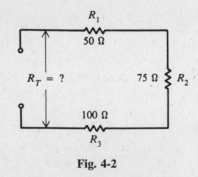

Fig. 4-2

Use Eq. (*4-1*) and add the values of the three resistors in series.

$$R_T = R_1 + R_2 + R_3 = 50 + 75 + 100 = 225 \ \Omega \qquad Ans.$$

The total voltage across a series circuit is equal to the sum of the voltages across each resistance of the circuit (Fig. 4-3), or

$$V_T = V_1 + V_2 + V_3 \qquad (4\text{-}2)$$

where V_T = total voltage, V
        V_1 = voltage across resistance R_1, V
        V_2 = voltage across resistance R_2, V
        V_3 = voltage across resistance R_3, V

Although Eqs. (*4-1*) and (*4-2*) were applied to circuits containing only three resistances, they are applicable to any number of resistances n; that is,

$$R_T = R_1 + R_2 + R_3 + \cdots + R_n \qquad (4\text{-}1a)$$

$$V_T = V_1 + V_2 + V_3 + \cdots + V_n \qquad (4\text{-}2a)$$

Ohm's law may be applied to an entire series circuit or to the individual parts of the circuit. When it is used on a particular part of a circuit, the voltage across that part is equal to the current in that part multiplied by the resistance of the part. For the circuit shown in Fig. 4-3,

$$V_1 = IR_1$$

$$V_2 = IR_2$$

$$V_3 = IR_3$$

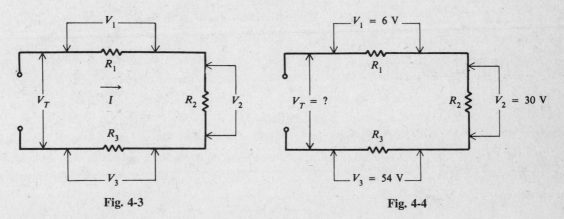

Fig. 4-3   Fig. 4-4

Example 4.2   A series circuit has 6 V across R_1, 30 V across R_2, and 54 V across R_3 (Fig. 4-4). What is the total voltage across the circuit?

Write Eq. (*4-2*) and add the voltage across each of the three resistances.

$$V_T = V_1 + V_2 + V_3 = 6 + 30 + 54 = 90 \text{ V} \qquad Ans.$$

To find the total voltage across a series circuit, multiply the current by the total resistance, or

$$V_T = IR_T \qquad (4\text{-}3)$$

where V_T = total voltage, V
        I = current, A
        R_T = total resistance, Ω

Remember that in a series circuit, the same current flows in every part of the circuit. *Do not* add the current in each part of the circuit to obtain I in the Eq. (*4-3*).

Example 4.3   A resistor of 45 Ω and a bell of 60 Ω are connected in series (Fig. 4-5). What voltage is required across this combination to produce a current of 0.3 A?

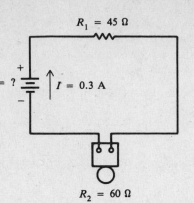

$R_1 = 45\ \Omega$

$V_T = ?$ $I = 0.3\ A$

$R_2 = 60\ \Omega$

Fig. 4-5

Step 1. Find the current I. The value of the current is the same in each part of a series circuit.

$$I = 0.3\ A \qquad \text{(Given)}$$

Step 2. Find the total resistance R_T. Add the two resistances.

$$R_T = R_1 + R_2 \tag{4-1}$$
$$= 45 + 60 = 105\ \Omega$$

Step 3. Find the total voltage V_T. Use Ohm's law.

$$V_T = IR_T \tag{4-3}$$
$$= 0.3(105) = 31.5\ V \qquad Ans.$$

Example 4.4 A 95-V battery is connected in series with three resistors: 20 Ω, 50 Ω, and 120 Ω (Fig. 4-6). Find the voltage across each resistor.

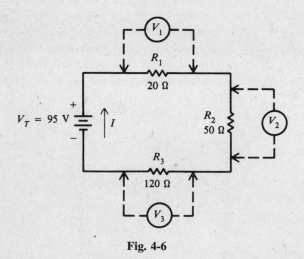

V_1

R_1

20 Ω

$V_T = 95\ V$ I

R_2
50 Ω V_2

R_3

120 Ω

V_3

Fig. 4-6

Step 1. Find the total resistance R_T.

$$R_T = R_1 + R_2 + R_3 \tag{4-1}$$
$$= 20 + 50 + 120 = 190\ \Omega$$

Step 2. Find the current I. Write Ohm's law,

$$V_T = IR_T \qquad\qquad (4\text{-}3)$$

from which we get

$$I = \frac{V_T}{R_T} = \frac{95}{190} = 0.5 \text{ A}$$

Step 3. Find the voltage across each part. In a series circuit, the current is the same in each part; that is, $I = 0.5$ A through each resistor.

$$V_1 = IR_1 = 0.5(20) = 10 \text{ V} \qquad Ans.$$

$$V_2 = IR_2 = 0.5(50) = 25 \text{ V} \qquad Ans.$$

$$V_3 = IR_3 = 0.5(120) = 60 \text{ V} \qquad Ans.$$

The voltages V_1, V_2, and V_3 found in Example 4.4 are known as *voltage drops* or *IR drops*. Their effect is to reduce the voltage that is available to be applied across the rest of the components in the circuit. The sum of the voltage drops in any series circuit is always equal to the voltage that is applied to the circuit. This relationship is expressed in Eq. (*4-2*), where the total voltage V_T is the same as the applied voltage, and can be verified in Example 4.4.

$$V_T = V_1 + V_2 + V_3$$

$$95 = 10 + 25 + 60$$

$$95 \text{ V} = 95 \text{ V} \qquad Check$$

POLARITY OF VOLTAGE DROPS

When there is a voltage drop across a resistance, one end must be more positive or more negative than the other end. The polarity of the voltage drop is determined by the direction of conventional current from a positive to a more negative potential. Current direction is through R_1 from point A to B (Fig. 4-7). Therefore the end of R_1 connected to point A has a more positive potential than point B. We say that the voltage across R_1 is such that point A is more positive than point B. Similarly, the voltage of point C is positive

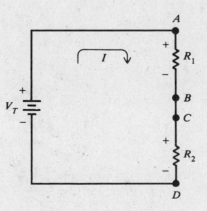

Fig. 4-7   Polarity of voltage drops

with respect to point D. Another way to look at polarity between any two points is that the one nearer to the positive terminal of the voltage source is more positive; also, the point nearer to the negative terminal of the applied voltage is more negative. Therefore, point A is more positive than B, while D is more negative than C (Fig. 4-7).

Example 4.5 Refer to Example 4.4. Ground the negative terminal of the 95-V battery (Fig. 4-6). Mark the polarity of voltage drops in the circuit (Fig. 4-8), and find the voltage values at points A, B, C, and D with respect to ground.

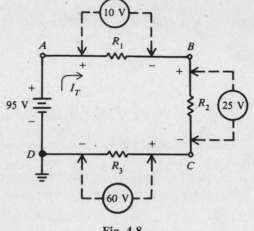

Fig. 4-8

Trace the complete circuit in the direction of current from the positive terminal of the battery to A, A to B, B to C, C to D, and D to the negative terminal. Mark plus (+) where current enters each resistor and minus (−) where current leaves each resistor (Fig. 4-8).

The voltage drops calculated in Example 4.4 are indicated (Fig. 4-8). Point A is the nearest point to the positive side of the terminal, so voltage at A is

$$V_A = +95 \text{ V} \qquad Ans.$$

There is a voltage drop of 10 V across R_1, so voltage at B is

$$V_B = 95 - 10 = +85 \text{ V} \qquad Ans.$$

There is a voltage drop of 25 V across R_2, so voltage at C is

$$V_C = 85 - 25 = +60 \text{ V} \qquad Ans.$$

There is a voltage drop of 60 V across R_3, so voltage at D is

$$V_D = 60 - 60 = 0 \text{ V} \qquad Ans.$$

Since we grounded the circuit at D, V_D must equal 0 V. If on tracing the voltage values, we find that V_D is not equal to 0 V, then we have made an error.

CONDUCTORS

A conductor is a material having many free electrons. Three good electrical conductors are copper, silver, and aluminum. Generally most metals are good conductors. Copper is the most common material used in electrical conductors. Second to copper is aluminum. Certain gases are also used as conductors under special conditions. For example, neon gas, mercury vapor, and sodium vapor are used in various kinds of lamps.

Conductors have very low resistance. A typical value for copper wire is less than 1 Ω for 10 feet (ft). The function of the wire conductor is to connect a source of applied voltage to a load resistance with a minimum *IR* voltage drop in the conductor so that most of the applied voltage can produce current in the load resistance.

Example 4.6   The resistance of two 10-ft lengths of copper wire conductors is about 0.05 Ω, which is very small compared with the 150-Ω resistance of the tungsten filament in the bulb shown in Fig. 4-9a. The conductors should have minimum resistance to light the bulb with full brilliance. When the current of 0.8 A flows in the bulb and series conductors, the *IR* voltage drop across the conductors is 0.04 V, with 109.96 V across the bulb (Fig. 4.9b). Practically all the applied voltage of 110 V is across the filament of the bulb.

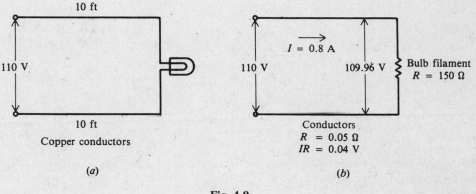

Fig. 4-9

Wire Measurement

Table 4-1 lists the standard wire sizes which correspond to the American Wire Gauge (AWG). The gauge numbers specify the size of round wire in terms of its diameter and cross-sectional circular area. Note the following:

1. As the gauge numbers increase from 1 to 40, the diameter and circular area decrease. Higher gauge numbers mean smaller wire sizes. Thus, No. 12 is a smaller wire than No. 4.

2. The circular area doubles for every three gauge sizes. For example, No. 12 wire has about twice the area of No. 15 wire.

3. The higher the gauge number and the smaller the wire, the greater the resistance of the wire for any given length. Therefore, for 1000 ft of wire, No. 12 has a resistance of 1.62 Ω while No. 4 has 0.253 Ω.

In typical house wiring applications, No. 14 or No. 12 wire is used for circuits where the current is not expected to exceed 15 A. Hookup wire for radio receiver circuits with current in milliamperes is about No. 22 wire. For this size, 0.5–1 A is the maximum current the wire can carry without excessive heating.

The cross-sectional area of round wire is measured in circular mils (abbreviated cmil or CM). A mil is one-thousandth of an inch (0.001 in). One circular mil is the cross-sectional area of a wire having a diameter of one mil. The number of circular mils in any circular area is equal to the square of the diameter in mils, or

$$\text{cmil} = \text{CM} = d^2 \qquad (4\text{-}4)$$

Table 4-1 Copper Wire Table

Gauge No.	Diameter, d, mil	Circular-mil Area, d^2	Ohms per 1000 ft of Copper Wire at 25°C*	Gauge No.	Diameter, d, mil	Circular-mil Area, d^2	Ohms per 1000 ft of Copper Wire at 25°C*
1	289.3	83 690	0.1264	21	28.46	810.1	13.05
2	257.6	66 370	0.1593	22	25.35	642.4	16.46
3	229.4	52 640	0.2009	23	22.57	509.5	20.76
4	204.3	41 740	0.2533	24	20.10	404.0	26.17
5	181.9	33 100	0.3195	25	17.90	320.4	33.00
6	162.0	26 250	0.4028	26	15.94	254.1	41.62
7	144.3	20 820	0.5080	27	14.20	201.5	52.48
8	128.5	16 510	0.6405	28	12.64	159.8	66.17
9	114.4	13 090	0.8077	29	11.26	126.7	83.44
10	101.9	10 380	1.018	30	10.03	100.5	105.2
11	90.74	8 234	1.284	31	8.928	79.70	132.7
12	80.81	6 530	1.619	32	7.950	63.21	167.3
13	71.96	5 178	2.042	33	7.080	50.13	211.0
14	64.08	4 107	2.575	34	6.305	39.75	266.0
15	57.07	3 257	3.247	35	5.615	31.52	335.0
16	50.82	2 583	4.094	36	5.000	25.00	423.0
17	45.26	2 048	5.163	37	4.453	19.83	533.4
18	40.30	1 624	6.510	38	3.965	15.72	672.6
19	35.89	1 288	8.210	39	3.531	12.47	848.1
20	31.96	1 022	10.35	40	3.145	9.88	1069

*20–25°C or 68–77°F is considered average room temperature.

Example 4.7   Find the area in circular mils of a wire with a diameter of 0.004 in.

First, convert the diameter to mils: 0.004 in = 4 mil. Then use Eq. (*4-4*) to find the cross-sectional area.

$$\text{CM} = d^2 = (4 \text{ mil})^2 = 16 \qquad Ans.$$

To prevent the conductors from short-circuiting to each other or to some other metal in the circuit, the wires are insulated. The insulator material should have very high resistance, be tough, and age without becoming brittle.

Resistivity

For any conductor, the resistance of a given length depends upon the resistivity of the material, the length of the wire, and the cross-sectional area of the wire according to the formula

$$R = \rho \frac{l}{A} \qquad\qquad (4\text{-}5)$$

where R = resistance of the conductor, Ω

l = length of the wire, ft

A = cross-sectional area of the wire, CM

ρ = specific resistance or resistivity, CM $\cdot$ Ω/ft

The factor ρ (Greek letter rho, pronounced "roe") permits different materials to be compared for resistance according to their nature without regard to different lengths or areas. Higher values of ρ mean more resistance.

Table 4-2 lists resistance values for different metals having the standard wire size of a 1-ft length with a cross-sectional area of 1 CM. Since silver, copper, gold, and aluminum have the lowest values of resistivity, they are the best conductors. Tungsten and iron have a much higher resistivity.

Table 4-2 Properties of Conducting Materials*

Material	ρ = Specific Resistance, at 20°C, CM · Ω/ft	Temperature Coefficient, Ω per °C, α
Aluminum	17	0.004
Carbon	†	−0.0003
Constantan	295	(average)
Copper	10.4	0.004
Gold	14	0.004
Iron	58	0.006
Nichrome	676	0.0002
Nickel	52	0.005
Silver	9.8	0.004
Tungsten	33.8	0.005

*Listings approximate only, since precise values depend on exact composition of material.
†Carbon has about 2500–7500 times the resistance of copper. Graphite is a form of carbon.

Example 4.8   What is the resistance of 500 ft of No. 20 copper wire?

From Table 4-1, the cross-sectional area for No. 20 wire is 1022 CM. From Table 4-2, ρ for copper is 10.4 CM · Ω/ft. Use Eq. (4-5) to find the resistance of 500 ft of wire.

$$R = \rho \frac{l}{A} = (10.4) \times \left(\frac{500}{1022} \right) = 5.09 \ \Omega \qquad Ans.$$

Example 4.9   What is the resistance of 500 ft of No. 23 copper wire?

From Table 4-1,

$$A = 509.5 \ CM$$

From Table 4-2,

$$\rho = 10.4 \ CM \cdot \Omega/ft$$

Substituting into Eq. (4-5),

$$R = \rho \frac{l}{A} = (10.4) \times \left(\frac{500}{509.5} \right) = 10.2 \ \Omega \qquad Ans.$$

Note from Examples 4.8 and 4.9 that the increase in gauge size of 3 from No. 20 to No. 23 gives one-half the circular area and doubles the resistance for the same wire length.

Temperature Coefficient

The temperature coefficient of resistance, α (Greek letter alpha), indicates how much the resistance changes for a change in temperature. A positive value for α means R increases with temperature; a negative α means R decreases; and a zero α means R is constant, not varying with changes in temperature. Typical values of α are listed in Table 4-2.

Although for a given material, α may vary slightly with temperature, an increase in wire resistance caused by a rise in temperature can be approximately determined from the equation

$$R_t = R_0 + R_0(\alpha \Delta T) \tag{4-6}$$

where R_t = higher resistance at higher temperature, Ω
R_0 = resistance at 20°C, Ω
α = temperature coefficient, $\Omega/°C$
ΔT = temperature rise above 20°C, °C

Note that carbon has a negative temperature coefficient (Table 4-2). In general, α is negative for all semiconductors such as germanium and silicon. A negative value for α means less resistance at higher temperatures. Therefore, the resistance of semiconductor diodes and transistors can be reduced considerably when they become hot with normal load current. Observe also that constantan has a value of zero for α (Table 4-2). Thus it can be used for precision wire-wound resistors, which do not change resistance when the temperature increases.

Example 4.10   A tungsten wire has a 10-Ω resistance at 20°C. Find its resistance at 120°C.
From Table 4-2,

$$\alpha = 0.005 \ \Omega/°C$$

The temperature rise is

$$\Delta T = 120 - 20 = 100°C$$

Substituting into Eq. (4-6),

$$R_t = R_0 + R_0(\alpha \Delta T) = 10 + 10(0.005 \times 100) = 10 + 5 = 15 \ \Omega \qquad Ans.$$

Because of the 100°C rise in temperature, the wire resistance is increased by 5 Ω, or 50 percent of its original value of 10 Ω.

TOTAL POWER IN A SERIES CIRCUIT

We found that Ohm's law could be used for total values in a series circuit as well as for individual parts of the circuit. Similarly, the formula for power may be used for total values.

$$P_T = IV_T \tag{4-7}$$

where P_T = total power, W
I = current, A
V_T = total voltage, V

The total power P_T produced by the source in a series circuit can also be expressed as the sum of the individual powers used in each part of the circuit.

$$P_T = P_1 + P_2 + P_3 + \cdots + P_n \tag{4-8}$$

where P_T = total power, W
P_1 = power used in first part, W
P_2 = power used in second part, W
P_3 = power used in third part, W
P_n = power used in nth part, W

Example 4.11 In the circuit shown (Fig. 4-10), find the total power P_T dissipated by R_1 and R_2.

Step 1. Find I by Ohm's law.

$$I = \frac{V_T}{R_T} = \frac{V_T}{R_1 + R_2} = \frac{60}{5 + 10} = 4\,A$$

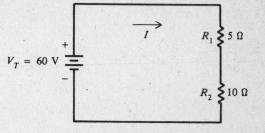

Step 2. Find the power used in R_1 and R_2.

$$P_1 = I^2 R_1 = 4^2(5) = 80\,W$$
$$P_2 = I^2 R_2 = 4^2(10) = 160\,W$$

Fig. 4-10

Step 3. Find the total power P_T by adding P_1 and P_2.

$$P_T = P_1 + P_2 = 80 + 160 = 240\,W \qquad Ans.$$

An alternative method is to use Eq. (4-7) directly.

$$P_T = IV_T$$
$$I = 4\,A$$
$$P_T = 4(60) = 240\,W \qquad Ans.$$

Calculated either way, the total power produced by the battery is 240 W and equals the power used by the load.

VOLTAGE DROP BY PROPORTIONAL PARTS

In a series circuit, each resistance provides a voltage drop V equal to its proportional part of the applied voltage. Stated as an equation,

$$V = \frac{R}{R_T} V_T \qquad (4\text{-}9)$$

where V = voltage, V
R = resistance, Ω
R_T = total resistance, Ω
R/R_T = proportional part of resistance
V_T = total voltage, V

A higher resistance R has a greater voltage drop than a smaller resistance in the same series circuit. Equal resistances have equal voltage drops.

Example 4.12 The circuit (Fig. 4-11) is an example of a proportional voltage divider. Find the voltage drop across each resistor by the method of proportional parts.

Write the formulas, using Eq. (4-9).

$$V_1 = \frac{R_1}{R_T} V_T \qquad V_2 = \frac{R_2}{R_T} V_T \qquad V_3 = \frac{R_3}{R_T} V_T$$

Find R_T.

$$R_T = R_1 + R_2 + R_3 = 20 + 30 + 50 = 100 \text{ k}\Omega$$

Substitute values.

$$V_1 = \frac{20}{100}100 = 20 \text{ V} \qquad Ans.$$

$$V_2 = \frac{30}{100}100 = 30 \text{ V} \qquad Ans.$$

$$V_3 = \frac{50}{100}100 = 50 \text{ V} \qquad Ans.$$

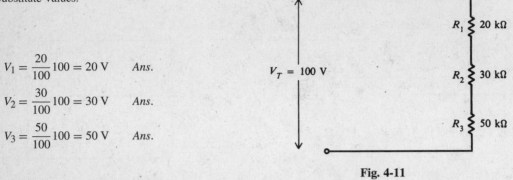

Fig. 4-11

The formula for the proportional method is derived from Ohm's law. For example, add V_1, V_2, and V_3 to obtain

$$V_1 + V_2 + V_3 = \frac{R_1}{R_T} V_T + \frac{R_2}{R_T} V_T + \frac{R_3}{R_T} V_T$$

Factor the right side of the equation.

$$V_1 + V_2 + V_3 = \frac{V_T}{R_T} (R_1 + R_2 + R_3)$$

Use the relationships

$$V_T = V_1 + V_2 + V_3$$
$$R_T = R_1 + R_2 + R_3$$

and substitute.

$$V_T = \frac{V_T}{R_T} R_T = V_T \qquad Check$$

Solved Problems

4.1   Find the voltage needed so that a current of 10 A will flow through the series circuit shown in Fig. 4-12a.

Step 1.   Find total resistance.

$$R_T = R_1 + R_2 + R_3$$
$$= 2 + 3 + 5 = 10 \ \Omega \qquad\qquad (4\text{-}1)$$

Step 2.   Find the voltage (we show the series circuit with R_T in Fig. 4-12b).

$$V_T = IR_T \qquad\qquad (4\text{-}3)$$
$$= 10(10) = 100 \text{ V} \qquad Ans.$$

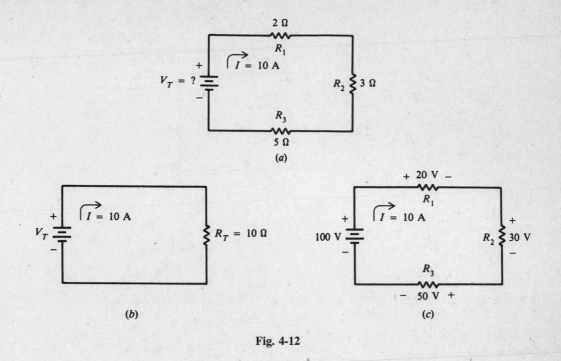

Fig. 4-12

4.2   Find the voltage across each resistor in the circuit of Fig. 4-12*a*. Show that the sum of the voltage drops equals the applied voltage of 100 V.

$$V_1 = IR_1 = 10(2) = 20 \text{ V} \qquad Ans.$$

$$V_2 = IR_2 = 10(3) = 30 \text{ V} \qquad Ans.$$

$$V_3 = IR_3 = 10(5) = 50 \text{ V} \qquad Ans.$$

See Fig. 4-12*c*. Remember that the polarity signs next to the resistors indicate the direction of voltage drops not current direction as the + and − next to the source indicate.

$$\text{Sum of voltage drops} = \text{applied voltage}$$

$$V_1 + V_2 + V_3 = V_T \qquad\qquad (4\text{-}2)$$

$$20 + 30 + 50 = 100$$

$$100 \text{ V} = 100 \text{ V} \qquad Check$$

4.3   In Fig. 4-13, a 12-V battery supplies a current of 2 A. If $R_2 = 2 \, \Omega$, find R_1 and V_1.

Step 1.   Find R_T. By Ohm's law,

$$R_T = \frac{V_T}{I} = \frac{12}{2} = 6 \, \Omega$$

Step 2.   Find R_1.

$$R_T = R_1 + R_2 \qquad\qquad (4\text{-}1)$$

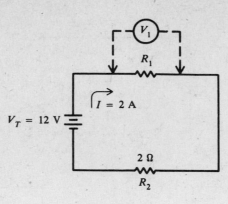

Fig. 4-13

Transposing,

$$R_1 = R_T - R_2 = 6 - 2 = 4\,\Omega \qquad Ans.$$

Step 3. Find V_1.

$$V_1 = IR_1 = 2(4) = 8\,\text{V} \qquad Ans.$$

An alternative method of solution is using voltage drops.

Step 1. Find V_1.

$$V_T = V_1 + V_2$$

Transposing,

$$V_1 = V_T - V_2 = 12 - V_2$$

But $\qquad\qquad\qquad V_2 = IR_2$

so $\qquad\qquad V_1 = 12 - IR_2 = 12 - 2(2) = 12 - 4 = 8\,\text{V} \qquad Ans.$

Step 2. Find R_1.

$$R_1 = \frac{V_1}{I} = \frac{8}{2} = 4\,\Omega \qquad Ans.$$

4.4 For the circuit in Fig. 4-14, find the voltage drop of R_3.

$$\text{Sum of voltage drops} = \text{applied voltage}$$
$$10 + 15 + V_3 + 8 + 10 = 60$$
$$43 + V_3 = 60$$
$$V_3 = 60 - 43 = 17\,\text{V} \qquad Ans.$$

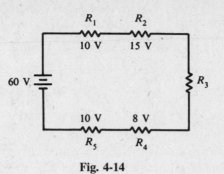

Fig. 4-14

4.5 A series circuit (Fig. 4-15a) uses ground as a common connection and a reference point for voltage measurement. (The ground connection is at 0 V.) Mark the polarity of the voltage drops across the resistances R_1, R_2, and find the voltage drops at points A and B with respect to ground.

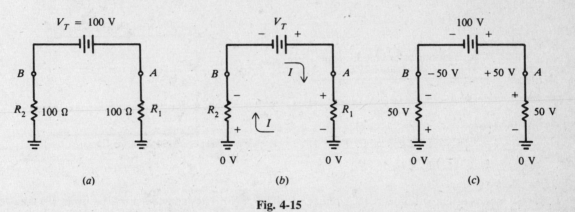

Fig. 4-15

Step 1. Mark the polarities. The current I flows from the positive terminal of the battery through R_1, through ground, up through R_2, and back to the negative terminal of the battery (Fig. 4-15b). Assign a $+$ sign where the current enters the resistance and a $-$ sign to the end where the current emerges (Fig. 4-15b). Mark the ground voltage 0 V as the reference to measure voltage drops.

Step 2. Find the total resistance, using Eq. (4-1).

$$R_T = R_1 + R_2 = 100 + 100 = 200 \ \Omega$$

Step 3. Find the current in the circuit.

$$I = \frac{V_T}{R_T} = \frac{100}{200} = 0.5 \ \text{A}$$

Step 4. Find the voltage drops.

$$V_1 = IR_1 = 0.5(100) = 50 \ \text{V}$$
$$V_2 = IR_2 = 0.5(100) = 50 \ \text{V}$$

Step 5. Find voltage polarity at points A and B. Point A is 50 V positive with respect to ground, while point B is 50 V negative with respect to ground (Fig. 4-15c). Point A is nearer to the positive terminal, while point B is nearer to the negative terminal.

Step 6.   Verify the voltage drops.

$$\text{Sum of voltage drops} = \text{applied voltage}$$
$$V_T = V_1 + V_2$$
$$100 = 50 + 50$$
$$100\,\text{V} = 100\,\text{V} \qquad Check$$

4.6   The terminal voltage of the motor (Fig. 4-16) should be not less than 223 V at a rated current of 20 A. Utility voltage variations produce a minimum of 228 V at the panel. What size branch circuit conductors are needed?

   Find the minimum allowable wire size for the voltage drop by calculating its resistance. The maximum voltage drop is $228 - 223 = 5$ V. Then the maximum wire resistance is $5/20 = 0.25\,\Omega/500\,\text{ft} = 0.50\,\Omega/1000\,\text{ft}$. From Table 4-1, No. 6 wire is satisfactory since it has $0.40\,\Omega/1000\,\text{ft}$ (No. 7 wire has $0.51\,\Omega/1000\,\text{ft}$).

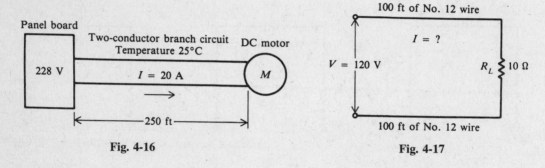

Fig. 4-16	**Fig. 4-17**

4.7   How much current will flow in the circuit (Fig. 4-17) if No. 12 conductors are (a) copper, (b) tungsten, and (c) nichrome? (Temperature is 20°C.)

(*a*)   Copper conductor:

$$\rho = 10.4 \qquad \text{(Table 4-2)}$$
$$A = 6530\,\text{CM} \qquad \text{(Table 4-1)}$$

Copper conductor resistance:

$$R = \rho\frac{l}{A} \qquad (4\text{-}5)$$
$$= \frac{10.4(200)}{6530} = 0.319\,\Omega$$

Total circuit resistance = conductor resistance + load resistance

$$R_T = R + R_L = 0.319 + 10 = 10.319\,\Omega$$
$$I = \frac{V}{R_T} = \frac{120}{10.319} = 11.6\,\text{A} \qquad Ans.$$

(*b*)   Tungsten conductor:

$$\rho = 33.8 \quad \text{(Table 4-2)}$$
$$A = 6530\,\text{CM} \quad \text{(equal diameter of copper conductor)}$$

$$R = \rho \frac{l}{A}$$ (4-5)

$$= \frac{33.8(200)}{6530} = 1.035 \ \Omega$$

$$R_T = R + R_L = 1.035 + 10 = 11.035 \ \Omega$$

$$I = \frac{V}{R_T} = \frac{120}{11.035} = 10.9 \ \text{A} \qquad Ans.$$

(c)   Nichrome conductor:

$$\rho = 676 \qquad \text{(Table 4-2)}$$

$$A = 6530 \ \text{CM} \quad \text{(equal diameter of copper conductor)}$$

$$R = \rho \frac{l}{A}$$ (4-5)

$$= \frac{676(200)}{6530} = 20.7 \ \Omega$$

$$R_T = R + R_L = 20.7 + 10 = 30.7 \ \Omega$$

$$I = \frac{V}{R_T} = \frac{120}{30.7} = 3.91 \ \text{A} \qquad Ans.$$

Notice that as the factor ρ increases, the circuit resistance increases and the circuit current decreases.

4.8   Five lamps are connected in series (Fig. 4-18). Each lamp requires 16 V and 0.1 A. Find the total power used.

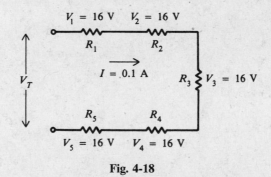

Fig. 4-18

The total voltage V_T equals the sum of the individual voltage across all parts in the series circuit.

$$V_T = V_1 + V_2 + V_3 + V_4 + V_5$$ (4-2)

$$= 16 + 16 + 16 + 16 + 16 = 80 \ \text{V}$$

The current through each resistance (lamp) is the single current in the series circuit.

$$I = 0.1 \ \text{A}$$

So the total power is

$$P_T = IV_T \qquad (4\text{-}7)$$

$$= 0.1(80) = 8 \text{ W} \qquad Ans.$$

Also total power is the sum of the individual powers.

For one lamp, $\qquad P_1 = V_1 I = 16(0.1) = 1.6\,\text{W}$

For five lamps, $\qquad P_T = 5P_1 = 5(1.6) = 8\,\text{W} \qquad Ans.$

4.9 Find I, V_1, V_2, P_2, and R_2 in the circuit shown (Fig. 4-19).

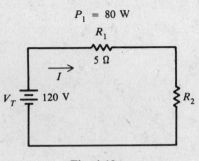

Fig. 4-19

Step 1. Find I.
Use the power formula, $P_1 = I_1^2 R_1$. Then

$$I_1^2 = \frac{P_1}{R_1} = \frac{80}{5} = 16$$

Taking the square root,

$$I_1 = \sqrt{16} = 4\,\text{A}$$

Since this is a series circuit,

$$I = I_1 = 4\,\text{A} \qquad Ans.$$

Step 2. Find V_1, V_2.

$$V_1 = IR_1 = 4(5) = 20\,\text{V} \qquad Ans.$$

$$V_2 = V_T - V_1 = 120 - 20 = 100\,\text{V} \qquad Ans.$$

Step 3. Find P_2, R_2.

$$P_2 = V_2 I = 100(4) = 400\,\text{W} \qquad Ans.$$

$$R_2 = \frac{V_2}{I} = \frac{100}{4} = 25\,\Omega \qquad Ans.$$

4.10   Three 20-kΩ resistors R_1, R_2, and R_3 are in series across an applied voltage of 120 V. What is the voltage drop across each resistor?

Since R_1, R_2, and R_3 are equal, each has exactly one-third the total resistance of the series circuit and one-third the total voltage drop. So

$$V = \frac{1}{3}120 = 40 \text{ V} \qquad Ans.$$

4.11   Use the voltage-divider method to find voltage drop (Fig. 4-20) across each resistor.

$$V_1 = \frac{R_1}{R_T}V_T = \frac{3}{10}10 = 3 \text{ V} \qquad Ans.$$

$$V_2 = \frac{R_2}{R_T}V_T = \frac{7}{10}10 = 7 \text{ V} \qquad Ans.$$

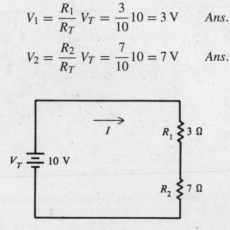

Fig. 4-20

An important advantage of using the voltage-divider formula is that we can find the voltage drops from V_T and the resistances without finding the current.

If we found the current first, then we could calculate the voltage drop by multiplying current and resistance. For example,

$$I = \frac{V_T}{R_T} = \frac{V_T}{R_1 + R_2} = \frac{10}{3+7} = \frac{10}{10} = 1 \text{ A}$$

Then
$$V_1 = IR_1 = 1(3) = 3 \text{ V} \qquad Check$$
$$V_2 = IR_2 = 1(7) = 7 \text{ V} \qquad Check$$

4.12   Compare the effect on voltage drop of a 1-Ω resistor and a 99-Ω resistor in series.

Because series voltage drops are proportional to the resistances, a very small resistance (1 Ω) has a very small effect in series with a much larger resistance (99 Ω). For example, if the applied voltage were 100 V, the voltage drop across the 1-Ω resistor would be 1 V [(1/100)(100) = 1 V], while across the 99-Ω resistor it would be 99 V [(99/100)(100) = 99 V].

4.13   A voltage of 5 V is to be made available from a 12-V source using a two-resistor voltage divider (Fig. 4-21). The current in the divider is to be 100 mA. Find the values for resistors R_1 and R_2.

Step 1.   Find R_T.

$$R_T = \frac{V_T}{I} = \frac{12}{100 \times 10^{-3}} = 120 \, \Omega$$

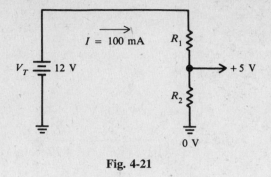

Fig. 4-21

Step 2. Find R_1. From Fig. 4-21, tracing the voltage across the resistors from ground at $0\,V$, we have $5\,V$ across R_2 ($V_2 = 5\,V$), so that there is $7\,V$ across R_1 ($V_1 = 7\,V$) for a total of $12\,V$ ($V_T = 12\,V$). Use the voltage-divider formula.

$$V_1 = \frac{R_1}{R_T} V_T \qquad (4\text{-}9)$$

Then

$$\frac{R_1}{R_T} = \frac{V_1}{V_T} = \frac{7}{12}$$

The ratio of the two resistors is known because the ratio of the two voltages is 7/12. So

$$R_1 = \frac{7}{12} R_T = \frac{7}{12} 120 = 70\,\Omega \qquad Ans.$$

Step 3. Find R_2.

$$R_T = R_1 + R_2$$

Transpose and substitute.

$$R_2 = R_T - R_1 = 120 - 70 = 50\,\Omega \qquad Ans.$$

Supplementary Problems

4.14 What is the total resistance of three 20-Ω resistors connected in series?   *Ans.* $60\,\Omega$

4.15 A car has a 3-V, 1.5-Ω dash light and a 3-V, 1.5-Ω taillight connected in series to a battery delivering 2 A (Fig. 4-22). Find the battery voltage and total resistance of the circuit.
Ans. $V_T = 6\,V$; $R_T = 3.0\,\Omega$

4.16 A 3-Ω, a 5-Ω, and a 4-Ω resistor are connected in series across a battery. The voltage drop across a 3-Ω resistor is 6 V. What is the battery voltage?   *Ans.* $24\,V$

4.17 If three resistors are connected in series across a 12-V battery and the voltage drop across one resistor is 3 V and the voltage drop across the second resistor is 7 V, what is the voltage drop across the third resistor?   *Ans.* $2\,V$

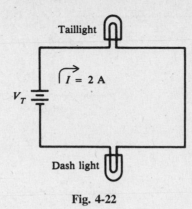

Fig. 4-22

4.18 A lamp using 10 V, a 10-Ω resistor drawing 4 A, and a 24-V motor are connected in series. Find the total voltage and the total resistance.   *Ans.* $V_T = 74$ V; $R_T = 18.5\ \Omega$

4.19 Find the missing values of current, voltage, and resistance in a high-voltage regulator circuit in a color television receiver (Fig. 4-23). The voltage drop across each resistor is used to supply voltage in other parts of the receiver.
Ans. $V_1 = 700$ V, $I_2 = 0.07$ mA; $V_2 = 105$ V, $I_3 = 0.07$ mA;
     $R_3 = 500$ kΩ, $V_T = 840$ V; $R_T = 12$ MΩ, $I = 0.07$ mA

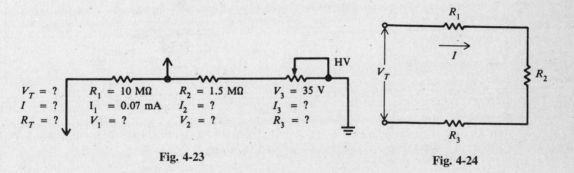

Fig. 4-23   Fig. 4-24

4.20 Given $I = 2$ A, $R_1 = 10\ \Omega$, $V_2 = 50$ V, and $V_3 = 40$ V, find V_1, V_T, R_2, R_3, and R_T (Fig. 4-24).
Ans. $V_1 = 20\ V$; $V_T = 110$ V; $R_2 = 25\ \Omega$; $R_3 = 20\ \Omega$; $R_T = 55\ \Omega$

4.21 A current of 3 mA flows through a resistor that is connected to a 1.5-V dry cell. If three additional 1.5-V cells are connected in series to the first cell, find the current flowing through the resistor.
Ans. $I = 0.012$ A $= 12$ mA

4.22 A voltage divider consists of a 3000-Ω, a 5000-Ω, and 10 000-Ω resistor in series. The series current is 15 mA. Find (a) the voltage drop across each resistance, (b) the total voltage, and (c) the total resistance.
Ans.   (a) $V_1 = 45$ V, $V_2 = 75$ V, $V_3 = 150$ V; (b) $V_T = 270$ V; (c) $R_T = 18\,000\ \Omega$

4.23 A dc circuit to a specialized transistor circuit can be represented as shown in Fig. 4-25. Find the total resistance and voltage between points A and B.   *Ans.* $R_T = 50$ kΩ; $V_{AB} = 30$ V

4.24 A 12-Ω spotlight in a theater is connected in series with a dimming resistor of 32 Ω (Fig. 4-26). If the voltage drop across the light is 31.2 V, find the missing values indicated in Fig. 4-26.

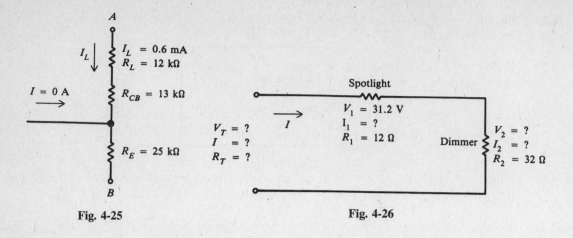

Fig. 4-25

Fig. 4-26

Ans. $I_1 = I_2 = I = 2.6\,\text{A};\; V_2 = 83.2\,\text{V};\; V_T = 114.4\,\text{V};\; R_T = 44\,\Omega$

4.25 Find all missing values of current, voltage, and resistance in the circuit shown in Fig. 4-27.

Ans. $V_3 = 30\,\text{V};$

$I = I_1 = I_2 = I_3 = 0.667\,\text{A};$

$R_1 = 30\,\Omega,$

$R_2 = 90\,\Omega,$

$R_3 = 45\,\Omega$

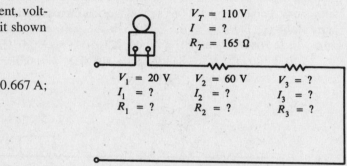

Fig. 4-27

4.26 Find voltage values at points A, B, C, and D shown in the circuit (Fig. 4-28) with respect to ground.
Ans. $V_A = +60\,\text{V};\; V_B = +50\,\text{V};\; V_C = +30\,\text{V};\; V_D = 0\,\text{V}$

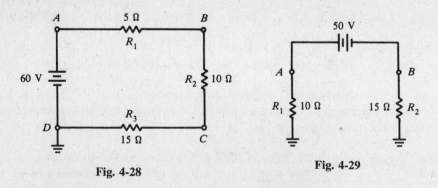

Fig. 4-28

Fig. 4-29

4.27 Find the voltage at points A and B with respect to ground (Fig. 4-29).
Ans. $V_A = +20\,\text{V};\; V_B = -30\,\text{V}$

4.28 A coil is wound with 4000 turns of No. 20 copper wire. If the average amount of wire in a turn is 3 in, how much is the total resistance of the coil? What will be its resistance if No. 25 wire is used instead? (The temperature is 25°C.)   *Ans.* $10.35\,\Omega$, $33.0\,\Omega$

4.29 Find the voltage drop across 1000 ft of No. 10 gauge copper wire connected to a 3-A load.
        Ans. 3.05 V

4.30 If total line length is 200 ft, find the smallest size copper wire that will limit the line drop to 5 V with
        115 V applied and a 6-A load.   *Ans.* No. 16 copper wire

4.31 A copper wire has a diameter of 0.031 96 in. Find (*a*) the circular-mil area, (*b*) its AWG size, and
        (*c*) the resistance of a 200-ft length.   *Ans.* (*a*) 1024 CM; (*b*) No. 20; (*c*) 2.07 Ω
        (for No. 20 wire)

4.32 What is the resistance of a 200-ft length of (*a*) No. 16 copper wire and (*b*) No. 20 aluminum wire?
        (Obtain diameter from Table 4-1.)   *Ans.*   (*a*) 0.805 Ω;   (*b*) 1.32 Ω

4.33 A copper conductor measures 0.8 Ω at 20°C. What is its resistance at 25°C?   *Ans.*   0.816 Ω

4.34 If a copper wire has a resistance of 4 Ω at 20°C, how much is its resistance at 75°C? If the wire is No.
        10, what is its length in feet?   *Ans.*   4.88 Ω, 4800 ft

4.35 Calculate the load current I (Fig. 4-30) for the wire IR drop of 24.6 V with a supply of 115 V. Also
        find the value of R_L.   *Ans.* $I = 30$ A; $R_L = 3.01$ Ω

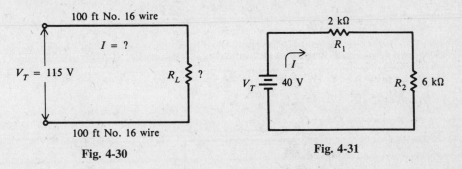

Fig. 4-30   Fig. 4-31

4.36 Two resistors form the base-bias voltage divider for an audio amplifier. The voltage drops across them
        are 2.4 V and 6.6 V in the 1.5-mA circuit. Find the power used by each resistor and the total power
        used in milliwatts (mW).   *Ans.* $P_1 = 3.6$ mW; $P_2 = 9.9$ mW; $P_T = 13.5$ mW

4.37 Find I, V_1, V_2, P_1, P_2, and P_T (Fig. 4-31).
        Ans. $I = 5$ mA; $V_1 = 10$ V; $V_2 = 30$ V; $P_1 = 50$ mW; $P_2 = 150$ mW; $P_T = 200$ mW

4.38 Find V_1, V_2, V_3, P_1, P_2, P_3, P_T, and R_3 (Fig. 4-32).
        Ans. $V_1 = 30$ V; $V_2 = 15$ V; $V_3 = 55$ V; $P_1 = 150$ mW; $P_2 = 75$ mW; $P_3 = 275$ mW;
           $P_T = 500$ mW; $R_3 = 11$ kΩ

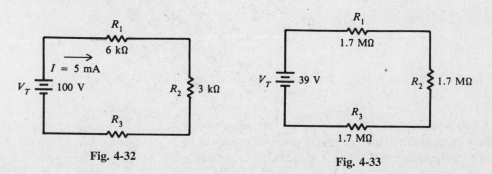

Fig. 4-32   Fig. 4-33

4.39   Find P_T, P_2, and V_3 (Fig. 4-33).   *Ans.* $P_T = 298\,\mu\text{W}$; $P_2 = 99.5\,\mu\text{W}$; $V_3 = 13\text{ V}$

4.40   A 90-Ω and a 10-Ω resistor are in series across a 3-V source. Find the voltage drop across each resistor by the voltage-divider method.   *Ans.*   2.7 V, 0.3 V

4.41   Eight 10-Ω resistances are in series across a 120-V source. What is the voltage drop across each resistance?   *Ans.*   15 V

4.42   A potentiometer can be considered a simple two-resistor voltage divider (Fig. 4-34). To what resistance point would the control arm have to be set in a 120-Ω potentiometer to obtain 2.5 V between the arm (point A) of the potentiometer and ground (point B)?   *Ans.*   25-Ω point from ground

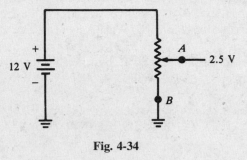

Fig. 4-34

4.43   Find the IR drop across each resistor in the following circuits by the voltage-divider method (Fig. 4-35).

Ans.   (*a*) $V_1 = 60\text{ V}$; $V_2 = 180\text{ V}$ (*b*) $V_1 = 25\text{ V}$; $V_2 = 50\text{ V}$; $V_3 = 35\text{ V}$ (*c*) $V_1 = 11.5\text{ V}$;
$V_2 = 23\text{ V}$; $V_3 = 34.5\text{ V}$; $V_4 = 46\text{ V}$

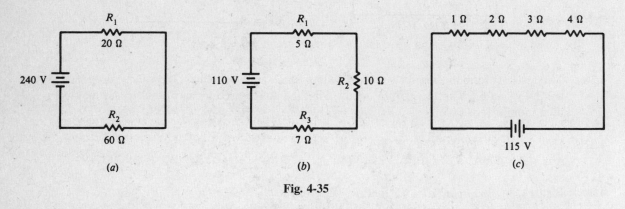

Fig. 4-35

Chapter 5

Direct-Current Parallel Circuits

VOLTAGE AND CURRENT IN A PARALLEL CIRCUIT

A *parallel circuit* is a circuit in which two or more components are connected across the same voltage source (Fig. 5-1). The resistors R_1, R_2, and R_3 are in parallel with each other and with the battery. Each parallel path is then a branch with its own individual current. When the total current I_T leaves the voltage source V, part I_1 of the current I_T will flow through R_1, part I_2 will flow through R_2, and the remainder I_3 through R_3. The branch currents I_1, I_2, and I_3 can be different. However, if a voltmeter (an instrument for measuring the voltage of a circuit) is connected across R_1, R_2, and R_3, the respective voltages V_1, V_2, and V_3 will be equal. Therefore,

$$V = V_1 = V_2 = V_3 \qquad (5\text{-}1)$$

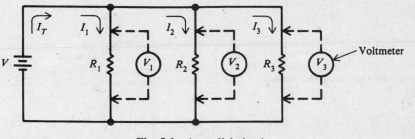

Fig. 5-1 A parallel circuit

The total current I_T is equal to the sum of all branch currents.

$$I_T = I_1 + I_2 + I_3 \qquad (5\text{-}2)$$

This formula applies for any number of parallel branches whether the resistances are equal or unequal.

By Ohm's law, each branch current equals the applied voltage divided by the resistance between the two points where the voltage is applied. Hence (Fig. 5-1), for each branch we have the following equations:

Branch 1:
$$I_1 = \frac{V_1}{R_1} = \frac{V}{R_1}$$

Branch 2:
$$I_2 = \frac{V_2}{R_2} = \frac{V}{R_2} \qquad (5\text{-}3)$$

Branch 3:
$$I_3 = \frac{V_3}{R_3} = \frac{V}{R_3}$$

With the same applied voltage, any branch that has less resistance allows more current through it than a branch with higher resistance.

Example 5.1 Two lamps each drawing 2 A and a third lamp drawing 1 A are connected in parallel across a 110-V line (Fig. 5-2). What is the total current?

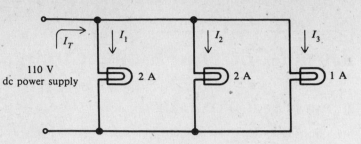

Fig. 5-2

The formula for total current is

$$I_T = I_1 + I_2 + I_3 \qquad (5\text{-}2)$$
$$= 2 + 2 + 1 = 5\,\text{A} \qquad Ans.$$

The total current is 5 A.

Example 5.2 Two branches R_1 and R_2 across a 110-V power line draw a total line current of 20 A (Fig. 5-3). Branch R_1 takes 12 A. What is the current I_2 in branch R_2?

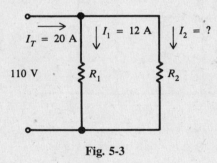

Fig. 5-3

Starting with Eq. (5-2), transpose to find I_2 and then substitute given values.

$$I_T = I_1 + I_2$$
$$I_2 = I_T - I_1$$
$$= 20 - 12 = 8\,\text{A} \qquad Ans.$$

The current in branch R_2 is 8 A.

Example 5.3 A parallel circuit consists of a coffee maker, a toaster, and a frying pan plugged into a kitchen appliance circuit on a 120-V line (Fig. 5-4a). What currents will flow in each branch of the circuit and what is the total current drawn by all the appliances?

First, draw the circuit diagram (Fig. 5-4b). Show the resistance for each appliance. There is a 120-V potential across each appliance. Then, using Eq. (5-3), apply Ohm's law to each appliance.

Coffee maker: $\qquad\qquad\qquad I_1 = \dfrac{V}{R_1} = \dfrac{120}{15} = 8\,\text{A} \qquad Ans.$

Toaster: $\qquad\qquad\qquad\quad I_2 = \dfrac{V}{R_2} = \dfrac{120}{15} = 8\,\text{A} \qquad Ans.$

Frying pan: $\qquad\qquad\qquad I_3 = \dfrac{V}{R_3} = \dfrac{120}{12} = 10\,\text{A} \qquad Ans.$

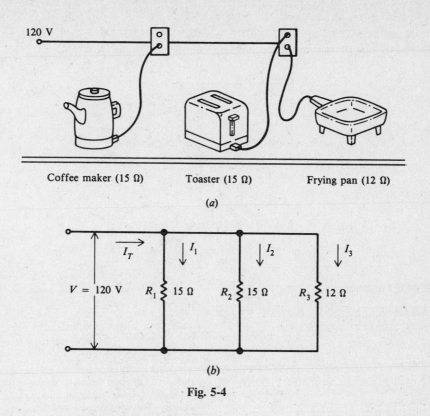

Coffee maker (15 Ω)   Toaster (15 Ω)   Frying pan (12 Ω)

(a)

(b)

Fig. 5-4

Now find total current, using Eq. (5-2).

$$I_T = I_1 + I_2 + I_3$$
$$= 8 + 8 + 10 = 26 \text{ A} \qquad Ans.$$

With this load of 26 A, a 20-A circuit breaker or fuse will open the circuit. This example shows the desirability of having two 20-A kitchen appliance circuits.

RESISTANCES IN PARALLEL

Total Resistance

The total resistance in a parallel circuit is found by applying Ohm's law: Divide the common voltage across the parallel resistances by the total line current.

$$R_T = \frac{V}{I_T} \qquad\qquad (5\text{-}4)$$

R_T is the total resistance of all the parallel branches across the voltage source V, and I_T is the sum of all the branch currents.

Example 5.4   What is the total resistance of the circuit shown in Fig. 5-4 (Example 5.3)?

In Example 5.3 the line voltage is 120 V and the total line current is 26 A. Therefore,

$$R_T = \frac{V}{I_T} = \frac{120}{26} = 4.62 \ \Omega \qquad Ans.$$

The total load connected to the 120-V line is the same as the single equivalent resistance of 4.62 Ω connected across the line (Fig. 5-5). The words *total resistance* and *equivalent resistance* are used interchangeably.

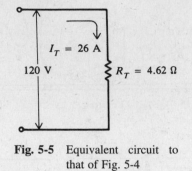

Fig. 5-5 Equivalent circuit to
that of Fig. 5-4

General Reciprocal Formula

The total resistance in parallel is given by the formula

$$\frac{1}{R_T} = \frac{1}{R_1} + \frac{1}{R_2} + \frac{1}{R_3} + \cdots + \frac{1}{R_n} \tag{5-5}$$

where R_T is the total resistance in parallel and R_1, R_2, R_3, and R_n are the branch resistances.

Example 5.5   Find the total resistance of a 2-Ω, a 4-Ω, and an 8-Ω resistor in parallel (Fig. 5-6).

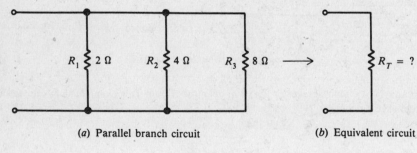

(*a*) Parallel branch circuit   (*b*) Equivalent circuit

Fig. 5-6

Write the formula for three resistances in parallel.

$$\frac{1}{R_T} = \frac{1}{R_1} + \frac{1}{R_2} + \frac{1}{R_3} \tag{5-5}$$

Substitute the resistance values.

$$\frac{1}{R_T} = \frac{1}{2} + \frac{1}{4} + \frac{1}{8}$$

Add fractions.

$$\frac{1}{R_T} = \frac{4}{8} + \frac{2}{8} + \frac{1}{8} = \frac{7}{8}$$

Invert both sides of the equation to solve for R_T.

$$R_T = \frac{8}{7} = 1.14 \ \Omega \qquad Ans.$$

Note that when resistances are connected in parallel, the total resistance is always *less* than the resistance of any single branch. In this case, $R_T = 1.14 \ \Omega$ is less than $R_1 = 2 \ \Omega$, $R_2 = 4 \ \Omega$, and $R_3 = 8 \ \Omega$.

Example 5.6   Add a fourth resistor of 2 Ω in parallel to the circuit in Fig. 5-6. What is the new total resistance and what is the net effect of adding another resistance in parallel?
Write the formula for four resistances in parallel.

$$\frac{1}{R_T} = \frac{1}{R_1} + \frac{1}{R_2} + \frac{1}{R_3} + \frac{1}{R_4} \qquad\qquad (5\text{-}5)$$

Substitute values.

$$\frac{1}{R_T} = \frac{1}{2} + \frac{1}{4} + \frac{1}{8} + \frac{1}{2}$$

Add fractions.

$$\frac{1}{R_T} = \frac{4}{8} + \frac{2}{8} + \frac{1}{8} + \frac{4}{8} = \frac{11}{8}$$

Invert.

$$R_T = \frac{8}{11} = 0.73 \ \Omega \qquad Ans.$$

Thus we see that the net effect of adding another resistance in parallel is a *reduction* of the total resistance from 1.14 to 0.73 Ω.

Simplified Formulas

The total resistance of *equal* resistors in parallel is equal to the resistance of one resistor divided by the number of resistors.

$$R_T = \frac{R}{N} \qquad\qquad (5\text{-}6)$$

where R_T = total resistance of equal resistors in parallel, Ω
R = resistance of one of the equal resistors, Ω
N = number of equal resistors

Example 5.7   Four lamps, each having a resistance of 60 Ω, are connected in parallel. Find the total resistance.
Given are

$$R = R_1 = R_2 = R_3 = R_4 = 60 \ \Omega$$

$$N = 4$$

Write Eq. (5-6) and substitute values.

$$R_T = \frac{R}{N} = \frac{60}{4} = 15 \ \Omega \qquad Ans.$$

When any two unequal resistors are in parallel, it is often easier to calculate the total resistance by multiplying the two resistances and then dividing the product by the sum of the resistances.

$$R_T = \frac{R_1 R_2}{R_1 + R_2} \qquad (5\text{-}7)$$

where R_T is the total resistance in parallel and R_1 and R_2 are the two resistors in parallel.

Example 5.8   Find the total resistance of a 6-Ω and an 18-Ω resistor in parallel.
   Given are $R_1 = 6\,\Omega$, $R_2 = 18\,\Omega$.
   Write Eq. (5-7) and substitute values.

$$R_T = \frac{R_1 R_2}{R_1 + R_2} = \frac{6(18)}{6 + 18} = \frac{108}{24} = 4.5\,\Omega \qquad Ans.$$

In some cases with two parallel resistors, it is useful to find what size R_x to connect in parallel with a known R in order to obtain a required value of R_T. To find the appropriate formula, we start with Eq. (5-7) and transpose the factors as follows:

$$R_T = \frac{R R_x}{R + R_x}$$

Cross-multiply. $$R_T R + R_T R_x = R R_x$$

Transpose. $$R R_x - R_T R_x = R_T R$$

Factor. $$R_x (R - R_T) = R_T R$$

Solve for R_x. $$R_x = \frac{R R_T}{R - R_T} \qquad (5\text{-}8)$$

Example 5.9   What value of resistance must be added in parallel with a 4-Ω resistor to provide a total resistance of 3 Ω (Fig. 5-7)?

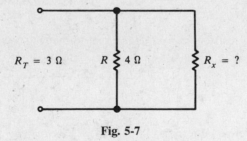

Fig. 5-7

Given are $R = 4\,\Omega$ and $R_T = 3\,\Omega$. Write Eq. (5-8) and substitute values.

$$R_x = \frac{R R_T}{R - R_T} = \frac{4(3)}{4 - 3} = \frac{12}{1} = 12\,\Omega \qquad Ans.$$

OPEN AND SHORT CIRCUITS

An "open" in any part of a circuit is, in effect, an extremely high resistance that results in no current flow in the circuit. When there is an open in the main line (the "X" in Fig. 5-8a), current to all the parallel branches is stopped. When there is an open in one branch (branch 2 in Fig. 5-8b), only that branch will have no current. However, current in branches 1 and 3 will continue to flow as long as they are connected to the voltage source.

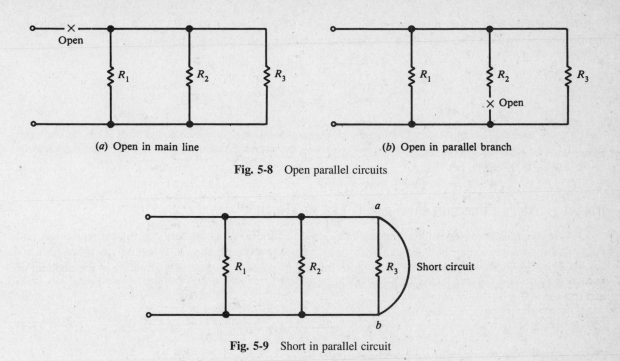

(a) Open in main line   (b) Open in parallel branch

Fig. 5-8 Open parallel circuits

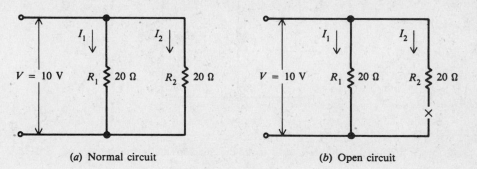

Fig. 5-9 Short in parallel circuit

A "short" in any part of a circuit is, in effect, an extremely low resistance. The result is that very high current will flow through the short circuit. Assume that a conducting wire at point a in Fig. 5-9 should accidentally contact the wire at point b. Since the wire is an excellent conductor, the short circuit offers a parallel path with practically zero resistance from points a to b. Almost all the current will flow in this path. Since the resistance of the short circuit is practically zero, the voltage drop across ab will be almost zero (by Ohm's law). Thus resistors R_1, R_2, and R_3 will not draw their normal current.

Example 5.10 Find the current in each parallel branch (Fig. 5-10a). If the resistor in the second branch burns out, causing an open circuit (Fig. 5-10b), find the new branch currents.

Use Eq. (5-3) and substitute values. With circuits normal (Fig. 5-10a),

(a) Normal circuit   (b) Open circuit

Fig. 5-10

$$I_1 = \frac{V}{R_1} = \frac{10}{20} = 0.5 \text{ A} \qquad Ans.$$

$$I_2 = \frac{V}{R_2} = \frac{10}{20} = 0.5 \text{ A} \qquad Ans.$$

With branch 2 open (Fig. 5-10*b*),

$$I_1 = \frac{V}{R_1} = \frac{10}{20} = 0.5 \text{ A} \qquad Ans.$$

$$I_2 = 0 \text{ A} \qquad Ans.$$

Branch 1 still operates normally at 0.5 A. This example shows the advantage of wiring components in parallel. An open circuit in one component merely opens the branch containing the component, while the other parallel branch keeps its normal voltage and current.

DIVISION OF CURRENT IN TWO PARALLEL BRANCHES

It is sometimes necessary to find the individual branch currents in a parallel circuit if the resistances and total current are known, but the voltage across the resistance bank is not known. When only two branches are involved, the current in one branch will be some fraction of the total current. This fraction is the quotient of the second resistance divided by the sum of the resistances.

$$I_1 = \frac{R_2}{R_1 + R_2} I_T \qquad (5\text{-}9)$$

$$I_2 = \frac{R_2}{R_1 + R_2} I_T \qquad (5\text{-}10)$$

where I_1 and I_2 are the currents in the respective branches. Notice that the equation for each branch current has the opposite R in the numerator. The reason is that each branch current is inversely proportional to the branch resistance. The denominator is the same in both equations, equal to the sum of the two branch resistances.

Example 5.11   Find the branch currents I_1 and I_2 for the circuit shown in Fig. 5-11.

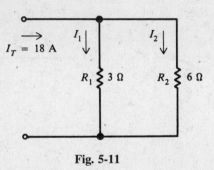

Fig. 5-11

Given are $I_T = 18$ A, $R_1 = 3\ \Omega$, and $R_2 = 6\ \Omega$. Write the equations and substitute values.

$$I_1 = \frac{R_2}{R_1 + R_2} I_T \qquad (5\text{-}9)$$

$$= \frac{6}{3+6} 18 = \frac{6}{9} 18 = 12 \text{ A} \qquad Ans.$$

$$I_2 = \frac{R_2}{R_1 + R_2} I_T = \frac{3}{9} 18 = 6 \text{ A} \qquad Ans. \qquad (5\text{-}10)$$

Since I_T and I_1 were known, we could have found I_2 simply by subtracting:

$$I_T = I_1 + I_2$$

$$I_2 = I_T - I_1 = 18 - 12 = 6 \text{ A} \qquad Ans.$$

CONDUCTANCES IN PARALLEL

Conductance is the opposite of resistance. The less the resistance, the higher the conductance. The symbol for conductance is G and its unit is siemens (S). G is the reciprocal of R, or

$$G = \frac{1}{R} \tag{5-11}$$

For example, 6 Ω resistance is equal to 1/6 S conductance.

Since conductance is equal to the reciprocal of resistance, the reciprocal resistance equation, Eq. (5-5), can be written for conductance as

$$G_T = G_1 + G_2 + G_3 + \cdots + G_n \tag{5-12}$$

where G_T is the total conductance in parallel and G_1, G_2, G_3, and G_n are the branch conductances.

Example 5.12　Find the total conductance of the circuit in Fig. 5-12. Then find the total resistance R_T and check the value with that computed in Example 5.5.

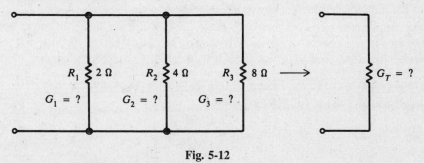

Fig. 5-12

This circuit is the same as that of Fig. 5-6 used in Example 5.5. Convert each branch resistance to conductance, using Eq. (5-11), and then add the values of conductance to obtain G_T.

$$G_1 = \frac{1}{R_1} = \frac{1}{2} = 0.5 \text{ S}$$

$$G_2 = \frac{1}{R_2} = \frac{1}{4} = 0.25 \text{ S}$$

$$G_3 = \frac{1}{R_3} = \frac{1}{8} = 0.125 \text{ S}$$

$$G_T = G_1 + G_2 + G_3 \tag{5-12}$$

$$= 0.5 + 0.25 + 0.125 = 0.875 \text{ S} \qquad Ans.$$

Finally

$$R_T = \frac{1}{G_T} = \frac{1}{0.875} = 1.14 \text{ Ω} \qquad Ans.$$

which agrees with the R_T value found in Example 5.5.

Ohm's law can be written in terms of conductance. Recall that

$$R_T = \frac{V}{I_T} \tag{5-4}$$

$$I_T = \frac{V}{R_T}$$

But $1/R_T = G_T$, so

$$I_T = V G_T \tag{5-13}$$

Example 5.13   If the source voltage across the parallel bank in Fig. 5-12 is 100 V, find the total current.
Given are $V = 100$ V and $G_T = 0.875$ S. Using Eq. (5-13),

$$I_T = V G_T = 100(0.875) = 87.5 \text{ A} \qquad \textit{Ans.}$$

POWER IN PARALLEL CIRCUITS

Since the power dissipated in the branch resistance must come from the voltage source, the total power equals the sum of the individual values of power in each branch.

$$P_T = P_1 + P_2 + P_3 + \cdots + P_n \tag{5-14}$$

where P_T is the total power and P_1, P_2, P_3, and P_n are the branch powers.
Total power can also be calculated by the equation

$$P_T = V I_T \tag{5-15}$$

where P_T is the total power, V is the voltage source across all parallel branches, and I_T is the total current.
The power P dissipated in each branch is equal to VI and equal to V^2/R.
In both parallel and series arrangements, the sum of the individual values of power dissipated in the circuit equals the total power generated by the source. The circuit arrangements cannot change the fact that all power in the circuit comes from the source.

Example 5.14   Find the power dissipated in each branch and the total power of the circuit in Fig. 5-13.

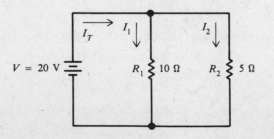

Fig. 5-13

First find the branch current and the power in each branch.

$$I_1 = \frac{V}{R_1} = \frac{20}{10} = 2\text{ A}$$

$$I_2 = \frac{V}{R_2} = \frac{20}{5} = 4\text{ A}$$

$$P_1 = VI_1 = 20(2) = 40\text{ W} \qquad Ans.$$

$$P_2 = VI_2 = 20(4) = 80\text{ W} \qquad Ans.$$

Then add these values for power in each branch to find P_T.

$$P_T = P_1 + P_2 \tag{5-14}$$

$$= 40 + 80 = 120\text{ W} \qquad Ans.$$

Another way to find P_T is to solve for I_T.

$$I_T = I_1 + I_2 = 2 + 4 = 6\text{A}$$

Then
$$P_T = VI_T \tag{5-15}$$

$$= 20(6) = 120\text{ W} \qquad Ans.$$

The 120 W of power supplied by the source is dissipated in the branch resistances.
There are still other ways to find power used in each branch and total power.

$$P_1 = \frac{V^2}{R_1} = \frac{(20)^2}{10} = 40\text{ W}$$

$$P_2 = \frac{V^2}{R_2} = \frac{(20)^2}{5} = 80\text{ W}$$

$$P_T = \frac{V^2}{R_T} = V^2 G_T = (20)^2(0.3) = 120\text{ W}$$

where
$$G_T = \frac{1}{R_T} = \frac{R_1 + R_2}{R_1 R_2} = \frac{10 + 5}{10(5)} = 0.3\text{ S}$$

Solved Problems

5.1   Write the word or words which most correctly complete the following statements.

(a)   The equivalent resistance R_T of parallel branches is _____ than the smallest branch resistance since all the branches must take _____ current from the source than any one branch.

(b)   When two resistances are connected in parallel, the voltage across each is the _____.

(c)   An open in one branch results in _____ current through that branch, but the other branches can have their _____ current.

(d) A short circuit has _____ resistance, resulting in _____ current.

(e) If each of two resistances connected in parallel dissipates 5 W, the total power supplied by the voltage source equals _____ W.

Ans.   (a) less, more;   (b) same;   (c) zero, normal;   (d) zero, excessive;   (e) 10

5.2 Branch circuits in a house wiring system are parallel circuits. A toaster, a coffee maker, and a frying pan are plugged into a kitchen appliance circuit across a 110-V line (Fig. 5-14). The current through the toaster is 8.3 A; through the coffee maker, 8.3 A; and through the frying pan, 9.6 A. Find (a) the total current from the main line, (b) the voltage across each appliance, and (c) the total resistance of the circuit.

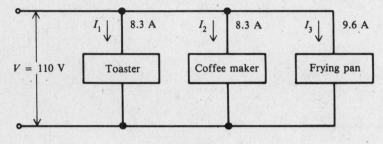

Fig. 5-14

(a) Find I_T.

$$I_T = I_1 + I_2 + I_3 \qquad\qquad (5\text{-}2)$$
$$= 8.3 + 8.3 + 9.6 = 26.2 \text{ A} \qquad Ans.$$

(b) Find V_1, V_2, V_3, using Eq. (5-1).

$$V = V_1 = V_2 = V_3 = 110 \text{ V} \qquad Ans.$$

(c) Find R_T.

$$R_T = \frac{V}{I_T} \qquad\qquad (5\text{-}4)$$
$$= \frac{110}{26.2} = 4.198 = 4.20 \,\Omega \qquad Ans.$$

5.3 Four 60-W lamps, each having the same resistance, are connected in parallel across a household terminal of 120 V, producing a line current of 2 A (Fig. 5-15a). The schematic diagram shows resistances that represent the lamps (Fig. 5-15b). What is (a) the equivalent resistance of the circuit, (b) the resistance R of each lamp, and (c) the current that each lamp draws?

(a)
$$R_T = \frac{V}{I_T} = \frac{120}{2} = 60 \,\Omega \qquad Ans.$$

(b)
$$R_T = \frac{R}{N}$$

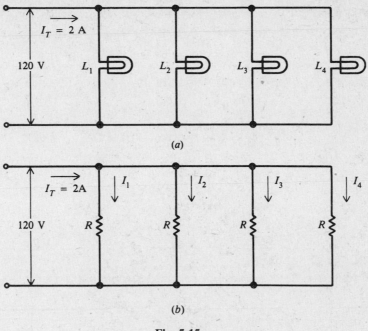

(a)

(b)

Fig. 5-15

so that

$$R = R_T N = 60(4) = 240 \ \Omega \qquad Ans.$$

(c)
$$I_1 = I_2 = I_3 = I_4 = \frac{I_T}{N} = \frac{2}{4} = 0.5 \ \text{A} \qquad Ans.$$

With equal resistance in each branch, the current in each branch is equal and the power consumed by each branch is equal.

5.4 For the circuit in Fig. 5-16, find (a) the total resistance, (b) each branch current, and (c) the total current.

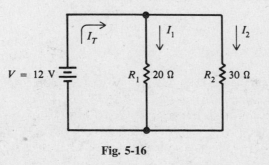

Fig. 5-16

(a) Since there are only two resistances in parallel, use the simplified formula, Eq. (5-7).

$$R_T = \frac{R_1 R_2}{R_1 + R_2} = \frac{20(30)}{20 + 30} = 12 \ \Omega \qquad Ans.$$

(*b*) Use Eq. (*5-3*).

$$I_1 = \frac{V}{R_1} = \frac{12}{20} = 0.6 \text{ A} \qquad Ans.$$

$$I_2 = \frac{V}{R_2} = \frac{12}{30} = 0.4 \text{ A} \qquad Ans.$$

(*c*) $$I_T = I_1 + I_2 = 0.6 + 0.4 = 1 \text{ A} \qquad Ans.$$

Or, as a check,

$$I_T = \frac{V}{R_T} = \frac{12}{12} = 1 \text{ A} \qquad Ans.$$

5.5 Find the total resistance R_T of each resistance arrangement in Fig. 5-17.

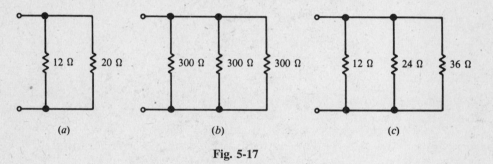

(*a*)   (*b*)   (*c*)

Fig. 5-17

(*a*) Use Eq. (*5-7*) for two parallel branches.

$$R_T = \frac{R_1 R_2}{R_1 + R_2} = \frac{12(20)}{12 + 20} = \frac{240}{32} = 7.5 \ \Omega \qquad Ans.$$

(*b*) Since all the resistances are equal, use Eq. (*5-6*).

$$R_T = \frac{R}{N} = \frac{300}{3} = 100 \ \Omega \qquad Ans.$$

(*c*) For three parallel branches with different resistances, use Eq. (*5-5*).

$$\frac{1}{R_T} = \frac{1}{R_1} + \frac{1}{R_2} + \frac{1}{R_3} = \frac{1}{12} + \frac{1}{24} + \frac{1}{36} = \frac{11}{72}$$

$$R_T = \frac{72}{11} = 6.55 \ \Omega \qquad Ans.$$

Note that the total resistance of a parallel circuit is always less than the smallest resistance of any individual resistor.

5.6 A spotlight of unknown resistance is placed in parallel with an automobile cigarette lighter of 75 Ω resistance (Fig. 5-18). If a current of 0.8 A flows when a voltage of 12 V is applied, find the resistance of the spotlight.

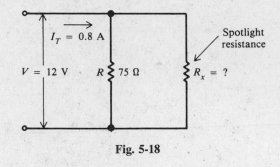

Fig. 5-18

Find R_T.

$$R_T = \frac{V}{I_T} \tag{5-4}$$

$$R_T = \frac{12}{0.8} = 15\ \Omega$$

Then use Eq. (5-8) to the unknown resistance.

$$R_x = \frac{RR_T}{R - R_T} = \frac{75(15)}{75 - 15} = \frac{1125}{60} = 18.8\ \Omega \qquad Ans.$$

Another way to find the answer is to use Ohm's law and the total current equation.

$$I_{\text{lighter}} = \frac{12}{75} = 0.16\ \text{A} \qquad Ans.$$

$$I_{\text{spot}} = 0.8 - 0.16 = 0.64\ \text{A} \qquad Ans.$$

$$R_{\text{spot}} = \frac{12}{0.64} = 18.75 = 18.8\ \Omega \qquad Ans.$$

5.7 (a) Derive Eq. (5-7) $R_T = R_1 R_2 / (R_1 + R_2)$ from the reciprocal formula for two parallel resistances.

(b) Derive a formula for R_T, given three parallel resistances.

(a)

$$\frac{1}{R_T} = \frac{1}{R_1} + \frac{1}{R_2} \tag{5-5}$$

Add fractions.

$$\frac{1}{R_T} = \frac{R_2}{R_1 R_2} + \frac{R_1}{R_1 R_2} = \frac{R_1 + R_2}{R_1 R_2}$$

Invert.

$$R_T = \frac{R_1 R_2}{R_1 + R_2} \qquad \text{which is Eq. (5-7)}$$

(b)   Use the formula

$$\frac{1}{R_T} = \frac{1}{R_1} + \frac{1}{R_2} = \frac{1}{R_3} \qquad (5\text{-}5)$$

Find the common denominator and combine numerators.

$$\frac{1}{R_T} = \frac{R_2 R_3 + R_1 R_3 + R_1 R_2}{R_1 R_2 R_3}$$

Invert.

$$R_T = \frac{R_1 R_2 R_3}{R_1 R_2 + R_1 R_3 + R_2 R_3} \qquad Ans.$$

5.8   Find the voltage required to send 2 A through a parallel combination of a 20-Ω, a 30-Ω, and a 40-Ω resistance (Fig. 5-19).

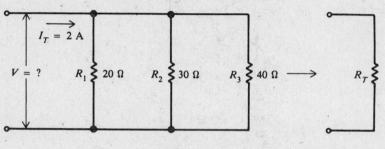

Fig. 5-19

Find R_T.

$$\frac{1}{R_T} = \frac{1}{R_1} + \frac{1}{R_2} + \frac{1}{R_3} \qquad (5\text{-}5)$$

$$= \frac{1}{20} + \frac{1}{30} + \frac{1}{40} = \frac{13}{120}$$

$$R_T = \frac{120}{13} = 9.23 \ \Omega$$

Then $V = I_T R_T = 2(9.23) = 18.5 \text{ V}$ Ans.

As a check,

$$I_1 = \frac{V}{R_1} = \frac{18.5}{20} = 0.925 \text{ A}$$

$$I_2 = \frac{V}{R_2} = \frac{18.5}{30} = 0.617 \text{ A}$$

$$I_3 = \frac{V}{R_3} = \frac{18.5}{40} = 0.463 \text{ A}$$

$$I_T = I_1 + I_2 + I_3 = 0.925 + 0.617 + 0.463 = 2.005 \approx 2 \text{ A}$$

which checks with the given value. (The sum of the currents is not exactly 2 A due to rounding off the individual branch currents.)

5.9   Two resistances are arranged in parallel (Fig. 5-20). Find the current in each resistance.

Use formulas for the division of current.

$$I_1 = \frac{R_2}{R_1 + R_2} I_T \qquad\qquad (5\text{-}9)$$

$$= \frac{18}{18 + 72} 30 = \frac{18}{90} 30 = 6\,\text{mA} \qquad Ans.$$

$$I_2 = \frac{R_1}{R_1 + R_2} I_T \qquad\qquad (5\text{-}10)$$

$$= \frac{72}{18 + 72} 30 = \frac{72}{90} 30 = 24\,\text{mA} \qquad Ans.$$

Check: $I_T = I_1 + I_2 = 6 + 24 = 30\,\text{mA}$, which agrees with the given value.

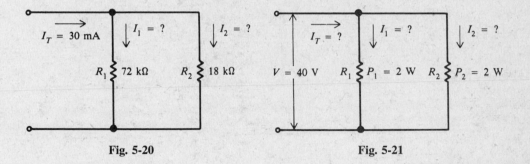

Fig. 5-20   **Fig. 5-21**

5.10   Two resistors, each dissipating 2 W, are connected in parallel across 40 V (Fig. 5-21). What is the current in each resistor? What is the total current drawn?

Find I_1, I_2, I_T.

$$I_1 = \frac{P_1}{V} = \frac{2}{40} = 0.05\,\text{A} \qquad Ans.$$

$$I_2 = \frac{P_2}{V} = \frac{2}{40} = 0.05\,\text{A} \qquad Ans.$$

$$I_T = I_1 + I_2 = 0.05 + 0.05 = 0.1\,\text{A} \qquad Ans.$$

Check:

$$P_T = P_1 + P_2 = 2 + 2 = 4\,\text{W}$$

$$I_T = \frac{P_T}{V} = \frac{4}{40} = 0.1\,\text{A}$$

which agrees with the previously calculated value.

5.11   The combined resistance of a coffee percolator and toaster in parallel is 24 Ω. Find the total power used if the line voltage is 120 V.

$$P_T = \frac{V^2}{R_T} = \frac{(120)^2}{24} = 600\,\text{W} \qquad Ans.$$

5.12 Find I_3 in the parallel current–divider circuit (Fig. 5-22).

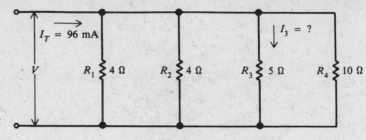

Fig. 5-22

Find R_T.

$$\frac{1}{R_T} = \frac{1}{R_1} + \frac{1}{R_2} + \frac{1}{R_3} + \frac{1}{R_4}$$

$$= \frac{1}{4} + \frac{1}{4} + \frac{1}{5} + \frac{1}{10} = \frac{32}{40} \qquad\qquad (5\text{-}5)$$

$$R_T = \frac{40}{32} = 1.25\,\Omega$$

Find V.

$$V = I_T R_T = 96(1.25) = 120\,\text{mV}$$

Then find I_3.

$$I_3 = \frac{V}{R_3} = \frac{120}{5} = 24\,\text{mA} \qquad Ans.$$

Supplementary Problems

5.13 Write the word or words which most correctly complete the following statements.

(a) There is only _____ voltage across all components in parallel.

(b) If a parallel circuit is open in the main line, the current is _____ in all the branches.

(c) For any number of conductances in parallel, their values are _____ to obtain G_T.

(d) When I_T divides into branch currents, each branch current is _____ proportional to the branch resistance.

(e) The sum of the _____ values of power dissipated in parallel resistances equals the _____ power produced by the source.

Ans.   (a) one;   (b) zero;   (c) added;   (d) inversely;   (e) individual, total

5.14 A 100-Ω and a 150-Ω resistor are connected in parallel. What is the total resistance?
Ans. $R_T = 60\,\Omega$

5.15 When the voltage across R_4 is 10 V, what is the source voltage in Fig. 5-23?
Ans. $V = 10\,\text{V}$

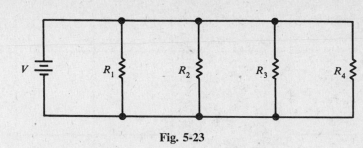

Fig. 5-23

5.16   Find the equivalent resistances in the circuits shown in Fig. 5-24.
   Ans.   (*a*) $R_T = 1\,\Omega$;   (*b*) $R_T = 2\,\Omega$;   (*c*) $R_T = 4.8\,\Omega$;   (*d*) $R_T = 3.6\,\Omega$

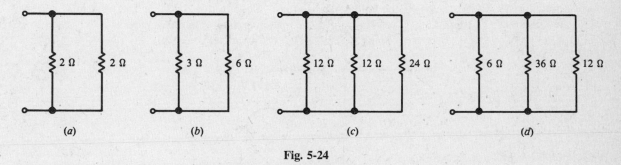

(*a*)   (*b*)   (*c*)   (*d*)

Fig. 5-24

5.17   Find the missing branch or total current as indicated in Fig. 5-25.
   Ans.   (*a*) $I_T = 3\,\text{A}$;   (*b*) $I_3 = 2\,\text{A}$

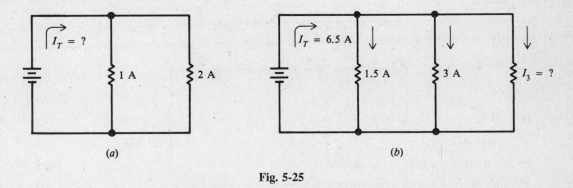

(*a*)   (*b*)

Fig. 5-25

5.18   Four equal resistances are connected in parallel across a 90-V source. If the resistances are $36\,\Omega$ for each branch, find the total resistance and the total current.   *Ans.* $R_T = 9\,\Omega$; $I_T = 10\,\text{A}$

5.19   Find the total resistance, each branch current, and total current (Fig. 5-26).
   Ans. $R_T = 2.67\,\Omega$; $I_1 = 2\,\text{A}$; $I_2 = 1\,\text{A}$; $I_T = 3\,\text{A}$

5.20   In the circuit shown (Fig. 5-27), find the total resistance, each branch current, and total current.
   Ans. $R_T = 4\,\Omega$; $I_1 = 20\,\text{A}$; $I_2 = 4\,\text{A}$; $I_3 = 1\,\text{A}$; $I_T = 25\,\text{A}$

5.21   If the 25-Ω resistor is removed from the circuit in Fig. 5-27, what is the total current and total resistance?   *Ans.* $I_T = 21\,\text{A}$; $R_T = 4.76\,\Omega$

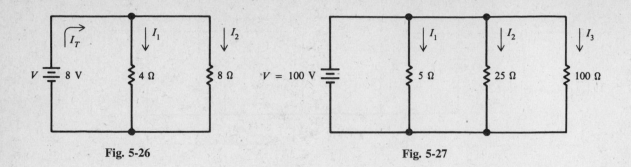

Fig. 5-26   Fig. 5-27

5.22   An ammeter (instrument that measures current) carries 0.05 A and is in parallel with a shunt resistor drawing 1.9 A (Fig. 5-28). If the voltage across the combination is 4.2 V, find (a) the total current, (b) resistance of the shunt, (c) resistance of the ammeter, and (d) total resistance.
Ans.   (a) $I_T = 1.95$ A;   (b) Shunt $R = 2.21\ \Omega$;   (c) Ammeter $R = 84.0\ \Omega$;   (d) $R_T = 2.15\ \Omega$

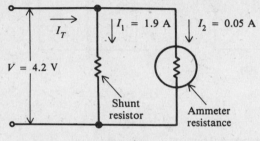

Fig. 5-28

5.23   Find the total resistance, each branch current, and total current (Fig. 5-29).
Ans. $R_T = 2.67\ \Omega$; $I_1 = 8$ A; $I_2 = 6$ A; $I_3 = 4$ A; $I_T = 18$ A

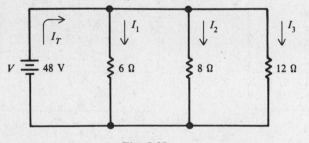

Fig. 5-29

5.24   A circuit consists of five identical resistances connected in parallel across a voltage source. If the total circuit current is 1 A, what is the current through each resistance?   *Ans.* $I = 0.2$ A

5.25   In the circuit of Fig. 5-30, find V if $I_3 = 0.2$ A. Then find I_T.   *Ans.* $V = 2$ V; $I_T = 0.4$ A

5.26   The ignition coil and the starting motor of a car are connected in parallel across a 12-V battery through an ignition switch (Fig. 5-31). Find (a) the total current drawn from the battery, (b) the voltage across the coil and the motor, and (c) the total resistance of the circuit.
Ans.   (a) $I = 105$ A;   (b) $V_1 = V_2 = 12$ V;   (c) $R_T = 0.114\ \Omega$

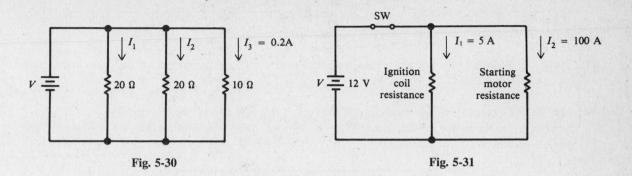

Fig. 5-30   Fig. 5-31

5.27 Two headlight lamps, each drawing 4 A, and two taillight lamps, each drawing 1 A, are wired in parallel across a 12-V storage battery. What is the total current drawn and the total resistance of the circuit? *Ans.* $I_T = 10 \text{ A}$; $R_T = 1.2 \, \Omega$

5.28 What is the value of a resistor that must be connected in parallel across a 100-kΩ resistance to reduce R_T to (*a*) 50 kΩ, (*b*) 25 kΩ, and (*c*) 10 kΩ? *Ans.* (*a*) $R_x = 100 \text{ k}\Omega$; (*b*) $R_x = 33.3 \text{ k}\Omega$; (*c*) $R_x = 11.1 \text{ k}\Omega$

5.29 What resistance must be connected in parallel with a 20-Ω and a 60-Ω resistor in parallel in order to provide a total resistance of 10 Ω? *Ans.* 30 Ω

5.30 Two resistances are connected in parallel. $R_1 = 24 \, \Omega$, $R_2 = 24 \, \Omega$, and $I_T = 6 \text{ A}$. Find the current in each branch. *Ans.* $I_1 = I_2 = 3 \text{ A}$

5.31 Find the current in each branch of a parallel circuit consisting of a 20-Ω percolator and a 30-Ω toaster if the total current is 10 A. *Ans.* I in percolator $= 6 \text{ A}$; I in toaster $= 4 \text{ A}$

5.32 Find the missing values in Fig. 5-32. *Ans.* $V = 4.5 \text{ V}$; $I_1 = 1.50 \text{ A}$; $I_2 = 1.13 \text{ A}$; $I_3 = 0.38 \text{ A}$

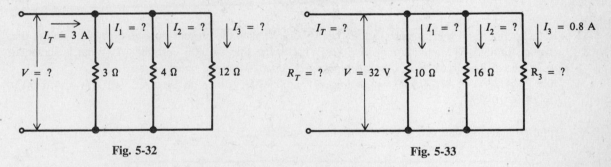

Fig. 5-32   Fig. 5-33

5.33 Find the missing values in Fig. 5-33.
Ans. $R_3 = 40 \, \Omega$; $R_T = 5.33 \, \Omega$; $I_1 = 3.2 \text{ A}$; $I_2 = 2 \text{ A}$; $I_T = 6 \text{ A}$

5.34 Find the total conductance in siemens for the following parallel branches: $G_1 = 6000 \, \mu\text{S}$, $G_2 = 7000 \, \mu\text{S}$, and $G_3 = 20\,000 \, \mu\text{S}$. *Ans.* $G_T = 33\,000 \, \mu\text{S}$

5.35 I_T is 12 mA for two branch resistances. R_1 is 10 kΩ and R_2 is 36 kΩ. Find I_1 and I_2 in this parallel current–divider circuit. *Ans.* $I_1 = 9.39 \text{ mA}$; $I_2 = 2.61 \text{ mA}$

5.36 What is the total power used by a 4.5-A electric iron, a 0.9-A fan, and a 2.4-A refrigerator motor if they are all connected in parallel across a 120-V line? *Ans.* $P_T = 936 \text{ W}$

5.37   Find the power drawn from a 12-V battery by a parallel circuit of two headlights, each drawing 4.2 A, and two taillights, each drawing 0.9 A.   *Ans.* $P_T = 122.4\,\text{W}$

5.38   Five 150-W light bulbs are connected in parallel across a 120-V power line. If one bulb opens, how many bulbs can light?   *Ans.*   Four

5.39   In Fig. 5-34 find (*a*) each branch current; (*b*) I_T; (*c*) R_T; and (*d*) P_1, P_2, P_3, and P_T.
    Ans.   (*a*) $I_1 = 30\,\text{mA}$, $I_2 = 14.6\,\text{mA}$, $I_3 = 60\,\text{mA}$;   (*b*) $I_T = 104.6\,\text{mA}$;   (*c*) $R_T = 1.15\,\text{k}\Omega$;
    (*d*) $P_1 = 3.60\,\text{W}$, $P_2 = 1.75\,\text{W}$, $P_3 = 7.20\,\text{W}$, $P_T = 12.6\,\text{W}$

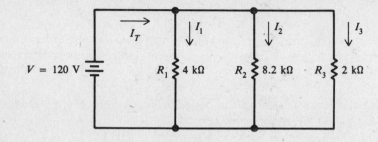

Fig. 5-34

5.40   Find R_2 in Fig. 5-35.   *Ans.* $R_2 = 1\,\text{k}\Omega$

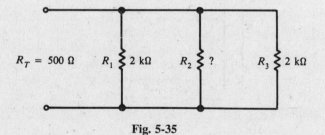

Fig. 5-35

5.41   Refer to Fig. 5-34 and assume that R_2 opens. (*a*) What is the current through R_2? (*b*) What is the current through R_1? and through R_3? (*c*) What is the line or total current? (*d*) What is the total resistance of the circuit? (*e*) How much power is generated by the battery?
    Ans.   (*a*) $I_2 = 0\,\text{A}$;   (*b*) $I_1 = 30\,\text{mA}$, $I_3 = 60\,\text{mA}$;   (*c*) $I_T = 90\,\text{mA}$;   (*d*) $R_T = 1.33\,\text{k}\Omega$;
    (*e*) $P_T = 10.8\,\text{W}$

5.42   Find I_2 and I_4 in the parallel current–divider circuit (Fig. 5-36).   *Ans.* $I_2 = 2.5\,\text{A}$; $I_4 = 1.67\,\text{A}$

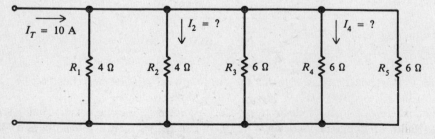

Fig. 5-36

Chapter 6

Batteries

THE VOLTAIC CELL

A *voltaic chemical cell* is a combination of materials used to convert chemical energy into electric energy. The chemical cell consists of two electrodes made of different kinds of metals or metallic compounds, and an electrolyte, which is a solution capable of conducting an electric current (Fig. 6-1a). A battery is formed when two or more cells are connected.

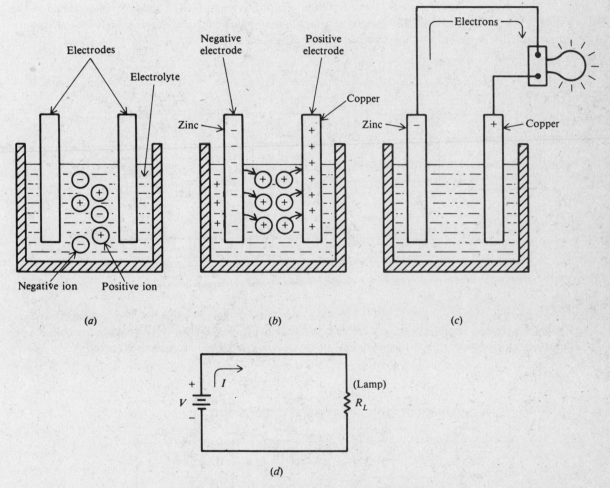

Fig. 6-1 Basic chemical action of a voltaic cell

An excellent example of a pair of electrodes is zinc and copper. Zinc contains an abundance of negatively charged atoms, while copper has an abundance of positively charged atoms. When plates of these metals are immersed in an electrolyte, chemical action between the two begins. The zinc electrode accumulates a much larger negative charge since it gradually dissolves into the electrolyte. The atoms which leave the zinc electrode are positively charged. They are attracted by the negatively charged ions (−) of the electrolyte, while they repel the positively charged ions (+) of the electrolyte toward the copper electrode (Fig. 6-1b). This causes

97

electrons to be removed from the copper, leaving it with an excess of positive charge. If a load such as a light bulb is connected across the terminals on the electrodes, the forces of attraction and repulsion will cause free electrons in the negative zinc electrode, connecting wires, and light bulb filament to move toward the positively charged copper electrode (Fig. 6-1c). The potential difference that results permits the cell to function as a source of applied voltage V (Fig. 6-1d).

The electrolyte of a cell may be liquid or a paste. If the electrolyte is a liquid, the cell is often called a *wet* cell. If the electrolyte is in a paste form, the cell is referred to as a *dry* cell.

SERIES AND PARALLEL CELLS

When cells are connected in series (Fig. 6-2), the total voltage across the battery of cells is equal to the sum of the voltage of each of the individual cells. In Fig. 6-2, the four 1.5-V cells in series provide a total battery voltage of 6 V. When cells are placed in series, the positive terminal of one cell is connected to the negative terminal of the other cell. The current flowing through such a battery of series cells is the same as for one cell because the same current flows through all the series cells.

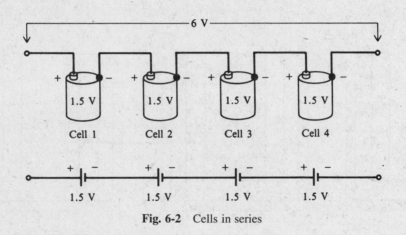

Fig. 6-2   Cells in series

To obtain a greater current, the battery has cells in parallel (Fig. 6-3). When cells are placed in parallel, all the positive terminals are connected together and all the negative terminals are connected together. Any point

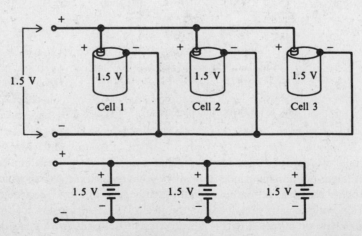

Fig. 6-3   Cells in parallel

on the positive side can serve as the positive terminal of the battery and any point on the negative side can be the negative terminal.

The total voltage output of a battery of three parallel cells is the same as that for a single cell (Fig. 6-3), but the available current is three times that of one cell. The parallel connection has the same effect of increasing the size of the electrodes and electrolyte in a single cell, which increases the current capacity.

Identical cells in parallel all supply equal parts of the current to the load. For example, of three identical parallel cells producing a load current of 270 mA, each cell contributes 90 mA.

PRIMARY AND SECONDARY CELLS

Primary cells are those which cannot be recharged or returned to good condition after their voltage output drops too low. Dry cells used in flashlights and transistor radios are examples of primary cells.

Secondary cells are those which are rechargeable. During recharging, the chemicals which provide electric energy are restored to their original condition. Recharging is done by passing direct current through a cell in a direction opposite to the direction of the current which the cell delivers to a circuit.

A cell is recharged by connecting it to a battery charger in "like-to-like" polarity (Fig. 6-4). Some battery chargers have a voltmeter and an ammeter which indicate the charging voltage and current.

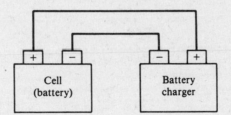

Fig. 6-4 Recharging a secondary cell with a battery charger

The most common example of a secondary cell is an automobile storage battery. Secondary cells or batteries are particularly useful for powering mobile equipment where a generator is available to keep them charged. Smaller, sealed secondary cells are used to power such portable equipment as shavers, electronic calculators, radios, and television receivers. These can be easily charged from ordinary house current by simple, low-cost chargers often built into the equipment or appliance itself.

TYPES OF BATTERIES

Lead–Acid Battery

The lead–acid battery consists of a number of lead–acid cells. Each cell has two groups of lead plates; one set is the positive terminal and the other is the negative terminal. All positive plates are connected together with a connecting strap (Fig. 6-5). All negative plates are similarly connected together. The positive and negative plates are *interlaced* so that alternately, there is a positive plate and a negative plate. Between the plates are sheets of insulating material called *separators*, made either of porous wood, perforated wood, or fiberglass. The separators prevent the positive and negative plates from touching each other and producing a short circuit, which would destroy the cell. The positive plate is treated chemically to form lead peroxide (a combination of lead and oxygen), and the negative electrode consists of porous, spongy lead. The two sets of plates with the separators between them are placed in a container filled with a dilute solution of sulfuric acid and water. The term *lead–acid battery* refers to the lead plates and sulfuric acid that are the principal components of the battery.

The voltage in this type of cell is slightly more than 2 V. Batteries used in modern automobiles contain six cells connected in series so that the output voltage from the battery is slightly more than 12 V.

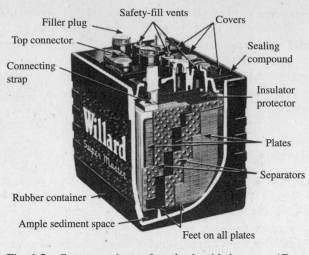

Fig. 6-5   Cutaway view of a lead–acid battery. (*From B. Grob, Basic Electronics, 4th ed., McGraw-Hill, New York, 1977, p. 247.*)

Older automobiles made before the mid-1950s used batteries in which three cells were connected in series to give an output voltage of slightly more than 6 V.

The storage battery can supply current for a much longer time than the average dry cell. When the storage battery is discharged and is no longer able to supply the current required by the circuit, the battery can be removed from the circuit and recharged by passing current through it in the opposite direction. Once the battery has been recharged, it can again be connected to the circuit and will supply current to the circuit.

In an automobile, the battery is connected to a device called an *alternator*. As long as the car is running at a reasonable speed, the alternator is both charging the battery and supplying the current needed to operate the car. However, when the car is operated at a slow speed or when it is stopped, the alternator is not turning fast enough to provide the electricity needed by the car. The battery then supplies this energy, causing it to slowly discharge.

When the battery discharges, some of the acid of the electrolyte combines with the active material on the plates (Fig. 6-6a). This chemical action changes the material in both plates to lead sulfate. When the battery is being charged by the alternator, the reverse action takes place, and the acid which was absorbed by the plates is returned to the electrolyte (Fig. 6-6b). The result is that the active material on the plates is changed back into the original (charged condition) lead peroxide and sponge lead, and the electrolyte is restored to its original strength.

Whenever a battery is charging, the chemical action produces hydrogen gas on one plate surface and oxygen gas on the other. These gases bubble to the surface and escape through the vent hole in the cap on the cell. Thus water (H_2O) is lost to the cell when the gases leave. The water that escapes must be replaced to maintain the proper electrolyte level. Only distilled water should be added to the cell. Otherwise, any impurity in the added water will combine chemically with the sulfuric acid on the plates and form a stable compound that will not enter into the charge or discharge action of the battery.

Carbon–Zinc Cell

This is one of the oldest and most widely used commercial types of dry cell. The carbon, in the form of a rod that is placed in the center of the cell, is the positive terminal. The case of the cell is made of zinc, which is the negative electrode (Fig. 6-7). Between the carbon electrode and the zinc case is the electrolyte of a chemical pastelike mixture. The cell is sealed to prevent the liquid in the paste from evaporating. The voltage of a cell of this type is about 1.5 V.

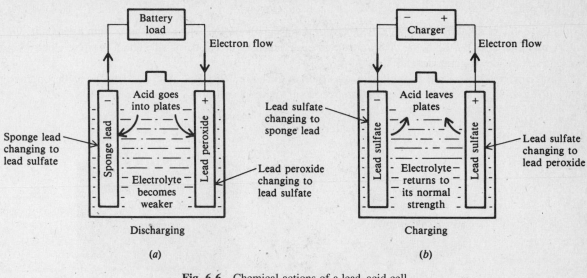

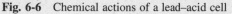

Fig. 6-6   Chemical actions of a lead–acid cell

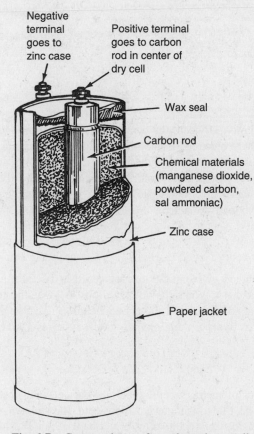

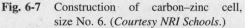

Fig. 6-7   Construction of carbon–zinc cell, size No. 6. (*Courtesy NRI Schools.*)

Alkaline Cell

The secondary alkaline cell is so called because it has an alkaline electrolyte of potassium hydroxide. One battery type that goes by the name *alkaline battery* has a negative electrode of zinc and a positive electrode of manganese dioxide. It generates 1.5 V.

The primary alkaline cell is similar in construction to the rechargeable type and has the same operating voltage (Fig. 6-8). This cell has extended life over a carbon–zinc cell of the same size.

Fig. 6-8   Manganese–alkaline battery.
(*From Grob*, p. 251.)

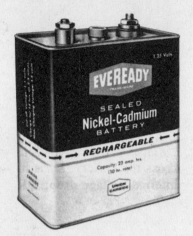

Fig. 6-9   Nickel–cadmium battery.
(*From Grob*, p. 253.)

Nickel–Cadmium Cell

In the secondary nickel–cadmium dry cell, the electrolyte is potassium hydroxide, the negative electrode is nickel hydroxide, and the positive electrode is cadmium oxide. The operating voltage is 1.25 V. These cells are manufactured in several sizes, including flat button shapes. The nickel–cadmium battery is the only dry battery that is a true storage battery with a reversible chemical reaction, allowing recharging many times (Fig. 6-9). It is a rugged device which gives dependable service under extreme conditions of shock, vibration, and temperature. Therefore, it is ideally suited for use in powering portable communication equipment such as a two-way radio.

Edison Cell

A lighter, more rugged secondary cell than the lead–acid cell is the Edison, or nickel–iron–alkaline, cell. It operates at a no-load voltage of 1.4 V. When the voltage drops to 1.0 V, the cell should be recharged. When fully charged, it has a positive plate of nickel and nickel hydrate and a negative plate of iron. Like the lead–acid cell, the Edison cell also produces hydrogen and oxygen gases. As a result, the electrolyte requires filling up with distilled water.

Mercury Cell

There are two different types of mercury cells. One is a flat cell that is shaped like a button, while the other is a cylindrical cell that looks like a standard flashlight cell. The advantage of the button-type cell is that several of them can be stacked inside one container to form a battery. A typical battery is made up of three flat cells (Fig. 6-10). A cell produces 1.35 V.

Mercury cells and batteries have a good shelf life and are very rugged. Because they produce a constant output voltage under different load conditions, they are used in many different products, including electric watches, hearing aids, test instruments, and alarm systems.

Fig. 6-10 A typical mercury battery.
(*Courtesy NRI Schools.*)

BATTERY CHARACTERISTICS

Internal Resistance

A battery is a dc voltage generator. All generators have internal resistance, R_i. In a chemical cell, the resistance of the electrolyte between electrodes is responsible for most of the cell's internal resistance (Fig. 6-11). Since any current in the battery must flow through the internal resistance, R_i is in series with the generated voltage V_B (Fig. 6-12a). With no current, the voltage drop across R_i is zero so that the full generated voltage V_B develops across the output terminals (Fig. 6-12a). This is the open-circuit voltage, or no-load voltage. If a load resistance R_L is connected across the battery, R_L is in series with R_i (Fig. 6-12b). When current I_L flows in this circuit, the internal voltage drop, $I_L R_i$, decreases the terminal voltage V_L of the battery so that $V_L = V_B - I_L R_i$.

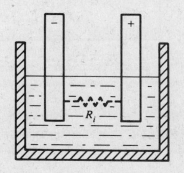

Fig. 6-11   Internal resistance in a cell

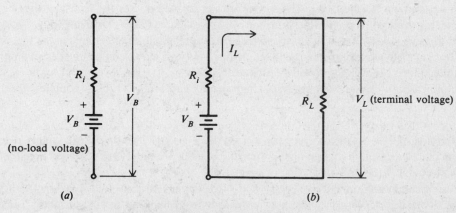

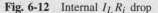

Fig. 6-12   Internal $I_L R_i$ drop

Example 6.1   A dry battery has an open-circuit, or no-load, voltage of 100 V (Fig. 6-13). If the internal resistance is 100 Ω and the load resistance is 600 Ω, find the voltage V_L across the output terminals.

The battery is marked 100 V because 100 V is its open-circuit voltage. With no load, the load current is zero. When load resistance R_L is added, there is a closed circuit, and the load current is calculated by Ohm's law.

$$I_L = \frac{V}{R_i + R_L} = \frac{100}{100 + 600} = \frac{100}{700} = 0.143 \text{ A}$$

The internal battery drop is

$$I_L R_i = 0.143(100) = 14.3 \text{ V}$$

so that the voltage at the battery's terminal is

$$V_L = V_B - I_L R_i = 100 - 14.3 = 85.7 \text{ V} \qquad Ans.$$

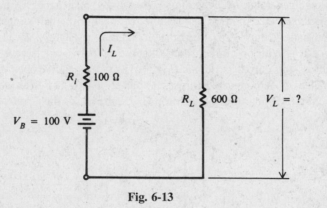

Fig. 6-13

Specific Gravity

The *specific gravity* of any liquid is a ratio comparing its weight with the weight of an equal volume of water. Pure sulfuric acid has a specific gravity of 1.835 since it weighs 1.835 times as much as water per unit volume.

The specific gravity of the electrolyte solution in a lead–acid cell ranges from 1.210 to 1.300 for new, fully charged batteries. The higher the specific gravity, the less internal resistance of the cell and the higher the possible load current. As the cell discharges, the water formed dilutes the acid and the specific gravity gradually decreases to about 1.150, at which time the cell is considered to be fully discharged. Specific gravity is measured with a *hydrometer* of the syringe type, which has a compressible rubber bulb at the top, a glass barrel, and a rubber hose at the bottom of the barrel. In taking readings with a hydrometer, the decimal point is usually omitted. For example, a specific gravity of 1.270 is read simply as "twelve-seventy." A hydrometer reading of 1210 to 1300 indicates full charge; about 1250 is half-charge; and 1150 to 1200 is complete discharge.

Capacity

The capacity of a battery is rated in ampere-hours (Ah). The capacity of a storage battery determines how long it will operate at a given discharge rate. For example, a 90-Ah battery must be recharged after 9 h of an average 10-A discharge.

A cell of a lead–acid automobile battery, when fully charged, has an initial voltage of about 2.1 V at no load, but discharges rapidly. The battery is "dead" after about 2 h of discharging under load condition. However, under normal use, this battery type is constantly recharged by the alternator in the automobile.

Shelf Life

The *shelf life* of a cell is that period of time during which the cell can be stored without losing more than approximately 10 percent of its original capacity. The *capacity* of a cell is its ability to deliver a given amount of current to the circuit in which it is used. The loss of capacity of a stored cell is primarily due to the drying out of its electrolyte (wet cell) and to chemical actions which change the materials within the cell. Since heat stimulates both these actions, the shelf life of a cell can be extended by keeping it in a cool, dry place.

Comparison of Types

Table 6-1 compares the types of cells described.

Table 6-1 Types of Cells

Name	Voltage	Wet or Dry	Primary or Secondary Type	Examples and Characteristics
Lead–acid cell	2.2	Wet	Secondary	Very low R_i and high current ratings; 6- and 12-V batteries
Carbon–zinc cell	1.5	Dry	Primary	*AA*, *A*, *B*, *C*, and *D* size cells; flashlight batteries; lowest price; short shelf life; low capacity
Manganese–alkaline cell	1.5	Dry	Both types	Manganese dioxide and zinc in hydroxide; currents above 300 mA
Nickel–cadmium cell	1.25	Dry	Secondary	Hydroxide electrolyte; constant voltage; reversible chemical reaction; used in rechargeable flashlights, portable power tools
Edison cell	1.4	Wet	Secondary	Nickel and iron in hydroxide; industrial uses
Mercury cell	1.35	Dry	Both types	Mercuric oxide and zinc in hydroxide; constant voltage, long shelf life; B batteries; miniature button cells for hearing aids, cameras, watches, calculators

Solved Problems

6.1 Write the word or words which most correctly complete the following statements.

(a) A _____ consists of _____ or more cells connected in series or parallel.

(b) A chemical cell consists basically of _____ electrodes of different kinds of metals or metallic compounds separated by an _____ .

(c) Cells which cannot be effectively recharged are called _____ cells.

(d) A cell or a battery is recharged by passing current through it in a direction _____ to the direction of its discharge current.

(e) In order to obtain higher voltages, cells are connected in _____ .

Ans.   (a) battery, two;   (b) two, electrolyte;   (c) primary;   (d) opposite;   (e) series

6.2 Match the type of cell in column 1 to its characteristic in column 2 (use a letter once only).

	Column 1		Column 2
1.	Lead–acid	(*a*)	Long shelf life
2.	Carbon–zinc	(*b*)	1.4-V voltage
3.	Nickel–cadmium	(*c*)	Automobile battery
4.	Edison cell	(*d*)	Inexpensive flashlight cell
5.	Mercury	(*e*)	Potassium hydroxide electrolyte
		(*f*)	3-V voltage

Ans. 1. (*c*) 2. (*d*) 3. (*e*) 4. (*b*) 5. (*a*)

6.3 A 6-V battery is temporarily short-circuited. The short-circuit current I_{SC} is 30 A. What is the internal resistance?

The battery rating of 6 V in this case is the open-circuit, or no-load, voltage. So

$$R_i = \frac{V}{I_{SC}} = \frac{6}{30} = 0.2\ \Omega \qquad Ans.$$

Note that the presence of internal resistance prevents the current from becoming very high.

6.4 A battery has a 12-V output on an open circuit, which drops to 11.5 V with a load current of 1 A. Find the internal resistance.

Open-circuit voltage = internal resistance drop + terminal voltage

$$V_B = I_L R_i + V_L$$

Solving of R_i,

$$I_L R_i = V_B - V_L$$

$$R_i = \frac{V_B - V_L}{I_L} = \frac{12 - 11.5}{1} = \frac{0.5}{1} = 0.5\ \Omega \qquad Ans.$$

We see that the internal resistance of any battery can be calculated by determining how much the output voltage drops for a specific amount of load current.

6.5 A discharged storage battery of three cells connected in series has an open-circuit voltage of 1.8 V per cell. Each cell has an internal resistance of 0.1 Ω. What is the minimum voltage of a charging battery to produce an initial charging rate of 10 A?

Charging battery voltage = battery voltage + internal resistance drop

$$\text{Battery voltage (cells in series)} = 3\ \text{cells} \times \frac{1.8\ \text{V}}{\text{cell}} = 5.4\ \text{V}$$

$$\text{Internal resistance drop} = I R_i = 10(3 \times 0.1) = 3\ \text{V}$$

Then,

$$\text{Charging battery voltage} = 5.4 + 3 = 8.4\ \text{V} \qquad Ans.$$

6.6   A lead–acid battery is rated at 200 Ah. Based on an 8-h discharge, what average load current can this battery supply?

In units, $\text{Capacity} = \text{amperes} \times \text{hours}$

Then, $\text{Load current (in amperes)} = \dfrac{\text{capacity}}{\text{hours}}$

and $\text{Load current} = \dfrac{200}{8} = 25\,\text{A}$ *Ans.*

6.7   What is the no-load voltage across four carbon–zinc cells in series?

$$\text{Voltage} = 4 \times \text{no-load voltage of a single cell}$$
$$= 4(1.5) = 6\,\text{V} \qquad \textit{Ans.}$$

6.8   What is the specific gravity of a solution with equal parts of sulfuric acid and water?

A solution with equal parts of sulfuric acid and water has a weight equally distributed between sulfuric acid and water (that is, each accounts for one-half the weight of the solution). If pure sulfuric acid has a specific gravity of 1.835, then

$$\text{Specific gravity of solution} = \frac{1}{2}(1.835) + \frac{1}{2}(1) = 0.918 + 0.500 = 1.418 \qquad \textit{Ans.}$$

Supplementary Problems

6.9   Write the word or words which most correctly complete the following statements.

(*a*) A cell which converts _____ energy into _____ energy is called a chemical cell.

(*b*) A cell in which the electrolyte is a liquid is commonly referred to as a _____ cell, while a cell in which the electrolyte is in a paste form is called a _____ cell.

(*c*) Cells which can be effectively recharged are called _____ cells.

(*d*) When charging a cell or a battery, its positive terminal is connected to the _____ terminal of the battery charger, and its negative terminal is connected to the _____ terminal of the charger.

(*e*) In order to obtain a greater current capacity, cells are connected in _____ .

Ans.   (*a*) chemical, electric;   (*b*) wet, dry;   (*c*) secondary;   (*d*) positive, negative;   (*e*) parallel

6.10   Match the type of cell in column 1 to its characteristic in column 2.

Column 1	Column 2
1. Lead–acid	(*a*) Nickel–iron–alkaline
2. Carbon–zinc	(*b*) 12-V battery
3. Nickel–cadmium	(*c*) Secondary dry battery
4. Edison cell	(*d*) 1.5-V primary cell
5. Mercury	(*e*) 5-V battery
	(*f*) Ideal for transistorized equipment

Ans.   1. (*b*)   2. (*d*)   3. (*c*)   4. (*a*)   5. (*f*)

6.11   Fill in the missing quantities (Fig. 6-14).

V_B, V	R_i, Ω	R_L, Ω	I_L, A	$I_L R_i$, V	V_L, V
100	?	?	2	?	80
6	0.2	1	?	?	?
12	?	5	?	?	2

Ans.

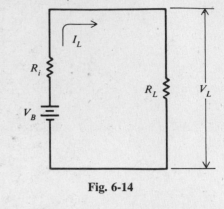

Fig. 6-14

V_B, V	R_i, Ω	R_L, Ω	I_L, A	$I_L R_i$, V	V_L, V
....	10	40	....	20	....
....	....	....	5	1	5
....	25	....	0.4	10	....

6.12   The terminal voltage V_L drops as the load current I_L increases. For a 12-V battery with internal resistance of 1 Ω, we vary the load resistance from a very high value to zero in order to observe how the terminal voltage varies with changing load current. Fill in the missing values of the table.

V_B, V	R_i, Ω	R_L, Ω	$R_T = R_L + R_i$, Ω	I_L, A	$I_L R_i$, V	$V_L = V_B - I_L R_i$, V
12	1	∞	?	?	?	?
12	1	9	?	?	?	?
12	1	5	?	?	?	?
12	1	3	?	?	?	?
12	1	1	?	?	?	?
12	1	0	?	?	?	?

Ans.

V_B, V	R_i, Ω	R_L, Ω	$R_T = R_L + R_i$, Ω	I_L, A	$I_L R_i$, V	$V_L = V_B - I_L R_i$, V
....	....	....	∞	0	0	12
....	....	....	10	1.2	1.2	10.8
....	....	....	6	2	2	10
....	....	....	4	3	3	9
....	....	....	2	6	6	6
....	....	....	1	12	12	0

6.13   For Problem 6.12 make a plot with terminal voltage V_L as the ordinate and load current I_L as the abscissa. Describe the plot.
Ans.   See Fig. 6-15. The plot is a straight line, where V_L is a maximum when the circuit is open (I_L is zero) and a minimum when the circuit is shorted (I_L is a maximum).

6.14   A 6-V lead–acid battery has an internal resistance of 0.02 Ω. How much current will flow if the battery has a short circuit?   *Ans.*   300 A

6.15   A new carbon–zinc cell has a voltage of 1.5 V. A battery made up of 30 cells connected in series ages so that its no-load voltage drops 15 percent. What is the no-load voltage of the cell and battery?
Ans.   1.28 V, 38.2 V

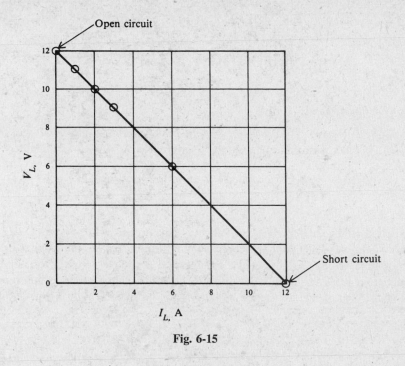

Fig. 6-15

6.16   What is the specific gravity of the electrolyte of a lead–acid battery with one-fourth part sulfuric acid and three-fourths part water? Would a hydrometer reading of that solution indicate full charge, half-charge, or discharge?   *Ans.*   1.209, full charge

6.17   How many cells are necessary to produce a battery with double the voltage and current rating of a single cell? Draw a schematic diagram.   *Ans.*   Four cells; see Fig. 6-16

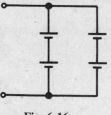

Fig. 6-16

6.18   Draw pictorial schematic diagrams showing two 12-V lead–acid batteries being charged by a 15-V source. Show the direction of current during charge.   *Ans.*   See Fig. 6-17

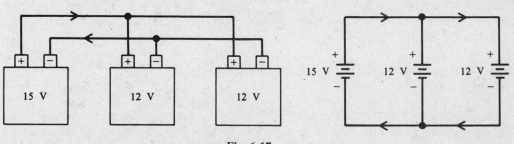

Fig. 6-17

Chapter 7

Kirchhoff's Laws

KIRCHHOFF'S VOLTAGE LAW (KVL)

Kirchhoff's *voltage law* states that *the voltage applied to a closed circuit equals the sum of the voltage drops in that circuit.* This fact was used in the study of series circuits and was expressed as follows:

$$\text{Voltage applied} = \text{sum of voltage drops}$$

$$V_A = V_1 + V_2 + V_3 \tag{7-1}$$

where V_A is the applied voltage and V_1, V_2, and V_3 are voltage drops.

Another way of stating KVL is that the algebraic sum of the voltage rises and voltage drops must be equal to zero. A voltage source or emf is considered a voltage rise; a voltage across a resistor is a voltage drop. Often for convenience in labeling, letter subscripts are shown for voltage sources and numerical subscripts for voltage drops. This form of the law can be written by transposing the right members of Eq. (7-1) to the left side.

$$\text{Voltage applied} - \text{sum of voltage drops} = 0$$

Substitute letters:

$$V_A - V_1 - V_2 - V_3 = 0$$

or

$$V_A - (V_1 + V_2 + V_3) = 0$$

Using a new symbol, Σ, the Greek capital letter sigma, we have

$$\Sigma V = V_A - V_1 - V_2 - V_3 = 0 \tag{7-2}$$

in which ΣV, the algebraic sum of all the voltages around any closed circuit, equals zero. Σ means "sum of."

We assign a $+$ sign to a voltage rise and a $-$ sign to a voltage drop for the $\Sigma V = 0$ formula (Fig. 7-1). In tracing voltage drops around a circuit, start at the negative terminal of the voltage source. The path from the negative terminal to the positive terminal through the source is a voltage rise. We continue to trace the circuit from the positive terminal through all resistors and back to the negative terminal of the source. In Fig. 7-1 if we start at point a, the negative terminal of the battery, and move around the circuit in the direction $abcda$, we go through V_A from $-$ to $+$ and $V_A = +100$ V. If we start at point b and move in the opposite direction $badcb$, we go through V_A from $+$ to $-$ and $V_A = -100$ V. The voltage drop across any resistance will be

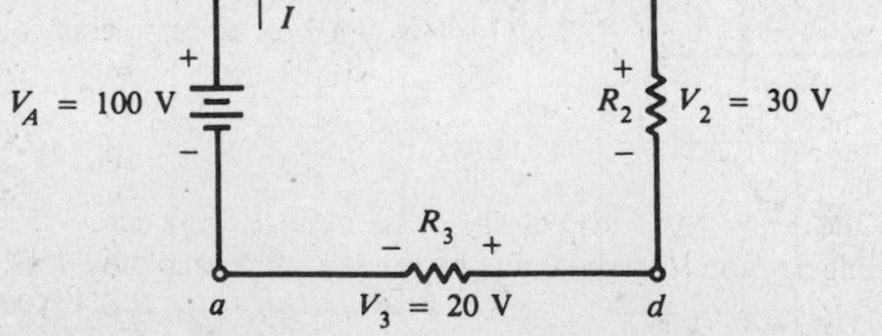

$$
\begin{aligned}
\Sigma V &= V_A - V_1 - V_2 - V_3 \\
&= 100 - 50 - 30 - 20 \\
&= 100 - 100 \\
&= 0
\end{aligned}
$$

Fig. 7-1 Illustration of $\Sigma V = 0$

110

negative $(-)$ if we trace it in the $+$ to $-$ direction. Thus in Fig. 7-1, if we trace the circuit in the direction $abcda$, $V_1 = -50$ V, $V_2 = -30$ V, and $V_3 = -20$ V. The voltage drop will be positive $(+)$ if we go through the resistance in the $-$ to $+$ direction. So in tracing the circuit in the direction $abcda$, we have

$$\Sigma V = 0$$

$$V_A - V_1 - V_2 - V_3 = 0$$

$$100 - 50 - 30 - 20 = 0$$

$$0 = 0$$

Example 7.1   Determine the direction of voltage around the circuit $abcda$ (Fig. 7-2), and then write the expression for voltages around the circuit.

Assume direction of current as shown. Mark the $+$ and $-$ polarities of each resistor.

V_A is a voltage source $(+)$.   (It is a voltage rise in the current direction assumed.)

V_1 is a voltage drop $(-)$.   (It is a decrease in the direction assumed.)

V_2 is a voltage drop $(-)$.   (It is a decrease in the direction assumed.)

V_B is a voltage source $(-)$.   (It is a decrease in voltage in the current direction assumed.)

V_3 is a voltage drop $(-)$.   (It is a decrease in the direction assumed.)

$$\Sigma V = 0$$

$$+V_A - V_1 - V_2 - V_B - V_3 = 0$$

Group the voltage rises and the voltage drops.

$$V_A - (V_1 + V_2 + V_3 + V_B)$$

Notice that the voltage drops include a voltage source V_B. Ordinarily a source would be positive. In this case the polarity of the source is acting against the assumed direction of current. Therefore, its effect is to decrease the voltage.

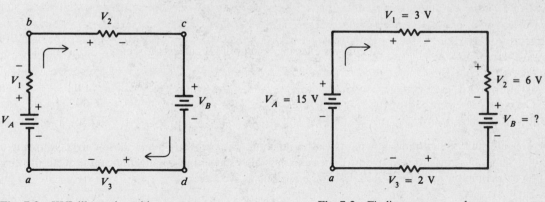

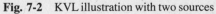
Fig. 7-2   KVL illustration with two sources

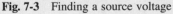
Fig. 7-3   Finding a source voltage

Example 7.2   Determine the voltage V_B (Fig. 7-3).

The direction of current flow is shown by the arrow. Mark the polarity of the voltage drops across the resistors. Trace the circuit in the direction of current flow starting at point a. Write the voltage equation around the circuit.

$$\Sigma V = 0 \qquad\qquad (7\text{-}2)$$

Use + and − rules for voltage rises and voltage drops.

$$V_A - V_1 - V_2 - V_B - V_3 = 0$$

Solve for V_B.

$$V_B = V_A - V_1 - V_2 - V_3 = 15 - 3 - 6 - 2 = 4 \text{ V} \qquad \textit{Ans.}$$

Since V_B was found to be positive, the assumed direction of current is in fact the actual direction of current.

KIRCHHOFF'S CURRENT LAW (KCL)

Kirchhoff's *current law* states that *the sum of the currents entering a junction is equal to the sum of the currents leaving the junction.* Suppose we have six currents leaving and entering a common junction or point, shown as P (Fig. 7-4). This common point is also called a *node*.

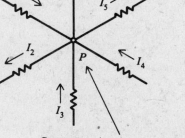

Common point, junction, or node

Fig. 7-4 Currents at a common point

Sum of all currents entering = sum of all currents leaving

Substitute letters:

$$I_1 + I_3 + I_4 + I_6 = I_2 + I_5$$

If we consider that the currents flowing toward a junction are positive (+) and those currents flowing away from the same junction are negative (−), then this law also states that the algebraic sum of all the currents meeting at a common junction is zero. Using the symbol Σ, we have

$$\Sigma I = 0 \qquad\qquad (7\text{-}3)$$

where ΣI, the algebraic sum of all the currents at the common point, is zero.

$$I_1 - I_2 + I_3 + I_4 - I_5 + I_6 = 0$$

If the negative terms are transposed to the right side of the equal sign, we would have the same form as the original equation.

Example 7.3   Write the equation for current I_1 for part a and part b of Fig. 7-5.

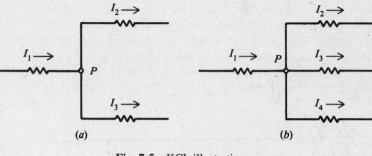

(a)   (b)

Fig. 7-5   KCL illustration

The algebraic sum of all currents at the node is zero. Entering currents are +; leaving currents are −.

(a) $+I_1 - I_2 - I_3 = 0$

$$I_1 = I_2 + I_3 \qquad Ans.$$

(b) $+I_1 - I_2 - I_3 - I_4 = 0$

$$I_1 = I_2 + I_3 + I_4 \qquad Ans.$$

Example 7.4   Find the unknown currents in part a and part b of Fig. 7-6.

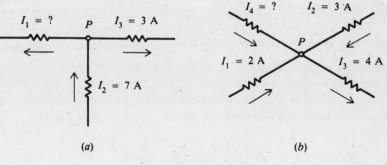

(a)   (b)

Fig. 7-6   Finding current

The algebraic sum of all currents at the node is zero. Entering currents are +; leaving currents are −.

(a) $-I_1 + I_2 - I_3 = 0$

$$I_1 = I_2 - I_3 = 7 - 3 = 4\,\text{A} \qquad Ans.$$

(b) $+I_1 + I_2 - I_3 + I_4 = 0$

$$I_4 = -I_1 - I_2 + I_3 = -2 - 3 + 4 = -1\,\text{A} \qquad Ans.$$

The negative sign for I_4 means that the assumed direction of I_4 is incorrect and that I_4 is actually flowing *away* from point P.

MESH CURRENTS

A simplification of Kirchhoff's laws is a method that makes use of *mesh* or *loop currents*. A *mesh* is any closed path of a circuit. It does not matter whether the path contains a voltage source. In solving a circuit with mesh currents, we must first decide which paths will be the meshes. Then we assign a mesh current to each mesh. For convenience, mesh currents are usually assigned in a clockwise direction. This direction is arbitrary, but the clockwise direction is usually assumed. Kirchhoff's voltage law is then applied about the path of each mesh. The resulting equations determine the unknown mesh currents. From these currents, the current or voltage of any resistor can be found.

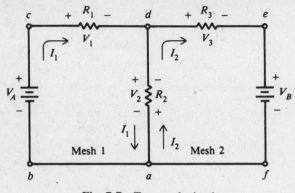

Fig. 7-7 Two-mesh circuit

In Fig. 7-7, we have a two-mesh circuit marked mesh 1 and mesh 2. Mesh 1 is path *abcda* and mesh 2 is path *adefa*. All voltage sources and resistances are known. A procedure for finding mesh currents I_1 and I_2 is as follows:

Step 1. After the meshes are selected, show the direction of mesh currents I_1 and I_2 in a clockwise direction. Mark the voltage polarity across each resistor, consistent with the assumed current. Remember that conventional current flow in a resistor produces positive polarity where the current enters.

Step 2. Apply Kirchhoff's voltage law, $\Sigma V = 0$, around each mesh. Trace each mesh in the direction of mesh current. Note that there are two different currents (I_1, I_2) flowing in opposite directions through the same resistor, R_2, which is common to both meshes. For this reason two sets of polarities are shown by R_2 (Fig. 7-7). Trace mesh 1 in direction *abcda*.

$$+V_A - I_1 R_1 - I_1 R_2 + I_2 R_2 = 0$$

$$+V_A - I_1(R_1 + R_2) + I_2 R_2 = 0$$

$$+I_1(R_1 + R_2) - I_2 R_2 = V_A \qquad (1)$$

Note that in the first expression $I_2 R_2$ is $+$ since we go through a voltage drop from $-$ to $+$.
    Trace mesh 2 in direction *adefa*.

$$-I_2 R_2 + I_1 R_2 - I_2 R_3 - V_B = 0$$

$$+I_1 R_2 - I_2(R_2 + R_3) = V_B \qquad (2)$$

Note that $I_1 R_2$ is a $+$ voltage drop since we go through a voltage drop from $-$ to $+$.

Step 3. Find I_1 and I_2 by solving Eqs. (*1*) and (2) simultaneously.

Step 4. When mesh currents are known, find all resistor voltage drops by using Ohm's law.

Step 5.   Check the solution of mesh currents by tracing mesh *abcdefa*.

$$V_A - I_1 R_1 - I_2 R_3 - V_B = 0$$

Example 7.5   Given $V_A = 58$ V, $V_B = 10$ V, $R_1 = 4\ \Omega$, $R_2 = 3\ \Omega$, and $R_3 = 2\ \Omega$ (Fig. 7-8a), find all mesh currents and voltage drops in the circuit.

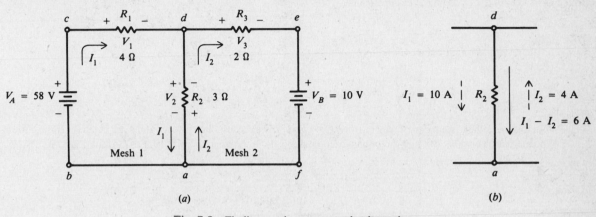

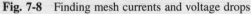

(a)   *(b)*

Fig. 7-8   Finding mesh currents and voltage drops

Step 1.   Choose the two loops or meshes shown. Show mesh current in the clockwise direction. Show polarity marks across each resistor.

Step 2.   Apply $\Sigma V = 0$ in mesh 1 and mesh 2 and trace the mesh in the direction of mesh current.

Mesh 1, *abcda*: $+58 - 4I_1 - 3I_1 + 3I_2 = 0$

$$+7I_1 - 3I_2 = 58 \tag{1}$$

Mesh 2, *adefa*: $3I_1 - 3I_2 - 2I_2 - 10 = 0$

$$3I_1 - 5I_2 = 10 \tag{2}$$

Note that mesh currents I_1 and I_2 flow through the common resistor R_2.

Step 3.   Find I_1 and I_2 by solving Eqs. (*1*) and (2) simultaneously.

$$7I_1 - 3I_2 = 58 \tag{1}$$

$$3I_1 - 5I_2 = 10 \tag{2}$$

Multiply Eq. (*1*) by 5 and multiply Eq. (2) by 3, getting Eqs. (*1a*) and (*2a*) and then subtract Eq. (*2a*) from Eq. (*1a*).

$$35I_1 - 15I_2 = 290 \tag{1a}$$

$$\underline{9I_1\ \ - 15I_2 =\ \ 30} \tag{2a}$$

$$26I_1\ \ \ \ \ \ \ \ \ = 260$$

$$I_1 = 10\,\text{A} \quad \textit{Ans.}$$

Substitute $I_1 = 10$ A in Eq. (I) to find I_2.

$$7I_1 - 3I_2 = 58$$

$$7(10) - 3I_2 = 58$$

$$-3I_2 = 58 - 70$$

$$I_2 = \frac{70 - 58}{3} = \frac{12}{3} = 4 \text{ A} \qquad Ans.$$

The current through branch da is

$$I_{da} = I_1 - I_2 = 10 - 4 = 6 \text{ A} \qquad Ans.$$

In this case, the assumed mesh current direction was correct because the current values are positive. If the current value were negative, the true direction would be opposite to the assumed direction of current. (See Fig. 7-8b.)

Step 4.   Find all voltage drops.

$$V_1 = I_1 R_1 = 10(4) = 40 \text{ V} \qquad Ans.$$

$$V_2 = (I_1 - I_2)R_2 = 6(3) = 18 \text{ V} \qquad Ans.$$

$$V_3 = I_2 R_3 = 4(2) = 8 \text{ V} \qquad Ans.$$

Step 5.   Check mesh current solution by tracing loop $abcdefa$ and applying KVL.

$$V_A - V_1 - V_3 - V_B = 0$$

$$58 - 40 - 8 - 10 = 0$$

$$58 - 58 = 0$$

$$0 = 0 \qquad Check$$

NODE VOLTAGES

Another method for solving a circuit with mesh currents uses the voltage drops to specify the currents at a node. Then node equations of currents are written to satisfy Kirchhoff's current law. By solving the node equations, we can calculate the unknown node voltages. A *node* is a common connection for two or more components. A *principal node* has three or more connections. To each node in a circuit a letter or number is assigned. A, B, G, and N are nodes, and G and N are principal nodes, or junctions (Fig. 7-9). A *node voltage* is the voltage of a given node with respect to one particular node called the *reference node*. Select node G connected to chassis ground as the reference node. Then V_{AG} is the voltage between nodes A and G, V_{BG} is the voltage between nodes B and G, and V_{NG} is the voltage between nodes N and G. Since the node voltage is always determined with respect to a specified reference node, the notations V_A for V_{AG}, V_B for V_{BG}, and V_N for V_{NG} are used.

With the exception of the reference node, equations using KCL can be written at each principal node. Thus the required number of equations is one less than the number of principal nodes. Since the circuit shown (Fig. 7-9) has two principal nodes (N and G), only one equation need be written at node N to find all voltage drops and currents in the circuit.

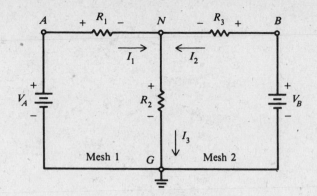

Fig. 7-9   Nodes in a two-mesh circuit

Assume that branch currents I_1 and I_2 enter node N, and I_3 leaves the node (Fig. 7-9). The selection of the direction of the currents is arbitrary. From KCL,

$$\Sigma I = 0$$

$$I_1 + I_2 - I_3 = 0$$

$$I_3 = I_1 + I_2 \tag{1}$$

By Ohm's law,

$$I_3 = \frac{V_N}{R_2} \tag{1a}$$

$$I_1 = \frac{V_A - V_N}{R_1} \tag{1b}$$

$$I_2 = \frac{V_B - V_N}{R_3} \tag{1c}$$

Substitute these expressions into Eq. (*1*).

$$\frac{V_N}{R_2} = \frac{V_A - V_N}{R_1} + \frac{V_B - V_N}{R_3} \tag{2}$$

If V_A, V_B, R_1, R_2, and R_3 are known, V_N can be calculated from Eq. (2). Then all voltage drops and currents in the circuit can be determined.

Example 7.6   The circuit of Fig. 7-8 (Example 7.5) solved by the method of branch currents is redrawn in Fig. 7-10. Solve by node-voltage analysis.

Step 1.   Assume the direction of currents shown (Fig. 7-10). Mark nodes A, B, N, and G. Mark the voltage polarity across each resistor consistent with assumed direction of current.

Step 2.   Apply KCL at principal node N and solve for V_N.

$$I_3 = I_1 + I_2$$

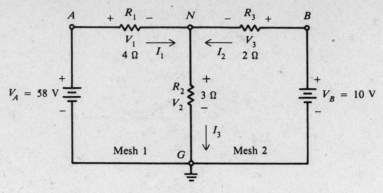

Fig. 7-10   Node-voltage analysis for same circuit as in Fig. 7-8

$$\frac{V_N}{R_2} = \frac{V_A - V_N}{R_1} + \frac{V_B - V_N}{R_3}$$

$$\frac{V_N}{3} = \frac{58 - V_N}{4} + \frac{10 - V_N}{2}$$

Clear fractions by multiplying each term by 12.

$$4V_N = 3(58 - V_N) + 6(10 - V_N)$$

$$4V_N = 174 - 3V_N + 60 - 6V_N$$

$$13V_N = 234$$

$$V_N = 18 \text{ V}$$

Step 3.   Find all voltage drops and currents.

$$V_1 = V_A - V_N = 58 - 18 = 40 \text{ V} \qquad Ans.$$

$$V_2 = V_N = 18 \text{ V} \qquad Ans.$$

$$V_3 = V_B - V_N = 10 - 18 = -8 \text{ V} \qquad Ans.$$

The negative value for V_3 means I_2 is flowing opposite to the assumed direction and the polarity of V_3 is the reverse of the signs shown across R_3 (Fig. 7-10).

$$I_1 = \frac{V_1}{R_1} = \frac{40}{4} = 10 \text{ A} \qquad Ans.$$

$$I_2 = \frac{V_3}{R_3} = \frac{-8}{2} = -4 \text{ A} \qquad Ans.$$

$$I_3 = I_1 + I_2 = 10 - 4 = 6 \text{ A} \qquad Ans.$$

$$I_3 = \frac{V_2}{R_2} = \frac{18}{3} = 6 \text{ A} \qquad Check$$

All calculated values agree with those of Example 7.5.

Solved Problems

7.1 Find the signs of the voltages when tracing the mesh *afedcba* and write the expression for KVL (Fig. 7-11).

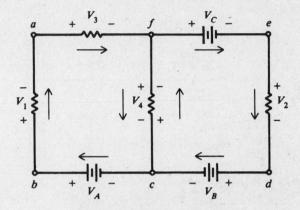

Fig. 7-11 Tracing two meshes

Assume directions of current flow as indicated. Mark the polarities across each resistor.

V_3 is $-$ since we go through a voltage drop $+$ to $-$.

V_C is $-$ since we go through a voltage rise $+$ to $-$.

V_2 is $-$ since we go through a voltage drop $+$ to $-$.

V_B is $-$ since we go through a voltage rise $+$ to $-$.

V_A is $+$ since we go through a voltage rise $-$ to $+$.

V_1 is $-$ since we go through a voltage drop $+$ to $-$.

$$\Sigma V = 0$$

$$-V_3 - V_C - V_2 - V_B + V_A - V_1 = 0$$

$$V_A - V_B - V_C - V_1 - V_2 - V_3 = 0$$

$$\underbrace{(V_A - V_B - V_C)}_{\text{Voltage rise}} - \underbrace{(V_1 + V_2 - V_3)}_{\text{Voltage drop}} = 0 \qquad \qquad \textit{Ans.}$$

7.2 Find I_3 and I_4 (Fig. 7-12).

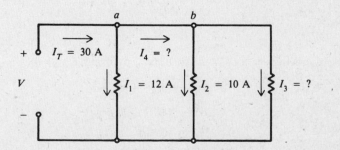

Fig. 7-12 Finding currents by KCL

Apply KCL, $\Sigma I = 0$ at node a.

$$30 - 12 - I_4 = 0$$

$$I_4 = 30 - 12 = 18 \text{ A} \qquad Ans.$$

Apply KCL, $\Sigma I = 0$ at node b.

$$18 - 10 - I_3 = 0$$

$$I_3 = 18 - 10 = 8 \text{ A} \qquad Ans.$$

Check solution.

$$I_T = I_1 + I_2 + I_3$$

$$30 = 12 + 10 + 8$$

$$30 = 30 \qquad Check$$

7.3   Solve the two-mesh circuit for all mesh currents (Fig. 7-13).

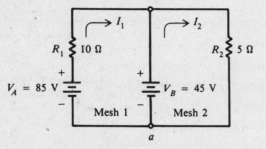

Fig. 7-13   Two meshes with voltage source in middle leg

Step 1.   Show mesh currents in clockwise direction.

Step 2.   Apply $\Sigma V = 0$ for mesh 1 and mesh 2 and trace each mesh from a in the direction of mesh current.

Mesh 1: $85 - 10I_1 - 45 = 0$

$$10I_1 = 40$$

$$I_1 = 4 \text{ A} \qquad Ans.$$

Mesh 2: $45 - 5I_2 = 0$

$$5I_2 = 45$$

$$I_2 = 9 \text{ A} \qquad Ans.$$

Step 3.   Check by tracing the loop of mesh 1 and 2 by using $\Sigma V = 0$.

$$V_A - I_1 R_1 - I_2 R_2 = 0$$

$$85 - 4(10) - 9(5) = 0$$

$$85 - 40 - 45 = 0$$

$$85 - 85 = 0 \qquad Check$$

7.4   Find all mesh currents and voltage drops for the two-mesh circuit shown in Fig. 7-14.

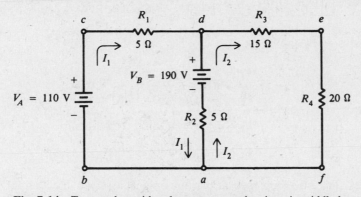

Fig. 7-14   Two meshes with voltage source and resistor in middle leg

Step 1.   Show the direction of mesh currents as indicated.

Step 2.   Apply $\Sigma V = 0$ for meshes 1 and 2, in the direction of mesh current.

Mesh 1, *abcda*: $110 - 5I_1 - 190 - 5I_1 + 5I_2 = 0$

$$-10I_1 + 5I_2 - 80 = 0$$

$$-10I_1 + 5I_2 = 80 \qquad\qquad (1)$$

Mesh 2, *adefa*: $5I_1 - 5I_2 + 190 - 15I_2 - 20I_2 = 0$

$$5I_1 - 40I_2 = -190 \qquad\qquad (2)$$

Step 3.   Find I_1 and I_2 by solving Eqs. (*1*) and (*2*) simultaneously.

$$-10I_1 + 5I_2 = \quad 80 \qquad\qquad (1)$$

$$5I_1 - 40I_2 = -190 \qquad\qquad (2)$$

Multiply Eq. (*2*) by 2 to get Eq. (*2a*); then add.

$$-10I_1 + \;5I_2 = 80 \qquad\qquad (1)$$
$$\underline{10I_1 - 80I_2 = -380} \qquad\qquad (2a)$$
$$-75I_2 = -300$$

$$I_2 = \frac{300}{75} = 4\,\text{A} \qquad \textit{Ans.}$$

Substitute $I_2 = 4\,\text{A}$ in Eq. (*1*) to find I_1.

$$-10I_1 + 5(4) = 80$$

$$-10I_1 = 60$$

$$I_1 = -6\,\text{A} \qquad \textit{Ans.}$$

The negative sign means that the assumed direction for I_1 was not correct. I_1 is actually going in a counterclockwise direction. In branch *ad*, I_1 and I_2 are going in the same direction.

Therefore,

$$I_{ad} = I_1 + I_2 = 6 + 4 = 10 \text{ A} \qquad Ans.$$

Step 4.   Find the voltage drops.

$$V_1 = I_1 R_1 = 6(5) = 30 \text{ V} \qquad Ans.$$

$$V_2 = (I_1 + I_2)R_2 = 10(5) = 50 \text{ V} \qquad Ans.$$

$$V_3 = I_2 R_3 = 4(15) = 60 \text{ V} \qquad Ans.$$

$$V_4 = I_2 R_4 = 4(20) = 80 \text{ V} \qquad Ans.$$

Step 5.   Check. Trace the loop *abcdefa* (use the original assumed direction for I_1 and I_2).

$$+V_A - I_1 R_1 - I_2 R_3 - I_2 R_4 = 0$$

$$110 - (-6)(5) - 4(15) - 4(20) = 0$$

$$110 + 30 - 60 - 80 = 0$$

$$140 - 140 = 0 \qquad Check$$

7.5   Find the voltage V_2 across R_2 by the method of node-voltage analysis (Fig. 7-15a).

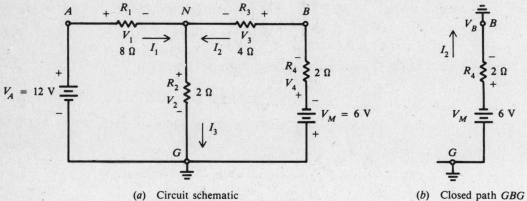

(a) Circuit schematic   (b) Closed path *GBG*

Fig. 7-15   Finding V_2 by the node-voltage method

Step 1.   Assume direction of currents shown. Mark voltage polarities. Show nodes A, B, N, G.

Step 2.   Apply $\Sigma I = 0$ at principal node N.

$$I_3 = I_1 + I_2 \qquad\qquad (1)$$

$$I_3 = \frac{V_2}{R_2} = \frac{V_N}{2} \qquad\qquad (1a)$$

$$I_1 = \frac{V_1}{R_1} = \frac{V_A - V_N}{R_1} = \frac{12 - V_N}{8} \qquad\qquad (1b)$$

$$I_2 = \frac{V_3}{R_3} = \frac{V_B - V_N}{R_3} = \frac{V_B - V_N}{4} \qquad\qquad (1c)$$

We are unable to determine V_B by inspection in Eq. (*1c*) because voltage drop V_4 is not given (Fig. 7-15*a*). So we use KVL to find V_B by tracing the complete circuit from G to B in the direction of I_2 (Fig. 7-15*b*). GBG is a complete path because V_B is the voltage at B with respect to ground.

$$-6 - 2I_2 - V_B = 0$$
$$V_B = -6 - 2I_2$$

Substitute expression for V_B into Eq. (*1c*),

$$I_2 = \frac{-6 - 2I_2 - V_N}{4}$$

from which we obtain

$$I_2 = \frac{-6 - V_N}{6}$$

Substitute the three expressions for current into Eq. (*1*).

$$\frac{V_N}{2} = \frac{12 - V_N}{8} + \frac{-6 - V_N}{6} \qquad (2)$$

Now Eq. (*2*) has one unknown, V_N.

Step 3. Find V_2 ($V_2 = V_N$). Multiply each member of Eq. (*2*) by 24.

$$12V_N = (36 - 3V_N) + (-24 - 4V_N)$$
$$19V_N = 12$$
$$V_N = \frac{12}{19} = 0.632 \text{ V}$$
$$V_2 = V_N = 0.632 \text{ V} \qquad Ans.$$

7.6 Write the mesh equations for the three-mesh circuit (Fig. 7-16). Do not solve.

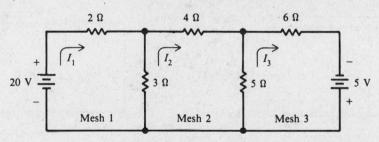

Fig. 7-16 A three-mesh circuit

Show mesh currents in clockwise direction. Trace loops in assumed direction of current, using KVL, $\Sigma V = 0$.

Mesh 1: $\qquad\qquad 20 - 2I_1 - 3I_1 + 3I_2 = 0 \qquad\qquad\qquad (1)$

Mesh 2: $\qquad -4I_2 - 5I_2 + 5I_3 - 3I_2 + 3I_1 = 0 \qquad\qquad (2)$

Mesh 3: $\qquad\qquad -6I_3 + 5 - 5I_3 + 5I_2 = 0 \qquad\qquad\qquad (3)$

Combine and rearrange terms in each equation.

Mesh 1: $\qquad 20 = 5I_1 - 3I_2 \qquad$ *Ans.* $\qquad\qquad\qquad\qquad$ (1a)

Mesh 2: $\qquad 0 = -3I_1 + 12I_2 - 5I_3 \qquad$ *Ans.* $\qquad\qquad\qquad$ (2a)

Mesh 3: $\qquad 5 = -5I_2 + 11I_3 \qquad$ *Ans.* $\qquad\qquad\qquad\qquad$ (3a)

A set with any number of simultaneous equations, for any number of meshes, can be solved by using determinants. Solution by determinants is shown in Chapter 8.

Supplementary Problems

7.7 Find the unknown quantities indicated in Fig. 7-17a and b. $\qquad$ *Ans.* $\quad$ (a) $I = 8$ A; $\quad$ (b) $V_B = 10$ V

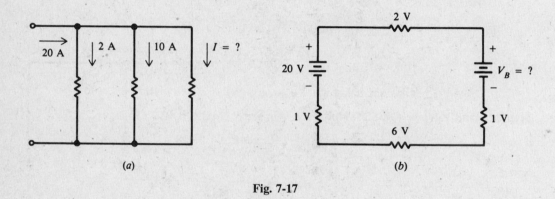

(a) $\qquad\qquad\qquad\qquad\qquad\qquad$ (b)

Fig. 7-17

7.8 Find the series current and voltage drops across R_1 and R_2 (Fig. 7-18).
$\quad$ *Ans.* $\quad I = 1$ A; $V_1 = 10$ V; $V_2 = 20$ V

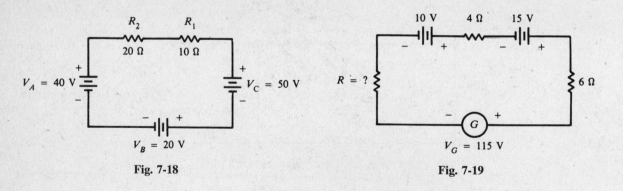

Fig. 7-18 $\qquad\qquad\qquad\qquad\qquad\qquad$ **Fig. 7-19**

7.9 A current of 6 A flows in the circuit (Fig. 7-19). Find the value of R. $\qquad$ *Ans.* $\quad R = 5\,\Omega$

7.10 Find I_2, I_3, and V_A (Fig. 7-20). $\qquad$ *Ans.* $\quad I_2 = 6$ A; $I_3 = 2$ A; $V_A = 152$ V

7.11 Find mesh currents I_1 and I_2 and all voltage drops by the mesh-current method (Fig. 7-21).
$\quad$ *Ans.* $\quad I_1 = 5$ A; $I_2 = 3$ A; $V_1 = 30$ V; $V_2 = 30$ V; $V_3 = 60$ V; $V_4 = 6$ V; $V_5 = 9$ V; $V_6 = 15$ V

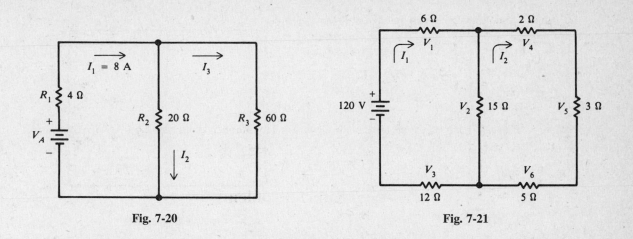

Fig. 7-20   Fig. 7-21

7.12   Find all currents through the resistances by the mesh-current method (Fig. 7-22).
Ans. $I_1 = 3$ A; $I_2 = 1$ A; $I_1 - I_2 = 2$ A (flowing from a to b)

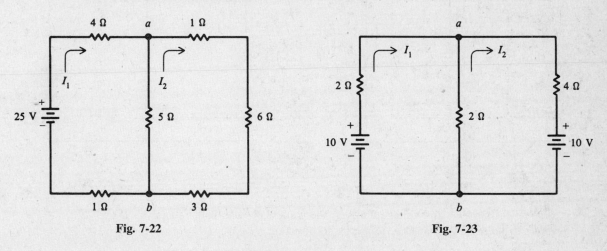

Fig. 7-22   Fig. 7-23

7.13   Find the current in each resistor, using the mesh-current method (Fig. 7-23).
Ans. $I_1 = 2$ A; $I_2 = -1$ A (current direction was assumed incorrectly), or $I_2 = 1$ A in counter-clockwise direction; $I_1 + I_2 = 3$ A (flowing from a to b)

7.14   Find currents I_1 and I_2 and the current through the 20-V battery using the mesh-current method (Fig. 7-24).   *Ans.* $I_1 = 2$ A; $I_2 = 5$ A; $I_2 - I_1 = 3$ A (flowing from b to a)

7.15   Find currents I_1 and I_2 and current through the resistor in series with the 20-V battery (Fig. 7-25). Use the mesh-current method.
Ans. $I_1 = -0.1$ A (direction assumed incorrectly. I_1 is actually going in the counterclockwise direction); $I_2 = 0.7$ A; $I_1 + I_2 = 0.8$ A (flowing from b to a)

7.16   Find currents I_1 and I_2 and current through the 20-Ω resistor common to meshes 1 and 2 (Fig. 7-26). Use the mesh-current method.   *Ans.* $I_1 = 0.6$ A; $I_2 = 0.4$ A; $I_1 - I_2 = 0.2$ A (flowing from a to b)

7.17   Find all the currents in the circuit shown by the mesh-current method (Fig. 7-27).
Ans. $I_1 = 6$ A; $I_2 = 7$ A; $I_2 - I_1 = 1$ A (flowing from b to a)

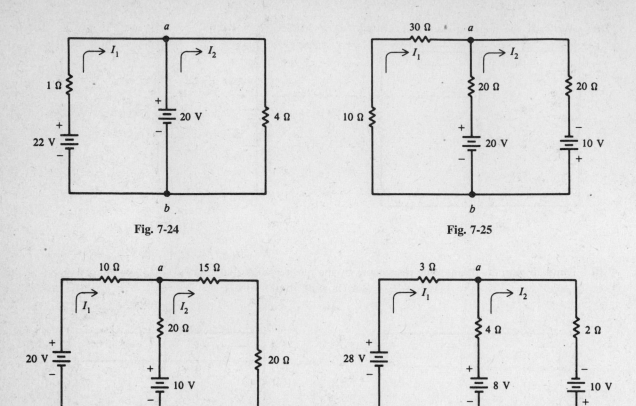

Fig. 7-24

Fig. 7-25

Fig. 7-26

Fig. 7-27

7.18   Find all the currents and voltage drops by the method of node-voltage analysis (Fig. 7-28).
Ans. $I_1 = 5$ A; $I_2 = -1$ A (opposite to direction shown); $I_3 = 4$ A; $V_1 = 60$ V; $V_2 = 24$ V; $V_3 = 3$ V

7.19   Find by the node-voltage method all currents and voltage drops (Fig. 7-29).
Ans. $I_1 = 1.42$ A; $I_2 = -1.10$ A (opposite to direction shown); $I_3 = 0.32$ A; $V_1 = 11.4$ V; $V_2 = 0.64$ V; $V_3 = 2.2$ V; $V_4 = 4.4$ V

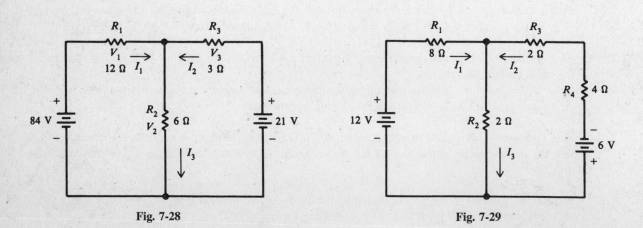

Fig. 7-28

Fig. 7-29

7.20 Write the mesh equations for the circuit (Fig. 7-30). Do not solve.
   Ans. $6I_1 - 2I_2 = 10$; $-2I_1 + 8I_2 - 2I_3 = 0$; $-2I_2 + 6I_3 = -4$

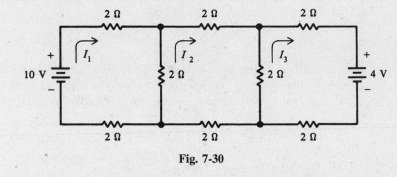

Fig. 7-30

7.21 Verify values of currents in circuit shown in Fig. 7-23 (Problem 7.13) by the node-voltage method.

7.22 Verify values of currents in circuit of Fig. 7-25 (Problem 7.15) by the node-voltage method.

7.23 Find the magnitude of all currents and show their directions at node N (Fig. 7-31). (*Hint*: $V_N = 1.67$ V.)

Ans.

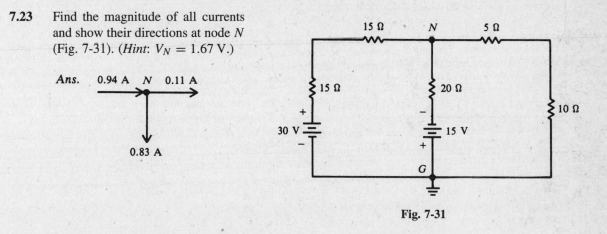

Fig. 7-31

7.24 If the 20-Ω resistor (Fig. 7-31) is replaced by a 30-Ω resistor, what is the nodal voltage V_N?
   Ans. $V_N = 3.75$ V

Chapter 8

Determinant Solutions for DC Networks

SECOND-ORDER DETERMINANTS

A *determinant* is an array of numbers or letters written in a square between vertical lines. The value of a determinant is found by multiplying the *elements* (numbers or letters) of the determinant in a specified way.

A second-order determinant has four elements arrayed in two rows and two columns, such as

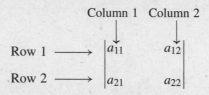

The four elements are a_{11}, a_{12}, a_{21}, and a_{22}. The first subscript indicates the row, while the second subscript indicates the column. Thus, the element a_{21} is in the second row and in the first column.

The value of the second-order determinant by definition is

$$\begin{vmatrix} a_{11} & a_{12} \\ a_{21} & a_{22} \end{vmatrix} = a_{11}a_{22} - a_{12}a_{21} \tag{8-1}$$

The rule for finding the value is to multiply together the elements on the diagonal lines, adding together those on lines that slope down to the right, and subtracting those on lines that slope down to the left.

Example 8.1 Find the value of the following second-order determinants.

(a) $\quad \begin{vmatrix} 3 & 4 \\ 2 & 5 \end{vmatrix} = (3)(5) - (4)(2) = 15 - 8 = 7 \qquad Ans.$

(b) $\quad \begin{vmatrix} -4 & 2 \\ -3 & -1 \end{vmatrix} = (-4)(-1) - (2)(-3) = 4 + 6 = 10 \qquad Ans.$

(c) $\quad \begin{vmatrix} 3 & 0 \\ -1 & 5 \end{vmatrix} = (3)(5) - (0)(-1) = 15 \qquad Ans.$

THIRD-ORDER DETERMINANTS

A third-order determinant has nine elements arrayed in three rows and three columns, such as

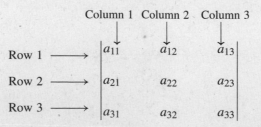

The value of the third-order determinant is

$$\begin{vmatrix} a_{11} & a_{12} & a_{13} \\ a_{21} & a_{22} & a_{23} \\ a_{31} & a_{32} & a_{33} \end{vmatrix} = a_{11} \begin{vmatrix} a_{22} & a_{23} \\ a_{32} & a_{33} \end{vmatrix} - a_{21} \begin{vmatrix} a_{12} & a_{13} \\ a_{32} & a_{33} \end{vmatrix} + a_{31} \begin{vmatrix} a_{12} & a_{13} \\ a_{22} & a_{23} \end{vmatrix} \qquad (8\text{-}2)$$

The rule for finding the value of a higher-order determinant is found by expanding the determinant in terms of any one of its rows or columns. The selection of a row or column is random. For illustration, the first column is selected for expanding. When a_{11} is selected, a sub array (second-order determinant) is formed by eliminating the row and column that intersects a_{11}. This sub array is called a *minor*.

$$a_{11} \text{ selected:} \quad \begin{vmatrix} a_{11} & a_{12} & a_{13} \\ a_{21} & a_{22} & a_{23} \\ a_{31} & a_{32} & a_{33} \end{vmatrix} \longrightarrow \begin{vmatrix} a_{22} & a_{23} \\ a_{32} & a_{33} \end{vmatrix}, \quad \text{minor for } a_{11}$$

Similarly,

$$a_{21} \text{ selected:} \quad \begin{vmatrix} a_{11} & a_{12} & a_{13} \\ a_{21} & a_{22} & a_{23} \\ a_{31} & a_{32} & a_{33} \end{vmatrix} \longrightarrow \begin{vmatrix} a_{12} & a_{13} \\ a_{32} & a_{33} \end{vmatrix}, \quad \text{minor for } a_{21}$$

$$a_{31} \text{ selected:} \quad \begin{vmatrix} a_{11} & a_{12} & a_{13} \\ a_{21} & a_{22} & a_{23} \\ a_{31} & a_{32} & a_{33} \end{vmatrix} \longrightarrow \begin{vmatrix} a_{12} & a_{13} \\ a_{22} & a_{23} \end{vmatrix}, \quad \text{minor for } a_{31}$$

The *cofactor* of the element a_{ij} is defined as the product of $(-1)^{i+j}$ and the minor determinant obtained by deleting the ith row and j^{th} column.

In the previous example, when a_{11} is selected, $(-1)^{i+j} = (-1)^{1+1} = (-1)^2 = 1$, so its cofactor is $+1 \begin{vmatrix} a_{22} & a_{23} \\ a_{32} & a_{33} \end{vmatrix}$.

When a_{21} is selected, $(-1)^{i+j} = (-1)^{1+2} = (-1)^3 = -1$, so its cofactor is $-1 \begin{vmatrix} a_{12} & a_{13} \\ a_{32} & a_{33} \end{vmatrix}$.

When a_{31} is selected, $(-1)^{i+j} = (-1)^{1+3} = (-1)^4 = 1$, so its cofactor is $+1 \begin{vmatrix} a_{12} & a_{13} \\ a_{22} & a_{23} \end{vmatrix}$.

A convenient way to remember the algebraic signs is to refer to a checker board arrangement:

$$\begin{vmatrix} + & - & + \\ - & + & - \\ + & - & + \end{vmatrix}$$

Start with the upper left-hand corner with the $+$ sign. All the other signs follow automatically, regardless of the number of elements in the determinant. Another way to remember the sign is when the sum of the row and column values is even, the sign is $+$; and when the sum is odd, the sign is $-$. Thus, for the minor of a_{11}, the sum is $1 + 1 = 2$, so the sign is $+$; and for a_{21}, the sum is $2 + 1 = 3$, so the sign is $-$.

Example 8.2 Find the value of the following third-order determinants by expanding the elements of the first column.

$$(a) \quad \begin{vmatrix} 1 & 2 & -1 \\ 3 & 1 & 1 \\ 1 & -1 & 2 \end{vmatrix} \qquad (b) \quad \begin{vmatrix} 1 & -3 & -1 \\ -3 & 3 & 0 \\ -1 & 0 & 5 \end{vmatrix}$$

(a) Refer to Eq. (8-2): $a_{11} = 1$, $a_{12} = 2$, $a_{13} = -1$, $a_{21} = 3$, $a_{22} = 1$, $a_{23} = 1$, $a_{31} = 1$, $a_{32} = -1$, and $a_{33} = 2$. Expand the first column of the determinant into its cofactors (second-order determinants).

$$\begin{vmatrix} 1 & 2 & -1 \\ 3 & 1 & 1 \\ 1 & -1 & 2 \end{vmatrix} = 1 \begin{vmatrix} 1 & 1 \\ -1 & 2 \end{vmatrix} - 3 \begin{vmatrix} 2 & -1 \\ -1 & 2 \end{vmatrix} + 1 \begin{vmatrix} 2 & -1 \\ 1 & 1 \end{vmatrix}$$

$$= 1\,[(1)(2) - (1)(-1)] - 3\,[(2)(2) - (-1)(-1)] + [(2)(1) - (-1)(1)]$$

$$= (2 + 1) - 3(4 - 1) + (2 + 1)$$

$$= 3 - 9 + 3$$

$$= -3 \qquad Ans.$$

(b) Refer to Eq. (8-2): $a_{11} = 1$, $a_{12} = -3$, $a_{13} = -1$, $a_{21} = -3$, $a_{22} = 3$, $a_{23} = 0$, $a_{31} = -1$, $a_{32} = 0$, and $a_{33} = 5$. Expand the first column of the determinant.

$$\begin{vmatrix} 1 & -3 & -1 \\ -3 & 3 & 0 \\ -1 & 0 & 5 \end{vmatrix} = 1 \begin{vmatrix} 3 & 0 \\ 0 & 5 \end{vmatrix} - (-3) \begin{vmatrix} -3 & -1 \\ 0 & 5 \end{vmatrix} + (-1) \begin{vmatrix} -3 & -1 \\ 3 & 0 \end{vmatrix}$$

$$= 1\,[(3)(5) - (0)(0)] + 3\,[(-3)(5) - (-1)(0)] - [(-3)(0) - (-1)(3)]$$

$$= 15 + 3(-15) - 3$$

$$= 15 - 45 - 3$$

$$= -33 \qquad Ans.$$

(c) Solve (b) by expanding the elements of the 2nd column. Refer to the checkerboard arrangement for algebraic signs. The answer should be the same.

$$\begin{vmatrix} 1 & -3 & -1 \\ -3 & 3 & 0 \\ -1 & 0 & 5 \end{vmatrix} = -(-3) \begin{vmatrix} -3 & 0 \\ -1 & 5 \end{vmatrix} + 3 \begin{vmatrix} 1 & -1 \\ -1 & 5 \end{vmatrix} - 0 \begin{vmatrix} 1 & -1 \\ -3 & 0 \end{vmatrix}$$

$$= 3(-15) + 3(5 - 1)$$

$$= -45 + 12$$

$$= -33 \qquad Ans. \qquad Check$$

(d) Solve (b) now by expanding the elements of the 3rd row. The answer should be the same.

$$\begin{vmatrix} 1 & -3 & -1 \\ -3 & 3 & 0 \\ -1 & 0 & 5 \end{vmatrix} = -1 \begin{vmatrix} -3 & -1 \\ 3 & 0 \end{vmatrix} - 0 \begin{vmatrix} 1 & -1 \\ -3 & 0 \end{vmatrix} + 5 \begin{vmatrix} 1 & -3 \\ -3 & 3 \end{vmatrix}$$

$$= -3 + 5(3 - 9)$$

$$= -3 - 30$$

$$= -33 \qquad Ans. \qquad Check$$

CRAMER'S RULE

Cramer's rule is a method to solve simultaneous linear equations by use of determinants. As an example, consider two equations with two unknowns, x and y, where A and B are constants. (A *linear equation* is an equation where the unknown terms, as $x, y, z, \ldots$ n, are only to the 1st degree. If the equation has the term x^2 or xy, it is of the 2nd degree and therefore not linear.)

$$a_{11}x + a_{12}y = A \tag{8-3a}$$

$$a_{21}x + a_{22}y = B \tag{8-3b}$$

Write the determinant of the coefficients of x and y, which we call Δ (the Greek capital letter delta).

$$\Delta = \begin{vmatrix} a_{11} & a_{12} \\ a_{21} & a_{22} \end{vmatrix} \tag{8-4}$$

If $\Delta = 0$, the equations have no solution. If $\Delta \neq 0$, the equations are solved by writing

$$x = \frac{N_x}{\Delta} = \frac{\begin{vmatrix} A & a_{12} \\ B & a_{22} \end{vmatrix}}{\Delta} \tag{8-5a}$$

Constant terms replace the coefficients of x. Coefficients of y.

$$y = \frac{N_y}{\Delta} = \frac{\begin{vmatrix} a_{11} & A \\ a_{21} & B \end{vmatrix}}{\Delta} \tag{8-5b}$$

Coefficients of x. Constant terms replace the coefficients of y.

The numerators for x and y, N_x and N_y, are also determinants and are formed from Δ by substituting A and B for the coefficients of the desired unknown. For example, in the numerator determinant for x, N_x, we substitute A and B for a_{11} and a_{21}, the coefficients of x.

Let's expand this example by considering three equations with three unknowns, $x, y,$ and z, where $A, B,$ and C are constants.

$$a_{11}x + a_{12}y + a_{13}z = A \tag{8-6a}$$
$$a_{21}x + a_{22}y + a_{23}z = B \tag{8-6b}$$
$$a_{31}x + a_{32}y + a_{33}z = C \tag{8-6c}$$

Write the determinant of the coefficients $x, y,$ and z, which again we call Δ.

$$\Delta = \begin{vmatrix} a_{11} & a_{12} & a_{13} \\ a_{21} & a_{22} & a_{23} \\ a_{31} & a_{32} & a_{33} \end{vmatrix} \tag{8-7}$$

If $\Delta = 0$, the equations have no solution. If $\Delta \neq 0$, the equations are solved by writing

$$x = \frac{N_x}{\Delta} = \frac{\begin{vmatrix} A & a_{12} & a_{13} \\ B & a_{22} & a_{23} \\ C & a_{32} & a_{33} \end{vmatrix}}{\Delta} \tag{8-8}$$

Constant terms replace the coefficients of x. Coefficients of y. Coefficients of z.

$$y = \frac{N_y}{\Delta} = \frac{\begin{vmatrix} a_{11} & A & a_{13} \\ a_{21} & B & a_{23} \\ a_{31} & C & a_{33} \end{vmatrix}}{\Delta}$$

Constant terms replace the coefficients of y

$$(8\text{-}9)$$

$$z = \frac{N_z}{\Delta} = \frac{\begin{vmatrix} a_{11} & a_{12} & A \\ a_{21} & a_{22} & B \\ a_{31} & a_{32} & C \end{vmatrix}}{\Delta}$$

Constant terms replace the coefficients of z

$$(8\text{-}10)$$

The numerators for x, y, and z are also determinants and are formed from Δ by substituting A, B, and C for the coefficients of the desired unknown.

Example 8.3 Refer to Example 7.5. Step 2 shows that the simultaneous equations for the two-mesh circuit in Fig. 7-8 are

$$7I_1 - 3I_2 = 58 \qquad\qquad (1)$$

$$3I_1 - 5I_2 = 10 \qquad\qquad (2)$$

Solve for I_1 and I_2 by use of determinants.

Step 1. Solve for Δ.

$$\Delta = \begin{vmatrix} a_{11} & a_{12} \\ a_{21} & a_{22} \end{vmatrix} \qquad\qquad (8\text{-}4)$$

$$\Delta = \begin{vmatrix} 7 & -3 \\ 3 & -5 \end{vmatrix} = (7)(-5) - (-3)(3) = -35 + 9$$

$$= -26$$

Step 2. Solve for I_1.

$$x = \frac{N_x}{\Delta} = \frac{\begin{vmatrix} A & a_{12} \\ B & a_{22} \end{vmatrix}}{\Delta} \qquad\qquad (8\text{-}5a)$$

Substituting I_1 for x,
$$I_1 = \frac{N_{I_1}}{\Delta} = \frac{\begin{vmatrix} 58 & -3 \\ 10 & -5 \end{vmatrix}}{-26} = \frac{(58)(-5) - (-3)(10)}{-26}$$

$$= \frac{-290 + 30}{-26} = \frac{-260}{-26} = 10 \text{ A} \qquad Ans.$$

Step 3. Solve for I_2.

$$y = \frac{N_y}{\Delta} = \frac{\begin{vmatrix} a_{11} & A \\ a_{21} & B \end{vmatrix}}{\Delta} \qquad\qquad (8\text{-}5b)$$

Substituting I_2 for y,
$$I_2 = \frac{N_{I_2}}{\Delta} = \frac{\begin{vmatrix} 7 & 58 \\ 3 & 10 \end{vmatrix}}{-26} = \frac{(7)(10) - (58)(3)}{-26}$$

$$= \frac{70 - 174}{-26} = \frac{-104}{-26} = 4 \text{ A} \qquad Ans.$$

Current value answers check with those of Example 7.5.

Example 8.4   In Solved Problem 7.6, three simultaneous equations are written for a three-mesh circuit (Fig. 8-1). Solve for the mesh currents by determinants.

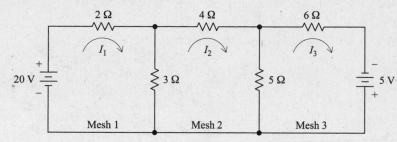

Fig. 8-1

$$5I_1 - 3I_2 = 20 \tag{1}$$

$$-3I_1 + 12I_2 - 5I_3 = 0 \tag{2}$$

$$-5I_2 + 11I_3 = 5 \tag{3}$$

Step 1.   Solve for the coefficient determinant Δ.

$$\Delta = \begin{vmatrix} a_{11} & a_{12} & a_{13} \\ a_{21} & a_{22} & a_{23} \\ a_{31} & a_{32} & a_{33} \end{vmatrix} \tag{8-7}$$

$$\Delta = \begin{vmatrix} 5 & -3 & 0 \\ -3 & 12 & -5 \\ 0 & -5 & 11 \end{vmatrix} = 5\begin{vmatrix} 12 & -5 \\ -5 & 11 \end{vmatrix} - (-3)\begin{vmatrix} -3 & 0 \\ -5 & 11 \end{vmatrix} + 0\begin{vmatrix} -3 & 0 \\ 12 & -5 \end{vmatrix}$$

$$= 5[(12)(11) - (-5)(-5)] + 3[(-3)(11) - (0)(-5)] + 0$$

$$= 5(107) + 3(-33)$$

$$= 436$$

Step 2.   Solve for I_1.

$$x = \frac{N_x}{\Delta} = \frac{\begin{vmatrix} A & a_{12} & a_{13} \\ B & a_{22} & a_{23} \\ C & a_{32} & a_{33} \end{vmatrix}}{\Delta} \tag{8-8}$$

Substituting I_1 for x, $$I_1 = \frac{N_{I_1}}{\Delta} = \frac{\begin{vmatrix} 20 & -3 & 0 \\ 0 & 12 & -5 \\ 5 & -5 & 11 \end{vmatrix}}{436}$$

$$= \frac{20\begin{vmatrix} 12 & -5 \\ -5 & 11 \end{vmatrix} - 0\begin{vmatrix} -3 & 0 \\ -5 & 11 \end{vmatrix} + 5\begin{vmatrix} -3 & 0 \\ 12 & -5 \end{vmatrix}}{436}$$

$$= \frac{20[(12)(11) - (-5)(-5)] - 0 + 5[(-3)(-5) - (0)(12)]}{436}$$

$$= \frac{20(107) + 5(15)}{436} = \frac{2215}{436}$$

$$I_1 = 5.08 \text{ A} \quad \textit{Ans.}$$

Step 3.   Solve for I_2.

$$y = \frac{N_y}{\Delta} = \frac{\begin{vmatrix} a_{11} & A & a_{13} \\ a_{21} & B & a_{23} \\ a_{31} & C & a_{33} \end{vmatrix}}{\Delta} \tag{8-9}$$

Substituting I_2 for y, $I_2 = \dfrac{N_{I_2}}{\Delta} = \dfrac{\begin{vmatrix} 5 & 20 & 0 \\ -3 & 0 & -5 \\ 0 & 5 & 11 \end{vmatrix}}{436}$

$$= \frac{5\begin{vmatrix} 0 & -5 \\ 5 & 11 \end{vmatrix} - (-3)\begin{vmatrix} 20 & 0 \\ 5 & 11 \end{vmatrix} + 0\begin{vmatrix} 20 & 0 \\ 0 & -5 \end{vmatrix}}{436}$$

$$= \frac{5(25) + 3(220)}{436} = \frac{785}{436}$$

$$I_2 = 1.80 \text{ A} \quad\quad Ans.$$

Step 4.   Solve for I_3.

$$z = \frac{N_z}{\Delta} = \frac{\begin{vmatrix} a_{11} & a_{12} & A \\ a_{21} & a_{22} & B \\ a_{31} & a_{32} & C \end{vmatrix}}{\Delta} \tag{8-10}$$

Substituting I_3 for z, $I_3 = \dfrac{N_{I_3}}{\Delta} = \dfrac{\begin{vmatrix} 5 & -3 & 20 \\ -3 & 12 & 0 \\ 0 & -5 & 5 \end{vmatrix}}{436}$

$$= \frac{5\begin{vmatrix} 12 & 0 \\ -5 & 5 \end{vmatrix} - (-3)\begin{vmatrix} -3 & 20 \\ -5 & 5 \end{vmatrix} + 0\begin{vmatrix} -3 & 20 \\ 12 & 0 \end{vmatrix}}{436}$$

$$= \frac{5(60) + 3(85)}{436} = \frac{555}{436}$$

$$I_3 = 1.27 \text{ A} \quad\quad Ans.$$

Step 5.   Check the solutions.

$$5I_1 - 3I_2 = 20 \tag{1}$$

$$5(5.08) - 3(1.80) = 20$$

$$25.40 - 5.40 = 20$$

$$20 = 20 \quad\quad Check$$

$$-3I_1 + 12I_2 - 5I_3 = 0 \tag{2}$$

$$-3(5.08) + 12(1.80) - 5(1.27) = 0$$

$$-15.24 + 21.60 - 6.35 = 0$$

$$0.01 = 0 \quad\quad Check, \text{ rounding error}$$

$$-5I_2 + 11I_3 = 5 \qquad\qquad (3)$$

$$-5(1.80) + 11(1.27) = 5$$

$$-9.00 + 13.97 = 5$$

$$4.97 = 5 \qquad \textit{Check}, \text{ rounding error}$$

DETERMINANT METHOD FOR SOLVING CURRENTS IN A TWO-MESH NETWORK

The determinant method for solving mesh currents is one whereby the determinant solution for mesh currents can be written *directly by inspecting* the network without writing first the proper simultaneous equations.

For a network of two meshes, the determinant of the network is

$$\Delta = \begin{vmatrix} R_{11} & -R_{12} \\ -R_{21} & R_{22} \end{vmatrix} \qquad\qquad (8\text{-}11)$$

where R_{11} = total resistance of mesh 1
R_{22} = total resistance of mesh 2
R_{12} = mutual (common) resistance between mesh 1 and mesh 2
R_{21} = mutual (common) resistance between mesh 2 and mesh 1

Because the mutual resistances are equal, $R_{12} = R_{21}$. Note that the sign for all mutual resistance is negative. The formulas to solve for mesh currents I_1 and I_2 are

$$I_1 = \frac{N_{I_1}}{\Delta} = \frac{\begin{vmatrix} V_1 & -R_{12} \\ V_2 & R_{22} \end{vmatrix}}{\Delta} \qquad\qquad (8\text{-}12)$$

$$I_2 = \frac{N_{I_2}}{\Delta} = \frac{\begin{vmatrix} R_{11} & V_1 \\ -R_{21} & V_2 \end{vmatrix}}{\Delta} \qquad\qquad (8\text{-}13)$$

where V_1 = net voltage for mesh 1
V_2 = net voltage for mesh 2

The numerator determinant for I_1, N_{I_1}, is formed by substituting the net voltage sources in column 1 of Δ; the numerator determinant for I_2, N_{I_2}, is formed by substituting the net voltage sources in column 2 of Δ. The polarity (+ or −) of V_1 and V_2 depends on the net voltage of the mesh as the mesh is traced in the direction of the assumed current.

Example 8.5 Find the determinant solution for mesh current I_1 and I_2 (Fig. 8-2). (This problem was solved in Example 8.3 by use of simultaneous equations.)

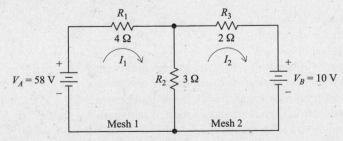

Fig. 8-2

Step 1. Assume all mesh currents in the same clockwise direction. I_1 and I_2 are shown clockwise.

Step 2. Find the determinant of the network Δ.

R_{11} = total resistance in mesh 1 = $R_1 + R_2 = 4 + 3 = 7 \ \Omega$

R_{22} = total resistance in mesh 2 = $R_2 + R_3 = 3 + 2 = 5 \ \Omega$

$R_{12} = R_{32}$ = mutual resistance between meshes 1 and 2 = $R_2 = 3 \ \Omega$

Substitute the proper values to find Δ.

$$\Delta = \begin{vmatrix} R_{11} & -R_{12} \\ -R_{21} & R_{22} \end{vmatrix} \tag{8-11}$$

$$= \begin{vmatrix} 7 & -3 \\ -3 & 5 \end{vmatrix} = (7)(5) - (-3)(-3) = 26$$

Step 3. Find the numerator determinant for I_1 and I_2.

V_1 = net voltage source for mesh 1 = $V_A = 58 \ \text{V}$

(V_1 is + since we go through a voltage source − to +)

V_2 = net voltage source for mesh 2 = $V_B = -10 \ \text{V}$

(V_2 is − since we go through a voltage source + to −)

Substitute the proper values to find N_{I_1} (see Eq. *8-12*) and N_{I_2} (see Eq. *8-13*).

$$N_{I_1} = \begin{vmatrix} V_1 & -R_{12} \\ V_2 & R_{22} \end{vmatrix} = \begin{vmatrix} 58 & -3 \\ -10 & 5 \end{vmatrix} = (58)(5) - (-3)(-10) = 260$$

$$N_{I_2} = \begin{vmatrix} R_{11} & V_1 \\ -R_{21} & V_2 \end{vmatrix} = \begin{vmatrix} 7 & 58 \\ -3 & -10 \end{vmatrix} = (7)(-10) - (58)(-3) = 104$$

Step 4. Find I_1 and I_2.

$$I_1 = \frac{N_{I_1}}{\Delta} = \frac{260}{26} = 10 \ \text{A} \qquad Ans. \tag{8-12}$$

$$I_2 = \frac{N_{I_2}}{\Delta} = \frac{104}{26} = 4 \ \text{A} \qquad Ans. \tag{8-13}$$

DETERMINANT METHOD FOR SOLVING CURRENTS IN A THREE-MESH NETWORK

For a network of three meshes, the determinant of the network is

$$\Delta = \begin{vmatrix} R_{11} & -R_{12} & -R_{13} \\ -R_{21} & R_{22} & -R_{23} \\ -R_{31} & -R_{32} & R_{33} \end{vmatrix} \tag{8-14}$$

where R_{11} = total resistance of mesh 1

R_{22} = total resistance of mesh 2

R_{33} = total resistance of mesh 3

R_{12} = mutual (common) resistance between mesh 1 and mesh 2

R_{21} = mutual (common) resistance between mesh 2 and mesh 1

R_{13} = mutual (common) resistance between mesh 1 and mesh 3

R_{31} = mutual (common) resistance between mesh 3 and mesh 1

R_{23} = mutual (common) resistance between mesh 2 and mesh 3

R_{32} = mutual (common) resistance between mesh 3 and mesh 2

Because the mutual resistances between two meshes are equal, $R_{12} = R_{21}$, $R_{13} = R_{31}$, and $R_{23} = R_{32}$. Note that the sign for all mutual resistances is negative. The formulas for mesh currents I_1, I_2, and I_3 are

$$I_1 = \frac{N_{I_1}}{\Delta} = \frac{\begin{vmatrix} V_1 & -R_{12} & -R_{13} \\ V_2 & R_{22} & -R_{23} \\ V_3 & -R_{32} & R_{33} \end{vmatrix}}{\Delta} \tag{8-15}$$

$$I_2 = \frac{N_{I_2}}{\Delta} = \frac{\begin{vmatrix} R_{11} & V_1 & -R_{13} \\ -R_{21} & V_2 & -R_{23} \\ -R_{31} & V_3 & R_{33} \end{vmatrix}}{\Delta} \tag{8-16}$$

$$I_3 = \frac{N_{I_3}}{\Delta} = \frac{\begin{vmatrix} R_{11} & -R_{12} & V_1 \\ -R_{21} & R_{22} & V_2 \\ -R_{31} & -R_{32} & V_3 \end{vmatrix}}{\Delta} \tag{8-17}$$

where V_1 = net voltage sources for mesh 1
$\quad\quad\ V_2$ = net voltage sources for mesh 2
$\quad\quad\ V_3$ = net voltage sources for mesh 3

The numerator determinants for the three mesh currents and the polarity of the net voltage sources are found in a similar manner to those for the two mesh currents.

Example 8.6   Find the determinant solution for mesh currents I_1, I_2, and I_3 (Fig. 8-3).

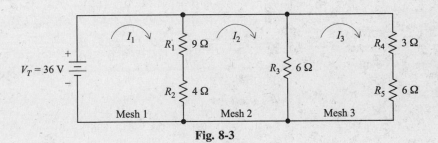

Fig. 8-3

Step 1.   Assume all mesh currents in the same clockwise direction.

Step 2.   Find the determinant of the network, Δ.

R_{11} = total resistance in mesh 1 = $R_1 + R_2 = 9 + 4 = 13\ \Omega$

R_{22} = total resistance in mesh 2 = $R_1 + R_2 + R_3 = 9 + 4 + 6 = 19\ \Omega$

R_{33} = total resistance in mesh 3 = $R_3 + R_4 + R_5 = 6 + 3 + 6 = 15\ \Omega$

$R_{12} = R_{21}$ = mutual resistance between meshes 1 and 2
$\quad\quad\quad = R_1 + R_2 = 9 + 4 = 13\ \Omega$

$R_{13} = R_{31}$ = mutual resistance between meshes 1 and 3 = $0\ \Omega$

$R_{23} = R_{32}$ = mutual resistance between meshes 2 and 3 = $R_3 = 6\ \Omega$

$$\Delta = \begin{vmatrix} R_{11} & -R_{12} & -R_{13} \\ -R_{21} & R_{22} & -R_{23} \\ -R_{31} & -R_{32} & R_{33} \end{vmatrix} \tag{8-14}$$

Substitute the value of resistances and expanding the 1st column.

$$\Delta = \begin{vmatrix} 13 & -13 & 0 \\ -13 & 19 & -6 \\ 0 & -6 & 15 \end{vmatrix} = 13 \begin{vmatrix} 19 & -6 \\ -6 & 15 \end{vmatrix} - (-13) \begin{vmatrix} -13 & 0 \\ -6 & 15 \end{vmatrix} + 0 \begin{vmatrix} -13 & 0 \\ 19 & -6 \end{vmatrix}$$

$$= 13(285 - 36) + 13(-195) = 13(249 - 195) = 13(54)$$

$$\Delta = 702$$

Step 3. Find the numerator determinant for I_1, I_2, and I_3.

V_1 = net voltage source for mesh 1 = V_T = 36 V

V_2 = net voltage source for mesh 2 = 0 V

V_3 = net voltage source for mesh 3 = 0 V

Substitute the proper values to find N_{I_1}, N_{I_2}, and N_{I_3}.

$$N_{I_1} = \begin{vmatrix} V_1 & -R_{12} & -R_{13} \\ V_2 & R_{22} & -R_{23} \\ V_3 & -R_{32} & R_{33} \end{vmatrix} = \begin{vmatrix} 36 & -13 & 0 \\ 0 & 19 & -6 \\ 0 & -6 & 15 \end{vmatrix} \qquad (8\text{-}15)$$

$$= 36 \begin{vmatrix} 19 & -6 \\ -6 & 15 \end{vmatrix} - 0 \begin{vmatrix} -13 & 0 \\ -6 & 15 \end{vmatrix} + 0 \begin{vmatrix} -13 & 0 \\ 19 & -6 \end{vmatrix}$$

$$= 36[(19)(15) - (-6)(-6)] = 36(249) = 8964$$

$$N_{I_2} = \begin{vmatrix} R_{11} & V_1 & -R_{13} \\ -R_{21} & V_2 & -R_{23} \\ -R_{31} & V_3 & R_{33} \end{vmatrix} = \begin{vmatrix} 13 & 36 & 0 \\ -13 & 0 & -6 \\ 0 & 0 & 15 \end{vmatrix} \qquad (8\text{-}16)$$

$$= 13 \begin{vmatrix} 0 & -6 \\ 0 & 15 \end{vmatrix} - (-13) \begin{vmatrix} 36 & 0 \\ 0 & 15 \end{vmatrix} + 0 \begin{vmatrix} 36 & 0 \\ 0 & -6 \end{vmatrix}$$

$$= 13(36)(15)$$

$$= 7020$$

$$N_{I_3} = \begin{vmatrix} R_{11} & -R_{12} & V_1 \\ -R_{21} & R_{22} & V_2 \\ -R_{31} & -R_{32} & V_3 \end{vmatrix} = \begin{vmatrix} 13 & -13 & 36 \\ -13 & 19 & 0 \\ 0 & -6 & 0 \end{vmatrix} \qquad (8\text{-}17)$$

$$= 13 \begin{vmatrix} 19 & 0 \\ -6 & 0 \end{vmatrix} - (-13) \begin{vmatrix} -13 & 36 \\ -6 & 0 \end{vmatrix} + 0 \begin{vmatrix} -13 & 36 \\ 19 & 0 \end{vmatrix}$$

$$= 13(36)(6)$$

$$= 2808$$

Step 4. Find I_1, I_2, and I_3.

$$I_1 = \frac{N_{I_1}}{\Delta} = \frac{8964}{702} = 12.77 \text{ A} \qquad Ans.$$

$$I_2 = \frac{N_{I_2}}{\Delta} = \frac{7020}{702} = 10 \text{ A} \qquad Ans.$$

$$I_3 = \frac{N_{I_3}}{\Delta} = \frac{2808}{702} = 4 \text{ A} \qquad Ans.$$

Step 5.   Check solutions.

In mesh 1 by KVL (Fig. 8-3),

$$I_1(R_1 + R_2) - I_2(R_1 + R_2) = 36$$
$$(12.77)(13) - 10(13) = 36$$
$$166 - 130 = 36$$
$$36 = 36 \qquad Check$$

In mesh 3 by KVL (Fig. 8-3),

$$I_2 R_3 - I_3(R_3 + R_4 + R_5) = 0$$
$$10(6) - 4(15) = 0$$
$$60 - 60 = 0$$
$$0 = 0 \qquad Check$$

Solved Problems

8.1   Evaluate the following second-order determinants.

(a) $\begin{vmatrix} 5 & 2 \\ 1 & 6 \end{vmatrix} = (5)(6) - (2)(1) = 30 - 2 = 28 \qquad Ans.$

(b) $\begin{vmatrix} 5 & -2 \\ 1 & -6 \end{vmatrix} = (5)(-6) - (-2)(1) = -30 + 2 = -28 \qquad Ans.$

(c) $\begin{vmatrix} 0 & 3 \\ 2 & -5 \end{vmatrix} = (0)(5) - (3)(2) = -6 \qquad Ans.$

(d) $\begin{vmatrix} 0 & 4 \\ 0 & -2 \end{vmatrix} = (0)(-2) - (4)(0) = 0 \qquad Ans.$

(e) $\begin{vmatrix} x^2 & x \\ y^2 & y \end{vmatrix} = x^2 y - xy^2 \qquad Ans.$

8.2   Evaluate the following third-order determinants. Expand the first column into its cofactors to form a series of second-order determinants.

(a) $\begin{vmatrix} 3 & 1 & 2 \\ 4 & 2 & -3 \\ 5 & 4 & 1 \end{vmatrix} = 3\begin{vmatrix} 2 & -3 \\ 4 & 1 \end{vmatrix} - 4\begin{vmatrix} 1 & 2 \\ 4 & 1 \end{vmatrix} + 5\begin{vmatrix} 1 & 2 \\ 2 & -3 \end{vmatrix}$

$$= 3[(2)(1) - (-3)(4)] - 4[(1)(1) - (2)(4)] + 5[(1)(-3) - (2)(2)]$$
$$= 3(2 + 12) - 4(1 - 8) + 5(-3 - 4)$$
$$= 3(14) - 4(-7) + 5(-7)$$
$$= 42 + 28 - 35$$
$$= 35 \qquad Ans.$$

(b) $\begin{vmatrix} 3 & -1 & -4 \\ 1 & -2 & 2 \\ 4 & -4 & 1 \end{vmatrix} = 3 \begin{vmatrix} -2 & 2 \\ -4 & 1 \end{vmatrix} - 1 \begin{vmatrix} -1 & -4 \\ -4 & 1 \end{vmatrix} + 4 \begin{vmatrix} -1 & -4 \\ -2 & 2 \end{vmatrix}$

$$= 3\,[(-2)(1) - (2)(-4)] - 1\,[(-1)(1) - (-4)(-4)] + 4\,[(-1)(2) - (-4)(-2)]$$

$$= 3(-2 + 8) - 1(-1 - 6) + 4(-2 - 8)$$

$$= 3(6) - 1(-17) + 4(-10)$$

$$= 18 + 17 - 40$$

$$= -5 \quad \text{Ans.}$$

(c)   Verify the answer to (b) by now expanding elements of the 2nd row.

$$\begin{vmatrix} 3 & -1 & -4 \\ 1 & -2 & 2 \\ 4 & -4 & 1 \end{vmatrix} = -1(1) \begin{vmatrix} -1 & -4 \\ -4 & 1 \end{vmatrix} + 1(-2) \begin{vmatrix} 3 & -4 \\ 4 & 1 \end{vmatrix} - 1(-2) \begin{vmatrix} 3 & -1 \\ 4 & -4 \end{vmatrix}$$

$$= -(-1 - 16) - 2(3 + 16) - 2(-12 + 4)$$

$$= 17 - 38 + 16$$

$$= -5 \quad \text{Ans.} \qquad \text{Check}$$

8.3   Solve the following simultaneous equations by the use of determinants.

(a) $x + y = 3$ ⠀⠀⠀⠀⠀⠀⠀⠀⠀⠀⠀⠀⠀⠀⠀⠀⠀⠀⠀⠀⠀(1)

⠀⠀⠀$2x + 3y = 1$ ⠀⠀⠀⠀⠀⠀⠀⠀⠀⠀⠀⠀⠀⠀⠀⠀⠀⠀⠀(2)

Step 1.   Solve for the coefficient determinant Δ.

$$\Delta = \begin{vmatrix} a_{11} & a_{12} \\ a_{21} & a_{22} \end{vmatrix} = \begin{vmatrix} 1 & 1 \\ 2 & 3 \end{vmatrix} = 3 - 2 = 1 \tag{8-4}$$

Step 2.   Solve for x and y.

$$x = \frac{N_x}{\Delta} = \frac{\begin{vmatrix} A & a_{12} \\ B & a_{22} \end{vmatrix}}{\Delta} = \frac{\begin{vmatrix} 3 & 1 \\ 1 & 3 \end{vmatrix}}{1} = 9 - 1 = 8 \quad \text{Ans.} \tag{8-5a}$$

$$y = \frac{N_y}{\Delta} = \frac{\begin{vmatrix} a_{11} & A \\ a_{21} & B \end{vmatrix}}{\Delta} = \frac{\begin{vmatrix} 1 & 3 \\ 2 & 1 \end{vmatrix}}{\Delta} = 1 - 6 = -5 \quad \text{Ans.} \tag{8-5b}$$

Step 3.   Check solutions of $x = 8$, $y = -5$ in Eqs. (1) and (2).

⠀⠀⠀(1)⠀⠀⠀⠀⠀$8 - 5 = 3$

⠀⠀⠀⠀⠀⠀⠀⠀⠀⠀⠀⠀$3 = 3 \quad \text{Check}$

⠀⠀(2)⠀⠀$2(8) + 3(-5) = 1$

⠀⠀⠀⠀⠀⠀⠀⠀$16 - 15 = 1$

⠀⠀⠀⠀⠀⠀⠀⠀⠀⠀$1 = 1 \quad \text{Check}$

8.3 (*b*)　$15I_1 - 10I_2 = 10$ (1)

$-10I_1 + 20I_2 = 0$ (2)

Step 1. Solve for Δ.

$$\Delta = \begin{vmatrix} a_{11} & a_{12} \\ a_{21} & a_{22} \end{vmatrix} = \begin{vmatrix} 15 & -10 \\ -10 & 20 \end{vmatrix} = (15)(20) - (-10)(-10) = 200 \qquad (8\text{-}4)$$

Step 2. Solve for I_1 and I_2.

$$I_1 = \frac{N_{I_1}}{\Delta} = \frac{\begin{vmatrix} A & a_{12} \\ B & a_{22} \end{vmatrix}}{\Delta} = \frac{\begin{vmatrix} 10 & -10 \\ 0 & 20 \end{vmatrix}}{200} = \frac{200}{200} = 1 \text{ A} \qquad Ans. \qquad (8\text{-}5)$$

$$I_2 = \frac{N_{I_2}}{\Delta} = \frac{\begin{vmatrix} a_{11} & A \\ a_{21} & B \end{vmatrix}}{\Delta} = \frac{\begin{vmatrix} 15 & 10 \\ -10 & 0 \end{vmatrix}}{200} = \frac{100}{200} = 0.5 \text{ A} \qquad Ans. \qquad (8\text{-}6)$$

Step 3. Check solutions.

(1)　　$15(1) - 10(0.5) = 10$

$15 - 5 = 10$

$10 = 10$　　*Check*

(2)　　$-10(1) + 20(0.5) = 0$

$-10 + 10 = 0$

$0 = 0$　　*Check*

8.4　Find the determinant solution for mesh currents I_1 and I_2 (Fig. 8-4).

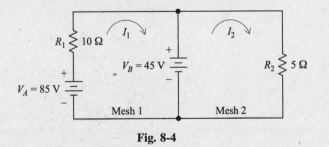

Fig. 8-4

Step 1.　Show mesh currents I_1 and I_2 in the clockwise direction.

Step 2.　Find the determinant of the network Δ.

$$R_{11} = R_1 = 10 \ \Omega$$

$$R_{22} = R_2 = 5 \ \Omega$$

$$R_{12} = R_{21} = 0 \ \Omega$$

$$\Delta = \begin{vmatrix} R_{11} & -R_{12} \\ -R_{21} & R_{22} \end{vmatrix} = \begin{vmatrix} 10 & 0 \\ 0 & 5 \end{vmatrix} = (10)(5) = 50 \tag{8-11}$$

Step 3. Find the numerator determinant for I_1 and I_2.

$$V_1 = V_A - V_B = 85 - 45 = 40 \text{ V}$$

$$V_2 = V_B = 45 \text{ V}$$

$$N_{I_1} = \begin{vmatrix} V_1 & -R_{12} \\ V_2 & R_{22} \end{vmatrix} = \begin{vmatrix} 40 & 0 \\ 45 & 5 \end{vmatrix} = (40)(5) = 200$$

$$N_{I_2} = \begin{vmatrix} R_{11} & V_1 \\ -R_{21} & V_2 \end{vmatrix} = \begin{vmatrix} 10 & 40 \\ 0 & 45 \end{vmatrix} = (10)(45) = 450$$

Step 4. Find I_1 and I_2.

$$I_1 = \frac{N_{I_1}}{\Delta} = \frac{200}{50} = 4 \text{ A} \qquad Ans. \tag{8-12}$$

$$I_2 = \frac{N_{I_2}}{\Delta} = \frac{450}{50} = 9 \text{ A} \qquad Ans. \tag{8-13}$$

8.5 Equation (8-2) showed the value of a third-order determinant by expanding the first column. The specific row or column selected for expanding is random. Show that by expanding the first row, the result is the same.

Expand the first column:

$$\begin{vmatrix} a_{11} & a_{12} & a_{13} \\ a_{21} & a_{22} & a_{23} \\ a_{31} & a_{32} & a_{33} \end{vmatrix} = a_{11} \begin{vmatrix} a_{22} & a_{23} \\ a_{32} & a_{33} \end{vmatrix} - a_{21} \begin{vmatrix} a_{12} & a_{13} \\ a_{32} & a_{33} \end{vmatrix} + a_{31} \begin{vmatrix} a_{12} & a_{13} \\ a_{22} & a_{23} \end{vmatrix} \tag{8-2}$$

$$= a_{11}(a_{22}a_{33} - a_{23}a_{32}) - a_{21}(a_{12}a_{33} - a_{13}a_{32}) + a_{31}(a_{12}a_{23} - a_{13}a_{22})$$

$$= a_{11}a_{22}a_{33} - a_{11}a_{23}a_{32} - a_{21}a_{12}a_{33} + a_{21}a_{13}a_{32} + a_{31}a_{12}a_{23} - a_{31}a_{13}a_{22} \quad (1)$$

Expand the first row:

$$\begin{vmatrix} a_{11} & a_{12} & a_{13} \\ a_{21} & a_{22} & a_{23} \\ a_{31} & a_{32} & a_{33} \end{vmatrix} = a_{11} \begin{vmatrix} a_{22} & a_{23} \\ a_{32} & a_{33} \end{vmatrix} - a_{12} \begin{vmatrix} a_{21} & a_{23} \\ a_{31} & a_{33} \end{vmatrix} + a_{13} \begin{vmatrix} a_{21} & a_{22} \\ a_{31} & a_{32} \end{vmatrix}$$

$$= a_{11}(a_{22}a_{33} - a_{23}a_{32}) - a_{12}(a_{21}a_{33} - a_{23}a_{31}) + a_{13}(a_{21}a_{32} - a_{22}a_{31})$$

$$= a_{11}a_{22}a_{33} - a_{11}a_{23}a_{32} - a_{12}a_{21}a_{33} + a_{12}a_{23}a_{31} + a_{13}a_{21}a_{32} - a_{13}a_{22}a_{31}$$

$$= a_{11}a_{22}a_{33} - a_{11}a_{23}a_{32} - a_{21}a_{12}a_{33} + a_{21}a_{13}a_{32} + a_{31}a_{12}a_{23} - a_{31}a_{13}a_{22} \quad (2)$$

by rearranging elements within the term, we show that by matching term by term, Eq. (1) = Eq. (2).

8.6 In Example 8.4, three mesh currents were found by solving three simultaneous equations by use of determinants. Write by inspection the determinant expression for all mesh currents. Figure 8-1 is repeated for convenience.

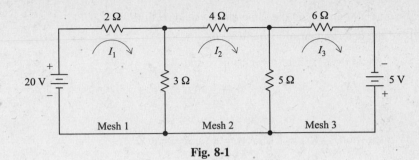

Fig. 8-1

By inspection, $R_{11} = 5\,\Omega$, $R_{22} = 12\,\Omega$, $R_{33} = 11\,\Omega$
$R_{12} = R_{21} = 3\,\Omega$, $R_{13} = R_{31} = 0\,\Omega$, $R_{23} = R_{32} = 5\,\Omega$
$V_1 = 20\,\text{V}$, $V_2 = 0\,\text{V}$, $V_3 = 5\,\text{V}$

$$\Delta = \begin{vmatrix} R_{11} & -R_{12} & -R_{13} \\ -R_{21} & R_{22} & -R_{23} \\ -R_{31} & -R_{32} & R_{33} \end{vmatrix} = \begin{vmatrix} 5 & -3 & 0 \\ -3 & 12 & -5 \\ 0 & -5 & 11 \end{vmatrix} \qquad Ans. \qquad (8\text{-}14)$$

$$I_1 = \frac{N_{I_1}}{\Delta} = \frac{\begin{vmatrix} V_1 & -R_{12} & -R_{13} \\ V_2 & R_{22} & -R_{23} \\ V_3 & -R_{32} & R_{33} \end{vmatrix}}{\Delta} = \frac{\begin{vmatrix} 20 & -3 & 0 \\ 0 & 12 & -5 \\ 5 & -5 & 11 \end{vmatrix}}{\Delta} \qquad Ans. \qquad (8\text{-}15)$$

$$I_2 = \frac{N_{I_2}}{\Delta} = \frac{\begin{vmatrix} R_{11} & V_1 & -R_{13} \\ -R_{21} & V_2 & -R_{23} \\ -R_{31} & V_3 & R_{33} \end{vmatrix}}{\Delta} = \frac{\begin{vmatrix} 5 & 20 & 0 \\ -3 & 0 & -5 \\ 0 & 5 & 11 \end{vmatrix}}{\Delta} \qquad Ans. \qquad (8\text{-}16)$$

$$I_3 = \frac{N_{I_3}}{\Delta} = \frac{\begin{vmatrix} R_{11} & -R_{12} & V_1 \\ -R_{21} & R_{22} & V_2 \\ -R_{31} & -R_{32} & V_3 \end{vmatrix}}{\Delta} = \frac{\begin{vmatrix} 5 & -3 & 20 \\ -3 & 12 & 0 \\ 0 & -5 & 5 \end{vmatrix}}{\Delta} \qquad Ans. \qquad (8\text{-}17)$$

Note that these current expressions are identical to those in Example 8.4. By inspection I_1, I_2, and I_3 can be explicitly expressed as a ratio of determinants without writing first the simultaneous equations.

8.7   Solve for all mesh currents by the use of determinants (Fig. 8-5).

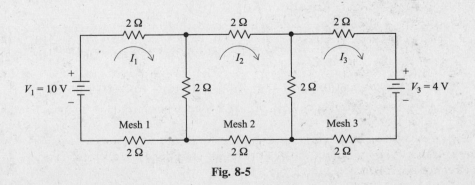

Fig. 8-5

Step 1. Assume all currents in the same clockwise direction.

Step 2. Find the determinant of the network Δ.

$$\Delta = \begin{vmatrix} R_{11} & -R_{12} & -R_{13} \\ -R_{21} & R_{22} & -R_{23} \\ -R_{31} & -R_{32} & R_{33} \end{vmatrix} \qquad (8\text{-}14)$$

By inspection, $R_{11} = 6\,\Omega, R_{22} = 8\,\Omega, R_{33} = 6\,\Omega$

$R_{12} = R_{21} = 2\,\Omega$

$R_{13} = R_{31} = 0\,\Omega$

$R_{23} = R_{32} = 2\,\Omega$

$V_1 = 10\,\text{V}$

$V_2 = 0\,\text{V}$

$V_3 = -4\,\text{V}$

Substituting R values and expanding the 1st column,

$$\Delta = \begin{vmatrix} 6 & -2 & 0 \\ -2 & 8 & -2 \\ 0 & -2 & 6 \end{vmatrix} = 6\begin{vmatrix} 8 & -2 \\ -2 & 6 \end{vmatrix} - (-2)\begin{vmatrix} -2 & 0 \\ -2 & 6 \end{vmatrix} + 0\begin{vmatrix} -2 & 0 \\ 8 & -2 \end{vmatrix}$$

$$= 6[(8)(6) - (-2)(-2)] + 2[(-2)(6) - (0)(-2)] + 0[(-2)(-2) - (0)(8)]$$

$$= 6(44) + 2(-12) = 264 - 24 = 240$$

Step 3. Find the numerator determinant for I_1, I_2, and I_3.

$$N_{I_1} = \begin{vmatrix} V_1 & -R_{12} & -R_{13} \\ V_2 & R_{22} & -R_{23} \\ V_3 & -R_{32} & R_{33} \end{vmatrix} \qquad \text{(from 8-15)}$$

$$= \begin{vmatrix} 10 & -2 & 0 \\ 0 & 8 & -2 \\ -4 & -2 & 6 \end{vmatrix} = 10\begin{vmatrix} 8 & -2 \\ -2 & 6 \end{vmatrix} - 0\begin{vmatrix} -2 & 0 \\ -2 & 6 \end{vmatrix} + (-4)\begin{vmatrix} -2 & 0 \\ 8 & -2 \end{vmatrix}$$

$$= 10[(8)(6) - (-2)(-2)] - 0 - 4[(-2)(-2) - (0)(8)]$$

$$= 10(44) - 4(4) = 440 - 16 = 424$$

$$N_{I_2} = \begin{vmatrix} R_{11} & V_1 & -R_{13} \\ -R_{21} & V_2 & -R_{23} \\ -R_{31} & V_3 & R_{33} \end{vmatrix} \qquad \text{(from 8-16)}$$

$$= \begin{vmatrix} 6 & 10 & 0 \\ -2 & 0 & -2 \\ 0 & -4 & 6 \end{vmatrix} = 6\begin{vmatrix} 0 & -2 \\ -4 & 6 \end{vmatrix} - (-2)\begin{vmatrix} 10 & 0 \\ -4 & 6 \end{vmatrix} + 0\begin{vmatrix} 10 & 0 \\ 0 & -2 \end{vmatrix}$$

$$= 6[(0)(6) - (-2)(-4)] + 2[(10)(6) - (0)(-4)] + 0$$

$$= 6(-8) + 2(60) = -48 + 120 = 72$$

$$N_{I_3} = \begin{vmatrix} R_{11} & -R_{12} & V_1 \\ -R_{21} & R_{22} & V_2 \\ -R_{31} & -R_{32} & V_3 \end{vmatrix} \qquad \text{(from 8-17)}$$

$$= \begin{vmatrix} 6 & -2 & 10 \\ -2 & 8 & 0 \\ 0 & -2 & -4 \end{vmatrix} = 6 \begin{vmatrix} 8 & 0 \\ -2 & -4 \end{vmatrix} - (-2) \begin{vmatrix} -2 & 10 \\ -2 & -4 \end{vmatrix} + 0 \begin{vmatrix} -2 & 10 \\ 8 & 0 \end{vmatrix}$$

$$= 6[(8)(-4) - (0)(-2)] + 2[(-2)(-4) - (10)(-2)] + 0$$

$$= 6(-32) + 2(28) = -192 + 56 = -136$$

Step 4. Find I_1, I_2, and I_3.

$$I_1 = \frac{N_{I_1}}{\Delta} = \frac{424}{240} = 1.77 \text{ A} \qquad Ans. \tag{8-15}$$

$$I_2 = \frac{N_{I_2}}{\Delta} = \frac{72}{240} = 0.30 \text{ A} \qquad Ans. \tag{8-16}$$

$$I_3 = \frac{N_{I_3}}{\Delta} = -\frac{136}{240} = -0.57 \text{ A} \qquad Ans. \tag{8-17}$$

(I_3 is counterclockwise)

Step 5. Check solutions.

In mesh 1 by KVL (Fig. 8-5),

$$10 = 6I_1 - 2I_2$$

$$10 = 6(1.77) - 2(0.30)$$

$$10 = 10.62 - 0.60$$

$$10 = 10.02 \qquad Check, \text{ rounding error}$$

In mesh 2 by KVL (Fig. 8-5),

$$0 = -2I_1 + 8I_2 - 2I_3$$

$$0 = -2(1.77) + 8(0.30) - 2(-0.57)$$

$$0 = -3.54 + 2.40 + 1.14$$

$$0 = 0 \qquad Check$$

8.8   (a) Solve for mesh currents I_1, I_2, and I_3 by determinants (Fig. 8-6).
         (b) What are the voltage drops across the resistors in mesh 1?

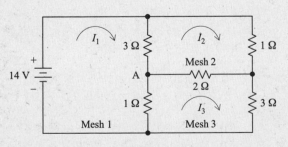

Fig. 8-6

(a) **Step 1.**   Write the determinant of the network Δ.

$$\Delta = \begin{vmatrix} R_{11} & -R_{12} & -R_{13} \\ -R_{21} & R_{22} & -R_{23} \\ -R_{31} & -R_{32} & R_{33} \end{vmatrix} = \begin{vmatrix} 4 & -3 & -1 \\ -3 & 6 & -2 \\ -1 & -2 & 6 \end{vmatrix} \qquad (8\text{-}14)$$

$$= 4\begin{vmatrix} 6 & -2 \\ -2 & 6 \end{vmatrix} - (-3)\begin{vmatrix} -3 & -1 \\ -2 & 6 \end{vmatrix} + (-1)\begin{vmatrix} -3 & -1 \\ 6 & -2 \end{vmatrix}$$

$$= 4(36 - 4) + 3(-18 - 2) - (6 + 6)$$

$$= 4(32) + 3(-20) - 12$$

$$= 128 - 60 - 12$$

$$\Delta = 56$$

Step 2.   Find the numerator determinant for I_1, I_2, and I_3.

Replace 1st column of Δ with values of V_1, V_2, V_3 to find N_{I_1}.

$$N_{I_1} = \begin{vmatrix} 14 & -3 & -1 \\ 0 & 6 & -2 \\ 0 & -2 & 6 \end{vmatrix}$$

$$= 14\begin{vmatrix} 6 & -2 \\ -2 & 6 \end{vmatrix} - 0\begin{vmatrix} -3 & -1 \\ -2 & 6 \end{vmatrix} + 0\begin{vmatrix} -3 & -1 \\ 6 & -2 \end{vmatrix}$$

$$= 14(36 - 4) = 14(32) = 448$$

Replace 2nd column of Δ with values of V_1, V_2, V_3 to find N_{I_2}.

$$N_{I_2} = \begin{vmatrix} 4 & 14 & -1 \\ -3 & 0 & -2 \\ -1 & 0 & 6 \end{vmatrix}$$

$$= 4\begin{vmatrix} 0 & -2 \\ 0 & 6 \end{vmatrix} - (-3)\begin{vmatrix} 14 & -1 \\ 0 & 6 \end{vmatrix} + (-1)\begin{vmatrix} 14 & -1 \\ 0 & -2 \end{vmatrix}$$

$$= 0 + 3(84) - (-28)$$

$$= 280$$

Replace 3rd column of Δ with values of V_1, V_2, V_3 to find N_{I_3}.

$$N_{I_3} = \begin{vmatrix} 4 & -3 & 14 \\ -3 & 6 & 0 \\ -1 & -2 & 0 \end{vmatrix}$$

$$= 4\begin{vmatrix} 6 & 0 \\ -2 & 0 \end{vmatrix} - (-3)\begin{vmatrix} -3 & 14 \\ -2 & 0 \end{vmatrix} + (-1)\begin{vmatrix} -3 & 14 \\ 6 & 0 \end{vmatrix}$$

$$= 0 + 3(28) - 1(-84) = 84 + 84$$

$$= 168$$

Step 3.   Find I_1, I_2, and I_3.

$$I_1 = \frac{N_{I_1}}{\Delta} = \frac{448}{56} = 8 \text{ A} \qquad Ans. \tag{8-15}$$

$$I_2 = \frac{N_{I_2}}{\Delta} = \frac{280}{56} = 5 \text{ A} \qquad Ans. \tag{8-16}$$

$$I_3 = \frac{N_{I_3}}{\Delta} = \frac{168}{56} = 3 \text{ A} \qquad Ans. \tag{8-17}$$

Step 4.   Check solution.

$$\text{At node A, } \Sigma I = 0$$

$$I_1 = I_2 + I_3$$

$$8 = 5 + 3$$

$$8 = 8 \qquad Check$$

(*b*)   Across 3 Ω-resistor:

$$V_{3\Omega} = 3(I_1 - I_2) = 3(8 - 5) = 9 \text{ V} \qquad Ans.$$

Across 1 Ω-resistor:

$$V_{1\Omega} = 1(I_1 - I_3) = 8 - 3 = 5 \text{ V} \qquad Ans.$$

$$14 = V_{3\Omega} + V_{1\Omega} = 9 + 5$$

$$14 = 14 \qquad Check$$

8.9   As the number of meshes increases, writing by inspection the explicit solutions of mesh currents by the ratio of determinants is far quicker and simpler than writing the simultaneous loop equations for each mesh. Write the general expression for finding the determinant of n-meshes of a DC network.

The determinant of order n is a square array of n rows by n columns. Following the pattern for 2nd order [Eq. (8-11)] and 3rd order [Eq. (8-14)] determinants, we can write

$$\Delta = \begin{vmatrix} +R_{11} & -R_{12} & -R_{13} & \cdots & -R_{1n} \\ -R_{21} & +R_{22} & -R_{23} & \cdots & -R_{2n} \\ -R_{31} & -R_{32} & +R_{33} & \cdots & -R_{3n} \\ \cdot & \cdot & \cdot & \cdots & \cdot \\ -R_{n1} & -R_{n2} & -R_{n3} & \cdots & +R_{nn} \end{vmatrix} \qquad Ans. \tag{8-18}$$

8.10   Given a network of $n = 4$ meshes, write the general expression for finding the value of I_2.

Step 1.   From Eq. *8-18*, Solved Problem 8.9, write the network determinant for four meshes.

$$\Delta = \begin{vmatrix} R_{11} & -R_{12} & -R_{13} & -R_{14} \\ -R_{21} & R_{22} & -R_{23} & -R_{24} \\ -R_{31} & -R_{32} & R_{33} & -R_{34} \\ -R_{41} & -R_{42} & -R_{43} & R_{44} \end{vmatrix} \tag{8-19}$$

Step 2. Replace the elements of the 2nd column by the net voltages of each mesh to form the numerator determinant of I_2.

$$I_2 = \frac{N_{I_2}}{\Delta} = \frac{\begin{vmatrix} R_{11} & V_1 & -R_{13} & -R_{14} \\ -R_{21} & V_2 & -R_{23} & -R_{24} \\ -R_{31} & V_3 & R_{33} & -R_{34} \\ -R_{41} & V_4 & -R_{43} & R_{44} \end{vmatrix}}{\Delta} \quad \text{Ans.} \qquad (8\text{-}20)$$

8.11   Add a fourth mesh to Fig. 8-6 to form a four-mesh network (Fig. 8-7). Write the expression for I_2. Do not solve.

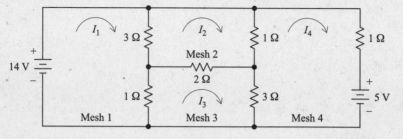

Fig. 8-7

Step 1. Find the network Δ. Refer to Δ for a three-mesh network in Solved Problem 8.8(a) and to Eq. (8-20). The new Δ for the four-mesh network becomes

$$\Delta = \begin{vmatrix} 4 & -3 & -1 & -R_{14} \\ -3 & 6 & -2 & -R_{24} \\ -1 & -2 & 6 & -R_{34} \\ -R_{41} & -R_{42} & -R_{43} & R_{44} \end{vmatrix} = \begin{vmatrix} 4 & -3 & -1 & 0 \\ -3 & 6 & -2 & -1 \\ -1 & -2 & 6 & -3 \\ 0 & -1 & -3 & 5 \end{vmatrix}$$

Step 2. Replace 2nd column elements by net voltages of each mesh to form N_{I_2} and write I_2 expression.

$$I_2 = \frac{N_{I_2}}{\Delta} = \frac{\begin{vmatrix} 4 & 14 & -1 & 0 \\ -3 & 0 & -2 & -1 \\ -1 & 0 & 6 & -3 \\ 0 & -5 & -3 & 5 \end{vmatrix}}{\Delta} \quad \text{Ans.}$$

By successive expanding of the elements of the first column, we reduce the order of the determinant from four to three and then to two. By reducing the order of the determinant eventually to two, you can solve explicitly for each mesh current. When $n = 4$, the expansion of the first column elements have signs of $+$, $-$, $+$, and $-$. (Compare to $n = 3$ where the signs are $+$, $-$, and $+$.)

Supplementary Problems

8.12   Evaluate the following second-order determinants.

(a) $\begin{vmatrix} 5 & -7 \\ 3 & -5 \end{vmatrix}$ (b) $\begin{vmatrix} 4 & -5 \\ 3 & 2 \end{vmatrix}$ (c) $\begin{vmatrix} 0 & 4 \\ 3 & 0 \end{vmatrix}$ (d)   Find x: $\begin{vmatrix} x & 3 \\ 1 & 6 \end{vmatrix} = 15$

   Ans.   (a) -4;   (b) 23;   (c) -12;   (d) $x = 3$

8.13 Solve the following simultaneous equations by the determinant method and check each solution.

    (a) $3x - 4y = 13$ (b) $16I_1 - 3I_2 = 10$ (c) $2I_1 + 3I_2 = 12$

         $5x + 6y = 9$ $8I_1 + 5I_2 = 18$ $3I_1 - I_2 = 7$

    Ans. (a) $x = 3$, $y = -1$; (b) $I_1 = 1$, $I_2 = 2$; (c) $I_1 = 3$, $I_2 = 2$

8.14 Evaluate the following third-order determinants.

$$(a)\ \begin{vmatrix} 4 & 4 & 2 \\ 3 & 1 & 1 \\ 2 & -5 & -1 \end{vmatrix} \quad (b)\ \begin{vmatrix} 4 & 5 & 2 \\ 1 & 4 & 1 \\ -5 & 3 & -1 \end{vmatrix} \quad (c)\ \begin{vmatrix} 6 & -4 & -2 \\ -4 & 15 & -6 \\ -2 & -6 & 11 \end{vmatrix} \quad (d)\ \begin{vmatrix} 1 & 3 & 0 \\ 2 & 0 & 1 \\ 0 & 4 & 3 \end{vmatrix}$$

    (e) Verify your answers for (a) and (c) by selecting a different row or column for expanding.

    Ans. (a) 2; (b) −2; (c) 442; (d) −22

8.15 Solve the following simultaneous equations by the determinant method and check each solution.

    (a) $2I_1 + 3I_2 + 5I_3 = 0$ (b) $I_1 + I_2 + 2I_3 = 3$ (c) $6I_1 - 4I_2 + 5I_3 = 10$

         $6I_1 - 2I_2 - 3I_3 = 3$ $2I_1 + I_2 + I_3 = 16$ $3I_1 + 2I_2 = 60$

         $8I_1 - 5I_2 - 6I_3 = 1$ $I_1 + 2I_2 + I_3 = 9$ $5I_1 + 4I_3 = 58$

    Ans. (a) $I_1 = 0.5$, $I_2 = 3$, $I_3 = -2$ $(\Delta = -40)$

            (b) $I_1 = 9$, $I_2 = 2$, $I_3 = -4$ $(\Delta = 4)$

            (c) $I_1 = 10$, $I_2 = 15$, $I_3 = 2$ $(\Delta = 46)$

8.16 Find the determinant solutions for mesh currents I_1 and I_2 in the following circuits. Assume the clockwise direction for current. Check solutions.

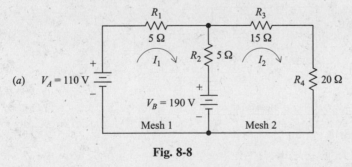

Fig. 8-8

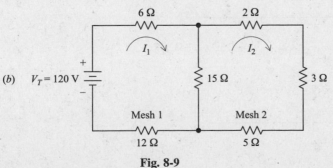

Fig. 8-9

(c)

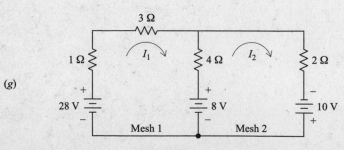

Fig. 8-10

(d)

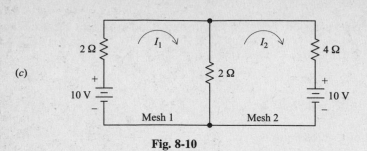

Fig. 8-11

(e)

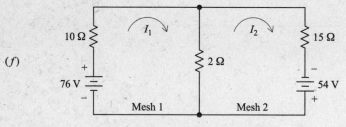

Fig. 8-12

(f)

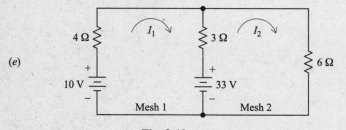

Fig. 8-13

(g)

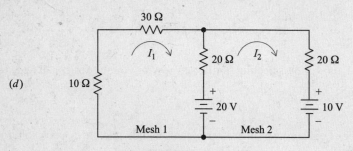

Fig. 8-14

Ans.   (*a*) $I_1 = -6$ A, $I_2 = 4$ A

(*b*) $I_1 = 5$ A, $I_2 = 3$ A

(*c*) $I_1 = 2$ A, $I_2 = -1$ A

(*d*) $I_1 = -0.3$ A, $I_2 = 0.1$ A

(*e*) $I_1 = -2$ A, $I_2 = 3$ A

(*f*) $I_1 = 7$ A, $I_2 = 4$ A

(*g*) $I_1 = 6$ A, $I_2 = 7$ A

8.17   Find the voltage drops across the following resistors in the circuits of Supplementary Problem 8.16.

(*a*)   15-ohm resistor (Fig. 8-9)

(*b*)   Common 2-ohm resistor (Fig. 8-10)

(*c*)   4-ohm and 6-ohm resistors (Fig. 8-12)

(*d*)   4-ohm resistor (Fig. 8-14)

Ans.   (*a*) 30 V;   (*b*) 6 V;   (*c*) 8 V, 18 V;   (*d*) 4 V

8.18   Find the determinant solution for mesh currents I_1, I_2, and I_3 in the following networks. Assume the clockwise direction for current. Check solutions.

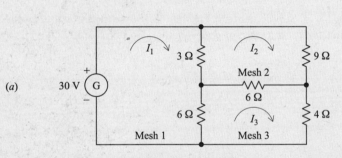

Fig. 8-15

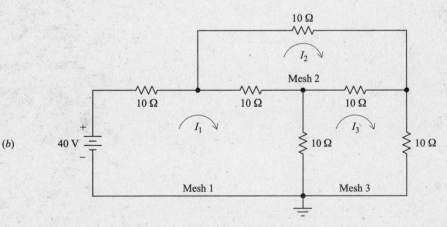

Fig. 8-16

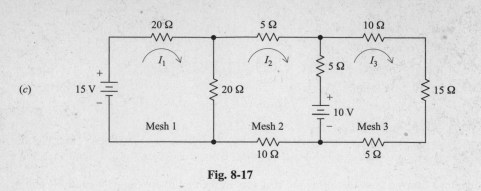

Fig. 8-17

Ans.   (*a*) $I_1 = 6\,\text{A}, \; I_2 = 2\,\text{A}, \; I_3 = 3\,\text{A}$
          (*b*) $I_1 = 2\,\text{A}, \; I_2 = 1\,\text{A}, \; I_3 = 1\,\text{A}$ ($\Delta = 16\,000$)
          (*c*) $I_1 = 0.357\,\text{A}, \; I_2 = -0.037\,\text{A}, \; I_3 = 0.280\,\text{A}$ ($\Delta = 41\,000$)

8.19   Find the voltage drops across the following resistors in Fig. 8-15 of Supplementary Problem 8.18.

(*a*)   9-ohm and 4-ohm resistors

(*b*)   6-ohm common resistor between mesh 2 and mesh 3

(*c*)   6-ohm common resistor between mesh 1 and mesh 3

Ans.   (*a*) $V_{9\Omega} = 18\,\text{V}, \; V_{4\Omega} = 12\,\text{V};$ (*b*) 6 V;   (*c*) 18 V

8.20   Add a fourth mesh to Fig. 8-17 in Supplementary Problem 8.18*c* to form a four-mesh network (Fig. 8-18). Write the expression for I_1 as a ratio of determinants. Do not solve.

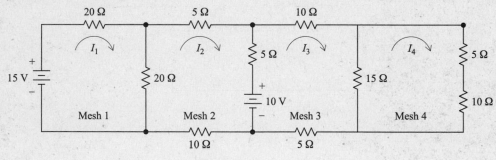

Fig. 8-18

Ans. $I_1 = \dfrac{\begin{vmatrix} 15 & -20 & 0 & 0 \\ -10 & 40 & -5 & 0 \\ 10 & -5 & 35 & -15 \\ 0 & 0 & -15 & 30 \end{vmatrix}}{\begin{vmatrix} 40 & -20 & 0 & 0 \\ -20 & 40 & -5 & 0 \\ 0 & -5 & 35 & -15 \\ 0 & 0 & -15 & 30 \end{vmatrix}}$

Chapter 9

Network Calculations

Y AND DELTA NETWORKS

The network in Fig. 9-1 is called a T ("tee") or Y ("wye") network because of its shape. T and Y are different names for the same network, except that in the Y network the R_a and R_b arms form the upper part of a Y.

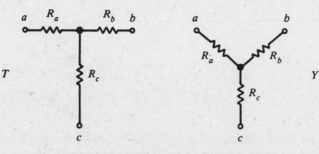

Fig. 9-1 Form of a T or Y network

The network in Fig. 9-2 is called a Π (pi) or Δ (delta) network because its shape resembles these Greek letters; Π and Δ are different names for the same network.

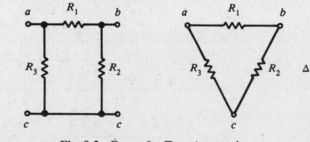

Fig. 9-2 Form of a Π or Δ network

In analyzing networks it is helpful to convert Y to Δ or Δ to Y to simplify the solution. The formulas for these conversions are derived from Kirchhoff's laws. Note that resistances in Y have subscript letters, R_a, R_b, and R_c, while the resistances in Δ are numbered R_1, R_2, and R_3.

Resistances are shown in a three-terminal network with three terminals a, b, and c. After the conversion formulas are used, one network is equivalent to the other because they have equivalent resistances across any one pair of terminals.

Δ to Y Conversion, or Π to T

See Fig. 9-3.

$$R_a = \frac{R_1 R_3}{R_1 + R_2 + R_3} \qquad (9\text{-}1)$$

$$R_b = \frac{R_1 R_2}{R_1 + R_2 + R_3} \qquad (9\text{-}2)$$

$$R_c = \frac{R_2 R_3}{R_1 + R_2 + R_3} \qquad (9\text{-}3)$$

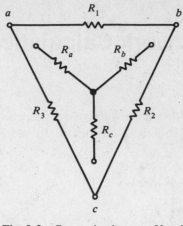

Fig. 9-3   Conversion between Y and
Δ networks

Rule 1:   The rule for Δ to Y conversion can be stated as follows: The resistance of any branch of the Y
network is equal to the product of the two adjacent sides of the Δ network divided by the sum of the
three Δ resistances.

Y to Δ Conversion, or T to Π

See Fig. 9-3.

$$R_1 = \frac{R_a R_b + R_b R_c + R_c R_a}{R_c} \qquad\qquad (9\text{-}4)$$

$$R_2 = \frac{R_a R_b + R_b R_c + R_c R_a}{R_a} \qquad\qquad (9\text{-}5)$$

$$R_3 = \frac{R_a R_b + R_b R_c + R_c R_a}{R_b} \qquad\qquad (9\text{-}6)$$

Rule 2:   The rule for Y to Δ conversion can be stated as follows: The resistance of any side of the Δ network
is equal to the sum of the Y network resistances multiplied two at a time, divided by the resistance
of the opposite branch of the Y network.

As an aid in using Eqs. (9-1)–(9-6) the following scheme is useful. Place the Y inside the Δ (Fig. 9-3). Note
that the Δ has three closed sides, while the Y has three open arms. Also note how each resistor in the open has
two adjacent resistors in the closed sides. For R_a, adjacent resistors are R_1 and R_3; for R_b, adjacent resistors
are R_1 and R_2; and for R_c, adjacent resistors are R_2 and R_3. Furthermore, each resistor can be considered
opposite to each other in the two networks. For example, open arm R_c is opposite to closed side R_1; R_b is
opposite to R_3; and R_a is opposite to R_2.

Example 9.1   A Δ network is shown in Fig. 9-4a. Find the resistances of an equivalent Y network (Fig. 9-4b) and draw
the network.

Place Y network within Δ network and find resistances by using Δ to Y conversion rule (see Fig. 9-4c).

$$R_a = \frac{R_1 R_3}{R_1 + R_2 + R_3} = \frac{4(6)}{10 + 6 + 4} = \frac{24}{20} = 1.2 \ \Omega \qquad Ans. \qquad (9\text{-}1)$$

$$R_b = \frac{R_1 R_2}{R_1 + R_2 + R_3} = \frac{4(10)}{20} = \frac{40}{20} = 2 \ \Omega \qquad Ans. \qquad (9\text{-}2)$$

$$R_c = \frac{R_2 R_3}{R_1 + R_2 + R_3} = \frac{10(6)}{20} = \frac{60}{20} = 3 \ \Omega \qquad Ans. \qquad (9\text{-}3)$$

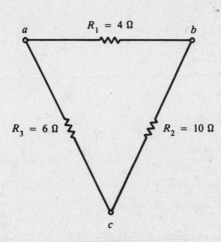

(a) Δ Known

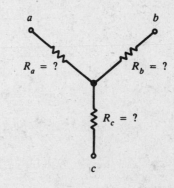

(b)   Y Unknown

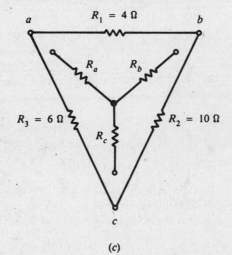

(c)

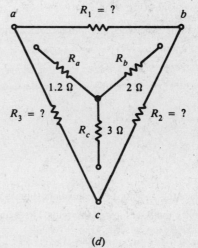

(d)

Fig. 9-4 Δ to Y conversion

The equivalent Y network is

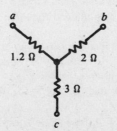

Example 9.2 Given the calculated resistances of the Y network in Example 9.1, $R_a = 1.2\ \Omega$, $R_b = 2\ \Omega$, and $R_c = 3\ \Omega$, confirm the values of equivalent resistances in the Δ network of $R_1 = 4\ \Omega$, $R_2 = 10\ \Omega$, and $R_3 = 6\ \Omega$.

Place Y network within Δ network as in previous example and find Δ resistance by using Y to Δ rule. (See Fig. 9-4d.)

$$R_1 = \frac{R_a R_b + R_b R_c + R_c R_a}{R_c} \qquad\qquad (9\text{-}4)$$

$$= \frac{1.2(2) + 2(3) + 3(1.2)}{3}$$

$$= \frac{2.4 + 6 + 3.6}{3} = \frac{12}{3} = 4\ \Omega \qquad Ans.$$

$$R_2 = \frac{R_a R_b + R_b R_c + R_c R_a}{R_a} \qquad\qquad (9\text{-}5)$$

$$= \frac{12}{1.2} = 10\ \Omega \qquad Ans.$$

$$R_3 = \frac{R_a R_b + R_b R_c + R_c R_a}{R_b} \qquad\qquad (9\text{-}6)$$

$$= \frac{12}{2} = 6\ \Omega \qquad Ans.$$

The results show that the Δ and Y networks (Fig. 9-4) are equivalent to each other when they have three resistance values obtained with the conversion formulas.

Example 9.3 Use network conversion to find the equivalent or total resistance R_T between a and d in a bridge circuit consisting of two deltas (Fig. 9-5a).

Step 1. Transform Δ network abc into its equivalent Y. Use the rule for Δ to Y conversion (Fig. 9-5b).

$$R_a = \frac{2(4)}{2 + 4 + 6} = \frac{8}{12} = 0.667\ \Omega$$

$$R_b = \frac{4(6)}{12} = \frac{24}{12} = 2\ \Omega$$

$$R_c = \frac{2(6)}{12} = \frac{12}{12} = 1\ \Omega$$

Step 2. Replace the Δ with its Y equivalent (Fig. 9-5c) in the original bridge circuit.

Step 3. Simplify the series–parallel circuit. First, combine series resistances. Resistances R_c and R_4 are in series, and R_b and R_5 are in series [Fig. 9-5d(1)].

$$R_c + R_4 = 1 + 5 = 6\ \Omega$$

$$R_b + R_5 = 2 + 4 = 6\ \Omega$$

Next combine the parallel branches, $R_c + R_4$ and $R_b + R_5$. Since the resistances are equal [Fig. 9-5d(2)],

$$R_p = \frac{6}{2} = 3\ \Omega$$

Finally, combine series resistances R_a and R_p [Fig. 9-5d(3)].

$$R_T = R_a + R_p = 0.667 + 3 = 3.67\ \Omega \qquad Ans.$$

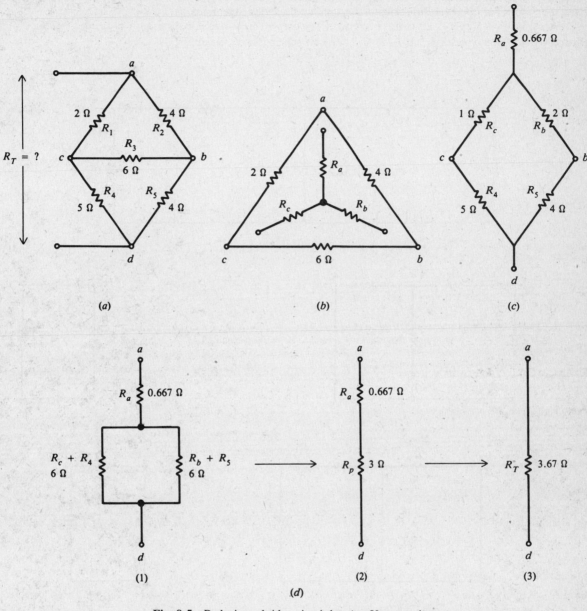

Fig. 9-5   Reducing a bridge circuit by Δ to Y conversion

SUPERPOSITION

The *superposition theorem* states that in a network with two or more sources the current or voltage for any component is the algebraic sum of the effects produced by each source acting independently. In order to use one source at a time, all other sources are removed from the circuit. A voltage source is removed by replacing it with a short circuit. A current source is removed by replacing it with an open circuit.

In order to superimpose currents and voltages, all the components must be linear and bilateral. *Linear* means that the current is proportional to the applied voltage; that is, the current and voltage obey Ohm's law, $I = V/R$. Then the currents calculated for different source voltages can be superimposed, that is, added algebraically. *Bilateral* means that the current is the same amount for opposite polarities of the source voltage. Then the values for opposite directions of current can be added algebraically.

Example 9.4   Find branch currents I_1, I_2, and I_3 by the superposition theorem (Fig. 9-6*a*).

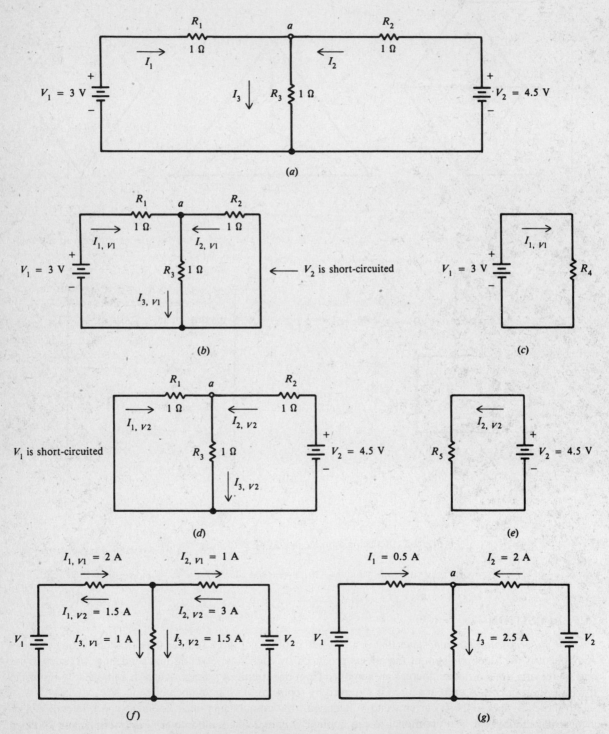

Fig. 9-6   Superposition applied to a two-mesh circuit

Step 1. Find the currents produced by voltage source V_1 *only*.

(*a*) Replace voltage source V_2 with a short circuit (Fig. 9-6*b*). Use the subscript V_1 after a comma to indicate that only source V_1 is supplying the circuit. For example, $I_{1,V1}$ is current I_1 due only to source V_1 and $I_{2,V1}$ is current I_2 due only to source V_1.

(*b*) Combine series and parallel resistances to reduce the circuit to a single source and a single resistance (Fig. 9-6*c*). Solve for currents produced by V_1.

$$R_4 = R_1 + \frac{R_2 R_3}{R_2 + R_3} = 1 + \frac{1(1)}{1+1} = 1 + 0.5 = 1.5 \ \Omega$$

$$I_{1,V1} = \frac{V_1}{R_4} = \frac{3}{1.5} = 2 \text{ A}$$

$I_{1,V1}$ will divide symmetrically at point *a* because of equal resistances R_2 and R_3 (Fig. 9-6*b*) so that

$$I_{2,V1} = -\frac{1}{2}I_{1,V1} = -\frac{1}{2}2 = -1 \text{ A}$$

$$I_{3,V1} = \frac{1}{2}I_{1,V1} = \frac{1}{2}2 = 1 \text{ A}$$

The negative sign is used to show that $I_{2,V1}$ actually leaves point *a* rather than enters point *a* as assumed.

Step 2. Find the currents produced by voltage source V_2 *only*.

(*a*) Replace voltage source V_1 with a short circuit (Fig. 9-6*d*). Use a subscript V_2 after a comma to indicate that only source V_2 is supplying the circuit. For example, $I_{1,V2}$ is current I_1 due only to source V_2.

(*b*) Reduce the circuit to a single source and a single resistance (Fig. 9-6*e*). Solve for currents produced by V_2.

$$R_5 = R_2 + \frac{R_1 R_3}{R_2 + R_3} = 1 + \frac{1(1)}{1+1} = 1 + 0.5 = 1.5 \ \Omega$$

$$I_{2,V2} = \frac{V_2}{R_5} = \frac{4.5}{1.5} = 3 \text{ A}$$

$I_{2,V2}$ will divide symmetrically at point *a* (Fig. 9-6*d*) so that

$$I_{3,V2} = \frac{1}{2}I_{2,V2} = \frac{1}{2}3 = 1.5 \text{ A}$$

$$I_{1,V2} = -\frac{1}{2}I_{2,V2} = -\frac{1}{2}3 = -1.5 \text{ A}$$

The negative sign shows that $I_{1,V2}$ actually leaves point *a* and does not enter point *a* as assumed.

Step 3. Add algebraically the individual currents to find the currents produced by both V_1 and V_2 (Fig. 9-6*f* and *g*).

$$I_1 = I_{1,V1} + I_{1,V2} = 2 - 1.5 = 0.5 \text{ A} \qquad \textit{Ans.}$$

$$I_2 = I_{2,V1} + I_{2,V2} = -1 + 3 = 2 \text{ A} \qquad \textit{Ans.}$$

$$I_3 = I_{3,V1} + I_{3,V2} = 1 + 1.5 = 2.5 \text{ A} \qquad \textit{Ans.}$$

THEVENIN'S THEOREM

Thevenin's theorem is a method used to change a complex circuit into a simple equivalent circuit. Thevenin's theorem states that any linear network of voltage sources and resistances, if viewed from any two points in the network, can be replaced by an equivalent resistance R_{Th} in series with an equivalent source V_{Th}. Figure 9-7a shows the original linear network with terminals a and b; Fig. 9-7b shows its connection to an external network or load; and Fig. 9-7c shows the Thevenin equivalent V_{Th} and R_{Th} that can be substituted for the linear network at the terminals a and b. The polarity of V_{Th} is such that it will produce current from a and b in the same direction as in the original network. R_{Th} is the Thevenin resistance across the network terminals a and b with each internal voltage source short-circuited. V_{Th} is the Thevenin voltage that would appear across the terminals a and b with the voltage sources in place and no load connected across a and b. For this reason, V_{Th} is also called the open-circuit voltage.

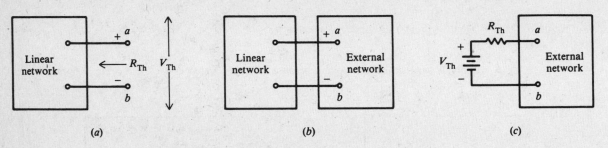

Fig. 9-7   Thevenin equivalent, V_{Th}, and series R_{Th}

Example 9.5   Find the Thevenin equivalent to the circuit at terminals a and b (Fig. 9-8a).

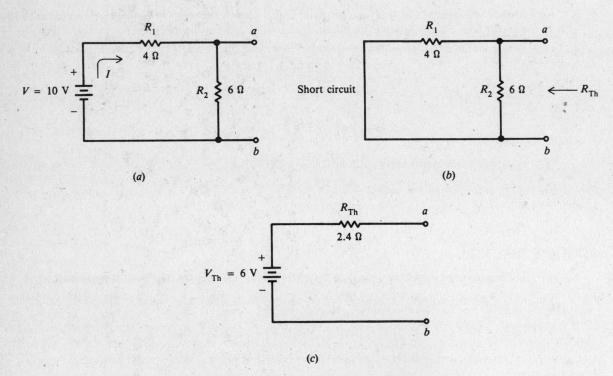

Fig. 9-8   Thevenin equivalent without load

Step 1.   Find R_{Th}. Short-circuit the voltage source $V = 10$ V (Fig. 9-8b). R_1 and R_2 are in parallel.

$$R_{Th} = \frac{R_1 R_2}{R_1 + R_2} = \frac{4(6)}{4+6} = \frac{24}{10} = 2.4\ \Omega \qquad Ans.$$

Step 2. Find V_{Th}. V_{Th} is the voltage across terminals a and b, which is the same as the voltage drop across resistance R_2.

$$I = \frac{V}{R_1 + R_2} = \frac{10}{4 + 6} = \frac{10}{10} = 1 \text{ A}$$

$$V_{Th} = V_2 = IR_2$$

Then $V_{Th} = 1(6) = 6 \text{ V}$ *Ans.*

The Thevenin equivalent is then as in Fig. 9-8c.

Example 9.6 To the circuit in Fig. 9-8a add a resistor load R_L of 3.6 Ω and find the current I_L through the load voltage V_L across the load.

The new circuit is as shown in Fig. 9-9a and with the Thevenin equivalent is as shown in Fig. 9-9b.

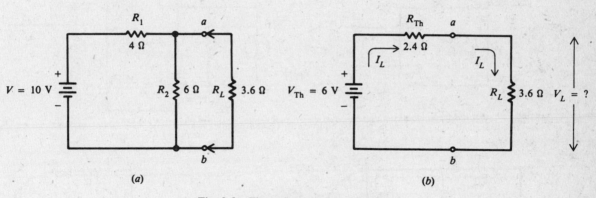

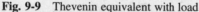

(a)   (b)

Fig. 9-9   Thevenin equivalent with load

$$I_L = \frac{V_{Th}}{R_{Th} + R_L} = \frac{6}{2.4 + 3.6} = \frac{6}{6} = 1 \text{ A}   \text{   } Ans.$$

$$V_L = I_L R_L = 1(3.6) = 3.6 \text{ V}   \text{   } Ans.$$

Note how the Thevenin equivalent has simplified the solution of the given two-mesh network. Further, if the load R_L were changed, we would not have to recalculate the entire network.

NORTON'S THEOREM

Norton's theorem is used to simplify a network in terms of currents instead of voltages. For current analysis, this theorem can be used to reduce a network to a simple parallel circuit with a current source, which supplies a total line current that can be divided among parallel branches.

If the current I (Fig. 9-10) is a 4-A source, it supplies 4 A no matter what is connected across the output terminals a and b. With nothing connected across a and b, all the 4 A flows through shunt R. When a load resistance R_L is connected across a and b, then the 4-A current divides according to the current-division rule for parallel branches.

The symbol for a current source is a circle with an arrow inside (Fig. 9-10) to show the direction of current. This direction must be the same as the current produced by the polarity of the corresponding voltage source. Remember that a source produces current flow out from the positive terminal.

Norton's theorem states that any network connected to terminals a and b [Fig. 9-11a(1)] can be replaced by a single current source I_N in parallel with a single resistance R_N [Fig. 9-11a(2)]. I_N is equal to the short-circuit current through the ab terminals (the current that the network would produce through a and b with a short

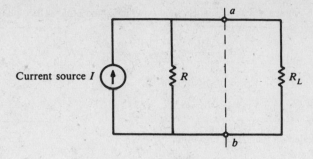

Fig. 9-10 I source with parallel R

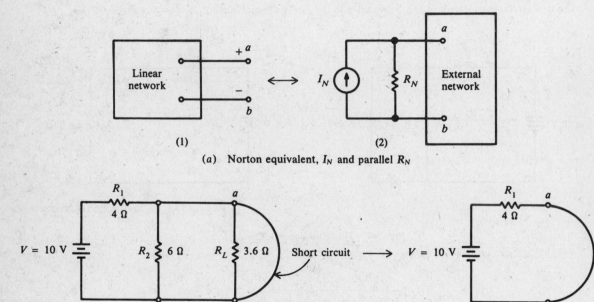

(1)   (2)

(a)   Norton equivalent, I_N and parallel R_N

(b)

Fig. 9-11a, b

circuit across these two terminsals). R_N is the resistance at terminals a and b, looking back from the open ab terminals. The value of the single resistor is the same for both the Norton and Thevenin equivalent circuits.

Example 9.7   Calculate the current I_L (see Fig. 9-9a) by Norton's theorem. (This was solved in Example 9.6 by Thevenin's theorem.)

Step 1.   Find I_N. Short-circuit across ab terminals (Fig. 9-11b). A short circuit across ab short-circuits R_L and the parallel R_2. Then the only resistance in the circuit is R_1 in series with source V.

$$I_N = \frac{V}{R_1} = \frac{10}{4} = 2.5 \text{ A}$$

Step 2.   Find R_N. Open terminals ab and short-circuit V (Fig. 9-11c). R_1 and R_2 are in parallel, so

$$R_N = \frac{4(6)}{4+6} = \frac{24}{10} = 2.4 \ \Omega$$

Note that R_N is the same as R_{Th}.

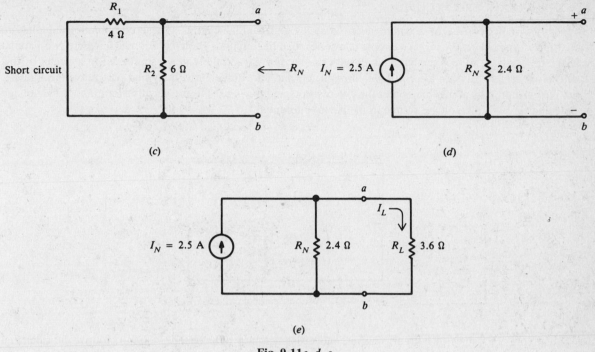

Fig. 9-11c, d, e

The Norton equivalent is then as in Fig. 9-11*d*. The arrow on the current source shows the direction of conventional current from terminal *a* to terminal *b*, as in the original circuit.

Step 3. Find I_L. Reconnect R_L to *ab* terminals (Fig. 9-11*e*). The current source still delivers 2.5 A, but now the current divides between the two branches R_N and R_L.

$$I_L = \frac{R_N}{R_N + R_L} I_N = \frac{2.4}{2.4 + 3.6} 2.5 = \frac{2.4}{6} 2.5 = 1 \text{ A} \qquad Ans.$$

This value is the same load current calculated in Example 9.6. Also, V_L can be calculated as $I_L R_L$, or (1 A)(3.6 Ω) = 3.6 V.

We therefore see that the Thevenin equivalent circuit (Fig. 9-12*a*) corresponds to the Norton equivalent circuit (Fig. 9-12*b*). So a general voltage source with a series resistance (Fig. 9-12*a*) can be converted to an equivalent current source with the same resistance in parallel (Fig. 9-12*b*). Divide the general source *V* by its series resistance *R* to find the value of *I* for the equivalent current source shunted by the same resistance *R*; that is, $I_N = V_{\text{Th}}/R_{\text{Th}}$.

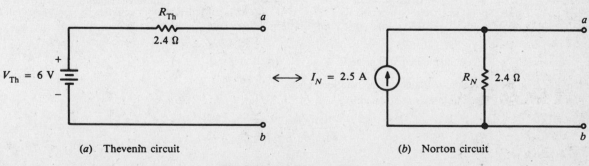

Fig. 9-12 Equivalent circuits

SERIES–PARALLEL CIRCUITS

Many circuits consist of a combination of series and parallel circuits. These combination circuits are called *series–parallel circuits*. An example of a series–parallel circuit is shown in Fig. 9-13, where two parallel resistors R_2 and R_3 are connected in series with the resistor R_1 and the voltage source V. In a circuit of this type, the current I_T divides after it flows through R_1, and part flows through R_2 and part flows through R_3. Then the current joins at the junction of the two resistors and flows back to the negative terminal of the voltage source and through the voltage source to the positive terminal.

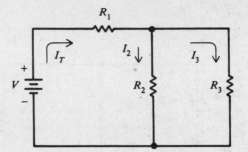

Fig. 9-13   A series–parallel circuit

In solving for values of current, voltage, and resistance in a series–parallel circuit, follow the rules that apply to a series circuit for the series part of the circuit, and follow the rules that apply to a parallel circuit for the parallel part of the circuit. Solving series–parallel circuits is simplified if all parallel and series groups are first reduced to single equivalent resistances and the circuits redrawn in simplified form. The redrawn circuit is called an *equivalent circuit*.

There are no general formulas for the solution of series–parallel circuits because there are so many different forms of these circuits.

Example 9.8   Find the total resistance, total circuit current, and branch currents of the circuit shown in Fig. 9-14a.

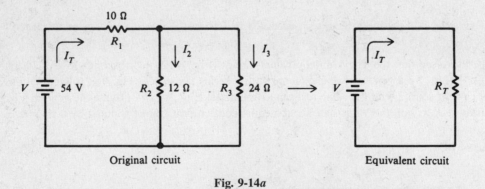

Original circuit   Equivalent circuit

Fig. 9-14a

It is best to solve combination circuits in steps:

Step 1.   Find the equivalent resistance of the parallel branch.

$$R_p = \frac{R_2 R_3}{R_2 + R_3} = \frac{12(24)}{12 + 24} = \frac{288}{36} = 8 \ \Omega$$

The equivalent circuit reduces to a series circuit (Fig. 9-14b).

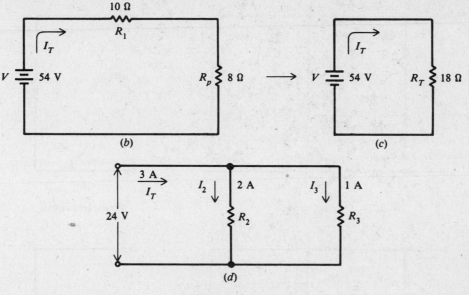

Fig. 9-14*b*, *c*, *d*

Step 2.   Find the resistance of the equivalent series circuit.

$$R_T = R_1 + R_p = 10 + 8 = 18\ \Omega \qquad Ans.$$

The equivalent circuit reduces to a single voltage source and a single resistance (Fig. 9-14*c*).

Step 3.   Find I_T. (I_T is the actual current being supplied in the original series–parallel circuit.)

$$I_T = \frac{V}{R_T} = \frac{54}{18} = 8\ \text{A} \qquad Ans.$$

Step 4.   Find I_2 and I_3. The voltage across R_2 and R_3 is equal to the applied voltage V less the voltage drop across R_1. See Fig. 9-14*d*.

$$V_2 = V_3 = V - I_T R_1 = 54 - (3 \times 10) = 24\ \text{V}$$

Then

$$I_2 = \frac{V_2}{R_2} = \frac{24}{12} = 2\ \text{A} \qquad Ans.$$

$$I_3 = \frac{V_3}{R_3} = \frac{24}{24} = 1\ \text{A} \qquad Ans.$$

or, by KCL,

$$I_T = I_1 = I_2 + I_3$$
$$I_3 = I_T - I_2 = 3 - 2 = 1\ \text{A}$$

Example 9.9   Find the total resistance R_T (Fig. 9-15*a*).

Step 1.   Add series resistances in each branch (Fig. 9-15*b*).

Branch *ab*: $R_1 + R_2 = 5 + 10 = 15\ \Omega$
Branch *cd*: $R_3 + R_4 = 6 + 9 = 15\ \Omega$
Branch *ef*: $R_5 + R_6 + R_7 = 8 + 5 + 2 = 15\ \Omega$

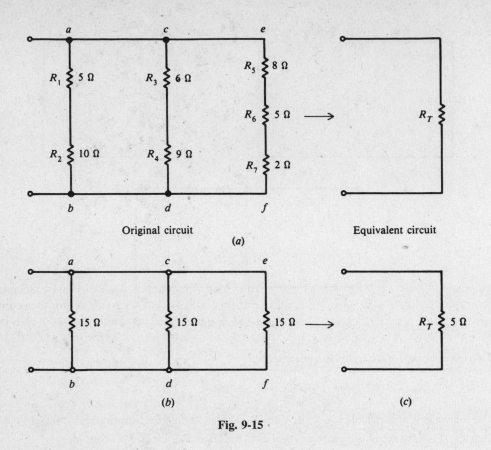

Fig. 9-15

Step 2. Find R_T. Each of three parallel resistors is 15 Ω. See Fig. 9-15c.

$$R_T = \frac{R}{N} = \frac{15}{3} = 5 \; \Omega \quad Ans.$$

WHEATSTONE BRIDGE CIRCUIT

The wheatstone bridge (Fig. 9-16) can be used to measure an unknown resistance R_x. Switch S_2 applies battery voltage to the four resistors in the bridge. To balance the bridge, the value of R_3 is varied. Balance is indicated by zero current in galvanometer G when switch S_1 is closed.

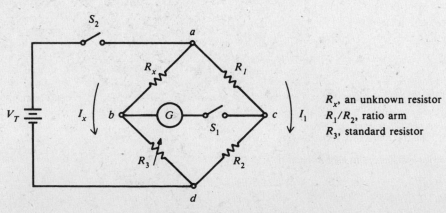

R_x, an unknown resistor
R_1/R_2, ratio arm
R_3, standard resistor

Fig. 9-16 Wheatstone bridge

When the bridge is balanced, points b and c must be at the same potential. Therefore

$$I_x R_x = I_1 R_1 \qquad (1)$$

$$I_x R_3 = I_1 R_2 \qquad (2)$$

Divide Eqs. (1) and (2). Note that I_x and I_1 cancel.

$$\frac{\not{I}_x R_x}{\not{I}_x R_3} = \frac{\not{I}_1 R_1}{\not{I}_1 R_2}$$

$$\frac{R_x}{R_3} = \frac{R_1}{R_2}$$

Solve for R_x.

$$R_x = \frac{R_1}{R_2} R_3 \qquad (9\text{-}7)$$

The ratio arm of the bridge is R_1/R_2. A rotary switching arrangement is used often in commercial bridges to adjust the ratio arm over a wide range of ratios. The bridge is balanced by varying R_3 for zero current in the meter. A decade box, where resistance can be varied in small ohmic steps up to as high as 10 KΩ, is commonly used as R_3. The value of R_x can be read directly from the calibrated scale of the rheostat when R_3 is adjusted for balance.

When current flows through the meter path bc, the bridge circuit is unbalanced and must be analyzed by Kirchhoff's laws or network theorems.

Example 9.10   An unknown resistance is to be measured by the Wheatstone bridge. If the ratio of R_1/R_2 is 1/100 and R_2 is 352 Ω when the bridge is balanced, find the value of the unknown resistance.
   Substitute known values into Eq. ($9\text{-}7$).

$$R_x = \frac{R_1}{R_2} R_3 = \frac{1}{100} 352 = 3.52 \ \Omega \qquad Ans.$$

MAXIMUM POWER TRANSFER

The maximum power is supplied by the voltage source and received by the load resistor if the value of the resistor equals the value of the internal resistance of the voltage source (Fig. 9-17). For maximum power transfer, then

$$R_L = R_i$$

and power received at the load is

$$P_L = I^2 R_L \qquad \text{where} \qquad I = \frac{V}{R_i + R_L}$$

Example 9.11   If a 10-V battery has an internal resistance of $R_i = 5 \ \Omega$, what is the maximum power that can be delivered to the load resistor (Fig. 9-18)?
   For maximum power transfer,

$$R_L = R_i = 5 \ \Omega$$

$$I = \frac{V}{R_i + R_L} = \frac{10}{5 + 5} = \frac{10}{10} = 1 \ \text{A}$$

$$P_L = I^2 R_L = 1^2 (5) = 5 \ \text{W} \qquad Ans.$$

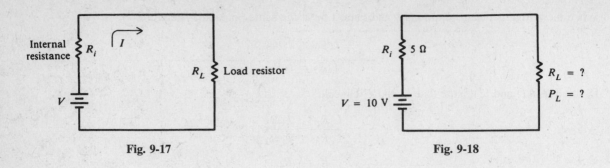

Fig. 9-17   Fig. 9-18

Example 9.12   For Example 9.11, prepare a table of values of power delivered to the load when the load resistance R_L is 1 Ω, 3 Ω, 4 Ω, 5 Ω, 6 Ω, 7 Ω, and 10 Ω.

Power Delivered to R_L

R_L, Ω	R_i, Ω	$I = \dfrac{V}{R_i + R_L}$, A	$P_L = I^2 R_L$, W
1	5	1.67	2.79
3	5	1.25	4.69
4	5	1.11	4.93
5	5	1.00	5.00 (max. power)
6	5	0.91	4.97
7	5	0.83	4.82
10	5	0.67	4.49

Notice that when $R_L = R_i = 5$ Ω, the maximum power of 5 W is transferred to the load.

LINE-DROP CALCULATIONS

The connecting wires are generally very short in electric circuits. Because the resistance of these short lengths is low, it was neglected in previous calculations. However, in home and factory electrical installations, where long lines of wires or feeders are used, the resistance of these long lengths must be included in all calculations.

The voltage drop across the resistance of the line wires is called the *line drop*. For example, if a generator delivers 120 V but the voltage available at a motor some distance away is only 117 V, then there has been a line drop in voltage of 3 V.

We must be careful about specifying the kind and size of wires used in any installation. If the wires are incorrectly chosen, then the line drop may be too large so that the voltage available to an electrical apparatus will be too low for proper operation (refer to the section on wire measurement in Chapter 4).

Example 9.13   A lamp bank consisting of three lamps, each drawing 1.5 A, is connected to a 120-V source (Fig. 9-19). Each line wire has a resistance of 0.25 Ω. Find the line drop, line power loss, and voltage available at the load.

Step 1.   Find the line current I_l.

$$I_l = I_1 + I_2 + I_3 = 1.5 + 1.5 + 1.5 = 4.5 \text{ A}$$

Step 2.   Find the resistance of the line wires R_l. Since the line wires are in series,

$$R_l = R_1 + R_2 = 0.25 + 0.25 = 0.5 \text{ Ω}$$

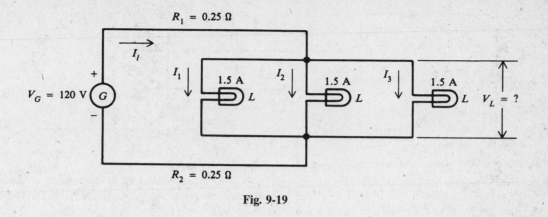

$R_1 = 0.25\ \Omega$

$V_G = 120\ \text{V}$ G

1.5 A L 1.5 A L 1.5 A L $V_L = ?$

$R_2 = 0.25\ \Omega$

Fig. 9-19

Step 3.   Find the line drop by ohm's law.

$$V_l = I_l R_l = 4.5(0.5) = 2.25\ \text{V} \qquad Ans.$$

Step 4.   Find the line power loss.

$$P_l = I_l^2 R_l = (4.5)^2(0.5) = 10.1\ \text{W} \qquad Ans.$$

Step 5.   Find voltage available at load.

$$V_L = V_G - V_l = 120 - 2.25 = 117.75 = 117.8\ \text{V} \qquad Ans.$$

Example 9.14   A bank of lathes is operated by individual motors in a machine shop (Fig. 9-20). The motors draw a total of 60 A at 110 V from the distributing panel box. What is the smallest size copper wire required for the two-wire line between the panel box and the switchboard, located 100 ft away, if the switchboard voltage is 115 V?

Step 1.   Find the line drop between the switchboard and the panel box.

$$V_l = V_G - V_L = 115 - 110 = 5\ \text{V}$$

Step 2.   Find the line resistance R_l for this drop.

$$V_l = I_l R_l$$

$$R_l = \frac{V_l}{I_l} = \frac{5}{60} = 0.0833\ \Omega$$

Step 3.   Find the circular-mil area of the wire that has this resistance. Since there are two wires, $l = 2L = 2(100) = 200$ ft. Use resistivity formula to solve for area A of wire.

$$R_l = \rho \frac{l}{A} \qquad\qquad (4\text{-}5)$$

$$A = \frac{\rho l}{R_l}$$

$$\rho = 10.4 \quad \text{(from Table 4-2)} \qquad l = 200\ \text{ft} \quad R_l = 0.0833\ \Omega$$

$$A = \frac{10.4(200)}{0.0833} = 25\ 000\ \text{CM}$$

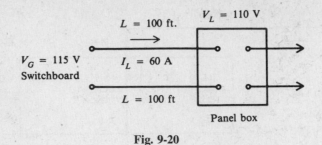

Fig. 9-20

Step 4.   Find the gauge number of the wire whose circular-mil area is larger than 25 000 CM. Refer to Table 4-1. Read down column 3, circular-mil area, until you get to a number just larger than 25 000 CM. The number is 26 250 CM. Read left to column 1 to find gauge No. 6.   *Ans.*

THREE-WIRE DISTRIBUTION SYSTEMS

The basic circuit described so far has been a two-wire circuit to a lamp, motor, or other load device. The basic two-wire circuit uses 120 V to supply the many household and factory load devices designed for that voltage. The three-wire circuit was developed to reduce the problems of voltage drop and power loss in the lines while still providing 120-V supply. The three wires of Fig. 9-21, including the grounded neutral, can be used for either 240 or 120 V. From either red or black high side to neutral, 120 V is available for separate branch circuits to the lights and outlets. Across the red and black wires, 240 V is available for high-power appliances such as a freezer or automatic clothes dryer. This 120/240-V three-wire distribution with a grounded neutral is called the *Edison system.*

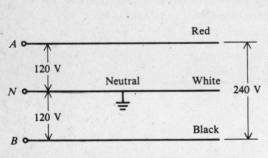

Fig. 9-21   Three-wire distribution system

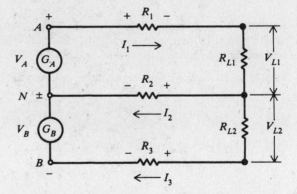

Fig. 9-22   Three-wire system with two voltage sources

The Edison System with Two Voltage Sources

This system (Fig. 9-22) has two direct-current generators, G_A and G_B, with load voltages V_{L1} and V_{L2}. R_1, R_2, and R_3 are the resistances of the line wires between the generators and the load. R_{L1} and R_{L2} are the load resistances. I_1, I_2, and I_3 are the currents in the line wires.

I_1 and I_3 will always flow in the direction indicated. I_2 may flow in either direction. The actual direction of flow will be determined by the sign of I_2. A + sign for I_2 means that it actually flows in the direction indicated, while a − sign means I_2 flows in the reverse direction.

By KCL and KVL, we can write

$$I_3 = I_1 - I_2 \qquad\qquad (1)$$

$$V_A - I_1 R_1 - I_1 R_{L1} - I_2 R_2 = 0 \tag{2}$$

$$V_B + I_2 R_2 - I_3 R_{L2} - I_3 R_3 = 0 \tag{3}$$

If the three wires are equal in size, as they usually are, their resistances are equal. So $R = R_1 = R_2 = R_3$. Equations (1), (2), and (3) can be solved to find the line currents.

$$I_1 = \frac{R V_B + V_A (2R + R_{L2})}{(R + R_{L1})(2R + R_{L2}) + R(R_{L2} + R)} \tag{9-8}$$

$$I_2 = \frac{I_1 (R_{L2} + R) - V_B}{2R + R_{L2}} \tag{9-9}$$

$$I_3 = I_1 - I_2 \tag{9-10}$$

Example 9.15   A three-wire 240/120-V circuit has line resistances of 0.5 Ω per line and load resistors $R_{L1} = 10\ \Omega$ and $R_{L2} = 5\ \Omega$ (Fig. 9-23). V_A and V_B are 120 V each. Find the currents and voltages at the loads.

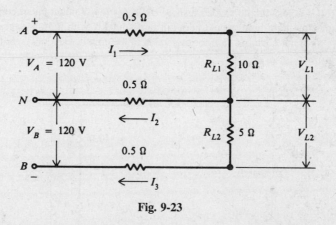

Fig. 9-23

Step 1.   Solve for I_1, I_2, and I_3. Substitute known values into the line current equations.

$$I_1 = \frac{R V_B + V_A (2R + R_{L2})}{(R + R_{L1})(2R + R_{L2}) + R(R_{L2} + R)} \tag{9-8}$$

$$= \frac{0.5(120) + 120(1.0 + 5)}{(0.5 + 10)(1.0 + 5) + 0.5(5 + 0.5)} = \frac{60 + 720}{63 + 2.75} = \frac{780}{65.75} = 11.9\ \text{A} \qquad Ans.$$

$$I_2 = \frac{I_1 (R_{L2} + R) - V_B}{2R + R_{L2}} \tag{9-9}$$

$$= \frac{11.9(5 + 0.5) - 120}{1.0 + 5} = \frac{-54.55}{6} = -9.1\ \text{A} \qquad Ans.$$

I_2 is thus flowing in opposite direction to that shown (Fig. 9-23).

$$I_3 = I_1 - I_2 \tag{9-10}$$

$$I_3 = 11.9 - (-9.1) = 11.9 + 9.1 = 21\ \text{A} \qquad Ans.$$

Note that because I_3 has the higher current, the heavier load is connected between the negative (lower) line and the neutral, and the neutral carries current away from the generator. If the 120-V loads were balanced on each side of the neutral ($R_{L1} = R_{L2}$), the neutral would carry no current ($I_2 = 0$).

Step 2. Find voltage across the loads by Ohm's law.

$$V_{L1} = I_1 R_{L1} = 11.9(10) = 119 \text{ V} \qquad Ans.$$

$$V_{L2} = I_3 R_{L2} = 21(5) = 105 \text{ V} \qquad Ans.$$

Example 9.16 Assume that the fuse in the neutral line is blown, opening the neutral line in Fig. 9-23. What now are the voltages across the two loads?

The connection is now reduced to a two-wire circuit (Fig. 9-24). By Ohm's law,

$$I = \frac{240}{0.5 + 10 + 5 + 0.5} = \frac{240}{16} = 15 \text{ A}$$

$$V_{L1} = I R_{L1} = 15(10) = 150 \text{ V} \qquad Ans.$$

$$V_{L2} = I R_{L2} = 15(5) = 75 \text{ V} \qquad Ans.$$

The high voltage of 150 V across the 10-Ω load resistor could destroy the equipment connected at that point in the circuit. If the load were lights, the light bulbs would burn out fairly quickly with a 31-V overage ($150 - 119 = 31$). For this reason, the neutral wire must not contain a fuse for protection of the circuit. Generally, electrical codes forbid fuses in the neutral line.

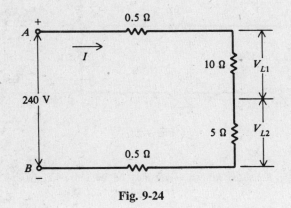

Fig. 9-24

Solved Problems

9.1 Reduce the bridge circuit (Fig. 9-25*a*) to a single equivalent input resistance at terminals *a* and *d*.

   Step 1. Convert Y network *bcd* to its equivalent Δ network. Use Rule 2 and visual aid. (See Fig. 9-25*b*.)

$$R_1 = \frac{10(10) + 10(10) + 10(10)}{10} = \frac{300}{10} = 30 \ \Omega$$

$$R_2 = \frac{300}{10} = 30 \ \Omega$$

$$R_3 = \frac{300}{10} = 30 \ \Omega$$

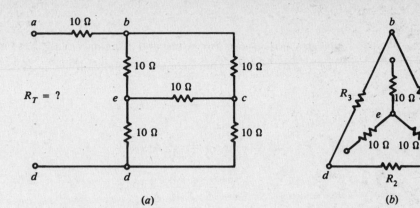

(a)   (b)

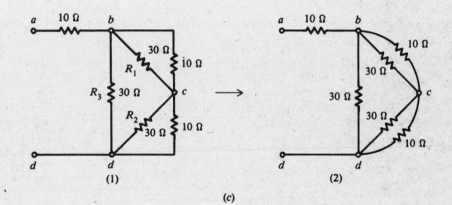

(1)   (2)

(c)

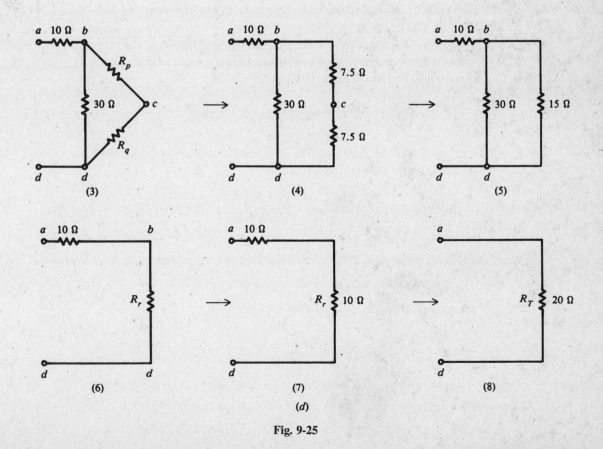

(3)   (4)   (5)

(6)   (7)   (8)

(d)

Fig. 9-25

Step 2.   Redraw Y as its Δ equivalent and connect it to the rest of the original circuit. (See Fig. 9-25c.)

Step 3.   Reduce the circuit. The two 10- and 30-Ω branches are in parallel. (See Fig. 9-25d.)

$$R_p = \frac{10(30)}{10 + 30} = \frac{300}{40} = 7.5\ \Omega$$

$$R_q = \frac{10(30)}{10 + 30} = \frac{300}{40} = 7.5\ \Omega$$

$$R_r = \frac{30(15)}{30 + 15} = \frac{450}{45} = 10\ \Omega$$

$$R_T = 10 + 10 = 20\ \Omega \qquad Ans.$$

9.2   Find the equivalent resistance R_T and output voltage V_o of a network with a bridged T form (Fig. 9-26a).

Step 1.   Transform the T (or Y) network into its equivalent Δ. Use the rule for Y to Δ conversion (Fig. 9-26b).

$$R_1 = \frac{2(2) + 2(2) + 2(2)}{2} = \frac{12}{2} = 6\ \Omega$$

$$R_2 = \frac{12}{2} = 6\ \Omega$$

$$R_3 = \frac{12}{2} = 6\ \Omega$$

Step 2.   Redraw T as its Δ equivalent and connect it to the remainder of the original circuit (Fig. 9-26c).

Step 3.   Redraw the circuit to show more clearly the two parallel branches, each containing two 6-Ω resistances. Then reduce the circuit until you get a single equivalent resistance R_T (Fig. 9-26c and e). Two 6-Ω resistances in parallel are equal to $6/2 = 3\ \Omega$.

$$R_T = \frac{6}{2} = 3\ \Omega \qquad Ans.$$

Step 4.   Solve for V_o by voltage distribution of 10 V. Look at circuit (3) in Fig. 9-26d.
By the voltage-division rule,

$$V_o = (\text{resistance ratio})(V_{ad}) = \left(\frac{3\ \Omega}{3\ \Omega + 3\ \Omega}\right)(10\ \text{V}) = \frac{3}{6}10 = 5\ \text{V} \qquad Ans.$$

9.3   For a two-delta bridge circuit (Fig. 9-27a), find the values of current through all the resistors.

Step 1.   Find the equivalent resistance between terminals a and d.

(a)   Transform Δ abc to its equivalent Y (Fig. 9-27b).

$$R_a = \frac{3(9)}{3 + 6 + 9} = \frac{27}{18} = 1.5\ \Omega \quad R_b = \frac{9(6)}{18} = \frac{54}{18} = 3\ \Omega \quad R_c = \frac{3(6)}{18} = \frac{18}{18} = 1\ \Omega$$

(b)   Connect the Y equivalent circuit to the original circuit (Fig. 9-27c).

(c)   Reduce the series–parallel circuit to its equivalent (Fig. 9-27d).

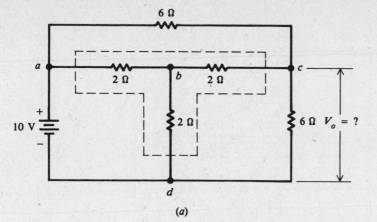

(a)

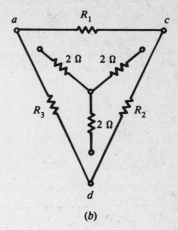

(b)

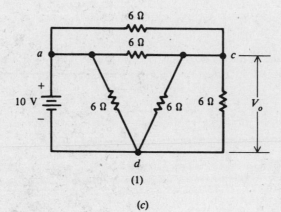

(1)

(c)

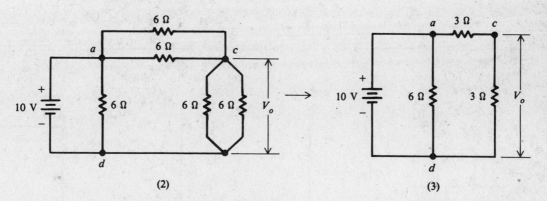

(2)   (3)

(d)

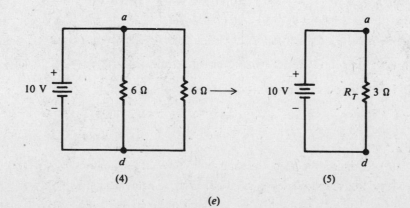

(4)   (5)

(e)

Fig. 9-26

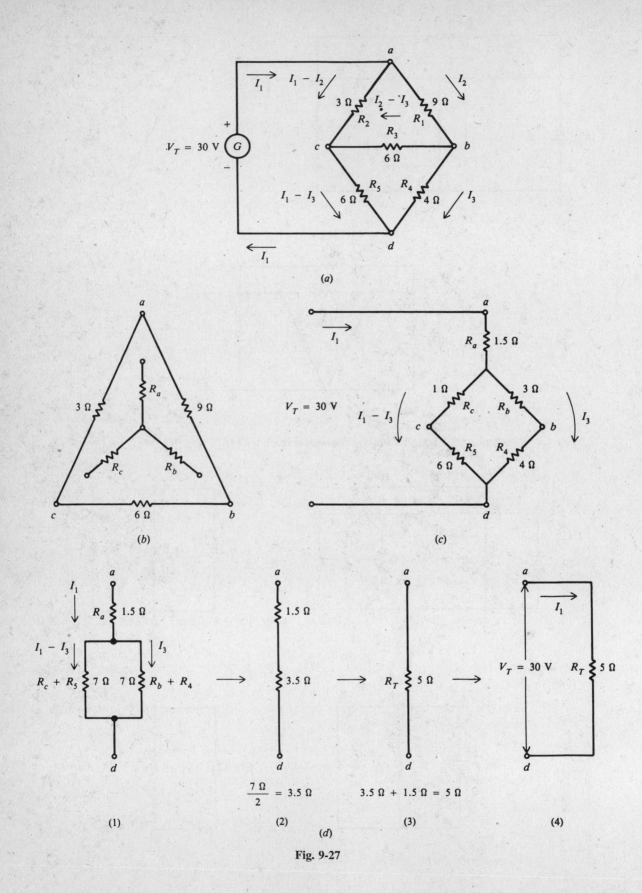

Fig. 9-27

Step 2. Find I_1 and I_3.

$$I_1 = \frac{V_T}{R_T} = \frac{30}{5} = 6\,\text{A}$$

I_1 divides equally into the two 7-Ω parallel branches so that

$$I_3 = \frac{7}{14}6 = \frac{6}{2} = 3\,\text{A}$$

Step 3. Solve for I_2 by KVL, $\sum V = 0$. Trace circuit *abca* in the clockwise direction (Fig. 9-27*a*).

$$-I_2 R_1 - (I_2 - I_3)R_3 + (I_1 - I_2)R_2 = 0$$

Substitute $R_1 = 9\,\Omega$, $R_2 = 3\,\Omega$, $R_3 = 6\,\Omega$, $I_1 = 6\,\text{A}$, and $I_3 = 3\,\text{A}$.

$$-9I_2 - (I_2 - 3)(6) + (6 - I_2)(3) = 0$$

$$-9I_2 - 6I_2 + 18 + 18 - 3I_2 = 0$$

$$-18I_2 + 36 = 0$$

$$-18I_2 = -36$$

$$I_2 = 2\,\text{A}$$

Step 4. Show currents through individual resistors (Fig. 9-27*a*).

$$I_1 - I_2 = 6 - 2 = 4\,\text{A} \qquad Ans.$$

$$I_2 = 2\,\text{A} \qquad Ans.$$

$$I_1 - I_3 = 6 - 3 = 3\,\text{A} \qquad Ans.$$

$$I_2 - I_3 = 2 - 3 = -1\,\text{A} \qquad Ans.$$

$$I_3 = 3\,\text{A} \qquad Ans.$$

The negative sign for $I_2 - I_3$ indicates that the current actually flows in the direction of c to b.

9.4 The superposition principle can be applied to a voltage-divider circuit with two sources (Fig. 9-28*a*). Find V_p.

The method is to calculate V_p contributed by each source separately and then add (superimpose) these voltage algebraically.

Step 1. Short-circuit G_2 and find $V_{p,G1}$ due only to source G_1 (Fig. 9-28*b*).

R_1 and R_2 form a series voltage divider for the V_1 source. $V_{p,G1}$ is the same as the voltage across R_2. To find $V_{p,G1}$ use the voltage-divider formula:

$$V_{p,G1} = \frac{R_2}{R_1 + R_2}V_1 = \frac{40}{20 + 40}240 = \frac{40}{60}240 = 160\,\text{V}$$

$V_{p,G1}$ is positive because V_1 is positive.

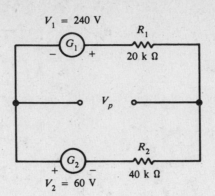

(a) Voltage divider circuit with two voltage sources

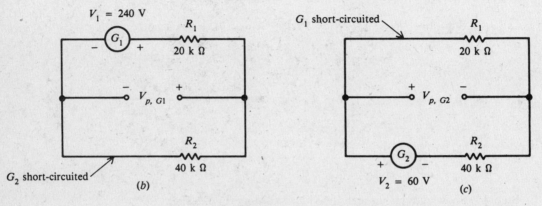

Fig. 9-28

Step 2. Short-circuit G_1 and find $V_{p,G2}$ due only to source G_2 (Fig. 9-28c).

R_1 and R_2 form a series voltage divider again, but here $V_{p,G2}$ is the same as the voltage across R_1. To find $V_{p,G2}$ we use the voltage-divider formula, but this time we have a negative voltage source G_2.

$$V_{p,G2} = \frac{R_1}{R_1 + R_2} V_2 = \frac{20}{20 + 40}(-60\,\text{V}) = \frac{20}{60}(-60\,\text{V}) = -20\,\text{V}$$

Step 3. To find V_p add the voltages calculated.

$$V_p = V_{p,G1} + V_{p,G2} = 160 - 20 = 140\,\text{V} \qquad Ans.$$

9.5 Find the current through the load resistor R_L in the two-generator source circuit (Fig. 9-29a) by superposition. R_1 and R_2 are the internal resistances of the generators.

Step 1. Find the current in R_L due to G_1 alone, designated $I_{L,G1}$.

(a) Short-circuit G_2 and reduce the circuit (Fig. 9-29b).

$$R_3 = R_1 + \frac{R_2 R_L}{R_2 + R_L} = 1 + \frac{1(10)}{1 + 10} = 1 + \frac{10}{11} = 1.91\,\Omega$$

$$I_{1,G1} = \frac{V_1}{R_3} = \frac{120}{1.91} = 62.8\,\text{A}$$

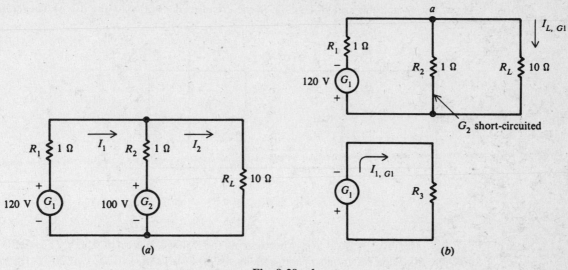

Fig. 9-29a, b

(b) Use current-divider formula to find $I_{L,G1}$ at point a.

$$I_{L,G1} = \frac{R_2}{R_2 + R_L} I_{1,G1} = \frac{1}{11} 62.8 = 5.71 \text{ A}$$

Step 2. Find the current in R_L due to G_2 alone, namely $I_{L,G2}$. Short-circuit G_1 and reduce the circuit (Fig. 9-29c).

$$R_4 = R_2 + \frac{R_1 R_L}{R_2 + R_L} = 1 + \frac{1(10)}{1 + 10} = 1.91 \ \Omega$$

$$I_{2,G2} = \frac{V_2}{R_4} = \frac{100}{1.91} = 52.4 \text{ A}$$

$$I_{L,G2} = \frac{R_1}{R_1 + R_L} I_{2,G2} = \frac{1}{11} 52.4 = 4.76 \text{ A}$$

Step 3. Add the individual currents algebraically.

$$I_L = I_{L,G1} + I_{L,G2} = 5.71 + 4.76 = 10.5 \text{ A} \qquad Ans.$$

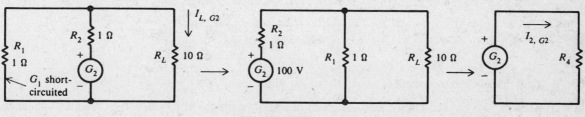

Fig. 9-29c

9.6   Find the load current I_L and the load voltage V_L in the circuit (Fig. 9-30*a*) by the use of Thevenin's theorem.

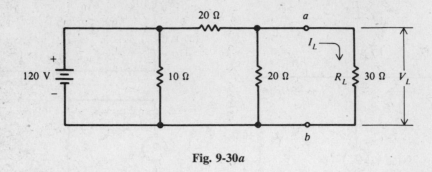

Fig. 9-30a

Step 1.   Find R_{Th}. Remove the load R_L. Short-circuit the voltage source of 120 V (Fig. 9-30*b*). Short-circuiting the battery also short-circuits the 10-Ω resistor, leaving two 20-Ω resistors in parallel.

$$R_{\text{Th}} = \frac{20}{2} = 10\ \Omega$$

Step 2.   Find V_{Th}. The two 20-Ω resistors are in series across the 120-V line (Fig. 9-30*a*). Since the voltage is the same across equal resistances and V_{Th} is the open-circuit voltage at *a* and *b* across the 20-Ω resistor,

$$V_{\text{Th}} = \frac{120}{2} = 60\ \text{V}$$

Step 3.   Draw the equivalent circuit with R_L and find I_L and V_L (Fig. 9-30*c*).

$$I_L = \frac{V_{\text{Th}}}{R_L + R_{\text{Th}}} = \frac{60}{30 + 10} = \frac{60}{40} = 1.5\ \text{A} \qquad Ans.$$

$$V_L = I_L R_L = 1.5(30) = 45\ \text{V} \qquad Ans.$$

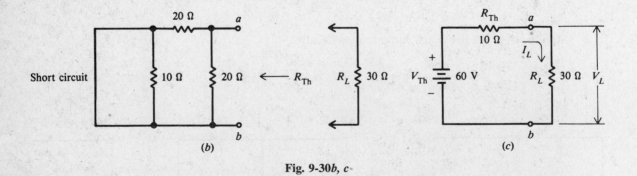

Fig. 9-30b, c

9.7   Find the Thevenin equivalent across R_L of the Wheatstone bridge network (Fig. 9-31*a*).

Step 1.   For greater clarity, move the voltage source inside the bridge and show the load outside the bridge (Fig. 9-31*b*).

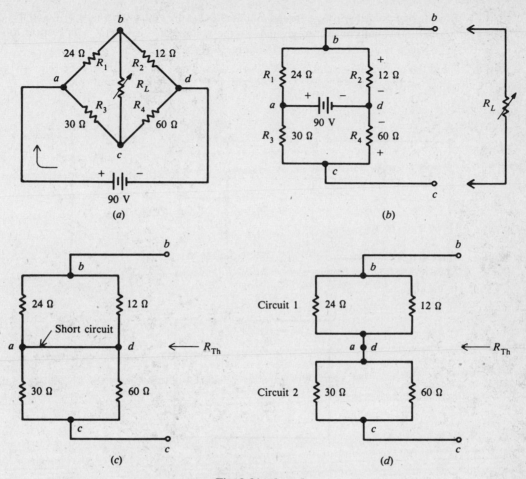

Fig. 9-31a, b, c, d

Step 2. Find R_{Th}. Short-circuit the 90-V source (Fig. 9-31c).

Redraw the circuit for simplicity (Fig. 9-31d). Find R_{eq1} for circuit 1, R_{eq2} for circuit 2, and then combine R_{eq1} and R_{eq2} in series to find R_{Th}.

$$R_{eq1} = \frac{24(12)}{24 + 12} = 8\ \Omega$$

$$R_{eq2} = \frac{30(60)}{30 + 60} = 20\ \Omega$$

$$R_{Th} = R_{eq1} + R_{eq2} = 8 + 20 = 28\ \Omega \qquad \textit{Ans.}$$

Step 3. Find V_{Th}. V_{Th} is the open-circuit voltage across terminals b and c and is equal to the algebraic sum of voltages across R_2 and R_4 (see Fig. 9-31e). By the voltage-division rule,

$$V_2 = \frac{12}{12 + 14}90 = \frac{1}{3}90 = 30\ V$$

$$V_4 = \frac{60}{60 + 30}90 = \frac{2}{3}90 = 60\ V$$

$$V_{Th} = V_4 - V_2 = 60 - 30 = 30\ V \qquad \textit{Ans.}$$

To show the polarity of V_{Th}, ground d and note that the voltage at b is $+30$ V and the voltage at c is $+60$ V with respect to ground. So the voltage at b is -30 V with respect to c. This can be seen if we ground point c. (See Fig. 9-31f.)

Step 4. Draw the equivalent circuit with R_L. Note the polarity of the source. (See Fig. 9-31g.)

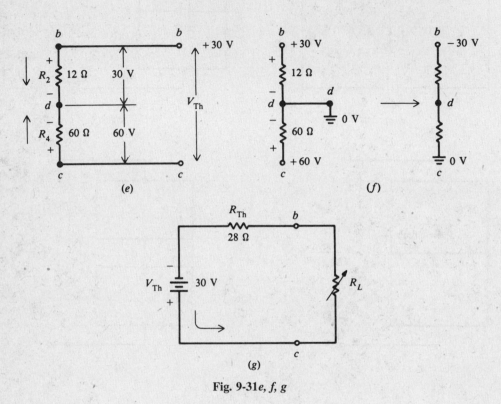

Fig. 9-31e, f, g

9.8 Convert the voltage source circuit (Fig. 9-32a) to its equivalent current source circuit. Prove that the two circuits are equivalent by calculating the voltage drop and current through a 10-Ω load resistor.

Step 1. Find the current source equivalent.

$$I = \frac{V}{R} = \frac{15}{5} = 3 \text{ A}$$

Shunt R is equal to the series R of 5 Ω. Therefore, the equivalent current source circuit is as shown in Fig. 9-32b.

Step 2. Add R_L at terminals a and b. Find I_L and V_L in each equivalent circuit and compare their values. From Fig. 9-32c:

$$I_L = \frac{V}{R + R_L} = \frac{15}{5 + 10} = 1 \text{ A}$$

$$V_L = I_L R_L = 1(10) = 10 \text{ V}$$

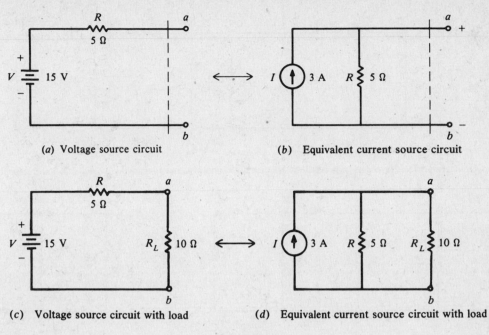

(a) Voltage source circuit　　　　　　　(b)　Equivalent current source circuit

(c)　Voltage source circuit with load　　　　(d)　Equivalent current source circuit with load

Fig. 9-32

From Fig. 9-32d:

$$I_L = \frac{R}{R + R_L} I = \frac{5}{5 + 10} 3 = \frac{1}{3} 3 = 1 \text{ A}$$

$$V_L = I_L R_L = 1(10) = 10 \text{ V}$$

The values of load current and load voltage are the same for each circuit.

9.9　　Conversion of voltage and current sources can often simplify circuits when there are two or more sources. Voltage sources are easier for series connections because we can add voltages, whereas current sources are easier for parallel connections because we can add currents. Find the current I_L through the middle load resistor R_L (Fig. 9-33a).

Step 1.　Convert voltage sources V_1 and V_2 into current sources.

$$I_1 = \frac{V_1}{R_1} = \frac{72}{9} = 8 \text{ A} \qquad \text{Shunt } R_1 = \text{series } R_1 = 9 \, \Omega$$

$$I_2 = \frac{V_2}{R_2} = \frac{24}{3} = 8 \text{ A} \qquad \text{Shunt } R_2 = \text{series } R_2 = 3 \, \Omega$$

Step 2.　Draw the equivalent current source circuit (see Fig. 9-33b). I_1 and I_2 can be combined for one equivalent current source I_T. Since they produce current in the same direction through R_L, they are added.

$$I_T = I_1 + I_2 = 8 + 8 = 16 \text{ A}$$

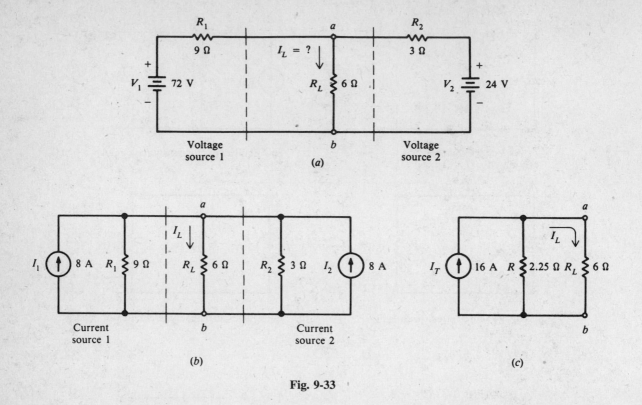

Fig. 9-33

The shunt R for the 16-A combined current source is the combined resistance of the 9-Ω R_1 and the 3-Ω R_2 in parallel. So

$$\text{Shunt } R = \frac{R_1 R_2}{R_1 + R_2} = \frac{9(3)}{9+3} = \frac{27}{12} = 2.25 \ \Omega$$

The circuit of Fig. 9-33b can be redrawn as shown in Fig. 9-33c.

Step 3. Find I_L. Use the current-divider formula for the 6- and 2.25-Ω branches.

$$I_L = \frac{2.25}{2.25 + 6} 16 = \frac{2.25}{8.25} 16 = 4.36 \ \text{A} \qquad \textit{Ans.}$$

9.10 Find the current I_L by converting the series current sources I_1 and I_2 into series voltage sources (Fig. 9-34a).

Step 1. Convert I_1 and I_2 into voltage sources.

$$V_1 = I_1 R_1 = 3(4) = 12 \ \text{V} \qquad \text{Shunt } R_1 = \text{series } R_1 = 4 \ \Omega$$

$$V_2 = I_2 R_2 = 4(2) = 8 \ \text{V} \qquad \text{Shunt } R_2 = \text{series } R_2 = 2 \ \Omega$$

Step 2. Draw the equivalent voltage source circuit (see Fig. 9-34b). The series voltages are added because they are series-aiding.

$$V_T = V_1 + V_2 = 12 + 8 = 20 \ \text{V}$$

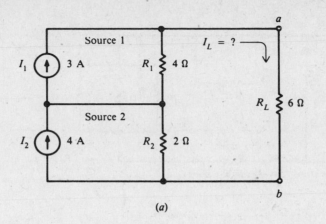

(a)

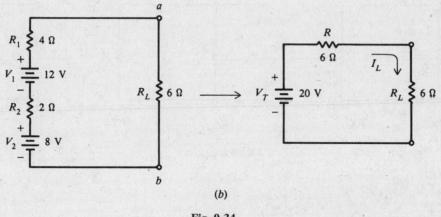

(b)

Fig. 9-34

The series resistances are added.

$$R = R_1 + R_2 = 4 + 2 = 6\ \Omega$$

Then by Ohm's law,

$$I_L = \frac{V_T}{R + R_L} = \frac{20}{6 + 6} = \frac{20}{12} = 1.67\ \text{A} \qquad Ans.$$

9.11 Find the total resistance R_T of the circuit shown in Fig. 9-35a.

> **Step 1.** Reduce the parallel resistances, R_2, R_3 and R_4, R_5 to a single resistance. Use formula for R/N. (See Fig. 9-35b.)

$$\frac{R_2}{2} = \frac{6}{2} = 3\ \Omega$$

$$\frac{R_4}{2} = \frac{10}{2} = 5\ \Omega$$

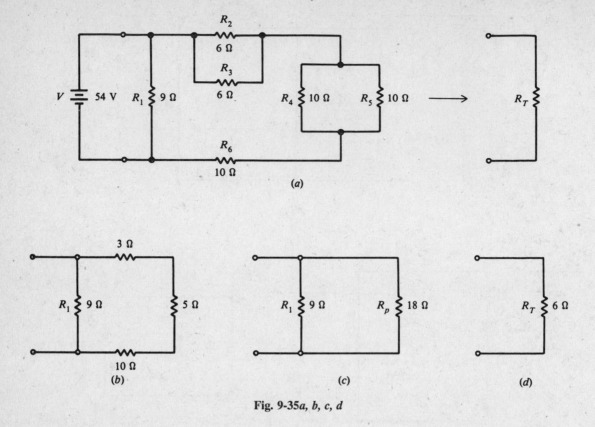

Fig. 9-35a, b, c, d

Step 2. Add series resistances. (See Fig. 9-35c.)

$$R_p = 10 + 5 + 3 = 18 \ \Omega$$

Step 3. Find R_T. (See Fig. 9-35d.)

$$R_T = \frac{R_1 R_p}{R_1 + R_p} = \frac{9(18)}{9 + 18} = 6 \ \Omega \qquad Ans.$$

9.12 For the circuit in Fig. 9-35a, determine the current values through all the resistors when the applied voltage is 54 V. Having found the equivalent resistances by steps in Problem 9.11, we shall work backward, starting with R_T and proceeding toward the original circuit.

Step 1. Find I_T (Fig. 9-35e).

$$I_T = \frac{54}{6} = 9 \ A$$

Step 2. Find I_1 and I_2 (Fig. 9-35f). By Ohm's law,

$$I_1 = \frac{V}{R_1} = \frac{54}{9} = 6 \ A$$

$$I_2 = \frac{V}{R_p} = \frac{54}{18} = 3 \ A$$

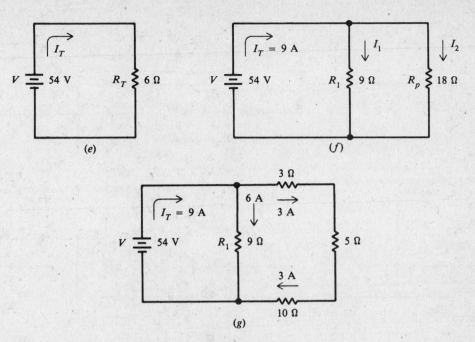

Fig. 9-35e, f, g

Step 3. Show the circuit equivalent. (See Fig. 9-35g.)

Step 4. Show the original circuit. (See Fig. 9-35h.)

Because the parallel resistances are equal, the current of 3 A divides equally into 1.5 A and 1.5 A.

$$I_1 = 6\,\text{A} \qquad I_4 = 1.5\,\text{A}$$

$$I_2 = 1.5\,\text{A} \qquad I_5 = 1.5\,\text{A} \qquad \textit{Ans.}$$

$$I_3 = 1.5\,\text{A} \qquad I_6 = 3\,\text{A}$$

9.13 Find the total resistance of the circuit (Fig. 9-36a).

Step 1. Reduce the circuit progressively from right to left. Add series resistances. (See Fig. 9-36b.)

$$R_A = R_6 + R_3 + R_7 = 30 + 40 + 50 = 120\,\Omega$$

Step 2. Combine parallel branches (Fig. 9-36c).

$$R_B = \frac{60\,(120)}{60 + 120} = 40\,\Omega$$

Step 3. Add series resistances (Fig. 9-36d).

$$R_C = 35 + 40 + 25 = 100\,\Omega$$

Step 4. Combine parallel branches (Fig. 9-36e).

$$R_D = \frac{300\,(100)}{300 + 100} = 75\,\Omega$$

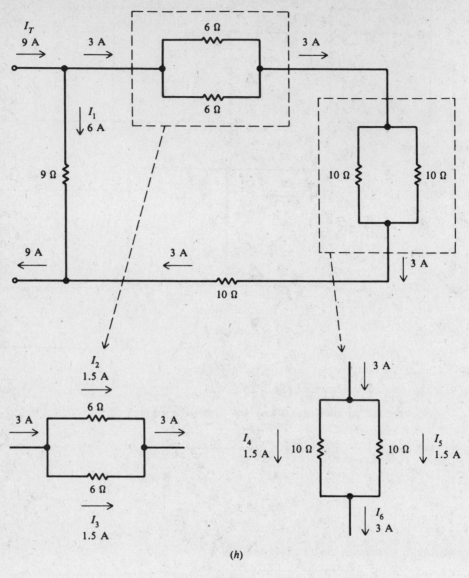

(h)

Fig. 9-35h

Step 5. Find R_T by combining series resistances (Fig. 9-36f).

$$R_T = 10 + 75 + 15 = 100 \,\Omega \qquad Ans.$$

9.14   In the Wheatstone bridge circuit (Fig. 9-37), the bridge is balanced. Calculate R_x, I_x, I_1, and each voltage.

Step 1.   Calculate R_x by Eq. (*9-7*).

$$R_x = \frac{R_1}{R_2} R_3 = \frac{1000}{10\,000} 42 = 4.2 \,\Omega \qquad Ans.$$

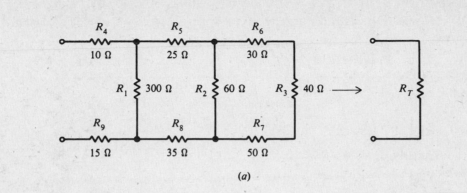

(a)

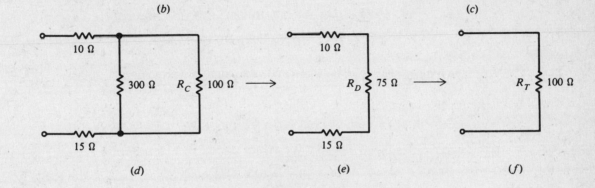

(b)   (c)

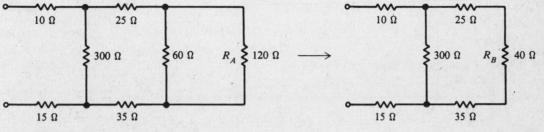

(d)   (e)   (f)

Fig. 9-36

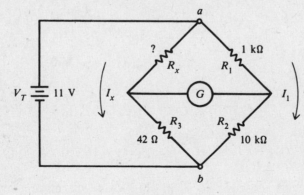

Fig. 9-37

Step 2. Calculate currents I_x and I_1. Express the voltage drop across terminals a and b.

$$I_x R_x + I_x R_3 = V_T$$

$$I_x(R_x + R_3) = V_T$$

$$I_x = \frac{V_T}{R_x + R_3} = \frac{11}{4.2 + 42} = \frac{11}{46.2} = 0.238 \text{ A} \qquad Ans.$$

Similarly, $I_1 R_1 + I_1 R_2 = V_T$

$$I_1(R_1 + R_2) = V_T$$

$$I_1 = \frac{V_T}{R_1 + R_2} = \frac{11}{1000 + 10\,000} = \frac{11}{11\,000} = 0.001 \text{ A} \qquad Ans.$$

Step 3. Find each voltage. By Ohm's law,

$$V_x = I_x R_x = 0.238(4.2) = 1 \text{ V} \qquad Ans.$$

$$V_1 = I_1 R_1 = 0.001(1000) = 1 \text{ V} \qquad Ans.$$

$$V_2 = I_1 R_2 = 0.001(10\,000) = 10 \text{ V} \qquad Ans.$$

$$V_3 = I_x R_3 = 0.238(42) = 10 \text{ V} \qquad Ans.$$

When the bridge is balanced, no current flows through the galvanometer, so that

$$V_x = V_1$$

$$1 \text{ V} = 1 \text{ V} \quad Check$$

and

$$V_3 = V_2$$

$$10 \text{ V} = 10 \text{ V} \quad Check$$

An alternative method to find each voltage is by the voltage-divider rule. With this method we need not solve for the currents.

$$V_1 = \frac{R_1}{R_1 + R_2} V_T = \frac{1}{1 + 10} 11 = \frac{1}{11} 11 = 1 \text{ V}$$

Then

$$V_2 = V_T - V_1 = 11 - 1 = 10 \text{ V}$$

Similarly,

$$V_x = \frac{R_x}{R_x + R_3} V_T = \frac{4.2}{4.2 + 42} 11 = \frac{4.2}{46.2} 11 = 1 \text{ V}$$

and

$$V_3 = V_T - V_x = 11 - 1 = 10 \text{ V}$$

9.15 Find the value of load resistance R_L that will provide the maximum power delivered to load (Fig. 9-38a). Also calculate maximum power P_L.

Step 1. Show the Thevenin equivalent circuit (Fig. 9-36b). R_1 and R_2 are in parallel.

$$R_{\text{Th}} = \frac{R_1 R_2}{R_1 + R_2} = \frac{5(20)}{5 + 20} = 4 \, \Omega$$

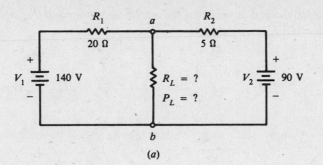

(a)

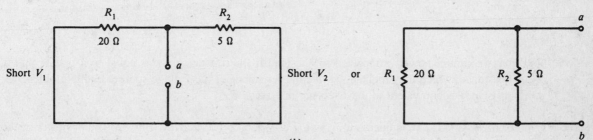

(b)

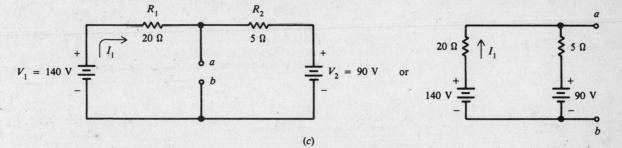

(c)

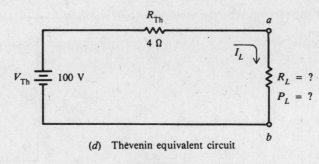

(d)   Thevenin equivalent circuit

Fig. 9-38

By Ohm's law (see Fig. 9-38c),

$$I_1 = \frac{V_1 - V_2}{R_1 + R_2} = \frac{140 - 90}{20 + 5} = \frac{50}{25} = 2 \text{ A}$$

$$V_{\text{Th}} = V_{\text{ab}} = V_1 - I_1 R_1 = 140 - 2(20) = 140 - 40 = 100 \text{ V}$$

The Thevenin equivalent circuit is shown in Fig. 9-38d.

Step 2. Find R_L and P_L. For maximum power to be delivered to the load, $R_L = R_{Th}$. Therefore,

$$R_L = 4\,\Omega \qquad Ans.$$

$$P_L = I_L^2 R_L = \left(\frac{V_{Th}}{R_{Th} + R_L}\right)^2 (R_L) = \left(\frac{100}{4+4}\right)^2 (4) = 625\,\text{W} \qquad Ans.$$

Or, more simply, since source voltage divides in half between R_{Th} and R_L,

$$V_L = 50\,\text{V} \quad \text{and} \quad P_L = \frac{V_L^2}{R_L} = \frac{50^2}{4} = 625\,\text{W}$$

Note that maximum power is received by the load if the resistance R_L is equal to a fixed value of series resistance, which may include the internal resistance of the voltage source. In this problem, the series resistance is equivalent to the Thevenin resistance R_{Th}.

9.16 A motor drawing 8333 W is operating from a 232-V source 100 ft away (Fig. 9-39). For power installations, the National Electrical Code permits a 5 percent drop. What is the line power loss? What is the minimum size copper wire that may be used for the line supplying the motor to avoid exceeding the 5 percent voltage drop?

Step 1. Find the minimum voltage at the motor load, V_L.

$$V_{\text{source}} = 232\,\text{V}$$

$$V_{\text{line drop}} = 0.05(232) = 11.6\,\text{V}$$

$$V_L = V_{\text{source}} - V_{\text{line drop}} = 232 - 11.6 = 220.4\,\text{V}$$

Step 2. Find the current drawn by the motor, I_L.

$$P_L = I_L V_L$$

$$I_L = \frac{P_L}{V_L} = \frac{8333}{220.4} = 37.8\,\text{A}$$

Step 3. Find the power loss P_l.

Line drop $V_l = 0.05(232) = 11.6\,\text{V}$

$I_L = 37.8\,\text{A}$

$P_l = V_l I_L$

$= 11.6(37.8) = 438\,\text{W} \qquad Ans.$

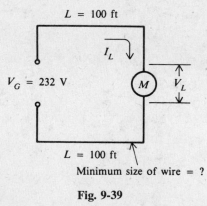

Fig. 9-39

Step 4. Find the resistance of the line wires, R_l.

$$V_l = I_L R_l$$

$$R_l = \frac{V_l}{I_L} = \frac{11.6}{37.8} = 0.307\,\Omega$$

Step 5. Find the circular-mil area of the wire that has this resistance.

$$R_l = \rho \frac{l}{A}$$ (4-5)

$$A = \rho \frac{l}{R_l}$$

$$\rho = 10.4 \quad \text{(Table 4-2)} \quad l = 2L = 200 \text{ ft} \quad R_l = 0.307 \ \Omega$$

$$A = \frac{10.4(200)}{0.307} = 6780 \text{ CM}$$

Step 6. Find the gauge number of the wire whose circular-mil area is larger than 6780 CM. Refer to Table 4-1; No. 11 wire is the minimum wire size that can be used to limit voltage drop to 5 percent.   *Ans.*

9.17 Find the voltage across the motor and across the lamp bank of the circuit (Fig. 9-40). The motor draws 4 A and the lamp banks 5 A. Resistances of the feeder line are indicated.

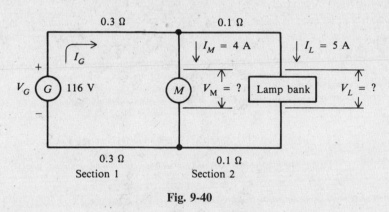

Fig. 9-40

Step 1. Find the current distribution.

$$I_G = I_M + I_L = 4 + 5 = 9 \text{ A}$$

In section 1 the current is 9 A, and in section 2 the current is 5 A.

Step 2. Find the line drop in each section.

Section 1: $$R_{11} = 0.3 + 0.3 = 0.6 \ \Omega$$
                          $$V_{11} = I_G R_{11} = 9(0.6) = 5.4 \text{ V}$$
Section 2: $$R_{12} = 0.1 + 0.1 = 0.2 \ \Omega$$
                          $$V_{12} = I_L R_{12} = 5(0.2) = 1 \text{ V}$$

Step 3. Find the load voltages.

$$V_M = V_G - V_{11} = 116 - 5.4 = 110.6 \text{ V} \qquad \text{Ans.}$$

$$V_L = V_M - V_{12} = 110.6 - 1 = 109.6 \text{ V} \qquad \text{Ans.}$$

9.18 Compare the line drop and load voltage of a three-wire system and a two-wire system. Use the circuit of Fig. 9-23 with the 10-Ω load for your calculations (see Fig. 9-41).

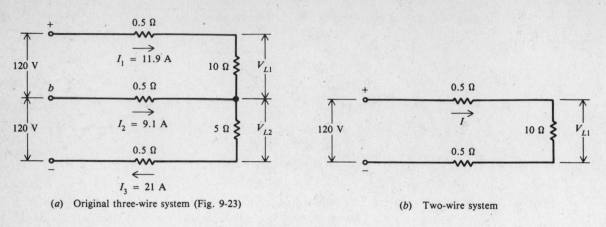

(a) Original three-wire system (Fig. 9-23)   (b) Two-wire system

Fig. 9-41

For three-wire system with respect to 10-Ω load (Fig. 9-41a):

$$V_{l1} = 11.9(0.5) = 5.95 \text{ V}$$
$$V_{l2} = 9.1(0.5) = 4.55 \text{ V}$$

Total line drop $= V_{l1} + V_{l2} = 5.95 + 4.55$
$$= 10.5 \text{ V}$$
$$V_{L1} = 11.9(10) = 119 \text{ V}$$

For two-wire system (Fig. 9-41b):

$$I = \frac{120}{0.5 + 10 + 0.5} = \frac{120}{11} = 10.9 \text{ A}$$

Total line drop $= 10.9(1) = 10.9 \text{ V}$

$$V_{L1} = 10.9(10) = 109 \text{ V}$$

Then,   Difference in line drop $= 10.9 - 10.5 = 0.4$ V   *Ans.*

Difference in load voltage $= 119 - 109 = 10$ V   *Ans.*

So we see that in this case a three-wire system has a line drop of 0.4 V less than that of a two-wire system and a higher load voltage by 10 V.

Supplementary Problems

9.19   Transform the Δ networks of Fig. 9-42a into Y networks. (*Hint*: Draw visual aid.)
Ans.   See Fig. 9-42b.

9.20   Transform the Y networks of Fig. 9-43a into Δ networks. (*Hint*: Draw visual aid.)
Ans.   See Fig. 9-43b.

9.21   Find the equivalent input resistance between terminals a and d for the bridge networks (Fig. 9-44).
Ans.   (a) $R_T = 10 \Omega$;   (b) $R_T = 11 \Omega$;   (c) $R_T = 5 \Omega$

9.22   If 50 V were applied between terminals a and d to the circuit shown in Fig. 9-44c, find the current in each resistor.   *Ans.* $I_{10 \Omega} = 4.5$ A; $I_{8 \Omega} = 5.5$ A; $I_{2 \Omega} = 0.5$ A; $I_{1 \Omega} = 5$ A; $I_{1.2 \Omega} = 5$ A

9.23   Find the equivalent resistance and output voltage V_o of a bridged T network (Fig. 9-45).
Ans. $R_T = 25 \Omega$; $V_o = 7.5$ V

9.24   Find the equivalent input resistance between terminals a and d (Fig. 9-46).   *Ans.* $R_T = 37 \Omega$

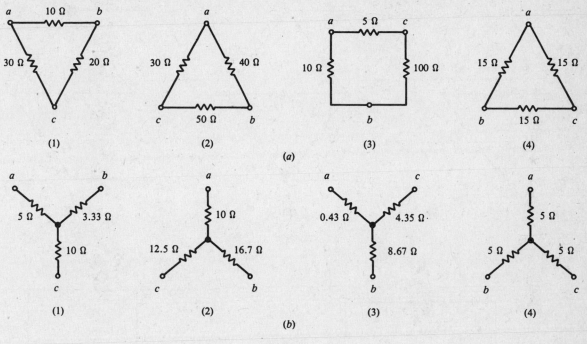

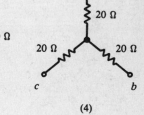

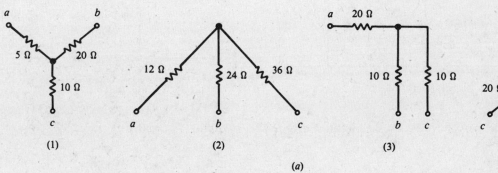

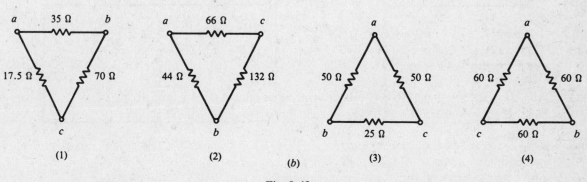

Fig. 9-42

Fig. 9-43

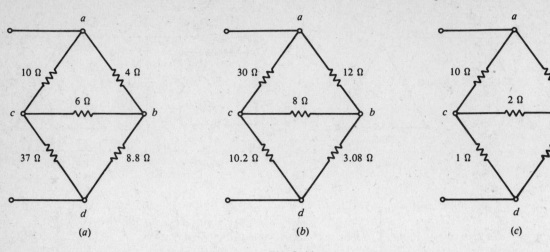

Fig. 9-44

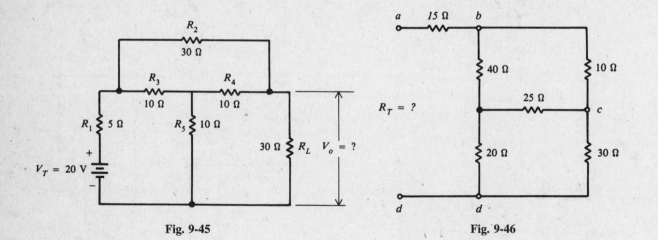

Fig. 9-45   Fig. 9-46

9.25   Determine the voltage V_p by superposition (Fig. 9-47).   *Ans.* $V_p = 30$ V

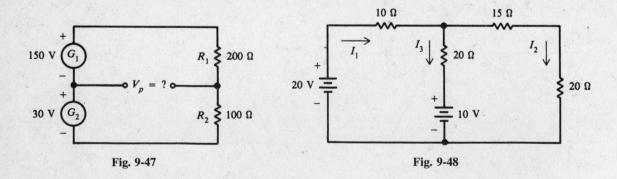

Fig. 9-47   Fig. 9-48

9.26   Solve for the indicated currents by using superposition (Fig. 9-48).
            Ans. $I_1 = 0.6$ A; $I_2 = 0.4$ A; $I_3 = 0.2$ A

9.27   Find the current in the load R_L by superposition (Fig. 9-49).
            Ans. $I_L = 14.8$ A (rounded from 14.84 A)

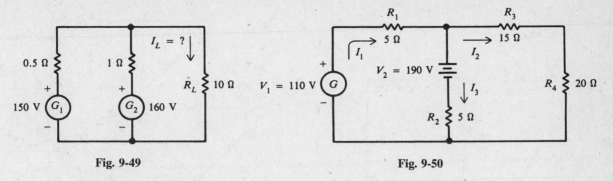

Fig. 9-49   Fig. 9-50

9.28 Find currents I_1, I_2, and I_3 in a two-mesh circuit by superposition (Fig. 9-50).
Ans. $I_1 = -6$ A (actual direction of current is opposite to the assumed direction); $I_2 = 4$ A; $I_3 = -10$ A (actual direction is opposite to the assumed direction)

9.29 Find the Thevenin equivalents to the circuits of Fig. 9-51.
Ans. (a) $R_{Th} = 1.2 \, \Omega$; $V_{Th} = 4.8$ V; (b) $R_{Th} = 1.6 \, \Omega$; $V_{Th} = 2.4$ V; (c) $R_{Th} = 0.89 \, \Omega$; $V_{Th} = 1.33$ V

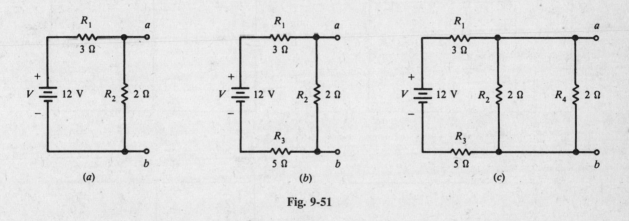

(a)   (b)   (c)

Fig. 9-51

9.30 Add a resistor load R_L of 5 Ω between terminals a and b to each circuit of Problem 9.29, and find the load current I_L and load voltage V_L.
Ans. (a) $I_L = 0.77$ A; $V_L = 3.87$ V; (b) $I_L = 0.36$ A; $V_L = 1.82$ V; (c) $I_L = 0.23$ A; $V_L = 1.13$ V

9.31 Find I_L and V_L by the Thevenin equivalent for the circuit of Fig. 9-52.
Ans. $I_L = 2$ A; $V_L = 20$ V

9.32 In the Wheatstone bridge network (Fig. 9-53), find the Thevenin equivalents R_{Th} and V_{Th}, and then find I_L and V_L. *Ans.* $R_{Th} = 21$; $V_{Th} = 30$ V; $I_L = 1$ A; $V_L = 9$ V

9.33 Find I_L and V_L (Fig. 9-54) by the Thevenin theorem.
Ans. $I_L = 3$ A; $V_L = 18$ V; ($R_{Th} = 1.71 \, \Omega$; $V_{Th} = 23.1$ V)

9.34 Find I_L and V_L (Fig. 9-55). *Ans.* $I_L = 1$ A; $V_L = 40$ V; ($R_{Th} = 6.67 \, \Omega$; $V_{Th} = 46.7$ V)

9.35 A voltage source has 24 V in series with 6 Ω (Fig. 9-56a). Draw the equivalent current source circuit.
Ans. See Fig. 9-56b.

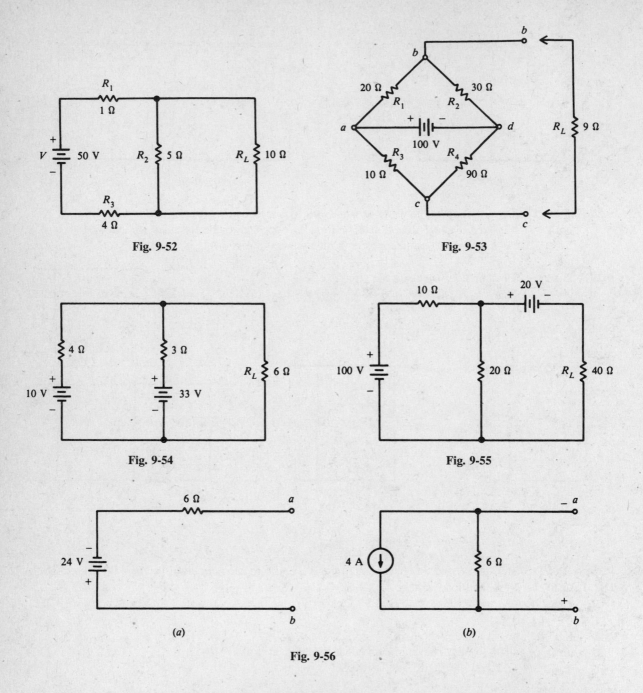

Fig. 9-52

Fig. 9-53

Fig. 9-54

Fig. 9-55

(a)   (b)

Fig. 9-56

9.36   Show the Norton equivalent circuit (Fig. 9-57a) and find I_L.   *Ans.*   See Fig. 9-57b. $I_L = 2.14$ A.

9.37   Find the Norton equivalent to the circuits of Fig. 9.58a, b, and c. (These are the same circuits for which you found the Thevenin equivalents in Problem 9.29.)
Ans.   (a) $I_N = 4$ A; $R_N = 1.2 \, \Omega$;   (b) $I_N = 1.5$ A; $R_N = 1.6 \, \Omega$;   (c) $I_N = 1.5$ A; $R_N = 0.89 \, \Omega$

9.38   Add a resistor load R_1 of 5 Ω between terminals a and b to each circuit of Problem 9.37. Calculate the load current I_L and load voltage V_L. Check your answers with those for Problem 9.30.
Ans.   (a) $I_L = 0.77$ A; $V_L = 3.87$ V;   (b) $I_L = 0.36$ A; $V_L = 1.82$ V;   (c) $I_L = 0.23$ A;
$V_L = 1.13$ V

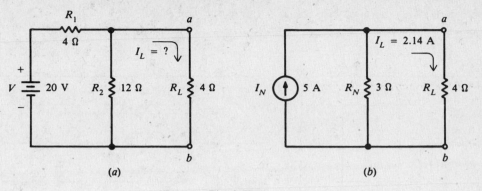

Fig. 9-57

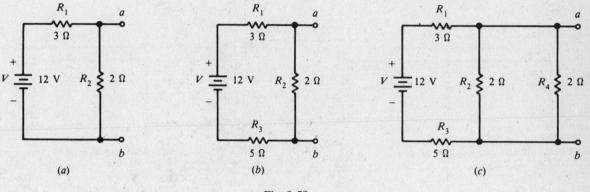

Fig. 9-58

9.39 Show the Thevenin equivalent (Fig. 9-59) and calculate V_L.
   Ans. $R_{\text{Th}} = 3 \ \Omega$; $V_{\text{Th}} = 22.5 \ \text{V}$; $V_L = 18 \ \text{V}$

9.40 In Fig. 9-59, solve for V_L by superposition.
   Ans. $V_L = 18 \ \text{V}$ ($I_{L,V1} = 1 \ \text{A}$, $I_{L,V2} = 0.5 \ \text{A}$, $I_L = 1.5 \ \text{A}$)

9.41 In Fig. 9-59, solve for V_L by the Norton equivalent theorem.
   Ans. $V_L = 18 \ \text{V}$ ($I_N = 7.5 \ \text{A}$, $R_N = 3 \ \Omega$)

9.42 Find the current through the load resistor R_L (Fig. 9-60).   *Ans.* $I_L = 0.2 \ \text{A}$

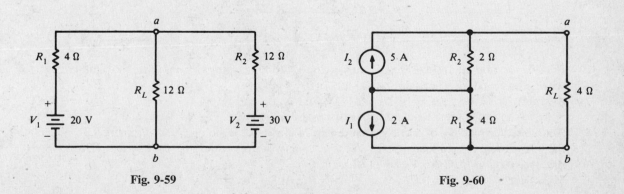

Fig. 9-59   **Fig. 9-60**

9.43   Find the total resistance of each circuit in Fig. 9-61a, b, c, and d.
       Ans.   (*a*) $R_T = 18\ \Omega$;   (*b*) $R_T = 10.6\ \Omega$;   (*c*) $R_T = 3.21\ \Omega$;   (*d*) $R_T = 2.86\ \Omega$

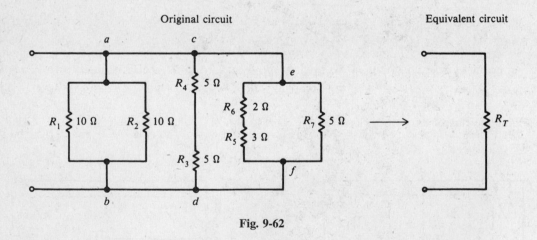

Fig. 9-61

9.44   Find the total resistance R_T (Fig. 9-62).   *Ans.* $R_T = 1.43\ \Omega$

9.45   Find the equivalent resistance of each resistance configuration (Fig. 9-63a, b, and c).
       Ans.   (*a*) $R_T = 2.86\ \Omega$;   (*b*) $R_T = 13.5\ \Omega$;   (*c*) $R_T = 15\ \Omega$

9.46   For the circuit (Fig. 9-64), find R_T, I_1, I_2, and I_3.
       Ans. $R_T = 10\ \Omega$; $I_1 = 15$ A; $I_2 = 10$ A; $I_3 = 5$ A

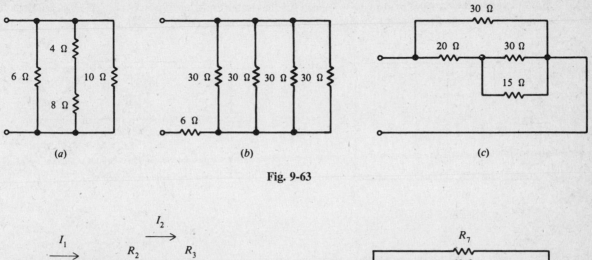

Fig. 9-63

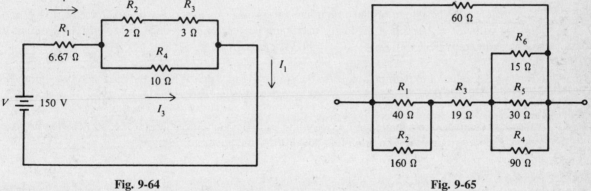

Fig. 9-64   Fig. 9-65

9.47   Find the total resistance of the circuit (Fig. 9-65).   *Ans.* $R_T = 30\,\Omega$

9.48   Find the total resistance of the circuit in Fig. 9-65 if the 60-Ω resistor were to burn out and open.
Ans. $R_T = 60\,\Omega$

9.49   An unknown resistance is to be checked by the Wheatstone bridge circuit (Fig. 9-66). When R_3 is
adjusted for 54 Ω, there is zero deflection on the galvanometer. Find R_x and each voltage.
Ans. $R_x = 1080\,\Omega$; $V_x = V_1 = 20\,\text{V}$; $V_2 = V_3 = 1\,\text{V}$

9.50   What load resistance R_L will produce maximum power at the load (Fig. 9-67) and what is the value
of that power?   *Ans.* $R_L = 0.075\,\Omega$; $P_L = 145\,\text{W}$

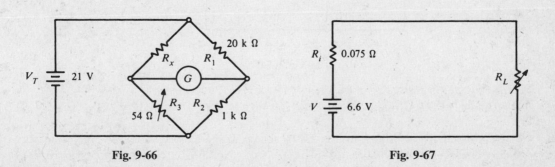

Fig. 9-66   Fig. 9-67

9.51 Determine the size of the load resistor needed for maximum transfer of power (Fig. 9-68). How much power then will be dissipated by the load?   *Ans.* $R_L = 2.4\ \Omega$; $P_L = 3.75$ W

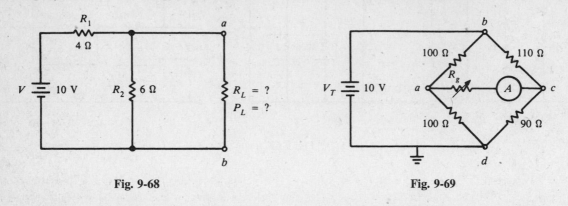

Fig. 9-68   **Fig. 9-69**

9.52 The unbalanced Wheatstone bridge circuit has a resistance R_g in series with an ammeter (Fig. 9-69). Find the value of R_g so that it dissipates maximum power. (*Hint*: Reduce the given bridge circuit to its Thevenin equivalent.)   *Ans.* $R_g = 99.5\ \Omega$ $(R_g = R_{\text{Th}})$

9.53 Calculate the value of maximum power for the circuit in Fig. 9-69. What is the reading of the ammeter? *Ans.* $P_L = 0.628$ mW; $I_L = 2.51$ mA $(V_{\text{Th}} = 0.5$ V, $R_{\text{Th}} = 99.5\ \Omega)$

9.54 A motor is connected to a generator by two wires, each having a resistance of 0.15 Ω. The motor takes 30 A at 211 V. What is the line drop, line power loss, and generator voltage? *Ans.* $V_l = 9$ V; $P_l = 270$ W; $V_G = 220$ V

9.55 Fixture wiring is often done with No. 16 wire, which has a resistance of 0.409 Ω for a 100-ft length. What is the loss in voltage from the house meter to an electric broiler using 10 A and located 100 ft from the meter? What is the power loss?   *Ans.* $V_l = 8.18$ V; $P_l = 81.8$ W

9.56 A generator is feeding current to a motor and lamp bank connected in parallel (Fig. 9-70). The feeder lines have the resistance shown. Find the voltage across the motor and lamp bank. *Ans.* $V_M = 113.8$ V; $V_L = 112$ V

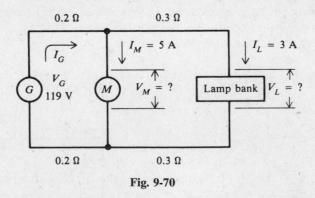

Fig. 9-70

9.57 Each lamp takes 0.5 A (Fig. 9-71). Find V_A and V_B.   *Ans.* $V_A = 112$ V; $V_B = 109.6$ V

9.58 A 20-kW motor load is 100 ft from a 230-V source. If the allowable drop is 5 percent, what is the smallest size copper wire that can be used?   *Ans.*   No. 8 wire (circular-mil area = 16 500)

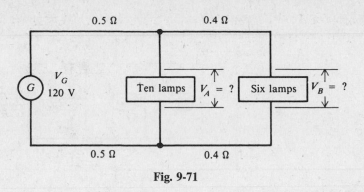

Fig. 9-71

9.59   A load 400 ft from a generator requires 80 A. The generator voltage is 115.6 V, and the load requires 110 V. What is the smallest size wire that may be used so that no less than 110 V will be across the load? (*Note*: Not only must voltage drop be considered, but the amperage capacity of the wire must not be exceeded.)   *Ans.*   No. 2 wire (circular-mil area = 59 400)

9.60   Find the currents in the three lines of the Edison system with two voltage sources (Fig. 9-72).
Ans. $I_1 = 6.64$ A; $I_2 = 1.65$ A; $I_3 = 4.99$ A

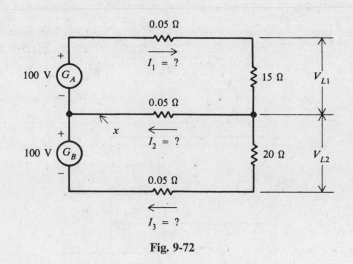

Fig. 9-72

9.61   If the neutral leg (Fig. 9-72) is broken at point X, possibly by a blown fuse, what is the current through the loads and the voltages across the loads?   *Ans.* $I = 5.70$ A; $V_{L1} = 85.5$ V; $V_{L2} = 114$ V

9.62   If the 15-Ω resistor (Fig. 9-72) is now replaced by a 20-Ω resistor, find the new values for the current in the three feeder lines (with equal source voltages, line resistances, and load resistances, the three-wire system is balanced).
Ans. $I_1 = 4.99$ A; $I_2 = 0$ A (current in neutral wire is zero for a balanced three-wire system); $I_3 = 4.99$ A

9.63   If each lamp (Fig. 9-73) requires a current of 1 A, find (*a*) the current in each of the three lines, (*b*) the IR drop in each line, and (*c*) the voltages V_1 and V_2. (*d*) Which line carries the heavier load?
Ans.   (*a*) $I_1 = 4$ A; $I_2 = 2$ A; $I_3 = 2$ A;   (*b*) Positive line, 3.2 V; neutral, 1.6 V; negative, 1.6 V; (*c*) $V_1 = 115.28$ V; $V_2 = 120$ V;   (*d*) Positive line

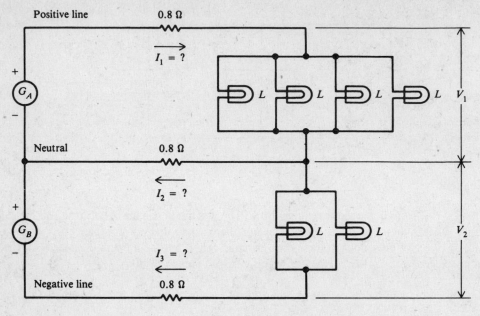

Fig. 9-73

9.64   In the three-wire dc distribution system (Fig. 9-74), each lamp bank consists of 50 lamps. Each lamp takes a power of 60 W when the voltage is 115 V. All three conductors are of the same size. Specify the size of copper wire to be used in order that the voltage at each bank be 115 V with all lamps turned on.   *Ans.*   No. 5 AWG

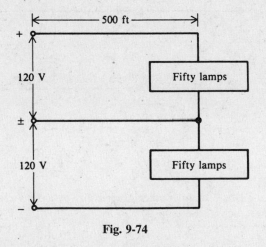

Fig. 9-74

Chapter 10

Magnetism and Electromagnetism

THE NATURE OF MAGNETISM

Most electrical equipment depends directly or indirectly upon magnetism. Without magnetism the electrical world we perceive today would not exist. There are few electrical devices used today that do not make use of magnetism.

Natural Magnets

The phenomenon of magnetism was discovered by the Chinese about 2637 B.C. The magnets used in their primitive compasses were called *lodestones* or *leading stones*. It is now known that lodestones were crude pieces of iron ore known as *magnetite*. Since magnetite has magnetic properties in its natural state, lodestones are classified as *natural* magnets. The only other natural magnet is the earth itself. All other magnets are human-made and are known as *artificial* magnets.

Magnetic Fields

Every magnet has two points opposite to each other which most readily attract pieces of iron. These points are called *poles* of the magnet: the north pole and the south pole. Just as like electric charges repel each other and opposite charges attract each other, like magnetic poles repel each other and unlike poles attract each other.

A magnet clearly attracts a bit of iron because of some force that exists around the magnet. This force is called the *magnetic field*. Although it is invisible to the naked eye, its force can be shown to exist by sprinkling small iron filings on a sheet of glass or paper over a bar magnet (Fig. 10-1a). If the sheet is tapped gently, the filings will move into a definite pattern which describes the field of force around the magnet. The field seems to be made up of *lines of force* that appear to leave the magnet at the north pole, travel through the air around the magnet, and continue through the magnet to the south pole to form a *closed loop* of force. The stronger the magnet, the greater the number of lines of force and the larger the area covered by the field.

In order to visualize the magnetic field without iron filings, the field is shown as lines of force in Fig. 10-1b. The direction of the lines outside the magnet shows the path a north pole would follow in the field, repelled away from the north pole of the magnet and attracted to its south pole.

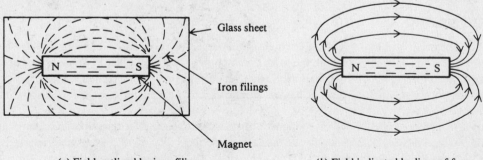

(a) Field outlined by iron filings   (b) Field indicated by lines of force

Fig. 10-1 Magnetic field of force around a bar magnet

Magnetic Flux ϕ

The entire group of magnetic field lines, which flow outward from the north pole of the magnet, is called the *magnetic flux*. The symbol for magnetic flux is the Greek lowercase letter ϕ (phi).

The SI unit of magnetic flux is the *weber* (Wb). One weber equals 1×10^8 magnetic field lines. Since the weber is a large unit for typical fields, the microweber (μWb) is used ($1\,\mu$Wb $= 10^{-6}$ Wb).

Example 10.1   If a magnetic flux ϕ has 3000 lines, find the number of microwebers.
Convert number of lines to microwebers.

$$\phi = \frac{3000\ \text{lines}}{1 \times 10^8\ \text{lines/Wb}} = \frac{3 \times 10^3}{10^8} = 30 \times 10^{-6}\ \text{Wb} = 30\,\mu\text{Wb} \qquad Ans.$$

Magnetic Flux Density B

The *magnetic flux density* is the magnetic flux per unit area of a section perpendicular to the direction of flux. The equation for magnetic flux density is

$$B = \frac{\phi}{A} \qquad\qquad (10\text{-}1)$$

where B = magnetic flux density in teslas (T)
ϕ = magnetic flux, Wb
A = area in square meters (m^2)

We see that the SI unit for B is webers per square meter (Wb/m^2). One weber per square meter is called a *tesla*.

Example 10.2   What is the flux density in teslas when there exists a flux of 600 μWb through an area of 0.0003 m^2?

Given $\phi = 600\,\mu\text{Wb} = 6 \times 10^{-4}\ \text{Wb}$

$$A = 0.0003\ \text{m}^2 = 3 \times 10^{-4}\ \text{m}^2$$

Substitute the values of ϕ and A in Eq. (10-1).

$$B = \frac{\phi}{A} = \frac{6 \times 10^{-4}\ \text{Wb}}{3 \times 10^{-4}\ \text{m}^2} = 2\ \text{T} \qquad Ans.$$

MAGNETIC MATERIALS

Magnetic materials are those materials which are attracted or repelled by a magnet and which can be magnetized themselves. Iron and steel are the most common magnetic materials. Permanent magnets are those of hard magnetic materials, such as cobalt steel, that retain their magnetism when the magnetizing field is removed. A temporary magnet is one that has *no* ability to retain a magnetized state when the magnetizing field is removed.

Permeability refers to the ability of a magnetic material to concentrate magnetic flux. Any material that is easily magnetized has high permeability. A measure of permeability for different materials in comparison with air or vacuum is called *relative* permeability. The symbol for relative permeability is μ_r (mu), where the subscript r stands for *relative*. μ_r is not expressed in units because it is a ratio of two flux densities, so the units cancel.

Classifying magnetic materials as either *magnetic* or *nonmagnetic* is based on the strong magnetic properties of iron. However, since weak magnetic materials can be important in some applications, classification includes three groups:

1.   *Ferromagnetic materials.* These include iron, steel, nickel, cobalt, and commercial alloys such as alnico and Permalloy. The *ferrites* are nonmagnetic materials that have ferromagnetic properties

of iron. A ferrite is a ceramic material. The permeability of ferrites is in the range of 50–3000. A common application is a ferrite core in the coils for RF (radio-frequency) transformers.

2. *Paramagnetic materials.* These include aluminum, platinum, manganese, and chromium. Relative permeability is slightly more than 1.

3. *Diamagnetic materials.* These include bismuth, antimony, copper, zinc, mercury, gold, and silver. Relative permeability is less than 1.

ELECTROMAGNETISM

In 1819, a Danish scientist named Oersted discovered a relation between magnetism and electric current. He found that an electric current flowing through a conductor produced a magnetic field around that conductor. In Fig. 10-2*a*, filings in a definite pattern of concentric rings around the conductor show the magnetic field of the current in the wire. Every section of the wire has this field of force around it in a plane perpendicular to the wire (Fig. 10-2*b*). The strength of the magnetic field around a conductor carrying current depends on the current. A high current will produce many lines of force extending far from the wire, while a low current will produce only a few lines close to the wire (Fig. 10-3).

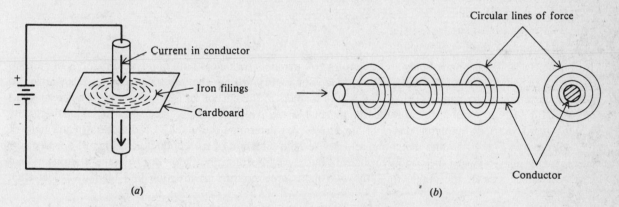

(a)   (b)

Fig. 10-2   Circular pattern of magnetic lines around current in a conductor

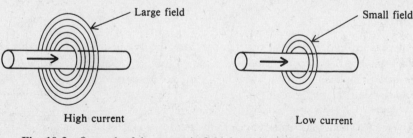

High current   Low current

Fig. 10-3   Strength of the magnetic field depends on the amount of current

Polarity of a Single Conductor

The *right-hand rule* is a convenient way to determine the relationship between the flow of current in a conductor (wire) and the direction of the magnetic lines of force around the conductor. Grasp the current-carrying wire in the right hand, wrapping the four fingers around the wire and extending the thumb along the wire. If the thumb points along the wire in the direction of current flow, the fingers will be pointing in the direction of the lines of force around the conductor (Fig. 10-4).

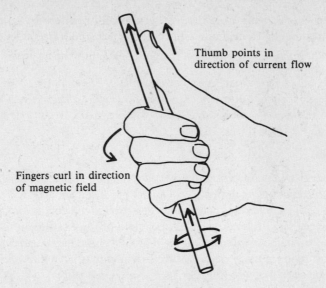

Thumb points in
direction of current flow

Fingers curl in direction
of magnetic field

Fig. 10-4   Right-hand rule

Magnetic Fields Aiding or Canceling

In Fig. 10-5, the magnetic fields are shown for two parallel conductors with opposite directions of current. The cross in the middle of the field of the conductor in Fig. 10-5a symbolizes the back of an arrow to indicate current into the paper. (Think of it as the feathers at the end of an arrow moving away from you.) The dot (Fig. 10-5b) symbolizes current moving out of the paper. (In this case, it is a point of the arrow facing toward you.) By applying the right-hand rule, you determine the clockwise direction of the field of the conductor in Fig. 10-5a and the counterclockwise field direction of the conductor in Fig. 10-5b. Because the magnetic lines between the conductors are in the *same* direction, the fields *aid* to make a stronger total field. On either side of the conductors, the two fields are *opposite* in direction and tend to cancel each other.

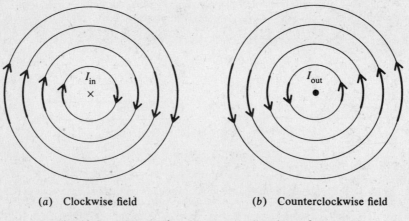

(*a*)   Clockwise field   (*b*)   Counterclockwise field

Fig. 10-5   Fields aiding with opposite directions of current

Magnetic Field and Polarity of a Coil

Bending a straight conductor into the form of a single loop has two results. First the magnetic field lines are more dense inside the loop, although the total number of lines is the same as for the straight conductor. Second, all the lines inside the loop are aiding in the same direction.

A coil of wire conductor is formed when there is more than one loop or turn. To determine the magnetic polarity of a coil, use the right-hand rule (Fig. 10-6). If the coil is grasped with the fingers of the right hand curled in the direction of current flow through the coil, the thumb points to the north pole of the coil.

Adding an iron core inside the coil increases the flux density. The polarity of the core is the same as that of the coil. The polarity depends on the direction of current flow and the direction of winding. Current flow is from the positive side of the voltage source, through the coil, and back to the negative terminal (Fig. 10-7). The north pole is found by using the right-hand rule.

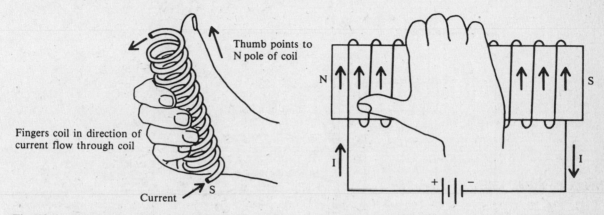

Fig. 10-6 Right-hand rule for coil of wire with several turns (solenoid)

Fig. 10-7 Right-hand rule to find the north pole of an electromagnet

Example 10.3   Determine the magnetic polarity of the pictured electromagnets (Fig. 10-8) by the right-hand rule.

The correct polarities are circled. Notice that A has the same direction of winding and current as in Fig. 10-7. In B, the battery polarity is opposite from A to reverse the direction of current. In C, the direction of winding is reversed from A; and in D, it is reversed from B.

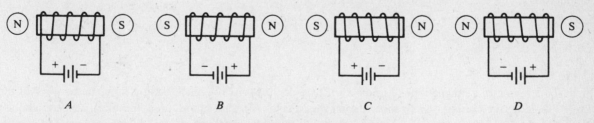

A B C D

Fig. 10-8   Finding the polarity of a coil

Electromagnet Applications

If a bar of iron or soft steel is placed in the magnetic field of a coil (Fig. 10-9), the bar will become magnetized. If the magnetic field is strong enough, the bar will be drawn into the coil until it is approximately centered within the magnetic field.

Electromagnets are widely used in electrical devices. One of the simplest and most common applications is in a relay. When the switch S is closed in a relay circuit (Fig. 10-10), current flows in the coil, causing a strong magnetic field around the coil. The soft iron bar in the lamp circuit is attracted toward the right end of the electromagnet and makes contact with the conductor at A. A path is thus completed for current in the lamp circuit. When the switch is opened, the current flow through the electromagnet ceases and the magnetic field collapses and disappears. Since the attraction for the soft iron bar by the electromagnet no longer exists,

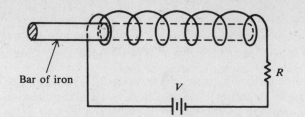

Fig. 10-9 A current-carrying coil magnetizes and attracts an iron bar placed in its field

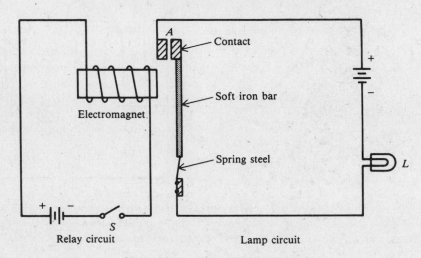

Fig. 10-10 A simple relay circuit

the iron bar is pulled away from the contact by the piece of spring steel to which it is attached. This opens the contacts at *A* and breaks the circuit for the lamp.

MAGNETIC UNITS

Ampere-Turns *NI*

The strength of a magnetic field in a coil of wire depends on how much current flows in the turns of the coil. The more the current, the stronger the magnetic field. Also, the more turns, the more concentrated are the lines of force. The product of the current times the number of turns of the coil, which is expressed in units called *ampere-turns* (At), is known as the *magnetomotive force* (mmf). As a formula,

$$F = \text{ampere-turns} = NI \qquad\qquad (10\text{-}2)$$

where F = magnetomotive force, At
        N = number of turns
        I = current, A

Example 10.4   Calculate the ampere-turns for a coil with 1500 turns and a 4-mA current.
    Use Eq. (*10-2*) and substitute $N = 150$ turns   and $I = 4 \times 10^{-3}$ A.

$$NI = 1500(4 \times 10^{-3}) = 6\,\text{At} \qquad Ans.$$

Field Intensity H

If a coil with a certain number of ampere-turns is stretched out to twice its original length, the intensity of the magnetic field, that is, the concentration of lines of force, will be half as great. The field intensity thus depends on how long the coil is. Expressed as an equation,

$$H = \frac{NI}{l} \qquad (10\text{-}3)$$

where H = magnetic field intensity, ampere-turns per meter (At/m)
NI = ampere-turns, At
l = length between poles of the coil, m

Equation (10-3) is for a solenoid. H is the intensity at the center of an air core. With an iron core, H is the intensity through the entire core and l is length or distance between poles of the iron core.

Example 10.5   (a) Find the field intensity of a 40-turn, 10-cm-long coil, with 3 A flowing in it (Fig. 10-11a). (b) If the same coil is stretched to 20 cm, with the wire length and current remaining the same, what is the new value of field intensity (Fig. 10-11b)? (c) The 10 cm coil in part (a) with the same 3 A flowing is now wound around an iron core that is 20 cm long (Fig. 10-11c). What is the field intensity?

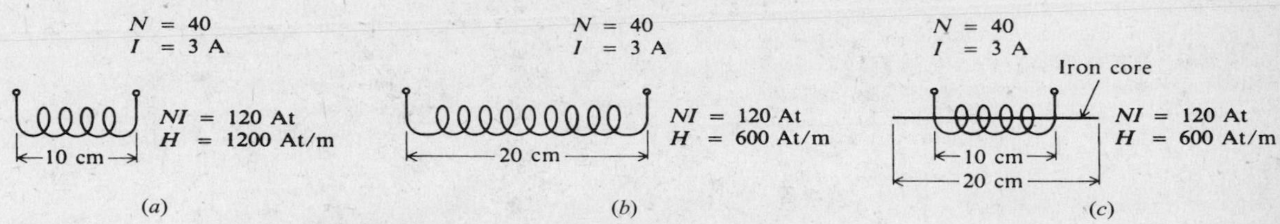

Fig. 10-11   Relation between mmf and field intensity with same value of mmf

(a)   Apply Eq. (10-3), where $N = 40$ turns, $l = 10$ cm $= 0.1$ m, and $I = 3$ A.

$$H = \frac{NI}{l} = \frac{40(3)}{0.1} = 1200 \text{ At/m} \qquad Ans.$$

(b)   The length l in Eq. (10-3) is between poles. The coil is stretched from 10 to 20 cm. Though the wire length is the same, the length between poles is 20 cm $= 0.2$ m. So

$$H = \frac{40(3)}{0.2} = 600 \text{ At/m} \qquad Ans.$$

Stretching out the coil to twice its original distance reduces the mmf by one-half.

(c)   The length l in Eq. (10-3) is 20 cm between the poles at the ends of the iron core although the winding is 10 cm long.

$$H = \frac{40(3)}{0.2} = 600 \text{ At/m} \qquad Ans.$$

Note that cases (b) and (c) have the same H value.

BH MAGNETIZATION CURVE

The *BH* curve (Fig. 10-12) is used to show how much flux density B results from increasing the amount of field intensity H. This curve is for two types of soft iron plotted for typical values. It shows that soft iron number 1 increases rapidly in B with an increase in H before it develops a "knee" and becomes saturated at $H = 2000$ At/m, $B = 0.2$ T. Past the knee an increase in H has little effect on the B value. Soft iron number 2 needs much more H to reach its saturation level at $H = 5000$ At/m, $B = 0.3$ T. Similar curves are obtained for all magnetic materials. Air, being nonmagnetic, has a very low *BH* profile (Fig. 10-12).

The permeability μ of a magnetic material is the ratio of B to H.

$$\mu = \frac{B}{H} \qquad (10\text{-}4)$$

Its average value is measured at the point where the knee is first established. Figure 10-12 illustrates that the normal or average permeability is as follows:

$$\mu \text{ for soft iron number } 1 = \frac{B}{H} = \frac{0.2}{2000} = 1 \times 10^{-4} \text{ (T} \cdot \text{m)/At}$$

$$\mu \text{ for soft iron number } 2 = \frac{B}{H} = \frac{0.3}{5000} = 6 \times 10^{-5} \text{ (T} \cdot \text{m)/At}$$

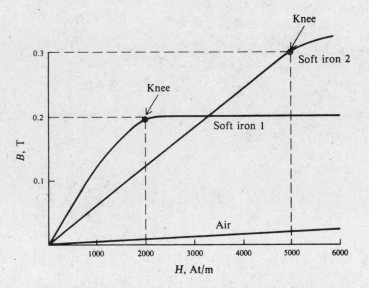

Fig. 10-12   Typical *BH* curve for two types of soft iron

In SI units, the permeability of air is $\mu_0 = 4\pi \times 10^{-7}$ or 1.26×10^{-6}. To calculate μ, the value of relative permeability μ_r must be multiplied by μ_0.

$$\mu = \mu_r \times \mu_0 \qquad (10\text{-}5)$$

Example 10.6   If a magnetic material has a relative permeability μ_r of 100, find its permeability μ.
Use Eq. (*10-5*) and substitute known values.

$$\mu = \mu_r \mu_0 = 100(1.26 \times 10^{-6})$$

$$= 126 \times 10^{-6} \text{ (T} \cdot \text{m)/At} \qquad Ans.$$

Hysteresis

When the current in a coil of wire reverses thousands of times per second, the hysteresis can cause a considerable loss of energy. *Hysteresis* means "a lagging behind"; that is, the magnetic flux in an iron core lags behind the increases or decreases of the magnetizing force.

The hysteresis loop is a series of curves that show the characteristics of a magnetic material (Fig. 10-13). Opposite directions of current result in the opposite directions of $+H$ and $-H$ for field intensity. Similarly, opposite polarities are shown for flux density as $+B$ or $-B$. The current starts at the center 0 (zero) when the material is unmagnetized. The dotted line is recognized as the magnetization curve illustrated in Fig. 10-12. Positive H values increase B to saturation at $+B_{max}$. Next H decreases to zero, but B drops to the value of B_r because of hysteresis. The current that produced the original magnetization now is reversed so that H becomes negative. B drops to zero and continues to $-B_{max}$. Then as the $-H$ values decrease, B is reduced to $-B_r$, when H is zero. Now with a positive swing of current, H becomes positive, producing saturation at $+B_{max}$ again. The hysteresis loop is now completed. The curve does not return to zero at the center because of hysteresis.

The value of $+B_r$ or $-B_r$, which is the flux density remaining after the magnetizing force is zero ($H = 0$), is called the *retentivity* of a magnetic material. The value of $-H_c$, which is the magnetizing force that must be applied in the reverse direction to reduce the flux density to zero ($B = 0$), is called the *coercive* force of the material.

The larger the area enclosed by the hysteresis loop, the greater the hysteresis loss.

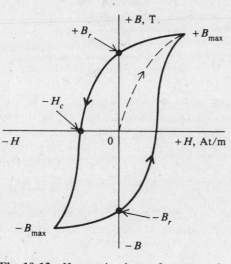

Fig. 10-13 Hysteresis loop for magnetic materials

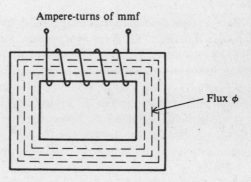

Fig. 10-14 Magnetic circuit with closed iron path

MAGNETIC CIRCUITS

A magnetic circuit can be compared with an electric current in which an emf produces a current flow. Consider a simple magnetic circuit (Fig. 10-14). The ampere-turns NI of the magnetomotive force produce the magnetic flux ϕ. Therefore, the mmf compares to emf or voltage and the flux ϕ compares to current. Opposition to the production of flux in a material is called its *reluctance*, which corresponds to resistance.

Reluctance $\mathcal{R}$

The symbol for reluctance is $\mathcal{R}$. Reluctance is inversely proportional to permeability. Iron has high permeability and therefore low reluctance. Air has low permeability and hence high reluctance.

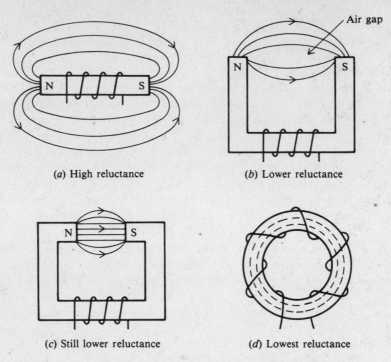

(a) High reluctance   (b) Lower reluctance

(c) Still lower reluctance   (d) Lowest reluctance

Fig. 10-15   Different physical forms of electromagnets

Different forms of electromagnets generally have different values of reluctance (Fig. 10-15). The air gap is the air space between the poles of a magnet. Since air has high reluctance, the size of the air gap affects the value of reluctance. The magnetic circuit in Fig. 10-15a has widely spaced poles in air so it has a high reluctance. In Fig. 10-15b, the reluctance has been decreased by bringing the two poles closer together. The field between N and S is more intense, assuming the same number of ampere-turns in the coils. In Fig. 10-15c, the air gap is smaller than that in Fig. 10-15b so the reluctance is lower. In Fig. 10-15d, there is no air gap in the toroid-shaped core so its reluctance is very low.

The shorter the air gap, the stronger the field in the gap. Since air is not magnetic and thus is unable to concentrate magnetic lines, a larger air gap only provides more space for the magnetic lines to spread out.

Ohm's Law for Magnetic Circuits

Ohm's law for magnetic circuits, corresponding to $I = V/R$, is

$$\phi = \frac{\text{mmf}}{\mathcal{R}}$$

(10-6)

where ϕ = magnetic flux, Wb
         mmf = magnetomotive force, At
         $\mathcal{R}$ = reluctance, At/Wb

Reluctance can be expressed as an equation as follows:

$$\mathcal{R} = \frac{l}{\mu A}$$

(10-7)

where $\mathcal{R}$ = reluctance, At/Wb
         l = length of coil, m
         μ = permeability of magnetic material, $(T \cdot m)/At$
         A = cross-sectional area of coil, m^2

Example 10.7   A coil has an mmf of 500 At and a reluctance of 2×10^6 At/Wb. Compute the total flux ϕ.
Write Ohm's law for magnetic circuits and substitute given values.

$$\phi = \frac{mmf}{\mathcal{R}} \tag{10-6}$$

$$= \frac{500 \text{ At}}{2 \times 10^6 \text{ At/Wb}} = 250 \times 10^{-6} \text{ Wb} = 250 \text{ } \mu\text{Wb} \qquad Ans.$$

Example 10.8   Starting with Eq. (10-6), show that $\mathcal{R} = l/\mu A$, which is Eq. (10-7).

$$\phi = \frac{mmf}{\mathcal{R}} \tag{10-6}$$

Also $$\phi = BA \tag{10-1}$$

Substitute $B = \mu H$ [Eq. (10-4)] and $H = NI/l$ [Eq. (10-3)] to obtain

$$\phi = BA = \mu HA = \frac{\mu NIA}{l} = NI\frac{\mu A}{l} = \frac{NI}{l/\mu A}$$

But Eq. (10-6) tells us that

$$\phi = \frac{mmf}{\mathcal{R}} = \frac{NI}{\mathcal{R}}$$

By comparing denominators of the two expressions for ϕ with the same numerator, we see that

$$\mathcal{R} = \frac{l}{\mu A}$$

which is Eq. (10-7).

ELECTROMAGNETIC INDUCTION

In 1831, Michael Faraday discovered the principle of electromagnetic induction. It states that if a conductor "cuts across" lines of magnetic force, or if lines of force cut across a conductor, an emf, or voltage, is induced across the ends of the conductor. Consider a magnet with its lines of force extending from the north to the south pole (Fig. 10-16). A conductor C, which can be moved between the poles, is connected to a galvanometer G used to indicate the presence of an emf. When the conductor is *not* moving, the galvanometer shows zero emf. If the wire conductor is moving *outside* the magnetic field at position 1, the galvanometer will still show zero. When the conductor is moved to the left to position 2, it cuts across the lines of magnetic force and the galvanometer pointer will deflect to A. This indicates that an emf was induced in the conductor because lines of force were cut. In position 3, the galvanometer pointer swings back to zero because no lines of force are being cut. Now reverse the direction of the conductor by moving it right through the lines of force back to position 1. During this movement, the pointer will deflect to B, showing that an emf has again been induced in the wire, but in the *opposite* direction. If the wire is held stationary in the middle of the field of force at position 2, the galvanometer reads zero. If the conductor is moved up or down *parallel* to the lines of force so that none is cut, no emf will be induced.

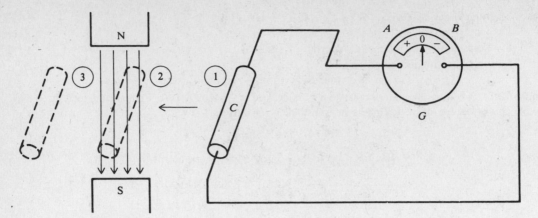

Fig. 10-16   When a conductor cuts lines of force, an emf is induced in the conductor

In summary,

1. When lines of force are cut by a conductor or lines of force cut a conductor, an emf or voltage, is induced in the conductor.

2. There must be relative motion between the conductor and the lines of force in order to induce an emf.

3. Changing the direction of cutting will change the direction of the induced emf.

The most important application of relative motion between conductor and magnetic field is made in electric generators. In a dc generator, fixed electromagnets are arranged in a cylindrical housing. Many conductors in the form of a coil are rotated on a core within the magnetic field so that these conductors are continually cutting the lines of force. As a result, voltage is induced in each of the conductors. Since the conductors are in series in the coil, the induced voltages add together to produce the output voltage of the generator.

Faraday's Law of Induced Voltage

The value of the induced voltage depends upon the number of turns of a coil and how fast the conductor cuts across the lines of force or flux. Either the conductor or the flux can move. The equation to calculate the value of the induced voltage is

$$v_{ind} = N \frac{\Delta \phi}{\Delta t} \qquad\qquad (10\text{-}8)$$

where v_{ind} = induced voltage, V
          N = number of turns in a coil
      $\Delta \phi / \Delta t$ = rate at which the flux cuts across the conductor, Wb/s

From Eq. (10-8) we see that v_{ind} is determined by three factors:

1. *Amount of flux.* The more the lines of force that cut across the conductor, the higher the value of induced voltage.

2. *Number of turns.* The more the turns in a coil, the higher the induced voltage.

3. *Time rate of cutting.* The faster the flux cuts a conductor or the conductor cuts the flux, the higher the induced voltage because more lines of force cut the conductor within a given period of time.

Example 10.9   The flux of an electromagnet is 6 Wb. The flux increases uniformly to 12 Wb in a period of 2 s. Calculate the voltage induced in a coil that has 10 turns if the coil is stationary in the magnetic field.

Write down known values.

$$\Delta\phi = \text{change in flux} = 12\,\text{Wb} - 6\,\text{Wb} = 6\,\text{Wb}$$

$$\Delta t = \text{change in time corresponding to the increase in flux} = 2\,\text{s}$$

Then
$$\frac{\Delta\phi}{\Delta t} = \frac{6}{2} = 3\,\text{Wb/s}$$

We are given that $N = 10$ turns. Substitute values into Eq. (*10-8*) and solve for v_{ind}.

$$v_{\text{ind}} = N\frac{\Delta\phi}{\Delta t} = 10(3) = 30\,\text{V} \qquad Ans.$$

Example 10.10　In Example 10-9, what is the value of induced voltage if the flux remains at 6 Wb after 2 s? Since there is no change in flux, $\Delta\phi = 0$. Using Eq. (*10-8*),

$$v_{\text{ind}} = N\frac{\Delta\phi}{\Delta t} = N\frac{0}{\Delta t} = N0 = 0\,\text{V} \qquad Ans.$$

That no voltage is induced in Example 10.10 confirms the principle that there must be relative motion between the conductor and the flux in order to induce a voltage. A magnetic field whose flux is increasing or decreasing in strength is, in effect, moving relative to any conductors in the field.

Lenz's Law

The polarity of the induced voltage is determined by Lenz's law. The induced voltage has the polarity that *opposes* the change causing the induction. When a current flows as a result of an induced voltage, this current sets up a magnetic field about the conductor such that this conductor magnetic field reacts with the external magnetic field, producing the induced voltage to oppose the change in the external magnetic field. If the external field increases, the conductor magnetic field of the induced current will be in the opposite direction. If the external field decreases, the conductor magnetic field will be in the same direction, thus sustaining the external field.

Example 10.11　A permanent magnet is moved into a coil and causes an induced current to flow in the coil circuit (Fig. 10-17a). Determine the polarity of the coil and the direction of the induced current.

By use of Lenz's law, the left end of the coil must be the N pole to oppose the motion of the magnet. Then the direction of the induced current can be determined by the right-hand rule. If the right thumb points to the left for the N pole, the fingers coil around the direction of current (Fig. 10-17b).

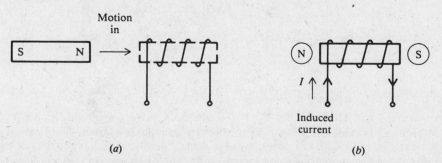

Fig. 10-17　Illustration of Lenz's law

INTERNATIONAL SYSTEM OF UNITS

Table 10-1 lists the SI units for magnetism.

Table 10-1 International System of Units for Magnetism

Term	Symbol	Unit (Abbreviation)
Flux	ϕ	Weber (Wb)
Flux density	B	Weber per square meter (Wb/m^2) = tesla (T)
Potential	mmf	Ampere-turn (At)
Field intensity	H	Ampere-turn per meter (At/m)
Reluctance	$\mathcal{R}$	Ampere-turn per weber (At/Wb)
Relative permeability	μ_r	None, pure number
Permeability	$\mu = \mu_r \times 1.26 \times 10^{-6}$	B/H = tesla per ampere-turn per meter [(T · m)/At]

Solved Problems

10.1   Match the term in column 1 with its closest meaning in column 2.

Column 1		Column 2
1. One weber	(a)	$B/\dot{H}$
2. Lenz's law	(b)	Ceramic material
3. Two north poles	(c)	Force of repulsion
4. v_{ind}	(d)	Inversely proportional to permeability
5. Field intensity	(e)	H/B
6. Electric generator	(f)	1×10^8 lines of force
7. Relative permeability	(g)	Application of electromagnetic induction
8. High permeability	(h)	$N\dfrac{\Delta\phi}{\Delta t}$
9. Ferrite	(i)	Respect to air
10. Reluctance	(j)	Polarity of induced voltage
	(k)	NI
	(l)	Force of attraction
	(m)	At/m
	(n)	Easily magnetized

Ans.   1. (f) 2. (j) 3. (c) 4. (h) 5. (m) 6. (g) 7. (i) 8. (n) 9. (b) 10. (d)

10.2   Describe the action that takes place when two like poles and when two unlike poles are placed near each other.

See Fig. 10-18. If the N poles of two magnets are placed near each other (Fig. 10-18a), the lines of force emanating from the N poles have the same direction and thus repel each other. This force of repulsion tends to move the two magnets apart. On the other hand, if the N and the S poles of two magnets are placed near each other (Fig. 10-18b), the adjacent lines of force are opposite in

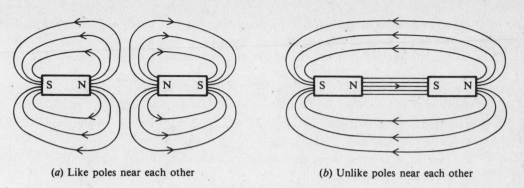

(a) Like poles near each other   (b) Unlike poles near each other

Fig. 10-18 Interaction of magnetic poles

direction and they link together to form long loops. These long, continuous lines tend to contract and this force of attraction pulls the two magnets together. So fields from like poles repel, tending to push the magnets apart, while fields from unlike poles attract, tending to pull the magnets together.

10.3 An example of magnetic attraction is the navigator's compass and the earth's magnetic field. The earth itself is a huge natural magnet. The earth has its magnetic south (S) pole near the geographic north (N) pole, and its magnetic north (N) pole near the geographic south (S) pole. The compass needle is a long, thin permanent magnet that is free to move on its central bearing point. The compass needle always lines up its magnetic field with the magnetic field of the earth, with its north end pointing toward the earth's magnetic south pole. The geographic N pole is located near the magnetic S pole. Show how a magnetic compass is used to indicate direction.

See Fig. 10-19.

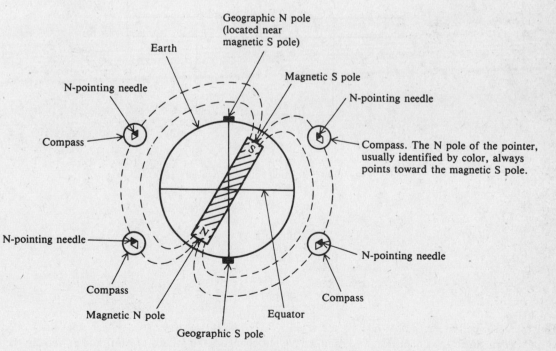

Fig. 10-19 The earth as a magnet

10.4 Would 1 A flowing through a 2-m length of wire made into a single loop produce more, the same, or less mmf if it were wound into a coil 2 cm in diameter and 4 cm long?

Basically flux is produced by current flowing in a wire. The mmf produced is the product of the current times the number of turns of the coil. Since 1 A is flowing through a coil of 1 turn in both cases, though the physical configuration of the coils is different, the mmf is the *same* at 1 At.

10.5 (a) Consider a coil with an air core (Fig. 10-20a). The coil is 5 cm long and has 8 turns. When the switch is closed, a current of 5 A flows in it. Find the mmf and H.

(b) If an iron core were slipped into the coil (Fig. 10-20b), what is now the mmf and H? What qualitative changes take place?

(c) The coil length remains the same, but the iron core is lengthened to 10 cm (Fig. 10-20c). What are the values of the mmf and H?

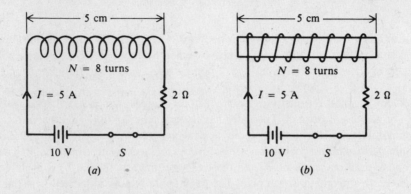

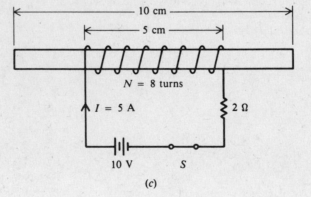

Fig. 10-20   Relation between mmf and H

(a) $$F = \text{mmf} = NI \qquad\qquad (10\text{-}2)$$
$$= (8 \text{ turns})(5 \text{ A}) = 400 \text{ At} \qquad Ans.$$

$$H = \frac{NI}{l} \qquad\qquad (10\text{-}3)$$

$$= \frac{400 \text{ At}}{5 \times 10^{-2} \text{ m}} = 8000 \text{ At/m} \qquad Ans.$$

(b) Since the quantities N, I, and l have not changed, the values of mmf and H remain the same as in part (a). So

$$\text{mmf} = 400 \text{ At} \qquad Ans.$$

$$H = 8000 \text{ At/m} \qquad Ans.$$

What changes is a large increase in flux density B. Suppose the iron core in air produced 50 lines of force. With the core placed into an energized coil, the number of lines in the core area might be 250 000. In this case, iron would have a permeability 5000 times that of air. Therefore, the use of an iron core instead of an air core increases the effectiveness of the magnet several thousand times. For this reason most electromagnets are made with iron cores.

(c) The lengthening of the iron core does not change the ampere-turns of the coil, so

$$\text{mmf} = 400 \,\text{At} \qquad Ans.$$

However, with a longer iron that is twice the initial length ($10\,\text{cm} = 2 \times 5\,\text{cm}$), the field intensity is reduced by half.

$$H = \frac{1}{2} \times 8000\,\text{At/m} = 4000\,\text{At/m} \quad \text{or} \quad H = \frac{NI}{l} = \frac{400}{0.1} = 4000\,\text{At/m} \qquad Ans.$$

10.6 An iron ring has a mean circumferential length of 40 cm and a cross-sectional area of $1\,\text{cm}^2$. It is wound uniformly with 500 turns of wire. Measurements made with a search coil around the ring show that the current in the windings is 0.06 A and the flux in the ring is 6×10^{-6} Wb. Find the flux density B, field intensity H, permeability μ, and relative permeability μ_r.

B is found by using Eq. (10-1).

$$B = \frac{\phi}{A} = \frac{6 \times 10^{-6}\,\text{Wb}}{10^{-4}\,\text{m}^2} = 6 \times 10^{-2}\,\text{T} \qquad Ans.$$

H is found by using Eq. (10-3).

$$H = \frac{NI}{l} = \frac{500 \times 0.06}{0.4} = 75\,\text{At/m} \qquad Ans.$$

μ is found from Eq. (10-4).

$$\mu = \frac{B}{H} = \frac{6 \times 10^{-2}}{75} = 8 \times 10^{-4}\,(\text{T} \cdot \text{m})/\text{At} \qquad Ans.$$

μ_r is found by Eq. (10-5).

$$\mu = \mu_r \mu_0$$

$$\mu_r = \mu/\mu_0 = \frac{8 \times 10^{-4}}{1.26 \times 10^{-6}} = 635 \qquad Ans.$$

The relative permeability is a pure number that has no unit of measurement.

10.7 Hysteresis loops of three different magnetic materials are shown in Fig. 10-21. Rank them in order from least to most hysteresis loss.

The smaller the area enclosed by the hysteresis loop, the lower the hysteresis loss. Hysteresis loss is likened to magnetic friction that must be overcome in magnetizing a material. Curve B, having the smallest area, has the least hysteresis loss. Loop B is characteristic of a temporary-magnet material. The coercive force is very small and hysteresis loss would be negligible. Next in area size is loop A, which is typical of a relatively permanent magnet material. And loop C with the largest area has the highest loss. This rectangular-shaped curve typifies permanent magnet material, such as alnico.

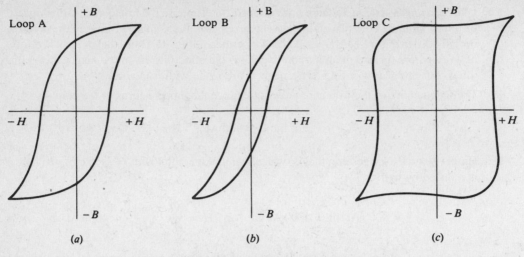

Fig. 10-21 Hysteresis loops

10.8 A core of annealed steel with a B of 0.72 T has a permeability μ of 8×10^{-3} (T·M)/At. If the length of the coil is 20 cm and the area of the core is 3 cm^2, find the reluctance of the path.

Given are $l = 20\,\text{cm} = 0.2\,\text{m}$, $A = 3\,\text{cm}^2 = 3 \times 10^{-4}\,\text{m}^2$, $\mu = 8 \times 10^{-3}$ (T·M)/At. Substitute these values in the formula for reluctance.

$$\mathcal{R} = \frac{I}{\mu A} \tag{10-7}$$

$$= \frac{0.2}{(8 \times 10^{-3})(3 \times 10^{-4})} = 83\,300\,\text{At/Wb} \qquad Ans.$$

10.9 If the magnetic circuit of Problem 10.8 has an air gap of 0.2 cm in addition to the 20 cm of annealed steel path, what is the reluctance of the air and how many ampere-turns would be needed to maintain a B of 0.72 T? Assume that the area of the air gap is the same as the area of the steel core.

The total reluctance of the magnetic circuit, $\mathcal{R}_T$, is the reluctance of the steel path plus the reluctance of the air gap. The μ_0 of air is 1.26×10^{-6} (T · m)/At. The reluctance $\mathcal{R}$ of the steel, as determined in Problem 10.8, is 83 300 At/Wb. The reluctance of the air gap is

$$\mathcal{R}_A = \frac{1}{\mu_0 A} \tag{10-7}$$

$$= \frac{2 \times 10^{-3}}{(1.26 \times 10^{-6})(3 \times 10^{-4})} = 5\,290\,000\,\text{At/Wb} \qquad Ans.$$

The total reluctance $\mathcal{R}_T$ is the sum of $\mathcal{R}$ and $\mathcal{R}_A$.

$$\mathcal{R}_T = \mathcal{R} + \mathcal{R}_A = 83\,300 + 5\,290\,000 = 5\,373\,300 = 5.37 \times 10^6\,\text{At/Wb}$$

To maintain a B of 0.72 T requires a total flux of

$$\phi = BA = 0.72(3 \times 10^{-4}) = 216 \times 10^{-6}\,\text{Wb}$$

The mmf in ampere-turns is found by the use of Eq. (*10-6*).

$$\phi = \frac{\text{mmf}}{\mathscr{R}_T}$$

from which

$$\text{mmf} = \mathscr{R}_T\phi = (5.37 \times 10^6)(216 \times 10^{-6}) = 1160\ \text{At} \qquad Ans.$$

10.10 Explain the terms of the induced voltage formula.

The equation is

$$v_{\text{ind}} = N\frac{\Delta\phi}{\Delta t} \qquad\qquad (10\text{-}8)$$

N, the number of turns, is a constant. More turns will provide more induced voltage, while fewer turns mean less voltage. Two factors are included in $\Delta\phi/\Delta t$. Its value can be increased by a higher value of $\Delta\phi$ or a smaller value of Δt. As an example, the value of 4 Wb/s for $\Delta\phi/\Delta t$ can be doubled by either increasing $\Delta\phi$ to 8 Wb or reducing Δt to 1/2s. Then $\Delta\phi/\Delta t$ is 8/1 or 4/(1/2), which equals 8 Wb/s in either case. For the opposite case, $\Delta\phi/\Delta t$ can be reduced by a smaller value of $\Delta\phi$ or a higher value of Δt.

10.11 The hysteresis loop for a magnetic material is shown by plotting a curve of flux density B for a periodically reversing magnetizing force H (Fig. 10-22). For this material what are its (*a*) permeability, (*b*) retentivity, and (*c*) coercive force?

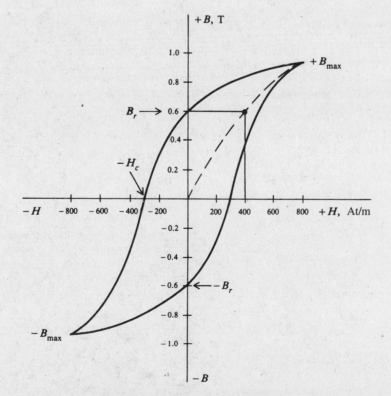

Fig. 10-22 Hysteresis loop, *BH* curve

(a) $\mu = B/H$, which is the slope of the curve from zero at the center (current is zero when the material is unmagnetized) to $+B_{\max}$. Since this curve is normally not a straight line, we approximate its slope from Fig. 10-22. Approximately,

$$\mu = \frac{B}{H} \approx \frac{0.6}{400} = 1.5 \times 10^{-3} \ (\text{T} \cdot \text{m})/\text{At} \qquad Ans.$$

(b) $+B_r$ or $-B_r$ is the flux density remaining after the magnetizing force H has been reduced to zero. This residual induction of a magnetic material is called its *retentivity*. From Fig. 10-22,

$$B_r = 0.6 \ \text{T} \qquad Ans.$$

(c)   The coercive force of the material is $-H_c$, which equals the magnetizing force that must be applied to reduce flux density to zero. From Fig. 10-22,

$$H_c = 300 \ \text{At/m} \qquad Ans.$$

Supplementary Problems

10.12   Match the term in column 1 with its closest meaning in column 2.

Column 1		Column 2
1. North and south poles	(a)	natural magnet
2. Ohm's law	(b)	NI
3. Magnetite	(c)	Iron
4. North pole	(d)	Value of B when $H = 0$
5. Relative permeability	(e)	Force of attraction
6. Induced voltage	(f)	μ_r less than 1
7. Ferromagnetic	(g)	mmf/$\mathcal{R}$
8. Retentivity	(h)	H value when $B = 0$
9. Diamagnetic	(i)	Lines of force cutting a conductor
10. mmf	(j)	B/H
	(k)	μ/μ_0
	(l)	Lines of force flowing from
	(m)	Force of repulsion
	(n)	Aluminum

Ans.   1. (e)   2. (g)   3. (a)   4. (l)   5. (k)   6. (i)   7. (c)   8. (d)   9. (f)   10. (b)

10.13   What is the flux density of a core having 20 000 lines and a cross-sectional area of 5 cm^2?
Ans. $B = 0.4$ T

10.14   Fill in the indicated values. All answers should be in SI units.

	ϕ	B	A			ϕ	B	A
(a)	35 μWb	?	0.001 m^2		(a)	....	0.035 T	....
(b)	?	0.8 T	0.005 m^2		(b)	400 μWb	....	....
(c)	10 000 lines	?	2 cm^2		(c)	....	0.5 T	....
(d)	90 μWb	?	0.003 m^2		(d)	....	0.03 T	....

Ans.

10.15   Draw the lines of force between the south poles of two bar magnets and indicate strong and weak fields.   *Ans.*   See Fig. 10-23.

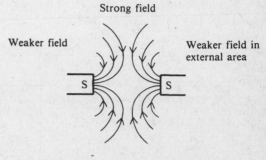

Fig. 10-23

10.16   Draw the lines of force for two parallel conductors having the same direction of current and indicate the strong and weak fields.   *Ans.*   See Fig. 10-24.

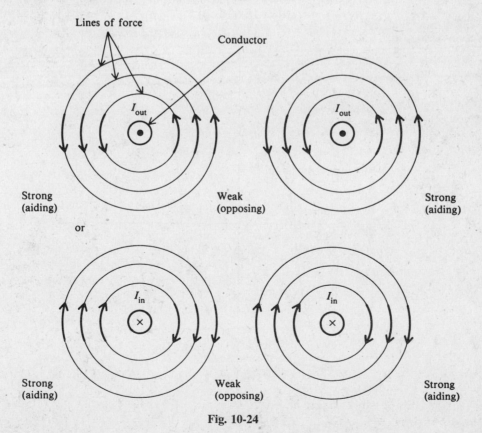

Fig. 10-24

10.17   Compute the ampere-turns rating of an electromagnet wound with 600 turns of wire when it is energized with 3 A of current.   *Ans.* $NI = 1800$ At

10.18   A core of annealed sheet steel is wound with 1500 turns of wire through which a current of 12 mA is flowing. If the length of the coil is 20 cm, find the magnetomotive force and field intensity.
Ans. $NI = 18$ At, $H = 90$ At/m

10.19   A coil has a field intensity of 300 At. Its length is doubled from 20 to 40 cm for the same *NI*. What is the new magnetic field intensity?   *Ans.* $H = 750$ At/m

10.20   An iron core has 250 times more flux density than air for the same field intensity. What is the value of μ_r?   *Ans.* $\mu_r = 250$

10.21   Fill in the indicated values. All answers are in SI units.

	B, T	H, At/m	μ, (T · m)/At	μ_r	*Ans.*	B, T	H, At/m	μ, (T · m)/At	μ_r
(a)	?	1200	650×10^{-6}	?	(a)	0.78	....	....	516
(b)	?	1000	?	200	(b)	0.25	....	252×10^{-6}	....
(c)	0.8	?	?	500	(c)	....	1270	630×10^{-6}	....
(d)	0.1	150	?	?	(d)	....	....	667×10^{-6}	529

10.22   A *BH* curve for soft iron is shown (Fig. 10-25). Find the value of permeability, retentivity, and coercive force.   *Ans.* $\mu = B/H \approx 0.4/200 = 2000 \times 10^{-6}$ (T · m)/At; $B_r = 0.4$ T; $H_c = 200$ At/m

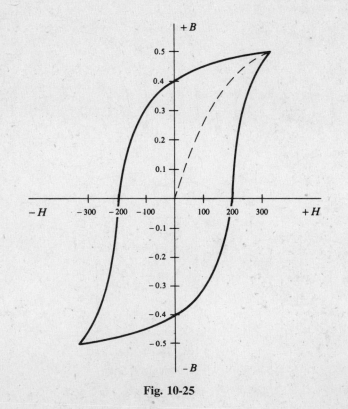

Fig. 10-25

10.23 When it has an annealed iron core, a coil has flux density of 1.44 T at a field intensity of 500 At/m. Find μ and μ_r.   *Ans.* $\mu = 2880 \times 10^{-6}$ (T · m)/At; $\mu_r = 2290$

10.24 The μ of an annealed iron core is 5600×10^{-6} (T·m)/At when the current is 80 mA. The coil consists of 200 turns on core 20 cm in length. Find H, B, and μ_r.
Ans. $H = 80$ At/m; $B = 0.45$ T; $\mu_r = 4440$

10.25 A coil of 100 turns is 8 cm in length. The current in the coil is 0.2 A. If the core is cast iron with a B of 0.13 T, find H, μ, and μ_r.
Ans. $H = 250$ At/m; $\mu = 520 \times 10^{-6}$ (T · m)/At; $\mu_r = 413$

10.26 If the core in Problem 10.25 has a cross-sectional area of 2 cm², find the reluctance and the mmf of this magnetic circuit.   *Ans.* $\mathcal{R} = 769\,000$ At/Wb; mmf = 20 At

10.27 A coil has 200 At (Fig. 10-26a) with a flux of 25 μWb in the iron core. Calculate the reluctance. If the reluctance of the path with an air gap were 800×10^6 At/Wb (Fig. 10-26b), how much mmf would be needed for the same flux of 25 μWb?   *Ans.* $\mathcal{R} = 8 \times 10^6$ At/Wb; mmf = 20 000 At

10.28 A magnetic flux of 1000 lines cuts across a coil of 800 turns in 2 μs. What is the voltage induced in the coil?   *Ans.* $v_{\text{ind}} = 4$ kV

10.29 In a stationary field coil of 500 turns, calculate the induced voltage produced by the following flux changes: (a) 4 Wb increasing to 6 Wb in 1 s; (b) 6 Wb decreasing to 4 Wb in 1 s; (c) 4000 lines of flux increasing to 5000 lines in 5 μs; (d) 4 Wb remaining the same over 1 s.
Ans.   (a) $v_{\text{ind}} = 1$ kV;   (b) $v_{\text{ind}} = 1$ kV;   (c) $v_{\text{ind}} = 1$ kV;   (d) $v_{\text{ind}} = 0$ V

10.30 A magnetic circuit has a 10-V battery connected to a 50-Ω coil of 500 turns with an iron core of 20 cm in length (Fig. 10-27). Find (a) mmf; (b) field intensity H; (c) flux density B in a core with μ_r of 600; and (d) the total flux ϕ at each pole with an area of 4 cm².
Ans.   (a) mmf = 100 At;   (b) $H = 500$ At/m;   (c) $B = 0.378$ T;   (d) $\phi = 1.51 \times 10^{-4}$ Wb

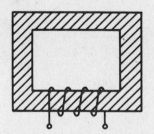

(a)

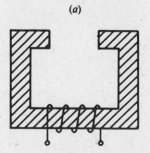

(b)

Fig. 10-26

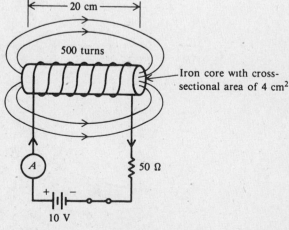

Fig. 10-27

10.31   If a conductor cuts 3.5 Wb in 0.25 s, what is the voltage induced in the conductor?
           Ans. $v_{ind} = 14$ V

10.32   If the iron core is removed from the coil in Problem 10.30, (*a*) What will the flux be in the air-core
           coil? (*b*) What value of induced voltage would be produced by this change in flux while the core
           is being moved out in 1/2 s? (*c*) What is the induced voltage after the core is removed and the flux
           remains constant?
           Ans.   (*a*) $\phi = 2.52 \times 10^{-7}$ Wb;   (*b*) $v_{ind} = 0.151$ V;   (*c*) $v_{ind} = 0$ V

10.33   The N pole of a permanent magnet is moved away from the coil (Fig. 10.28*a*). What is the polarity
           of the coil and the direction of induced current?   *Ans.*   See Fig. 10-28*b*.

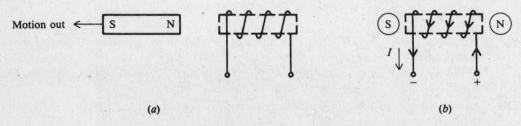

(*a*)   (*b*)

Fig. 10-28

10.34   A *BH* magnetization curve for soft iron has the following values:

B, T	H, At/m
0.126	1000
0.252	2000
0.378	3000
0.428	4000
0.441	5000

(*a*) What is the value of μ? (*b*) Find μ_r. (*c*) At what value of H does the *BH* curve begin to saturate?
Ans.   (*a*) $\mu \approx 1.25 \times 10^{-4}$ (T · m)/At;   (*b*) $\mu_r = 100$;   (*c*) $H = 4000$ At/m

Chapter 11

Direct-Current Generators and Motors

MOTORS AND GENERATORS

A *motor* is a machine that converts electric energy into rotary mechanical energy. Motors turn washing machines, dryers, fans, furnace blowers, and much of the machinery found in industry. A *generator*, on the other hand, is a machine that converts rotary mechanical energy into electric energy. The mechanical energy might be supplied by a waterfall, steam, wind, gasoline or diesel engine, or an electric motor.

Components

The main parts of direct-current motors and generators are basically the same (Fig. 11-1).

Fig. 11-1   Main parts of a dc motor. (*From B. Grob, Basic Electronics, McGraw-Hill, New York, 1977, 4th ed., p. 338*)

Armature

In a motor, the armature receives current from an external electrical source. This causes the armature to turn. In a generator, the armature is rotated by an external mechanical force. The voltage generated in the armature is then connected to an external circuit. In brief, the motor armature receives current from an external circuit (the power supply), and the generator armature delivers current to an external circuit (the load). Since the armature rotates, it is also called a *rotor*.

229

Commutator

A dc machine has a commutator to convert the alternating current flowing in its armature into the direct current at its terminals (in the case of the generator). The commutator (Fig. 11-1) consists of copper segments with one pair of segments for each armature coil. Each commutator segment is insulated from the others by mica. The segments are mounted around the armature shaft and are insulated from the shaft and armature iron. Two stationary brushes are mounted on the frame of the machine so that they contact opposite segments of the commutator.

Brushes

These graphite connectors are stationary and spring-mounted to slide or "brush" against the commutator on the armature shaft. Thus, brushes provide a connection between the armature coils and the external load.

Field Winding

This electromagnet produces the flux cut by the armature. In a motor, current for the field is provided by the same source that supplies the armature. In a generator, the field-current source may come from a separate source called an *exciter* or from its own armature output.

SIMPLE DC GENERATOR

A simple dc generator consists of an armature coil with a single turn of wire. This armature coil cuts across the magnetic field to produce voltage. If a complete path is present, current will move through the circuit in the direction shown by the arrows (Fig. 11-2a). In this position of the coil, commutator segment 1 is in contact with brush 1, while commutator segment 2 is in contact with brush 2. As the armature rotates a half turn in a clockwise direction, the contacts between the commutator segments and the brushes are reversed (Fig. 11-2b). Now, segment 1 is in contact with brush 2 and segment 2 is in contact with brush 1. Because of this commutator action, that side of the armature coil which is in contact with either of the brushes is always cutting across the magnetic field in the same direction. Therefore, brushes 1 and 2 have constant polarity, and a pulsating direct current is delivered to the external load circuit.

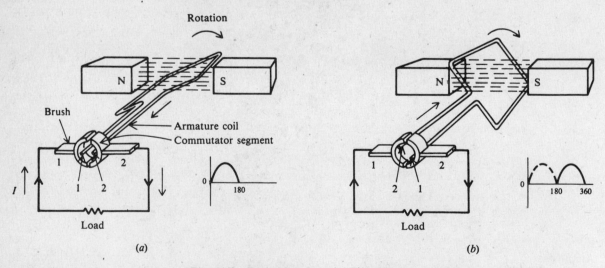

Fig. 11-2   Basic operation of a dc generator

Example 11.1   A dc generator with a single coil produces a pulsating dc output. By using more coils and combining their output, a smoother waveform can be obtained. Draw a voltage output waveform that results when a second coil is added to the armature and placed perpendicular to the first coil.

See Fig. 11-3. Notice that a voltage is included at all times. Although the current still pulsates, the output is smoother. In practical generators, many coils are wound around the armature to produce a still smoother dc output.

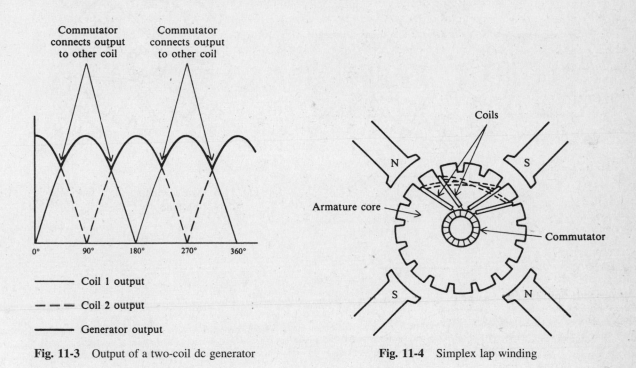

Fig. 11-3	Output of a two-coil dc generator

Fig. 11-4	Simplex lap winding

ARMATURE WINDINGS

The armature coils used in large dc machines are usually wound in their final shape before being put on the armature. The sides of the preformed coil are placed in the slots of the laminated armature core. There are two ways the coils can be connected, *lap* winding and *wave* winding.

In a *simplex lap winding*, the ends of each coil are connected to adjacent commutator segments (Fig. 11-4). In this way all the coils are connected in series. In a *duplex lap winding* there are in effect two separate sets of coils, each set connected in series (Fig. 11-5). The two sets of coils are connected to each other only by the brushes. Similarly, *a triplex lap winding* is in effect three separate sets of series-connected coils. In a simplex lap winding, a single brush short-circuits the two ends of a single coil.

In a *wave winding*, the ends of each coil are connected to commutator segments two pole spans apart (Fig. 11-6). Instead of short-circuiting a single coil, a brush will short-circuit a small group of coils in series.

The area in a generator where no voltage can be induced in an armature coil is called the *commutating* or *neutral plane*. This plane is midway between adjacent north and south field poles. The brushes are always set so that they short-circuit the armature coils passing through the neutral plane while, at the same time, the output is taken from the other coils.

Example 11.2	Explain the commutating action in a simplex lap–wound armature that has 22 coils.

See Fig. 11-7. An armature with 22 coils is connected to 22 commutator segments. There are two brushes. The + brush is short-circuiting armature coil 11, while the − brush is short-circuiting armature coil 22. There is no voltage induced in either of these coils. The two coil groups, 1–10 and 12–21, are connected in parallel by the brushes because the voltages in both coil groups have the same polarity. The brushes also connect the generated voltage to the external load circuit. While the brush is short-circuiting one armature coil, it is receiving the voltage and current induced in the other armature coils because one end of two different coils is connected to the same commutator segment (e.g., coil 21 and coil 22).

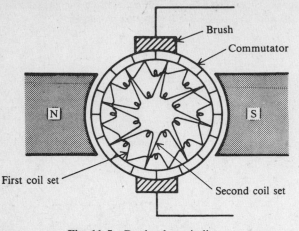

Fig. 11-5 Duplex lap winding

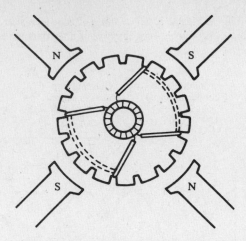

Fig. 11-6 Wave winding for a four-pole dc machine

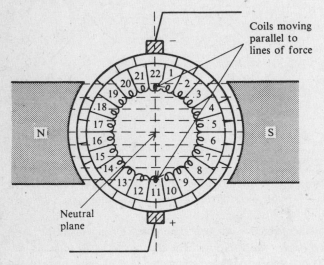

Fig. 11-7 Brush–commutator action in a simplex lap–wound armature

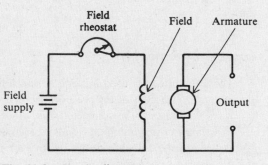

Fig. 11-8 Circuit diagram of separately excited generator

FIELD EXCITATION

DC generators take their names from the type of field excitation used. When the generator's field is supplied or "excited" from a separate dc source, such as a battery, it is called a *separately excited* generator (Fig. 11-8). When a generator supplies its own excitation, it is called a *self-excited* generator. If its field is connected in parallel with the armature circuit, it is called a *shunt* generator (Fig. 11-9*a*). When the field is in series with the armature, the generator is called a *series* generator (Fig. 11-9*b*). If both shunt and series fields are used, the generator is called a *compound* generator. Compound generators may be connected *short-shunt* (Fig. 11-9*c*), with the shunt field in parallel only with the armature, or *long-shunt* (Fig. 11-9*d*), with the shunt field in parallel with both the armature and series field. When the series field is so connected that its ampere-turns act in the same direction as those of the shunt field, the generator is said to be a *cumulative-compound* generator. Field rheostats are adjustable resistances placed in the field circuits to vary the field flux and therefore the emf generated by the generator.

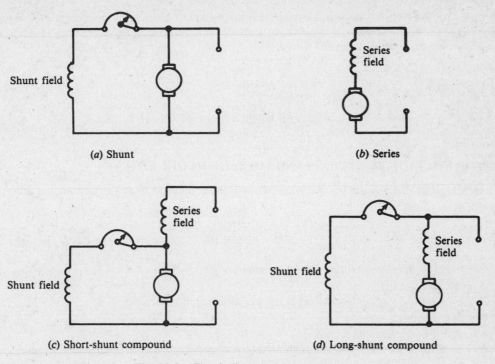

(a) Shunt

(b) Series

(c) Short-shunt compound

(d) Long-shunt compound

Fig. 11-9   Circuit diagrams of dc generators

The compound generator is used more extensively than other types of generators because it can be designed so that it has a wide variety of characteristics.

DC GENERATOR EQUIVALENT CIRCUIT

Voltage and current relationships of a dc generator equivalent circuit (Fig. 11-10) are, according to Ohm's law,

$$V_{ta} = V_g - I_a r_a \qquad (11\text{-}1)$$
$$V_t = V_g - I_a(r_a + r_s) \qquad (11\text{-}2)$$
$$I_L = I_a - I_f \qquad (11\text{-}3)$$

where V_{ta} = armature terminal voltage, V
V_g = armature generated voltage, V
I_a = armature current, A
V_t = generator terminal voltage, V
r_a = armature-circuit resistance, Ω
r_s = series-field resistance, Ω
r_f = shunt-field resistance, Ω
I_L = line current, A
I_f = shunt-field current, A

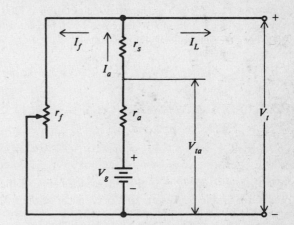

Fig. 11-10   DC generator equivalent circuit

Example 11.3   A dc generator has a 100-kW, 250-V rating. What do these ratings mean?

The generator can continuously deliver 100 kW of power to an external load. The terminal voltage V_t of the generator is 250 V when it is delivering its rated current.

Example 11.4   A 100-kW, 250-V dc generator has an armature current of 400 A, armature resistance (including brushes) of 0.025 Ω, and series-field resistance of 0.005 Ω. It is driven at 1200 revolutions per minute (rpm) by a constant-speed motor. Compute the armature generated voltage.

From Eq. *(11-2)*,

$$V_g = V_t + I_a(r_a + r_s)$$

$$= 250 + 400(0.025 + 0.005) = 250 + 12 = 262 \text{ V} \qquad Ans.$$

GENERATOR VOLTAGE EQUATIONS AND VOLTAGE REGULATION

The average generated voltage V_g of a generator may be calculated from the formula

$$V_g = \frac{pZ\phi n}{60b \times 10^8} \qquad (11\text{-}4)$$

where V_g = average generated voltage of a dc generator, V
   p = number of poles
   Z = total number of conductors on armature (also called inductors)
   ϕ = flux per pole
   n = speed of the armature, rpm
   b = number of parallel paths through armature, depending on type of armature winding

For any generator, all factors in Eq. *(11-4)* are fixed values except ϕ and n. Hence Eq. *(11-4)* may be simplified to

$$V_g = k\phi n \qquad (11\text{-}5)$$

where $k = \dfrac{pZ}{60b \times 10^8}$

Equation *(11-5)* indicates that the value of an induced emf in any circuit is proportional to the rate at which the flux is being cut. Thus, if ϕ is doubled while n remains constant, V_g is doubled. Similarly, if n is doubled, ϕ remaining constant, V_g will be doubled.

Example 11.5   When a generator is being driven at 1200 rpm, the generated voltage is 120 V. What will be the generated voltage (*a*) if the field flux is decreased by 10 percent with the speed remaining fixed and (*b*) if the speed is reduced to 1000 rpm, the field flux staying unchanged?

(*a*)
$$V_{g1} = k\phi_1 n_1 \quad \text{or} \quad k = \frac{V_{g1}}{\phi_1 n_1} \qquad (11\text{-}5)$$

$$V_{g2} = k\phi_2 n_1 = \frac{V_{g1}}{\phi_1 n_1}\phi_2 n_1 = V_{g1}\frac{\phi_2}{\phi_1} = 120\frac{1.00}{1.00 - 0.10} = 120(0.90) = 108 \text{ V} \qquad Ans.$$

(*b*)
$$V_{g2} = k\phi_1 n_2 = \frac{V_{g1}}{\phi_1 n_1}\phi_1 n_2 = V_{g1}\frac{n_2}{n_1} = 120\frac{1000}{1200} = 100 \text{ V} \qquad Ans.$$

Voltage regulation is the difference between the no-load (NL) and full-load (FL) terminal voltage of a generator and is expressed as a percentage of the full-load value.

$$\text{Voltage regulation} = \frac{\text{NL voltage} - \text{FL voltage}}{\text{FL voltage}} \qquad (11\text{-}6)$$

Low-percentage regulation, characteristic of lighting circuits, means that the generator's terminal voltage is nearly the same at full load as it is at no load.

Example 11.6   A shunt generator has a full-load terminal voltage of 120 V. When the load is removed, the voltage increases to 150 V. What is the percentage voltage regulation?

$$\text{Voltage regulation} = \frac{\text{NL voltage} - \text{FL voltage}}{\text{FL voltage}} = \frac{150 - 120}{120} = \frac{30}{120} = 0.25 = 25\% \quad Ans.$$

LOSSES AND EFFICIENCY OF A DC MACHINE

The losses of generators and motors consist of copper losses in the electric circuits and mechanical losses due to the rotation of the machine. Losses include:

1. Copper losses
   - (a) Armature I^2R losses
   - (b) Field losses
     - (1) Shunt field I^2R
     - (2) Series field I^2R
2. Mechanical or rotational losses
   - (a) Iron losses
     - (1) Eddy-current loss
     - (2) Hysteresis loss
   - (b) Friction losses
     - (1) Bearing friction
     - (2) Brush friction
     - (3) Windage or air friction loss

Copper losses are present because power is used when a current is made to flow through a resistance. As the armature rotates in a magnetic field, the emf induced in the iron parts causes eddy currents to flow which heat the iron and thus represent wasted energy. Hysteresis loss also results when a magnetic material is first magnetized in one direction and then in an opposite direction. Other rotational losses are caused by bearing friction, the friction of the brushes riding on the commutator, and air friction or windage.

Efficiency is the ratio of the useful power output to total power input.

$$\text{Efficiency} = \frac{\text{output}}{\text{input}} \tag{11-7}$$

Also,

$$\text{Efficiency} = \frac{\text{input} - \text{losses}}{\text{input}} = \frac{\text{output}}{\text{output} + \text{losses}} \tag{11-8}$$

Efficiency is usually expressed as a percentage.

$$\text{Efficiency (\%)} = \frac{\text{output}}{\text{input}} \times 100$$

Example 11.7   A shunt generator has an armature-circuit resistance of 0.4 Ω, a field-circuit resistance of 60 Ω, and a terminal voltage of 120 V when it is supplying a load current of 30 A (Fig. 11-11). Find the (a) field current, (b) armature current, and (c) copper losses at the above load. (d) If the rotational losses are 350 W, what is the efficiency at the above load?

(a) $I_f = \dfrac{V_t}{r_f} = \dfrac{120}{60} = 2\,\text{A}$ *Ans.*

(b) $I_a = I_L + I_f = 30 + 2 = 32\,\text{A}$ *Ans.*

(c)   Armature loss $= I_a^2 r_a = 32^2(0.4) = 410\,\text{W}$
Shunt-field loss $= I_f^2 r_f = 2^2(60) = 240\,\text{W}$
Copper loss = armature loss + shunt-field loss = $410 + 240 = 650\,\text{W}$ *Ans.*

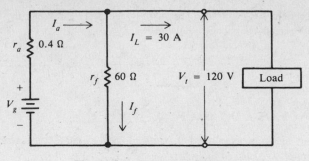

Fig. 11-11   DC shunt generator

(d)
$$\text{Efficiency} = \frac{\text{output}}{\text{output} + \text{losses}} \qquad (11\text{-}8)$$

$\text{Output} = P = V_t I_L = 120(30) = 3600 \text{ W}$

$\text{Total losses} = \text{copper losses} + \text{rotational losses} = 650 + 350 = 1000 \text{ W}$

$\text{Efficiency (\%)} = \dfrac{3600}{3600 + 1000} 100 = \dfrac{3600}{4600} 100 = 0.783(100) = 78.3\% \qquad Ans.$

DIRECT-CURRENT MOTOR

Motor Principle

Although the mechanical construction of dc motors and generators is very similar, their functions are different. The function of a generator is to generate a voltage when conductors are moved through a field, while that of a motor is to develop a turning effort, or *torque*, to produce mechanical rotation.

Direction of Armature Rotation

The *left-hand rule* is used to determine the direction of rotation of the armature conductors. The left-hand rule for motors is as follows: With the forefinger, middle finger, and thumb of the left hand mutually perpendicular, point the forefinger in the direction of the field and the middle finger in the direction of the current in the conductor; the thumb will point in the direction in which the conductor tends to move (Fig. 11-12a). In a single-turn rectangular coil parallel to a magnetic field (Fig. 11-12b), the direction of current in the left-hand conductor is *out* of the paper, while in the right-hand conductor it is *into* the paper. Therefore, the left-hand conductor tends to move upward with a force F_1, and the right-hand conductor tends to move downward with an equal force F_2. Both forces act to develop a torque which turns the coil in a clockwise direction. A single-coil motor (Fig. 11-12b) is impractical because it has dead centers and the torque developed is pulsating. Good results are obtained when a large number of coils are used as in a four-pole motor (Fig. 11-13). As the armature rotates and the conductors move away from under a pole into the neutral plane, the current is reversed in them by the action of the commutator. Thus, the conductors under a given pole carry current in the same direction at all times.

Torque

The torque T developed by a motor is proportional to the strength of the magnetic field and to the armature current.

$$T = k_t \phi I_a \qquad (11\text{-}9)$$

where T = torque, ft-lb

k_t = constant depending on physical dimensions of motor

ϕ = total number of lines of flux entering the armature from one N pole

I_a = armature current, A

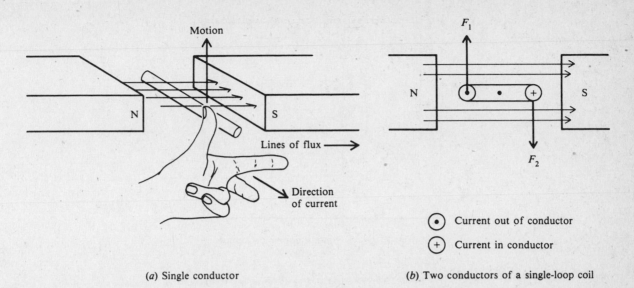

(a) Single conductor   (b) Two conductors of a single-loop coil

Fig. 11-12   Applications of left-hand rule for motors

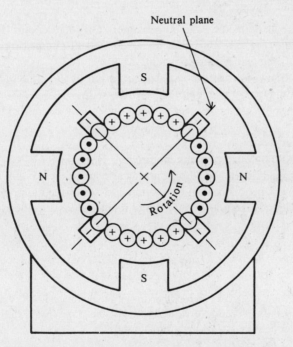

Fig. 11-13   Armature-current directions in a four-pole
motor for counterclockwise rotation

DC MOTOR EQUIVALENT CIRCUIT

Voltage and current relationships of a dc motor equivalent circuit (Fig. 11-14) are as follows:

$$V_{ta} = V_g + I_a r_a \qquad\qquad (11\text{-}10)$$

$$V_t = V_g + I_a(r_a + r_s) \qquad\qquad (11\text{-}11)$$

$$I_L = I_a + I_f \qquad\qquad (11\text{-}12)$$

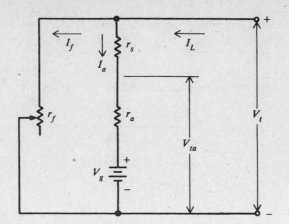

Fig. 11-14 DC motor equivalent circuit

where V_{ta} = armature terminal voltage, V
         V_g = counter emf, V
         I_a = armature current, A
         V_t = motor terminal voltage, V

and r_a, r_s, r_f, I_L, and I_f are as defined for the dc generator equivalent circuit (Fig. 11-10). A comparison of a generator equivalent circuit (Fig. 11-10) with a motor equivalent circuit (Fig. 11-14) shows that the only difference is the direction of line and armature current.

The counter emf of a motor, V_g, is generated by the action of the armature conductors cutting lines of force. If in a shunt motor, Eq. (*11-11*) is multiplied by I_a ($r_s = 0$),

$$V_t I_a = V_g I_a + I_a^2 r_a \qquad (11\text{-}13)$$

$V_t I_a$ is the power supplied to the armature of the motor; $I_a^2 r_a$ is the power lost as heat in the armature current; and $V_t I_a$ is the power developed by the armature. But this armature power is not the useful output since some of it must be used to overcome the mechanical or rotational losses of the motor. The rated output of the motor is equal to the input ($V_t I_L$) less the heat losses ($I^2 R$) and rotational losses. The common unit for motor output is *horsepower* (hp), where

$$\text{Horsepower} = \frac{\text{watts}}{746} \qquad (11\text{-}14)$$

Example 11.8   (*a*) Find the counter emf of a motor when the terminal voltage is 240 V and the armature current is 50 A. The armature resistance is 0.08 Ω. The field current is negligible. (*b*) What is the power developed by the motor armature? (*c*) What is the power delivered to the motor in kilowatts?

(*a*) $V_t = V_g + I_a r_a \qquad r_s = 0$
     $V_g = V_t - I_a r_a = 240 - 50(0.08) = 240 - 4 = 236$ V   *Ans.*
(*b*)   Power developed $= V_g I_a = 236(50) = 11\,800$ W

     $\text{Horsepower} = \dfrac{\text{watts}}{746} = \dfrac{11\,800}{746} = 15.8$ hp   *Ans.*
(*c*)   Power delivered $= V_t I_L = 240(50) = 12\,000$ W $= 12$ kW   *Ans.*

SPEED OF A MOTOR

Speed is designated by the number of revolutions of the shaft with respect to time and is expressed in units of revolutions per minute (rpm). A reduction of the field flux of a motor causes the motor speed to increase.

Conversely, an increase in the field flux causes the motor speed to decrease. Because the speed of a motor varies with field excitation, a convenient means for controlling the speed is to vary the field flux by adjusting the resistance in the field circuit.

If a motor is able to maintain a nearly constant speed for varying loads, the motor is said to have a good speed regulation. Speed regulation is usually expressed as a percentage as follows:

$$\text{Speed regulation} = \frac{\text{no-load speed} - \text{full-load speed}}{\text{full-load speed}} \qquad (11\text{-}15)$$

Example 11.9 A 220 V shunt motor has an armature resistance of 0.2 Ω. For a given load on the motor, the armature current is 25 A. What is the immediate effect on the torque developed by the motor if the field flux is reduced by 2 percent?

The torque developed when $I_a = 25$ A is

$$T_1 = k_t \phi I_a = 25 k_t \phi \qquad (11\text{-}9)$$

and the counter emf is

$$V_{g1} = V_t - I_a r_a = 220 - 25(0.2) = 215 \text{ V}$$

If ϕ is reduced by 2 percent, the value of V_g is also reduced by 2 percent since $V_g = k \phi n$ and the speed n cannot change instantly. Hence, the new counter emf is

$$V_{g2} = 0.98(215) = 210.7 \text{ V}$$

The new armature current is

$$I_{a2} = \frac{V_t - V_{g2}}{r_a} = \frac{220 - 210.7}{0.2} = 46.5 \text{ A}$$

and the new torque developed is

$$T_2 = k_t (0.98) \phi (46.5) = 45.6 k_t \phi$$

The torque increase is

$$\frac{T_2}{T_1} = \frac{45.6 k_t \phi}{25 k_t \phi} = 1.82 \text{ times} \qquad Ans.$$

Thus a decrease in flux by 2 percent increases the torque of a motor 1.82 times. This increased torque causes the armature speed to increase to a higher value, at which the increased counter emf ($V_g \propto n$) limits the armature current to a value just high enough to carry the load at the higher speed.

Example 11.10 The no-load speed of a dc shunt motor is 1200 rpm. When the motor carries its rated load, the speed drops to 1140. What is the speed regulation?

$$\text{Speed regulation} = \frac{\text{NL speed} - \text{FL speed}}{\text{FL speed}} \qquad (11\text{-}15)$$

$$= \frac{1200 - 1140}{1140} = 0.053 = 5.3\% \qquad Ans.$$

MOTOR TYPES

Shunt Motor

This is the most common type of dc motor. It is connected in the same way as the shunt generator (Fig. 11-15a). Its characteristic speed-load and torque-load curves (Fig. 11-15b) show that the torque increases linearly with an increase in armature current, while the speed drops slightly as the armature current is increased. The basic speed is the full-load speed. Speed adjustment is made by inserting resistance in the field circuit

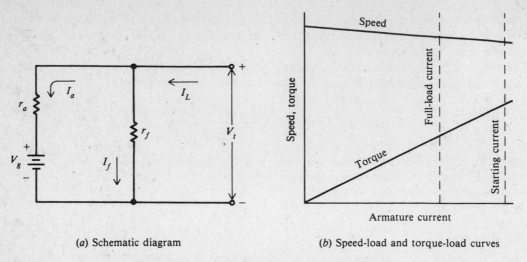

(a) Schematic diagram

(b) Speed-load and torque-load curves

Fig. 11-15 Characteristics of a typical shunt motor

with a field rheostat. At one setting of the rheostat, the motor speed remains practically constant for all loads. Starters used with dc motors limit the armature starting current to 125–200 percent of full-load current. Care must be taken never to open the field circuit of a shunt motor that is running unloaded because the motor speed will increase without limit until the motor destroys itself.

Series Motor

The field of this type of motor is connected in series with the armature (Fig. 11-16a). The speed varies from a very high speed at light load to a lower speed at full load (Fig. 11-16b). The series motor is suitable for starting with heavy, connected loads (driving cranes and winches) because at high armature currents, it develops a high torque and operates at low speed (Fig. 11-16b). At no load, the speed of a series motor will increase without limit until the motor destroys itself (Fig. 11-16b). Large series motors are therefore generally connected directly to their load rather than by belts and pulleys.

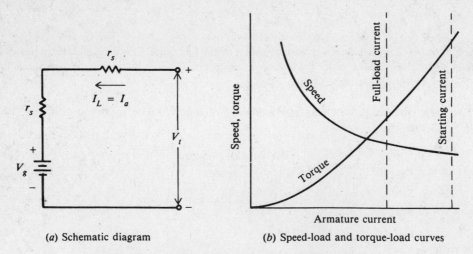

(a) Schematic diagram

(b) Speed-load and torque-load curves

Fig. 11-16 Characteristics of a typical series motor

Compound Motor

It combines the operating characteristics of the shunt and series motors (Fig. 11-17a and b). The compound motor may be operated safely at no load. As load is added, its speed decreases, and torque is greater compared with that of a shunt motor (Fig. 11-18).

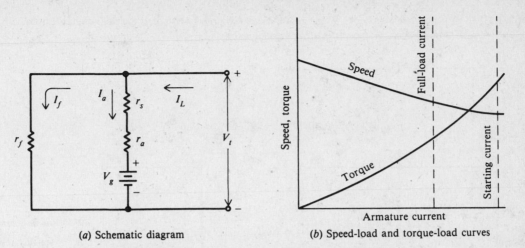

(a) Schematic diagram　　　　(b) Speed-load and torque-load curves

Fig. 11-17　Characteristics of a typical compound motor

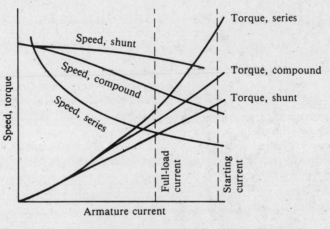

Fig. 11-18　Comparative characteristics for shunt, series, and compound dc motors

STARTING REQUIREMENTS FOR MOTORS

There are two starting requirements for motors:

1. Both motor and supply lines are protected from flow of excessive current during the starting period by placing external resistance in series with the armature circuit.

2. Motor-starting torque should be made as large as possible to bring the motor up to full speed in minimum time.

The amount of starting resistance needed to limit the armature starting current to the desired value is

$$R_s = \frac{V_t}{I_s} - r_a \qquad (11\text{-}16)$$

where　R_s = starting resistance, Ω
　　　　V_t = motor voltage, V
　　　　I_s = desired armature starting current, A
　　　　r_a = armature resistance, Ω

Example 11.11   A shunt motor on a 240-V line has an armature current of 75 A. If the field-circuit resistance is 100 Ω, find the field current, line current, and power input to the motor (Fig. 11-19).

$$I_f = \frac{V_t}{r_f} = \frac{240}{100} = 2.4 \text{ A} \qquad Ans.$$

$$I_L = I_f + I_a = 2.4 + 75 = 77.4 \text{ A} \qquad Ans.$$

$$P_{1N} = V_t I_L = 240(77.4) = 18\,576 \text{ W} = 18.6 \text{ kW} \qquad Ans.$$

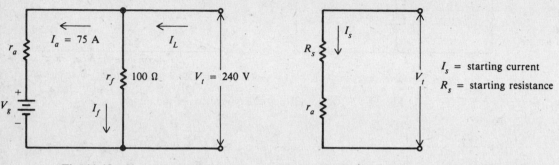

Fig. 11-19   Shunt motor   **Fig. 11-20**   Equivalent dc motor circuit at start-up

Example 11.12   A 10-hp shunt motor with an armature resistance of 0.5 Ω is connected directly to a 220-V supply line.
What is the resulting current if the armature is held stationary? Neglect the field current. If the full-load armature current is 40 A and it is desired to limit the starting current to 150 percent of this value, find the starting resistance that must be added in series with the armature.

At start-up, when a motor armature is stationary, no counter emf is being generated. The only factor limiting the current being drawn from the supply, therefore, is the armature-circuit resistance (Fig. 11-20). At motor start-up with $R_s = 0$, $V_g = 0$, and negligible shunt current,

$$I = \frac{V_t}{r_a} = \frac{220}{0.5} = 440 \text{ A} \qquad Ans.$$

which is far above the normal full-load armature current for a motor of this size. The result will be probable damage to brushes, commutator, and windings. With R_s added in series in the armature circuit,

$$R_s = \frac{V_t}{I_s} - r_a \qquad\qquad (11\text{-}16)$$

$$= \frac{220}{40(1.5)} - 0.5 = 3.67 - 0.5 = 3.17 \ \Omega \qquad Ans.$$

Solved Problems

11.1   A generator has an emf of 520 V, has 2000 armature conductors or inductors, a flux per pole of 1 300 000 lines, a speed of 1200 rpm, and the armature has four paths. Find the number of poles.

From Eq. (*11-4*),

$$p = \frac{V_g \left(60b \times 10^8\right)}{Z\phi n} = \frac{520 \left[60(4) \times 10^8\right]}{\left(2 \times 10^3\right)\left(1.3 \times 10^6\right)\left(1.2 \times 10^3\right)} = 4 \text{ poles} \qquad Ans.$$

11.2   A shunt-field winding of a 240-V generator has a resistance of 50 Ω (Fig. 11-21). How much field-rheostat resistance must be added to limit the field current to 3 A when the generator is operating at rated voltage?

Solve Ohm's law for r: $I_f = \dfrac{V_t}{r_f + r}$

$$r = \frac{V_t - I_f r_f}{I_f} = \frac{240 - 3(50)}{3} = 30\ \Omega \qquad Ans.$$

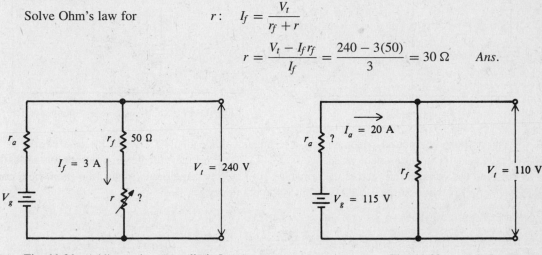

Fig. 11-21   Adding resistance to limit I_f **Fig. 11-22**

11.3   The terminal voltage of a shunt generator is 110 V when the generated voltage is 115 V and the armature current is 20 A (Fig. 11-22). What is the armature resistance?

The generated voltage minus the voltage drop across the armature equals the terminal voltage.

$$V_g - I_a r_a = V_t \qquad\qquad (11\text{-}2)$$

Solve for r_a: $r_a = \dfrac{V_g - V_t}{I_a} = \dfrac{115 - 110}{20} = 0.25\ \Omega \qquad Ans.$

11.4   The terminal voltage of a 75-kW shunt generator is 600 V at rated load. The resistance of the shunt field is 120 Ω and the armature resistance is 0.2 Ω (Fig. 11-23). Find the generated emf.

The rated current is

$$I_L = \frac{P}{V_t} = \frac{75\,000}{600} = 125\ \text{A}$$

$$I_f = \frac{V_t}{r_f} = \frac{600}{120} = 5\ \text{A}$$

$$I_a = I_f + I_L = 5 + 125 = 130\ \text{A}$$

$$V_g = V_t + I_a r_a = 600 + 130(0.2) = 626\ \text{V} \qquad Ans.$$

11.5   A shunt generator requires 50-hp input from its prime mover when it delivers 150 A at 240 V. Find the efficiency of the generator.

$$\text{Output} = 240(150) = 36\,000\ \text{W}$$

$$\text{Input} = 50(746) = 37\,300\ \text{W}$$

$$\text{Efficiency (\%)} = \frac{\text{output}}{\text{input}} \times 100 \qquad\qquad (11\text{-}7)$$

$$= \frac{36\,000}{37\,300}\,100 = 0.965(100) = 96.5\% \qquad Ans.$$

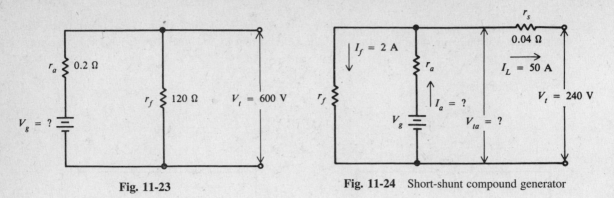

Fig. 11-23   **Fig. 11-24**   Short-shunt compound generator

11.6   A short-shunt compound generator has a terminal voltage of 240 V when the line current is 50 A (Fig. 11-24). The series-field resistance is 0.04 Ω. (*a*) Find the voltage drop across the series field. (*b*) Find the voltage drop across the armature. (*c*) Find the armature current if the shunt-field current is 2 A. (*d*) If the losses are 2000 W, what is the efficiency?

(*a*) $I_L r_s = 50(0.04) = 2 \text{ V}$ *Ans.*

(*b*) $V_{ta} = V_t + I_L r_s = 240 + 2 = 242 \text{ V}$ *Ans.*

(*c*) $I_a = I_f + I_L = 2 + 50 = 52 \text{ A}$ *Ans.*

(*d*)   Output $= V_t I_L = 240(50) = 12\,000 \text{ W}$

$$\text{Efficiency (\%)} = \frac{\text{output}}{\text{output} + \text{losses}} \times 100 \qquad (11\text{-}8)$$

$$= \frac{12\,000}{12\,000 + 2000} 100 = \frac{12\,000}{14\,000} 100 = 0.857(100) = 85.7\% \qquad \textit{Ans.}$$

11.7   From the following data on a shunt generator (Fig. 11-25), find the efficiency at full load:

$$\text{Rated power output} = 10 \text{ kW}$$

$$\text{Rated voltage} = 230 \text{ V}$$

$$\text{Armature resistance} = 0.6 \text{ Ω}$$

$$\text{Field resistance} = 182 \text{ Ω}$$

$$\text{Rotational losses at full load} = 700 \text{ W}$$

First find the generator currents and then the copper losses.

$$I_L = \frac{\text{power out}}{V_t} = \frac{10\,000}{230} = 43.48 \text{ A}$$

$$I_f = \frac{V_t}{r_f} = \frac{230}{182} = 1.26 \text{ A}$$

$$I_a = I_f + I_L = 1.26 + 43.48 = 44.74 \text{ A}$$

Copper losses:

$$\begin{aligned}
\text{Armature:} \quad & I_a^2 r_a = (44.74)^2 (0.6) = 1201 \text{ W} \\
\text{Field:} \quad & I_f^2 r_f = (1.26)^2 (182) = \underline{289 \text{ W}} \\
& \text{Total copper losses} = 1490 \text{ W}
\end{aligned}$$

Total losses:

$$\begin{array}{ll} \text{Copper:} & 1490 \text{ W} \\ \text{Rotational:} & \underline{700 \text{ W}} \\ \text{Total} = & 2190 \text{ W} \end{array}$$

$$\text{Efficiency (\%)} = \frac{\text{output}}{\text{output} + \text{losses}} \times 100 \qquad (11\text{-}8)$$

$$= \frac{10\,000}{10\,000 + 2190}100 = \frac{10\,000}{12\,190}100 = 0.820(100) = 82.0\% \qquad Ans.$$

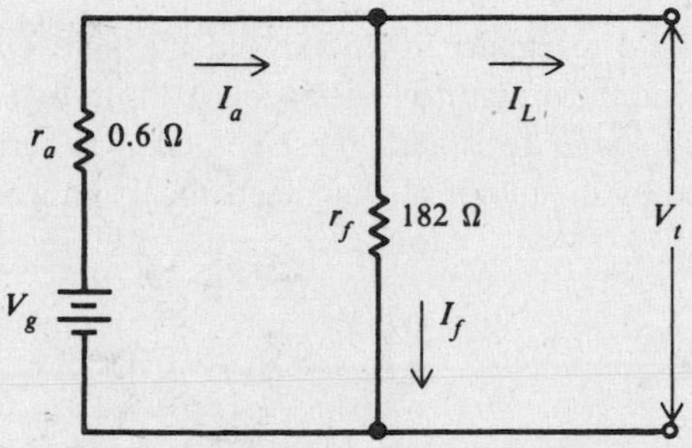

Fig. 11-25

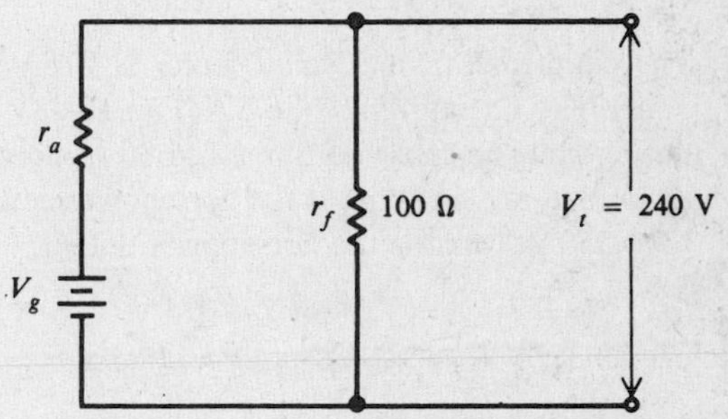

Fig. 11-26

11.8 A shunt motor draws 6 kW from a 240-V line (Fig. 11-26). If the field resistance is 100 Ω, find I_L, I_f, and I_a.

$$P_{1N} = V_t I_L$$

$$I_L = \frac{P_{1N}}{V_t s} = \frac{6000}{240} = 25 \text{ A} \qquad Ans.$$

$$I_f = \frac{V_t}{r_f} = \frac{240}{100} = 2.4 \text{ A} \qquad Ans.$$

$$I_a = I_L - I_f = 25 - 2.4 = 22.6 \text{ A} \qquad Ans.$$

11.9 A shunt motor connected to a 120-V line runs at a speed of 1200 rpm when the armature current is 20 A (Fig. 11-27). The armature resistance is 0.05 Ω. Assuming constant field flux, what is the speed when the armature current is 60 A?

Speed is directly proportional to counter emf.

$$V_{g1} = V_t - I_{a1}r_a = 120 - 20(0.05) = 119 \text{ V}$$

$$V_{g2} = V_t - I_{a2}r_a = 120 - 60(0.05) = 117 \text{ V}$$

$$\frac{V_{g1}}{V_{g2}} = \frac{n_1}{n_2}$$

$$n_2 = \frac{V_{g2}}{V_{g1}}n_1 = \frac{117}{119}1200 = 1180 \text{ rpm} \qquad Ans.$$

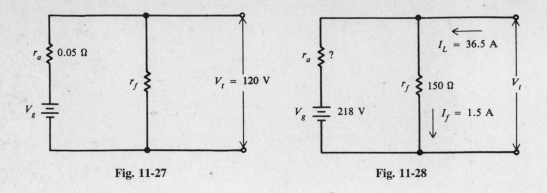

Fig. 11-27   Fig. 11-28

11.10 The counter emf of a shunt motor is 218 V, the field resistance is 150 Ω and the field current is 1.5 A. The line current is 36.5 A (Fig. 11-28). (*a*) Find the armature resistance. (*b*) If the line current during start-up must be limited to 55 A, how much starter resistance must be added in series with the armature? (*c*) What is the horsepower developed by the motor? If the mechanical and iron losses total 550 W, what is the horsepower output?

(*a*) $V_t = I_f r_f = 1.5(150) = 225$ V

$I_a = I_L - I_f = 36.5 - 1.5 = 35$ A

$I_a r_a + V_g = V_t$

$r_a = \dfrac{V_t - V_g}{I_a} = \dfrac{225 - 218}{35} = \dfrac{7}{35} = 0.2 \ \Omega$ *Ans.*

(*b*)   Neglecting the field current,

$$R_s = \frac{V_t}{I_s} - r_a = \frac{225}{55} - 0.2 \qquad\qquad (11\text{-}16)$$

$$= 4.09 - 0.20 = 3.89 \ \Omega \qquad Ans.$$

(*c*)   The horsepower of the motor is the horsepower developed by the armature. The power output is the horsepower available at the motor shaft.

$$V_g I_a = 218(35) = 7630 \text{ W}$$

$$1 \text{ hp} = 746 \text{ W}$$

So
$$\text{hp} = \frac{7630}{746} = 10.2 \qquad Ans.$$

The horsepower output is the horsepower developed by the armature less the power needed to overcome the mechanical or rotational losses of the motor.

$$\text{hp output} = \frac{7630 - 550}{746} = \frac{7080}{746} = 9.5 \qquad Ans.$$

11.11 The efficiency at rated load of a 100-hp 600-V shunt motor is 85 percent (0.85) (Fig. 11-29). The field resistance is 190 Ω and the armature resistance is 0.22 Ω. The full-load speed is 1200 rpm. Find (*a*) the rated line current, (*b*) the field current, (*c*) the armature current at full load, and (*d*) the counter emf at full load.

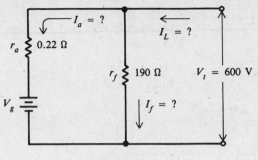

Fig. 11-29

(a) Efficiency $= \dfrac{\text{output}}{\text{input}}$

$$\text{Input} = \frac{\text{output}}{\text{efficiency}} = \frac{100(746)}{0.85} = 87\,765\ \text{W}$$

$V_t I_L = \text{input} = 87\,765\ \text{W}$

$$I_L = \frac{87\,765}{V_t} = \frac{87\,865}{600} = 146.3\ \text{A} \qquad Ans.$$

(b) $I_f = \dfrac{V_t}{r_f} = \dfrac{600}{190} = 3.16\ \text{A} \qquad Ans.$

(c) $I_a = I_L - I_f = 146.3 - 3.2 = 143.1\ \text{A} \qquad Ans.$

(d) $V_g = V_t - I_a r_a = 600 - 143.1(0.22) = 600 - 31.5 = 568.5\ \text{V} \qquad Ans.$

11.12 A long-shunt compound motor has an armature current of 12 A, armature resistance of 0.05 Ω and a series-field resistance of 0.15 Ω (Fig. 11-30). The motor is connected to a 115-V supply. Find (a) the counter emf and (b) the horsepower developed in the armature.

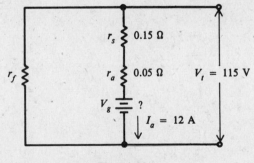

Fig. 11-30

(a) $V_t = V_g + I_a(r_a + r_s)$ $\qquad\qquad\qquad\qquad\qquad\qquad\qquad\qquad\qquad\qquad$ (11-11)

$V_g = V_t - I_a(r_a + r_s) = 115 - 12(0.05 + 0.15) = 115 - 12(0.2) = 115 - 2.4 = 112.6\ \text{V} \qquad Ans.$

(b) Developed hp $= \dfrac{V_g I_a}{746} = \dfrac{112.6(12)}{746} = 1.8\ \text{hp} \qquad Ans.$

11.13 At full load a 15-hp motor draws 55 A from a 240-V line. (a) What is the motor efficiency? (b) What is the motor efficiency at no load?

(a)   Motor input $= I_L V_t = 55(240) = 13\,200$ W
      Motor output $= 15$ hp $= 15(746) = 11\,190$ W

$$\text{Motor efficiency (\%)} = \frac{\text{output}}{\text{input}} \times 100 = \frac{11\,190}{13\,200}100 = 0.848(100) = 84.8\% \qquad Ans.$$

(b)   The output of a motor is considered the power delivered to a load. At no load, motor output is zero. Therefore,

$$\text{Motor efficiency} = 0\% \qquad Ans.$$

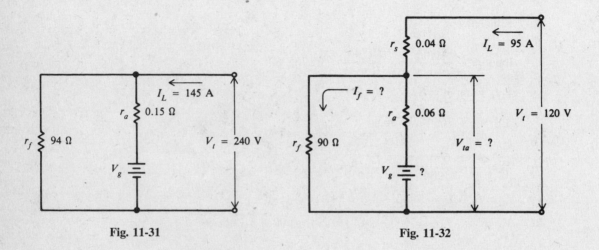

Fig. 11-31   Fig. 11-32

11.14   At rated load the rotational losses (iron losses plus mechanical losses) of a 240-V shunt motor are 900 W (Fig. 11-31). The field resistance is 94 Ω and the armature-circuit resistance is 0.15 Ω. The rated motor current is 145 A. Find (a) the field copper loss, (b) the armature copper loss, (c) the rated horsepower output, and (d) the efficiency.

(a) $$I_f = \frac{V_t}{r_f} = \frac{240}{94} = 2.25 \text{ A}$$

$$\text{Field copper loss} = I_f^2 r_f = (2.55)^2(94) = 611 \text{ W} \qquad Ans.$$

(b) $I_a = I_L - I_f = 145 - 2.25 = 142.5$ A   Ans.

$$\text{Armature copper loss} = I_a^2 r_a = (142.5)^2(0.15) = 3046 \text{ W} \qquad Ans.$$

(c)   Total copper losses $= 611 + 3046 = 3657$ W
      Rotational losses $= \underline{\quad 900 \text{ W}}$
      Total losses $= \overline{4557 \text{ W}}$

   Output $=$ input $-$ total losses

$$= V_t I_L - \text{total losses} = 240(145) - 4557 = 34\,800 - 4557 = 30\,243 \text{ W}$$

$$\text{Rated hp output} = \frac{30\,243}{746} = 40.5 \qquad Ans.$$

(d)   Efficiency (%) = $\dfrac{\text{output}}{\text{input}} \times 100 = \dfrac{30\,243}{34\,800}100 = 0.869(100) = 86.9\%$ *Ans.*

11.15   A 10-hp short-shunt compound motor is supplied by a 120-V source (Fig. 11-32). The full load current is 95 A. The shunt-field resistance is 90 Ω, the armature resistance is 0.06 Ω and the series-field resistance is 0.04 Ω. Find (a) the shunt-field current, (b) the armature current, (c) the counter emf, (d) the efficiency at full load, (e) the full-load copper losses, and (f) the rotational losses.

(a) $V_{ta} = V_t - I_L r_s = 120 - 95(0.04) = 116.2$ V

   $I_f = \dfrac{V_{ta}}{r_f} = \dfrac{116.2}{90} = 1.29$ A   *Ans.*

(b) $I_a = I_L - I_f = 95 - 1.29 = 93.7$ A   *Ans.*

(c) $V_g = V_{ta} - I_a r_a = 116.2 - 93.7(0.06) = 110.6$ V   *Ans.*

(d)   Efficiency = $\dfrac{\text{output}}{\text{input}} = \dfrac{10 \text{ hp}}{V_t I_L} = \dfrac{10(746)}{120(95)} = \dfrac{74\,60}{11\,400} = 0.654$

   Efficiency (%) = 0.654(100) = 65.4%   *Ans.*

(e)   Shunt-field copper loss = $I_f^2 r_f = (1.29)^2(90) = $ 150 W
   Series-field copper loss = $I_L^2 r_s = (95)^2(0.04) = $ 361 W
   Armature copper loss = $I_a^2 r_a = (93.7)^2(0.06) = $ 527 W
                        Total copper loss = $\overline{1038 \text{ W}}$ *Ans.*

(f)   Total losses = input − output = 11 400 − 7460 = 3940 W
   Total losses = total copper losses + rotational losses = 3940
   Rotational losses = 3940 − 1038 = 2902 W   *Ans.*

Supplementary Problems

11.16   How many amperes will a 60-kW 240-V dc generator deliver at full load?   *Ans.*   250 A

11.17   What is the full-load kilowatt output of a dc generator if the full-load line current is 30 A and the terminal voltage is 115 V?   *Ans.*   3.45 kW

11.18   A shunt generator generates 100 V when its speed is 800 rpm. What emf does it generate if the speed is increased to 1200 rpm, the field flux remaining constant?   *Ans.*   150 V

11.19   If the generated voltage of a generator is 120 V and the *IR* drop in the armature circuit is 5 V, what is the terminal voltage?   *Ans.*   115 V

11.20   A 240-V shunt generator has a field resistance of 100 Ω. What is the field current when the generator operates at rated voltage?   *Ans.*   2.4 A

11.21   A shunt generator is rated at 200 kW at 240 V. (a) What is the full-load current? (b) If the field resistance is 120 Ω, what is the field current? (c) What is the full-load armature current?
   Ans.   (a) 833.3 A;   (b) 2 A;   (c) 835.3 A

11.22   In a 50-kW 240-V shunt generator, 260 V is generated in the armature when the generator delivers rated current at rated voltage. The shunt-field current is 4 A. What is the resistance in the armature circuit?   *Ans.* $r_a = 0.049$ Ω

11.23 A shunt generator has a field resistance of 50 Ω in series with a rheostat. When the terminal voltage of the generator is 110 V, the field current is 2 A. How much resistance is cut in on the shunt-field rheostat? *Ans.* 5 Ω

11.24 Find the efficiency at full load of a 50-kW generator when the input is 80 hp. *Ans.* 83.8%

11.25 The losses of a 20-kW generator at full load are 5000 W. What is its efficiency? *Ans.* 80%

11.26 The full-load losses of a 20-kW 230-V shunt generator are as follows:

$$\text{Field } I^2R \text{ loss} = 200 \text{ W}$$

$$\text{Armature } I^2R \text{ loss} = 1200 \text{ W}$$

$$\text{Windage and friction losses} = 400 \text{ W}$$

$$\text{Iron loss} = 350 \text{ W}$$

Find the efficiency at full load. *Ans.* 90.3%

11.27 A short-shunt compound generator delivers 210 A to load at 250 V. Its shunt-field resistance is 24.6 Ω, its shunt-field rheostat resistance is 6.4 Ω, its series-field resistance is 0.038 Ω, and its armature resistance is 0.094 Ω. Find the copper losses in (*a*) the shunt-field winding, (*b*) the shunt-field rheostat, (*c*) the series field, and (*d*) the armature winding. (*e*) If the rotational losses at full load are 800 W, find the efficiency.
Ans. (*a*) 1704 W; (*b*) 443 W; (*c*) 1676 W; (*d*) 4480 W; (*e*) 85.2%

11.28 The voltage of a 110-V generator rises to 120 V when the load is removed. What is the percent of regulation of the generator? *Ans.* 9.1%

11.29 Indicate direction of rotation of the motor armature in Fig. 11.33*a* and *b*.
Ans. (*a*) clockwise; (*b*) counterclockwise

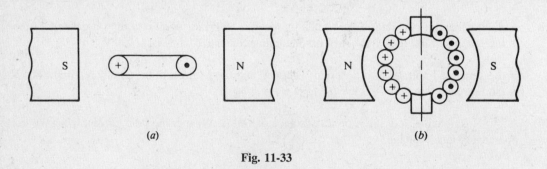

(*a*)   (*b*)

Fig. 11-33

11.30 Find the armature current of a shunt motor when the terminal voltage is 110 V, the counter emf is 108 V, and the armature-circuit resistance is 0.2 Ω. *Ans.* 10 A

11.31 A certain shunt motor is connected to a 240-V line. The armature-circuit resistance is 0.1 Ω. When the armature current is 50 A, what is the counter emf? *Ans.* 235 V

11.32   A shunt motor draws a current of 38 A from a 120-V source. The field-circuit resistance is 50 Ω and the armature-circuit resistance is 0.25 Ω. Find (a) the field current, (b) the armature current, (c) the counter emf, and (d) the counter emf at start-up.
Ans.   (a) 2.4 A;   (b) 35.6 A;   (c) 111.1 V;   (d) 0 V

11.33   A 10-hp motor has a shunt-field resistance of 110 Ω and a field current of 2 A. What is the applied voltage?   *Ans.*   220 V

11.34   What horsepower is developed by a motor when the armature current is 18 A, the applied voltage is 130 V, and the counter emf is 124 V?   *Ans.*   3 hp

11.35   A motor has a no-load speed of 900 rpm and a full-load speed of 855 rpm. What is the speed regulation?
Ans.   5.3%

11.36   The armature resistance of a shunt motor is 0.05 Ω. When the motor is connected across 120 V, it develops a counter emf of 111 V. Find (a) the IR drop in the armature circuit, (b) the armature current, (c) the armature current if the armature were stationary, and (d) the counter emf when the armature current is 155 A.   *Ans.*   (a) 9 V;   (b) $I_a = 180$ A;   (c) $I_a = 2400$ A;   (d) $V_g = 112.2$ V

11.37   The power input to a shunt motor is 5810 W for a given load on the motor. The terminal voltage is 220 V, the $I_a R_a$ drop is 5.4 V, and the armature resistance is 0.25 Ω. Find (a) the counter emf, (b) the power taken by the field, and (c) the field current.
Ans.   (a) $V_g = 214.6$ V;   (b) 1056 W;   (c) $I_f = 4.8$ A

11.38   A 10-hp short-shunt compound motor is supplied by a 120-V source. The full-load current is 86 A. The shunt-field resistance is 90 Ω, the armature resistance is 0.07 Ω, and the series-field resistance is 0.06 Ω. Find (a) the shunt-field current, (b) the armature current, (c) the counter emf, (d) the efficiency at full load, (e) the full-load copper losses, and (f) the rotational losses.
Ans.   (a) $I_f = 1.28$ A;   (b) $I_a = 84.7$ A;   (c) $V_g = 108.9$ V;   (d) 72.3%;   (e) 1093 W;   (f) 1767 W

11.39   If the 10-hp motor of Problem 11.38 is now connected by *long-shunt* and the parameters given remain the same, find the same quantities.
Ans.   (a) $I_f = 1.33$ A;   (b) $I_a = 84.7$ A;   (c) $V_g = 109.0$ V;   (d) 72.3%;   (e) 1091 W;   (f) 1769 W

Chapter 12

Principles of Alternating Current

GENERATING AN ALTERNATING VOLTAGE

An *ac voltage* is one that continually changes in magnitude and periodically reverses in polarity (Fig. 12-1). The zero axis is a horizontal line across the center. The vertical variations on the voltage wave show the changes in magnitude. The voltages above the horizontal axis have positive (+) polarity, while voltages below the horizontal axis have negative (−) polarity.

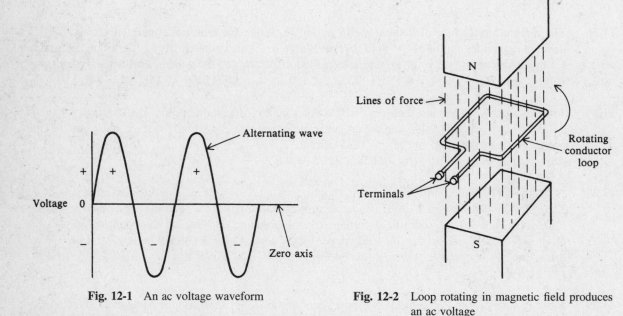

Fig. 12-1 An ac voltage waveform

Fig. 12-2 Loop rotating in magnetic field produces an ac voltage

An ac voltage can be produced by a generator, called an *alternator* (Fig. 12-2). In the simplified generator shown, the conductor loop rotates through the magnetic field and cuts lines of force to generate an induced ac voltage across its terminals. One complete revolution of the loop around the circle is a *cycle*. Consider the position of the loop at each quarter turn during a cycle (Fig. 12-3). At position A, the loop is traveling parallel to the magnetic flux and therefore cuts no lines of force. The induced voltage is zero. At top position B, the loop cuts across the field at 90° to produce maximum voltage. When it reaches C, the conductor is again moving parallel to the field and cannot cut across the flux. The ac wave from A to C is $\frac{1}{2}$ cycle of revolution, called an *alternation*. In D, the loop cuts across the flux again for maximum voltage, but here the flux is cut in the opposite direction (left to right) from B (right to left). Thus the polarity at D is negative. The loop completes the last quarter turn in the cycle where it returns to position A, the point where it started. The cycle of voltage values is repeated in positions $A'B'C'D'A''$ as the loop continues to rotate (Fig. 12-3). A cycle includes the variations between two successive points having the same value and varying in the same direction. For example, 1 cycle can be shown also between B and B' (Fig. 12-3).

ANGULAR MEASUREMENT

Because the cycles of voltage correspond to rotation of the loop around a circle, parts of the circle are expressed in angles. The complete circle is 360°. One half cycle, or one alternation, is 180°. A quarter turn

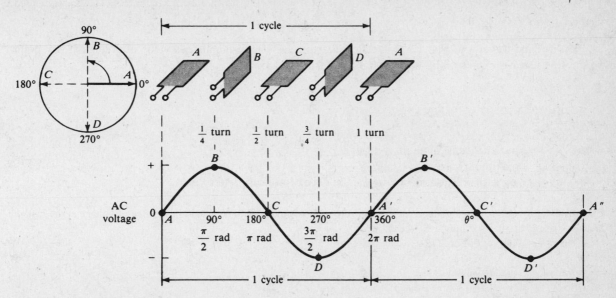

Fig. 12-3 Two cycles of alternating voltage generated by rotating loop. (*From B. Grob, Basic Electronics, 4th ed., McGraw-Hill, New York, 1977, p. 313.*)

is 90°. Degrees are also expressed in *radians* (rad). One radian is equal to 57.3°. A complete circle has 2π rad; therefore

$$360° = 2\pi \text{ rad}$$

Then
$$1° = \frac{\pi}{180} \text{ rad} \qquad (12\text{-}1)$$

or
$$1 \text{ rad} = \frac{180°}{\pi} \qquad (12\text{-}2)$$

In a two-pole generator (Fig. 12-2), the rotation of the armature coil through 360 geometric degrees (1 revolution) will always generate 1 cycle (360°) of ac voltage. But in a four-pole generator, an armature rotation through only 180 geometric degrees will generate 1 ac cycle or 180 electrical degrees. Therefore, the degree markings along the horizontal axis of ac voltage or current refer to *electrical* degrees rather than *geometric* degrees.

Example 12.1 How many radians are there in 30°?
Use Eq. (*12-1*) to convert from degrees into radians.

$$30° = 30° \times \frac{\text{equivalent rad}}{1°} = 30° \times \frac{\pi/180 \text{ rad}}{1°} = \frac{\pi}{6} \text{ rad} \qquad Ans.$$

Example 12.2 How many degrees are there in $\pi/3$ rad?
Use Eq. (*12-2*) to convert from radians into degrees.

$$\frac{\pi}{3} \text{ rad} = \frac{\pi}{3} \text{ rad} \times \frac{\text{equivalent}°}{1 \text{ rad}} = \frac{\pi}{3} \text{ rad} \times \frac{180°/\pi}{1 \text{ rad}} = 60° \qquad Ans.$$

In most handheld calculators, there is a selector switch to designate angles either in degrees or radians (DEG or RAD), so it normally will not be necessary to convert the angles. However, it is useful to know how the angle conversions can be done.

SINE WAVE

The voltage waveform (Fig. 12-3) is called a *sine wave*. The instantaneous value of voltage at any point on the sine wave is expressed by the equation

$$v = V_M \sin \theta \tag{12-3}$$

where v = instantaneous value of voltage, V
$\quad\quad\quad V_M$ = maximum value of voltage, V
$\quad\quad\quad \theta$ = angle of rotation, degrees (θ is the Greek lowercase letter theta)

Example 12.3   A sine wave voltage varies from zero to a maximum of 10 V. What is the value of voltage at the instant that the cycle is at 30°? 45°? 60°? 90°? 180°? 270°?
Substitute 10 for V_M in Eq. (*12-3*):

$$v = 10 \sin \theta$$
At 30°: $v = 10 \sin 30° = 10(0.5) = 5$ V   *Ans.*
At 45°: $v = 10 \sin 45° = 10(0.707) = 7.07$ V   *Ans.*
At 60°: $v = 10 \sin 60° = 10(0.866) = 8.66$ V   *Ans.*
At 90°: $v = 10 \sin 90° = 10(1) = 10$ V   *Ans.*
At 180°: $v = 10 \sin 180° = 10(0) = 0$ V   *Ans.*
At 270°: $v = 10 \sin 270° = 10(-1) = -10$ V   *Ans.*

ALTERNATING CURRENT

When a sine wave of alternative voltage is connected across a load resistance, the current that flows in the circuit is also a sine wave (Fig. 12-4).

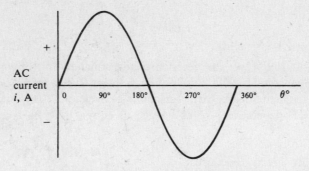

Fig. 12-4   One cycle of alternating current

Example 12.4   The ac sine wave voltage (Fig. 12-5*a*) is applied across a load resistance of 10 Ω (Fig. 12-5*b*). Show the resulting sine wave of alternating current.
The instantaneous value of current is $i = v/R$. In a pure resistance circuit, the current waveform follows the polarity of the voltage waveform. The maximum value of current is

$$I_M = \frac{V_M}{R} = \frac{10}{10} = 1 \text{ A}$$

In the form of an equation $i = I_M \sin \theta$. (See Fig. 12-6.)   *Ans.*

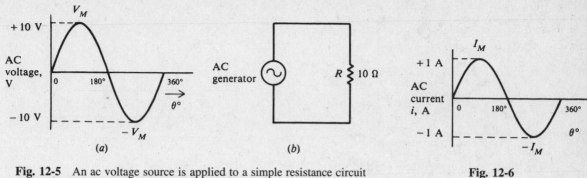

Fig. 12-5 An ac voltage source is applied to a simple resistance circuit   **Fig. 12-6**

FREQUENCY AND PERIOD

The number of cycles per second is called *frequency*. It is indicated by the symbol f and is expressed in hertz (Hz). One cycle per second equals one hertz. Thus 60 cycles per second (formerly abbreviated cps) equals 60 Hz. A frequency of 2 Hz (Fig. 12-7b) is twice the frequency of 1 Hz (Fig. 12-7a).

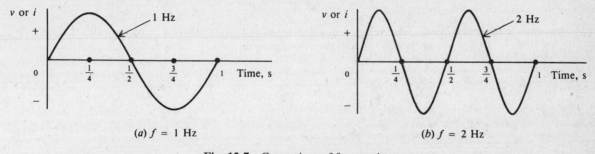

Fig. 12-7 Comparison of frequencies

The amount of time for the completion of 1 cycle is the *period*. It is indicated by the symbol T for time and is expressed in seconds (s). Frequency and period are reciprocals of each other.

$$f = \frac{1}{T} \qquad\qquad (12\text{-}4)$$

$$T = \frac{1}{f} \qquad\qquad (12\text{-}5)$$

The higher the frequency, the shorter the period.

The angle of 360° represents the time for 1 cycle, or the period T. So we can show the horizontal axis of the sine wave in units of either electrical degrees or seconds (Fig. 12-8).

Example 12.5 An ac current varies through one complete cycle in 1/100 s. What are the period and frequency? If the current has a maximum value of 5 A, show the current waveform in units of degrees and milliseconds.

$$T = \frac{1}{100}\,\text{s} \quad\text{or}\quad 0.01\,\text{s} \quad\text{or}\quad 10\,\text{ms} \qquad \textit{Ans.}$$

$$f = \frac{1}{T} \qquad\qquad (12\text{-}4)$$

$$= \frac{1}{1/100} = 100\,\text{Hz} \qquad \textit{Ans.}$$

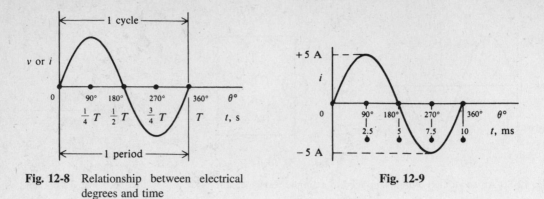

Fig. 12-8 Relationship between electrical degrees and time

Fig. 12-9

See Fig. 12-9 for the waveform.

The *wavelength* λ (Greek lowercase lambda) is the length of one complete wave or cycle. It depends upon the frequency of the periodic variation and its velocity of transmission. Expressed as a formula,

$$\lambda = \frac{\text{velocity}}{\text{frequency}} \qquad (12\text{-}6)$$

For electromagnetic radio waves, the velocity in air or a vacuum is 186 000 mi/s, or 3×10^8 m/s, which is the speed of light. Equation (12-6) is written in the familiar form

$$\lambda = \frac{c}{f} \qquad (12\text{-}7)$$

where   λ = wavelength, m
c = speed of light, 3×10^8 m/s, a constant
f = radio frequency, Hz

Example 12.6   TV Channel 2 has a frequency of 60 Hz. What is its wavelength?
Convert $f = 60$ MHz to $f = 60 \times 10^6$ Hz and substitute into Eq. (12-7).

$$\lambda = \frac{c}{f} = \frac{3 \times 10^8}{60 \times 10^6} = 5\,\text{m} \qquad Ans.$$

PHASE RELATIONSHIPS

The *phase angle* between two waveforms of the same frequency is the angular difference at a given instant of time. As an example, the phase angle between waves B and A (Fig. 12-10a) is 90°. Take the instant of time at 90°. The horizontal axis is shown in angular units of time. Wave B starts at maximum value and reduces to zero value at 90°. Wave B reaches its maximum value 90° ahead of wave A, so wave B leads wave A by 90°. This 90° phase angle between waves B and A is maintained throughout the complete cycle and all successive cycles. At any instant of time, wave B has the value that wave A will have 90° later. Wave B is a cosine wave because it is displaced 90° from wave A, which is a sine wave. Both waveforms are called *sinusoids*.

PHASORS

To compare phase angles or phases of alternating voltages and currents, it is more convenient to use phasor diagrams corresponding to the voltage and current waveforms. A *phasor* is a quantity that has magnitude

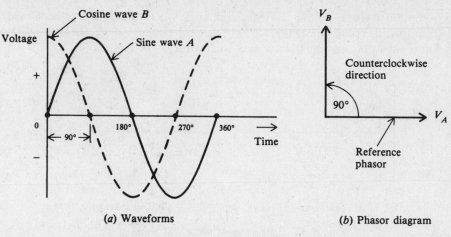

(a) Waveforms   (b) Phasor diagram

Fig. 12-10   Wave B leads wave A by a phase angle of 90°

and direction. The terms *phasor* and *vector* are used for quantities that have direction. However, a phasor quantity varies with time, while a vector quantity has direction in space. The length of the arrow in a phasor diagram indicates the magnitude of the alternating voltage. The angle of the arrow with respect to the horizontal axis indicates the phase angle. One waveform is chosen as the reference. Then the second waveform can be compared with the reference by means of the angle between the phasor arrows. For example, the phasor V_A represents the voltage wave A with a phase angle of 0° (Fig. 12-10b). The phasor V_B is vertical (Fig. 12-10b) to show the phase angle of 90° with respect to phasor V_A, which is the reference. Since lead angles are shown in the counterclockwise direction from the reference phasor, V_B *leads* V_A by 90° (Fig. 12-10b).

Generally, the reference phasor is horizontal, corresponding to 0°. If V_B were shown as the reference (Fig. 12-11b), V_A would have to be 90° clockwise in order to have the same phase angle. In this case V_A *lags* V_B by 90°. There is no fundamental difference between V_B leading V_A by 90° (Fig. 12-11a) or V_A lagging V_B by 90° (Fig. 12-11b).

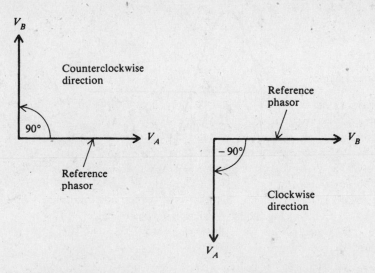

(a) V_B leads V_A by 90°   (b) V_A lags V_B by 90°

Fig. 12-11   Leading and lagging phase angles

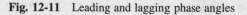

When two waves are in phase (Fig. 12-12a), the phase angle is zero. Then the amplitudes add (Fig. 12-12b). When two waves are exactly out of phase (Fig. 12-13a), the phase angle is 180°. Their amplitudes are opposing (Fig. 12-13b). Equal values of opposite phase cancel each other.

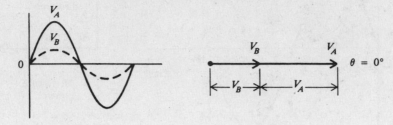

(a) Waveforms   (b) Phasor diagram

Fig. 12-12   Two waves in phase with angle of 0°

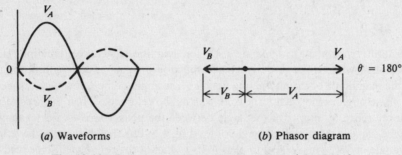

(a) Waveforms   (b) Phasor diagram

Fig. 12-13   Two waves in opposite phase with angle of 180°

Example 12.7   What is the phase angle between waves A and B (Fig. 12-14)? Draw the phasor diagram first with wave A as reference and then with wave B as reference.

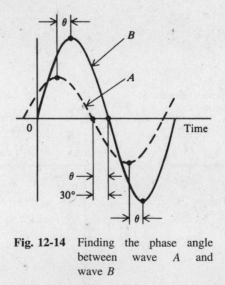

Fig. 12-14   Finding the phase angle between wave A and wave B

The phase angle is the angular distance between corresponding points on waves A and B. Convenient corresponding points are the maximum, minimum, and zero crossing of each wave. At the zero crossings on the horizontal axis (Fig. 12-14), the phase angle $\theta = 30°$. Since wave A reaches its zero crossing before wave B does, A leads B.

Wave A as reference: V_B lags V_A by $30°$.

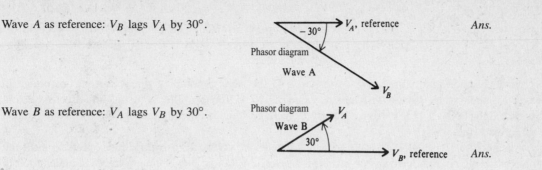

Wave B as reference: V_A lags V_B by $30°$.

Though the phasors are not drawn to scale, V_A is drawn smaller than V_B because the maximum value of wave A is less than that of wave B.

CHARACTERISTIC VALUES OF VOLTAGE AND CURRENT

Since an ac sine wave voltage or current has many instantaneous values throughout the cycle, it is convenient to specify magnitudes for comparing one wave with another. The peak, average, or root-mean-square (rms) value can be specified (Fig. 12-15). These values apply to current or voltage.

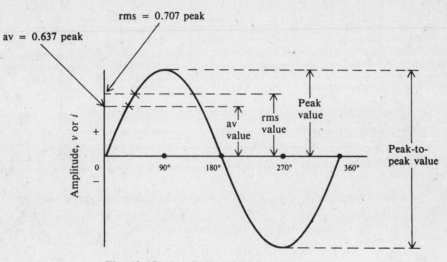

Fig. 12-15 Amplitude values for ac sine wave

The *peak value* is the maximum value of V_M or I_M. It applies to either the positive or negative peak. The *peak-to-peak (p-p) value* may be specified and is double the peak value when the positive and negative peaks are symmetrical.

The *average value* is the arithmetic average of all values in a sine wave for 1 half-cycle. The half-cycle is used for the average because over a full cycle the average value is zero.

$$\text{Average value} = 0.637 \times \text{peak value} \qquad (12\text{-}8)$$

or

$$V_{av} = 0.637\,V_M$$

$$I_{av} = 0.637\,I_M$$

The *root-mean-square value* or *effective value* is 0.707 times the peak value.

$$\text{rms value} = 0.707 \times \text{peak value} \qquad (12\text{-}9)$$

or
$$V_{rms} = 0.707\, V_M$$
$$I_{rms} = 0.707\, I_M$$

The rms value of an alternating sine wave corresponds to the same amount of direct current or voltage in heating power. An alternating voltage with an rms value of 115 V, for example, is just as effective in heating the filament of a light bulb as 115 V from a steady dc voltage source. For this reason, the rms value is also called the effective value.

Unless indicated otherwise, all sine wave ac measurements are given in rms values. The letters V and I are used to denote rms voltage and current. For instance, $V = 220$ V (an ac power-line voltage) is understood to mean 220 V rms.

Use Table 12-1 as a convenient way to convert from one characteristic value to another.

Table 12-1 Conversion Table for AC Sine Wave Voltage and Current

Multiply the Value	By	To Get the Value
Peak	2	Peak-to-peak
Peak-to-peak	0.5	Peak
Peak	0.637	Average
Average	1.570	Peak
Peak	0.707	rms (effective)
rms (effective)	1.414	Peak
Average	1.110	rms (effective)
rms (effective)	0.901	Average

Example 12.8   If the peak voltage for an ac wave is 60 V, what are its average and rms values?

$$\text{Average value} = 0.637 \times \text{peak value} \qquad\qquad (12\text{-}8)$$
$$= 0.637(60) = 38.2 \text{ V} \qquad Ans.$$
$$\text{rms value} = 0.707 \times \text{peak value} \qquad\qquad (12\text{-}9)$$
$$= 0.707(60) = 42.4 \text{ V} \qquad Ans.$$

Example 12.9   It is often necessary to convert from rms to peak value. Develop the formula.
Start with

$$\text{rms value} = 0.707 \times \text{peak value} \qquad\qquad (12\text{-}9)$$

Then invert:

$$\text{Peak value} = \frac{1}{0.707} \times \text{rms value} = 1.414 \times \text{rms value}$$
$$V_M = 1.414\, V_{rms}$$
or
$$I_M = 1.414\, I_{rms}$$

Verify this relationship by referring to Table 12-1.

Example 12.10   A commercial ac power-line voltage is 240 V. What are the peak and peak-to-peak voltages?

AC measurements are given in rms values unless noted otherwise. From Table 12-1,

$$V_M = 1.414 V_{\text{rms}} = 1.414(240) = 339.4 \text{ V} \qquad Ans.$$

$$V_{p\text{-}p} = 2V_M = 2(339.4) = 678.8 \text{ V} \qquad Ans.$$

RESISTANCE IN AC CIRCUITS

In an ac circuit with only resistance, the current variations are in phase with the applied voltage (Fig. 12-16). This in-phase relationship between V and I means that such an ac circuit can be analyzed by the same methods used for dc circuits. Therefore, Ohm's laws for dc circuits are applicable also to ac circuits with resistance only. The calculations in ac circuits are generally in rms values, unless otherwise specified. For the series circuit (Fig. 12-16a), $I = V/R = 110/10 = 11$ A. The rms power dissipation is $P = I^2 R = 11^2(10) = 1210$ W.

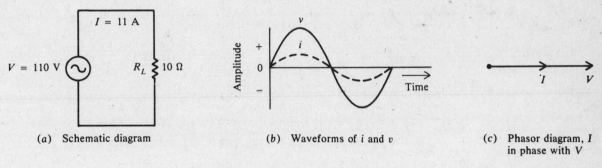

(a)   Schematic diagram   (b)   Waveforms of i and v (c)   Phasor diagram, I in phase with V

Fig. 12-16   AC circuit with only resistance

Example 12.11   A 110-V ac voltage is applied across 5- and 15-Ω resistances in series (Fig. 12-17a). Find the current and voltage drop across each resistance. Draw the phasor diagram.

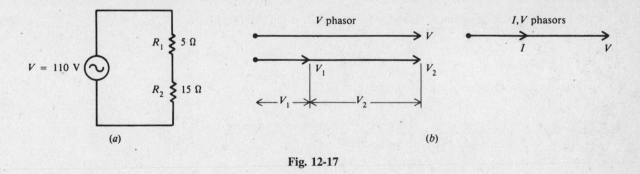

(a)   (b)

Fig. 12-17

Use Ohm's law,

$$R_T = R_1 + R_2 = 5 + 15 = 20 \ \Omega$$

$$I = \frac{V}{R_T} = \frac{110}{20} = 5.5 \text{ A} \qquad Ans.$$

$$V_1 = IR_1 = 5.5(5) = 27.5 \text{ V} \qquad Ans.$$

$$V_2 = IR_2 = 5.5(15) = 82.5 \text{ V} \qquad Ans.$$

Since the ac voltages V_1 and V_2 are in phase, phasors V_1 and V_2 are added to obtain phasor V. See Fig. 12-17*b*. The length of each phasor is proportional to its magnitude. I is in phase with V.

Solved Problems

12.1   Find the instantaneous current when $\theta = 30°$ and $225°$ for the ac current wave (Fig. 12-18) and locate these points on the waveform.

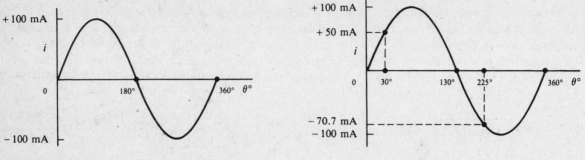

Fig. 12-18   Alternating-current waveform   **Fig. 12-19**

It is seen that $I_M = 100$ mA. The current wave is then

$$i = I_M \sin\theta = 100 \sin\theta$$

At $\theta = 30°$: $i = 100 \sin 30° = 100(0.5) = 50$ mA   *Ans.*
At $\theta = 225°$: $i = 100 \sin 225° = 100(-0.707) = -70.7$ mA   *Ans.*

See Fig. 12-19.

12.2   Many ac waves (e.g., sine wave, square wave) can be produced by a device called a *signal generator*. This unit usually can generate an ac voltage with a frequency as low as 20 Hz or as high as 200 MHz. Three basic control knobs are function, frequency, and amplitude. The operator selects the controls to produce a sine wave (function) at 100 kHz (frequency) with 5 V amplitude (maximum value). Draw 2 cycles of the ac voltage generated. Show both degrees and time units on the horizontal axis.

To obtain time units, solve for the period T, using Eq. (*12-5*).

$$T = \frac{1}{f} = \frac{1}{100 \times 10^3} = 10 \times 10^{-6} \text{ s} = 10 \,\mu\text{s}$$

Now draw the sine wave of voltage (Fig. 12-20).

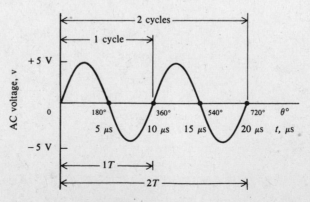

Fig. 12-20   AC voltage wave

12.3   Calculate the time delay for a phase angle of 45° at a frequency of 500 MHz.

Find the period that corresponds to the time for 1 cycle of 360°, and then find the proportional part of the period that corresponds to 45°.

$$T = \frac{1}{f} \tag{12-5}$$

$$= \frac{1}{500} = 2 \times 10^{-3} = 2 \text{ ms}$$

At $\theta = 45°$: $t = \frac{45°}{360°}(2 \text{ ms}) = 0.25 \text{ ms}$ *Ans.*

12.4   The sine wave of an alternating current shows a maximum value of 80 A. What value of dc current will produce the same heating effect?

If an ac wave produces as much heat as 1 A of direct current, we say that the ac wave is as effective as 1 A of direct current. So

$$I_{\text{dc}} = I_{\text{rms}} = 0.707\, I_M \tag{12-9}$$

$$= 0.707(80) = 56.6 \text{ A} \quad \textit{Ans.}$$

12.5   If an ac voltage has a peak value of 155.6 V, what is the phase angle at which the instantaneous voltage is 110 V?

Write $v = V_M \sin \theta$ \hspace{2cm} (12-3)

Solve for θ: $\sin \theta = \dfrac{v}{V_M}$

$$\theta = \arcsin \frac{v}{V_M} = \arcsin \frac{110}{155.6} = 0.707 = 45° \quad \textit{Ans.}$$

12.6   The frequency of the audio range extends from 20 Hz to 20 kHz. Find the range of period and wavelength for this sound wave over the range of audio frequencies.

Range of T: $T = \dfrac{1}{f}$ \hspace{2cm} (12-5)

At 20 Hz: $T = \dfrac{1}{20} = 0.05 \text{ s} = 50 \text{ ms}$

At 20 kHz: $T = \dfrac{1}{20 \times 10^3} = 0.05 \text{ ms}$

So T is from 0.05 to 50 ms   *Ans.*

Range of λ: $\lambda = \dfrac{c}{f}$ \hspace{2cm} (12-7)

where c = speed of light at 3×10^8 m/s

At 20 Hz: $$\lambda = \frac{3 \times 10^8}{20} = 15 \times 10^6 \text{ m}$$

At 20 kHz: $$\lambda = \frac{3 \times 10^8}{20 \times 10^3} = 15 \times 10^3 \text{ m}$$

So λ is from 15×10^3 to 15×10^6 m   *Ans.*

12.7   Find the phase angle for the following ac waves (Fig. 12-21) and draw their phasor diagrams.

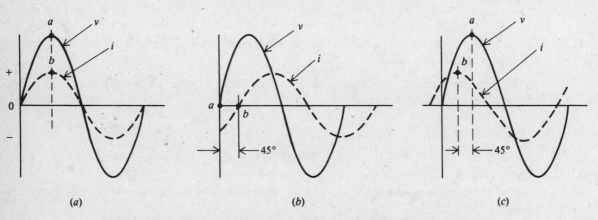

Fig. 12-21   Finding the phase angle between voltage and current waveforms

To determine the phase angle, select a corresponding point on each wave. The maximum and zero crossing corresponding points are convenient. The angular difference of the two points is the phase angle. Then compare the two points to decide if one wave is in phase with, leading, or lagging the other wave.

In Fig. 12-21*a*, curves *v* and *i* reach their maximum values at the same instant, so they are in phase (the phasor diagram as shown).

Phasor diagram: *Ans.*
                                    I   *V*

In Fig. 12-21*b*, curve *v* reaches its zero value at *a*, 45° before curve *i* is zero at the corresponding point *b*, so *v* leads *i* by 45° (phasor diagram as shown).

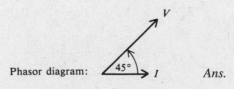

Phasor diagram: *Ans.*

In Fig. 12-21*c*, curve *i* reaches its maximum at *b* before curve *v* reaches its maximum at *a* so *i* leads *v* by 45° (phasor diagram as shown).

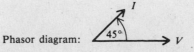

Phasor diagram: *Ans.*

12.8 A, B, and C are three sine ac voltage waveforms of the same frequency. Sine wave A leads sine wave B by a phase angle of 60° and lags sine wave C by 130°. What is the phase angle between wave B and wave C? Which wave is leading?

Draw the sine waves with phase angles described. A convenient way to draw or measure the phase angle between two sine waves is to compare their zero crossings. Wave A is drawn as the reference beginning at 0° (Fig. 12-22). Wave B is drawn beginning at 60° to indicate that wave A leads wave B by 60°. Wave C is shown beginning at −130° to show that A lags C by that angle. Compare the zero crossings on the horizontal axis of waves B and C as they move toward the positive cycle. B crosses the axis upward at 60°, while C does so at 230°. The phase angle is the phase difference between 230° and 60°, or 170°. Since B crosses the axis before wave C, B leads C.

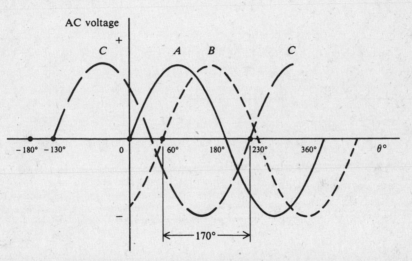

Fig. 12-22   Measuring phase angle between sine waves

12.9   Alternating-current and voltage meters are always calibrated to read effective values. An ac voltmeter indicates that the voltage across a resistive load is 40 V. What is the peak voltage across this load?

From Table 12-1,

$$V_M = 1.414 V \quad \text{(effective or rms value is understood for } V\text{)}$$

$$= 1.414(40) = 56.6\,\text{V} \quad Ans.$$

12.10   The current through an incandescent lamp is measured with an ac ammeter and found to be 0.95 A. What is the average value of this current?

From Table 12-1,

$$I_{av} = 0.901 I \quad \text{(effective or rms value is understood for } I\text{)}$$

$$= 0.901(0.95) = 0.86\,\text{A} \quad Ans.$$

12.11   Find V, period T, frequency f, and peak-to-peak voltage $V_{p\text{-}p}$ of the voltage waveform shown (Fig. 12-23).

Use Eq. (*12-9*).

$$V = 0.707 V_M = 0.707(48\,\mu\text{V}) = 33.9\,\mu\text{V} \quad Ans.$$

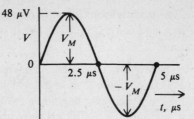

Fig. 12-23 Finding characteristics of
a sine wave

The time for 1 cycle is 5 μs. Therefore,

$$T = 5\,\mu\text{s} = 5 \times 10^{-6}\,\text{s} \qquad Ans.$$

$$f = \frac{1}{T} \tag{12-4}$$

$$= \frac{1}{5 \times 10^{-6}} = 200 \times 10^{3}\,\text{Hz} = 200\,\text{kHz} \qquad Ans.$$

The ac wave is symmetric with respect to the horizontal axis. So

$$V_{p\text{-}p} = 2V_{M} = 2(48) = 96\,\mu\text{V} \qquad Ans.$$

12.12 Any waveform that is not a sine or cosine wave is a *nonsinusoidal* waveform. Common examples are the rectangular and sawtooth waves (Fig. 12-24). What are the peak-to-peak voltages for these particular waves?

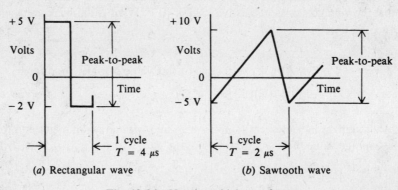

(a) Rectangular wave   (b) Sawtooth wave

Fig. 12-24 Nonsinusoidal waveforms

Peak-to-peak amplitudes, measured between the maximum and minimum peak values, are generally used for measuring nonsinusoidal wave shapes since they often have nonsymmetrical peaks. By inspection of Fig. 12-24,

Rectangular wave: $V_{p\text{-}p} = 5 + 2 = 7\,\text{V} \qquad Ans.$

Sawtooth wave: $V_{p\text{-}p} = 10 + 5 = 15\,\text{V} \qquad Ans.$

12.13 Calculate the frequency of the nonsinusoidal waveforms shown in Fig. 12-24.

The period T for a complete cycle is 4 μs (Fig. 12-24a) and 2 μs (Fig. 12-24b).

$$f = \frac{1}{T} \qquad \text{also for periodic nonsinusoids}$$

Rectangular wave: $\qquad\qquad$ $f = \dfrac{1}{4\,\mu s} = 0.25\,\text{MHz} \qquad Ans.$

Sawtooth wave: $\qquad\qquad$ $f = \dfrac{1}{2\,\mu s} = 0.5\,\text{MHz} \qquad Ans.$

12.14 A 120-V ac voltage is applied across a 20-Ω resistive load (Fig. 12-25). Find values of I, V_M, $V_{p\text{-}p}$, V_{av}, I_M, $I_{p\text{-}p}$, I_{av}, and P.

By Ohm's law,

$$I = \frac{V}{R_L} = \frac{120}{20} = 6\,\text{A} \qquad Ans.$$

Use Table 12-1 to calculate voltage and current values.

$$V_M = 1.414V = 1.414(120) = 169.7\,\text{V} \qquad Ans.$$

$$V_{p\text{-}p} = 2V_M = 2(169.7) = 339.4\,\text{V} \qquad Ans.$$

$$V_{av} = 0.637V_M = 0.637(169.7) = 108.3\,\text{V} \qquad Ans.$$

$$I_M = 1.414I = 1.414(6) = 8.5\,\text{A} \qquad Ans.$$

$$I_{p\text{-}p} = 2I_M = 2(8.5) = 17.0\,\text{A} \qquad Ans.$$

$$I_{av} = 0.637I_M = 0.637(8.5) = 5.4\,\text{A} \qquad Ans.$$

$$P = I^2 R_L = 6^2(20) = 720\,\text{W} \qquad Ans.$$

or $\qquad\qquad$ $P = \dfrac{V^2}{R_L} = \dfrac{120^2}{20} = 720\,\text{W} \quad$ or $\quad P = VI = 120(6) = 720\,\text{W} \qquad Ans.$

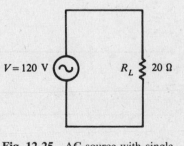

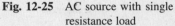

Fig. 12-25 AC source with single
resistance load

12.15 A 20-Ω electric iron and a 100-Ω lamp are connected in parallel across a 120-V 60-Hz ac line (Fig. 12-26). Find the total current, the total resistance, and the total power drawn by the circuit, and draw the phasor diagram.

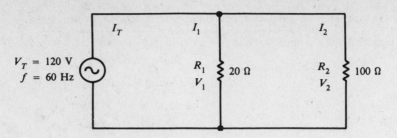

Fig. 12-26 AC source in parallel circuit

For a parallel circuit, $V_T = V_1 = V_2 = 120\,\text{V}$

$$I_1 = \frac{V_1}{R_1} = \frac{120}{20} = 6\,\text{A} \quad I_2 = \frac{V_2}{R_2} = \frac{120}{100} = 1.2\,\text{A}$$

Then

$$I_T = I_1 + I_2 = 6 + 1.2 = 7.2\,\text{A} \qquad Ans.$$

$$R_T = \frac{V_T}{I_T} = \frac{120}{7.2} = 16.7\,\Omega \qquad Ans.$$

In a purely resistive set of branch currents, the total current I_T is in phase with the total voltage V_T. The phase angle is therefore equal to $0°$.

$$P = V_T I_T \cos\theta = 120(7.2)(\cos 0°) = 120(7.2)(1) = 864\,\text{W} \qquad Ans.$$

Since the voltage in a parallel circuit is constant, the voltage is used as the reference phasor. The currents I_1 and I_2 are drawn in the same direction as the voltage because the current through pure resistances is in phase with the voltage. The I_1 phasor is shown longer than I_2 because its current value is higher (see the phasor diagram).

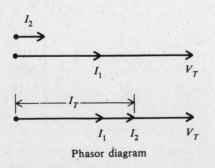

Phasor diagram

12.16   A series–parallel ac circuit has two branches across the 60-Hz 120-V power line (Fig. 12-27). Find I_1, I_2, I_3, V_1, V_2, and V_3. (Double-ended arrows are sometimes used to indicate direction for ac current.)

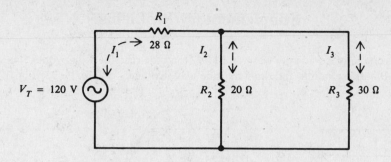

Fig. 12-27　AC source in series–parallel circuit

Proceed to solve the ac circuit with resistance only in the same manner as a dc circuit.

Step 1.　Simplify the circuit to a single resistance R_T.

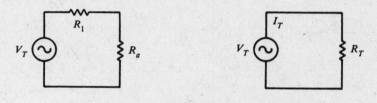

$$R_a = \frac{R_2 R_3}{R_2 + R_3} = \frac{20(30)}{20 + 30} = \frac{600}{50} = 12\ \Omega$$

$$R_T = R_1 + R_a = 12 + 28 = 40\ \Omega$$

Step 2.　Solve for the total current I_T.

$$I_T = \frac{V_T}{R_T} = I_1 = \frac{120}{40} = 3\ \text{A} \qquad Ans.$$

Step 3.　Solve for the branch currents I_2 and I_3.

$$I_2 = \frac{R_3}{R_2 + R_3} I_T = \frac{30}{50}3 = 1.8\ \text{A} \qquad Ans.$$

$$I_3 = I_T - I_2 = 3 - 1.8 = 1.2\ \text{A} \qquad Ans.$$

Step 4.　Solve for the branch voltages V_2 and V_3.

$$V_1 = I_1 R_1 = 3(28) = 84\ \text{V} \qquad Ans.$$

$$V_2 = V_3 = I_2 R_2 = 1.2(20) = 36\ \text{V} \qquad Ans.$$

Step 5.　Verify answer on voltage division.

$$V_T = V_1 + V_2$$

$$120 = 84 + 36$$

$$120\ \text{V} = 120\ \text{V} \qquad Check$$

Supplementary Problems

12.17   The peak voltage of an ac sine wave is 100 V. Find the instantaneous voltage at 0, 30, 60, 90, 135, and 245°. Plot these points and draw the sine wave voltage.   *Ans.*   See Fig. 12-28.

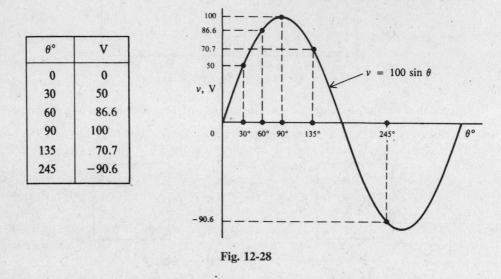

$\theta°$	V
0	0
30	50
60	86.6
90	100
135	70.7
245	−90.6

Fig. 12-28

12.18   If an ac voltage wave has an instantaneous value of 90 V at 30°, find the peak value.
Ans. $V_M = 180$ V

12.19   An ac wave has an effective value of 50 mA. Find the maximum value and the instantaneous value at 60°.   *Ans.* $I_M = 70.7$ mA; $i = 61.2$ mA

12.20   An electric stove draws 7.5 A from a 120-V dc source. What is the maximum value of an alternating current which will produce heat at the same rate? Find the power drawn from the ac line.
Ans. $I_M = 10.6$ A; $P = 900$ W

12.21   Calculate V, $V_{p\text{-}p}$, T, and f for the sine wave voltage in Fig. 12-29.
Ans. $V = 38.2\,\mu$V; $V_{p\text{-}p} = 108\,\mu$V; $T = 2\,\mu$s; $f = 0.5$ MHz

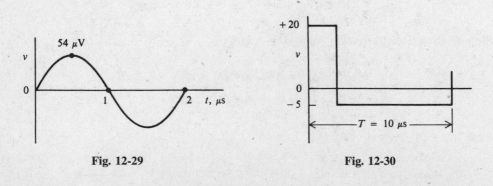

Fig. 12-29   **Fig. 12-30**

12.22   What is the peak-to-peak voltage and frequency of the nonsymmetrical rectangular waveform in Fig. 12-30?   *Ans.* $V_{p\text{-}p} = 25$ V; $f = 0.1$ MHz

12.23   Find the instantaneous voltage at 45° in a wave whose peak value is 175 V.   *Ans.* $V = 123.7$ V

12.24   Find the peak value of an ac wave if the instantaneous current is 35 A at 30°.   *Ans.* $I_M = 70$ A

12.25   Find the phase angle at which an instantaneous voltage of 36.5 V appears in a wave whose peak value is 125 V.   *Ans.* $\theta = 17°$

12.26   What is the period of an ac voltage that has a frequency of (*a*) 50 Hz, (*b*) 95 kHz, and (*c*) 106 kHz?   *Ans.*   (*a*) $T = 0.02$ s;   (*b*) $T = 0.0105$ ms;   (*c*) $T = 0.00943$ ms or 9.43 μs

12.27   Find the frequency of an ac current when its period is (*a*) 0.01 s, (*b*) 0.03 ms, and (*c*) 0.006 ms.   *Ans.*   (*a*) $f = 100$ Hz;   (*b*) $f = 33.3$ kHz;   (*c*) $f = 166.7$ kHz

12.28   What is the wavelength of radio station WMAL which broadcasts FM (frequency modulation) at a frequency of 107.3 kHz?   *Ans.* $\lambda = 2796$ m

12.29   What is the wavelength of an ac wave whose frequency is (*a*) 60 Hz, (*b*) 1 kHz, (*c*) 30 kHz, and (*d*) 800 kHz?   *Ans.*   (*a*) $\lambda = 5 \times 10^6$ m;   (*b*) $\lambda = 3 \times 10^5$ m;   (*c*) $\lambda = 10\ 000$ m;   (*d*) $\lambda = 375$ m

12.30   Find the frequency of a radio wave whose wavelength is (*a*) 600 m, (*b*) 2000 m, (*c*) 3000 m, and (*d*) 6000 m.   *Ans.*   (*a*) $f = 500$ Hz;   (*b*) $f = 150$ kHz;   (*c*) $f = 100$ kHz;   (*d*) $f = 50$ kHz

12.31   Determine the phase angle for each ac wave shown (Fig. 12-31) and draw its phasor. One cycle is shown for each wave. Show *I* as the reference phasor.

       Ans. (*a*) *v* and *i* are in phase   Phasor diagram: $\theta = 0°$

       (*b*) *v* leads *i* by 180° or *i* lags *v* by 180°   Phasor diagram: $\theta = 180°$

       (*c*) *i* leads *v* by 90° or *v* lags *i* by 90°   Phasor diagram: $\theta = -90°$

       (*d*) *i* leads *v* by 90° or *v* lags *i* by 90°   Phasor diagram: $\theta = -90°$

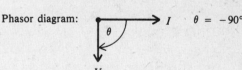

       (*e*) *v* leads *i* by 130° or *i* lags *v* by 130°   Phasor diagram: $\theta = 130°$

12.32   The ac power line delivers 120 V to your home. This is the voltage as measured by an ac voltmeter. What is the peak value of this voltage?   *Ans.* $V_M = 169.7$ V

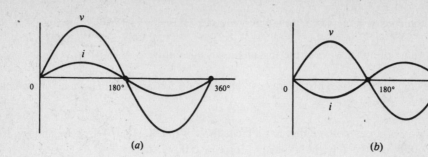

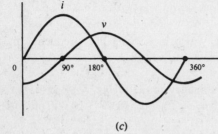
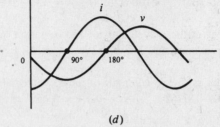

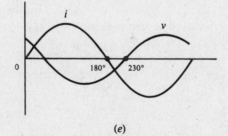

Fig. 12-31

12.33 An industrial oven draws 8.5 A from a 12-V dc source. What is the maximum value of an ac current which will heat at the same rate?   *Ans.* $I_M = 12.0$ A

12.34 Find the values indicated.

	Peak Value	rms Value	Average Value	Phase Angle	Instantaneous Value
(a)	45 A	?	?	45°	?
(b)	?	220 V	?	60°	?
(c)	?	?	10 A	30°	?
(d)	200 V	?	?	60°	?
(e)	?	110 V	?	75°	?

	Peak Value	rms Value	Average Value	Phase Angle	Instantaneous Value
(f)	?	?	?	15°	75.1 V
(g)	100 V	?	?	?	86.6 V
(h)	?	?	20 A	?	15.7 A
(i)	?	30 A	?	?	30 A
(j)	?	?	100.1 V	?	136.1 V

Ans.

	Peak Value	rms Value	Average Value	Phase Angle	Instantaneous Value
(a)	....	31.8 A	28.7 A	....	31.8 A
(b)	311.1 V	....	198.2 V	....	269.4 V
(c)	15.7 A	11.1 A	....	....	7.85 A
(d)	....	141.4 V	127.4 V	....	173.2 V
(e)	155.6 V	....	99.1 V	....	150.3 V
(f)	290.2 V	205.2 V	184.9 V	....	....
(g)	....	141.4 V	63.7 V	60°	....
(h)	31.4 A	22.2 A	....	30°	....
(i)	42.4 A	....	27.0 A	45°	....
(j)	157.1 V	111.1 V	....	60°	....

12.35 An ac ammeter reads 22 A rms current through a resistive load, and a voltmeter reads 385 V rms drop across the load. What are the peak values and the average values of the alternating current and voltage?   *Ans.* $I_M = 31.1$ A; $V_M = 545$ V; $I_{av} = 19.8$ A; $V_{av} = 347$ V

12.36 An ac power line delivers 240 V to a sidewalk heating cable that has a total resistance of 5 Ω. Find I, V_M, V_{p-p}, V_{av}, I_M, I_{p-p}, I_{av}, and P.
    Ans. $I = 48$ A; $V_M = 339$ V; $V_{p-p} = 678$ V; $V_{av} = 216$ V; $I_M = 67.9$ A; $I_{p-p} = 135.8$ A; $I_{av} = 43.3$ A; $P = 11\,520$ W

12.37 An electric soldering iron draws 0.8 A from a 120-V 60-Hz power line. What is its resistance? How much power will it consume? Draw the phasor diagram.   *Ans.* $R = 150$ Ω; $P = 96$ W

Phasor diagram:

12.38 Find the current and power drawn from a 110-V 60-Hz line by a tungsten lamp whose resistance is 275 Ω. Draw the phasor diagram.   *Ans.* $I = 0.4$ A; $P = 44$ W

Phasor diagram:

12.39 A circuit has a 5-MΩ resistor R_1 in series with a 15-MΩ resistor R_2 across a 200-V ac source. Calculate I, V_1, V_2, P_1, and P_2.
Ans. $I = 10\,\mu$A; $V_1 = 50$ V; $V_2 = 150$ V; $P_1 = 0.5$ mW; $P_2 = 1.5$ mW

12.40 For the ac series–parallel circuit (Fig. 12-32), find the total current, the current through each resistance, and the voltage across each resistance.
Ans. $I_T = I_1 = 24$ A; $I_2 = 12$ A; $I_3 = 12$ A; $V_1 = 96$ V; $V_2 = V_3 = 24$ V

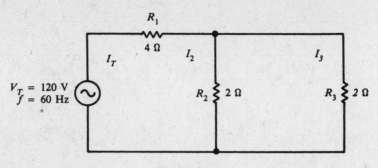

Fig. 12-32

12.41 A series–parallel ac circuit has two branches across the 60-Hz 120-V power line. One branch has a 20-Ω R_1 in series with a 10-Ω R_2. The other branch has a 30-MΩ R_3 in series with a 10-MΩ R_4. Calculate V_1, V_2, V_3, and V_4.   *Ans.* $V_1 = 80$ V; $V_2 = 40$ V; $V_3 = 90$ V; $V_4 = 30$ V

12.42 An ac circuit has a 5-MΩ resistor R_1 in parallel with a 10-MΩ resistor R_2 across a 200-V source. Find I_1, I_2, V_1, V_2, P_1, and P_2.   *Ans.* $I_1 = 40\,\mu$A; $I_2 = 20\,\mu$A; $V_1 = V_2 = 200$ V; $P_1 = 8$ mW; $P_2 = 4$ mW

Chapter 13

Inductance, Inductive Reactance, and Inductive Circuits

INDUCTION

The ability of a conductor to induce voltage in itself when the current changes is its *self-inductance*, or simply *inductance*. The symbol for inductance is L, and its unit is the *henry* (H). One henry is the amount of inductance that permits one volt to be induced when the current changes at the rate of one ampere per second (Fig. 13-1). The formula for inductance is

$$L = \frac{v_L}{\Delta i / \Delta t} \qquad (13\text{-}1)$$

where L = inductance, H
          v_L = induced voltage across the coil, V
         $\Delta i / \Delta t$ = rate of change of current, A/s

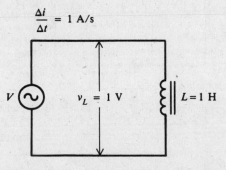

Fig. 13-1   The inductance of a coil is 1 H
            when a change of 1 A/s induces
            1 V across the coil

The self-induced voltage v_L from Eq. (*13-1*) is

$$v_L = L \frac{\Delta i}{\Delta t} \qquad (13\text{-}2)$$

Example 13.1   What is the value of inductance of a coil that induces 20 V when the current through the coil changes from 12 to 20 A in 2 s?
     We are given that

$$v_L = 20 \text{ V} \qquad \Delta i = 20 - 12 = 8 \text{ A} \qquad \Delta t = 2 \text{ s}$$

So

$$\frac{\Delta i}{\Delta t} = \frac{8}{2} = 4 \text{ A/s}$$

$$L = \frac{v_L}{\Delta i / \Delta t} \qquad (13\text{-}1)$$

$$= \frac{20}{4} = 5 \text{ H} \qquad Ans.$$

275

Example 13.2   A coil has an inductance of 50 μH. What voltage is induced across the coil when the rate of change of the current is 10 000 A/s?

$$v_L = L\frac{\Delta i}{\Delta t} \tag{13-2}$$

$$= (50 \times 10^{-6})(10^4) = 0.5 \text{ V} \qquad Ans.$$

When the current in a conductor or coil changes, the varying flux can cut across any other conductor or coil located nearby, thus inducing voltages in both. A varying current in L_1, therefore, induces voltage across L_1 and across L_2 (Fig. 13-2). When the induced voltage v_{L2} produces current in L_2, its varying magnetic field induces voltage in L_1. Hence, the two coils L_1 and L_2 have *mutual inductance* because current change in one coil can induce voltage in the other. The unit of mutual inductance is the henry, and the symbol is L_M. Two coils have L_M of 1 H when a current change of 1 A/s in one coil induces 1 V in the other coil.

The schematic symbol for two coils with mutual inductance is shown in Fig. 13-3.

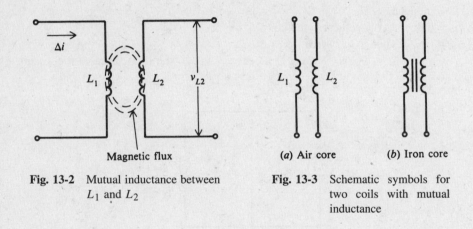

Fig. 13-2   Mutual inductance between L_1 and L_2

Fig. 13-3   Schematic symbols for two coils with mutual inductance

CHARACTERISTICS OF COILS

Physical Characteristics

A coil's inductance depends on how it is wound, the core material on which it is wound, and the number of turns of wire with which it is wound.

1. Inductance L increases as the number of turns of wire N around the core increases. Inductance increases as the square of the turns increases. For example, if the number of turns is doubled (2×), inductance increases 2^2 or 4×, assuming the area and length of the coil remain constant.

2. Inductance increases as the relative permeability μ_r of the core material increases.

3. As the area A enclosed by each turn increases, the inductance increases. Since the area is a function of the square of the diameter of the coil, inductance increases as the square of the diameter.

4. Inductance decreases as the length of the coil increases (assuming the number of turns remains constant).

Example 13.3   An approximate formula in SI units for the inductance of a coil where the length is at least 10 times the diameter is

$$L = \mu_r\frac{N^2 A}{l}(1.26 \times 10^{-6}), \text{ H}$$

Note that this formula follows the proportional relationship described. Find L when $\mu_r = 200$, $N = 200$ turns, $A = 1 \times 10^{-4}$ m², and $l = 0.1$ m.

$$L = 200 \frac{200^2(1 \times 10^{-4})}{0.1}(1.26 \times 10^{-6}) = 10 \times 10^{-3} \text{ H} = 10 \text{ mH} \quad Ans.$$

Core Losses

Losses in the magnetic core are due to *eddy-current* losses and *hysteresis* losses. Eddy currents flow in a circular path within the core material itself and dissipate as heat in the core. The loss is equal to $I^2 R$, where R is the resistance of the path through the core. The higher the frequency of alternating current in the inductance, the higher the eddy currents and the greater the eddy-current loss.

Hysteresis losses arise from the additional power needed to reverse the magnetic field in magnetic materials with an alternating current. Hysteresis losses generally are less than eddy-current losses.

To reduce eddy-current losses while sustaining flux density, the iron core can be made of laminated sheets insulated from each other, insulated powdered-iron granules pressed into one solid, or ferrite material. Air-core coils have practically no losses from eddy currents or hysteresis.

INDUCTIVE REACTANCE

Inductive reactance X_L is the opposition to ac current due to the inductance in the circuit. The unit of inductive reactance is the ohm. The formula for inductive reactance is

$$X_L = 2\pi f L \qquad (13\text{-}3)$$

Since $2\pi = 2(3.14) = 6.28$, Eq. (*13-3*) becomes

$$X_L = 6.28 f L$$

where X_L = inductive reactance, Ω
    f = frequency, Hz
    L = inductance, H

If any two quantities are known in Eq. (*13-3*), the third can be found.

$$L = \frac{X_L}{6.28 f} \qquad (13\text{-}4)$$

$$f = \frac{X_L}{6.28 L} \qquad (13\text{-}5)$$

In a circuit containing *only* inductance (Fig. 13-4), Ohm's law can be used to find current and voltage by substituting X_L for R.

$$I_L = \frac{V_L}{X_L} \qquad (13\text{-}6)$$

$$X_L = \frac{V_L}{I_L} \qquad (13\text{-}7)$$

$$V_L = I_L X_L \qquad (13\text{-}8)$$

where I_L = current through the inductance, A
    V_L = voltage across the inductance, V
    X_L = inductive reactance, Ω

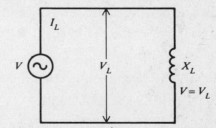

Fig. 13-4   Circuit with only X_L

Example 13.4 A resonant tank circuit consists of a 20-mH coil operating at a frequency of 950 kHz. What is the inductive reactance of the coil?

$$X_L = 6.28fL \tag{13-3}$$

$$= 6.28(950 \times 10^3)(20 \times 10^{-3}) = 11.93 \times 10^4 = 119\,300\ \Omega \quad \text{Ans.}$$

Example 13.5 What must the inductance of a coil be in order that it have a reactance of 942 Ω at a frequency of 60 kHz?

$$L = \frac{X_L}{6.28f} \tag{13-4}$$

$$= \frac{942}{6.28(60 \times 10^3)} = 2.5 \times 10^{-3} = 2.5\ \text{mH} \quad \text{Ans.}$$

Example 13.6 A tuning coil in a radio transmitter has an inductance of 300 μH. At what frequency will it have an inductive reactance of 3768 Ω?

$$f = \frac{X_L}{6.28L} \tag{13-5}$$

$$= \frac{3768}{6.28(300 \times 10^{-6})} = 2 \times 10^6 = 2\ \text{MHz} \quad \text{Ans.}$$

Example 13.7 A choke coil of negligible resistance is to limit the current through it to 50 mA when 25 V is applied across it at 400 kHz. Find its inductance.
    Find X_L by Ohm's law and then find L.

$$X_L = \frac{V_L}{I_L} \tag{13-7}$$

$$= \frac{25}{50 \times 10^{-3}} = 500\ \Omega \quad \text{Ans.}$$

$$L = \frac{X_L}{6.28f} \tag{13-4}$$

$$= \frac{500}{6.28(400 \times 10^3)} = 0.199 \times 10^{-3} = 0.20\ \text{mH} \quad \text{Ans.}$$

Example 13.8 The primary coil of a power transformer has an inductance of 30 mH with negligible resistance (Fig. 13-5). Find its inductive reactance at a frequency of 60 Hz and the current it will draw from a 120-V line.
    Find X_L by using Eq. (*13-4*) and then I_L by using Ohm's law [Eq. (*13-6*)].

$$X_L = 6.28fL = 6.28(60)(30 \times 10^{-3}) = 11.3\ \Omega \quad \text{Ans.}$$

$$I_L = \frac{V_L}{X_L} = \frac{120}{11.3} = 10.6\ \text{A} \quad \text{Ans.}$$

INDUCTORS IN SERIES OR PARALLEL

    If inductors are spaced sufficiently far apart so that they do not interact electromagnetically with each other, their values can be combined just like resistors when connected together. If a number of

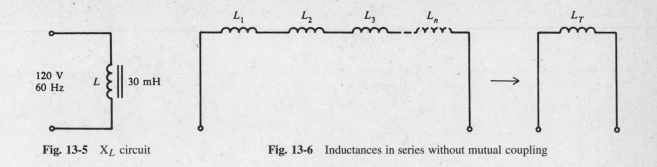

Fig. 13-5 X_L circuit   **Fig. 13-6** Inductances in series without mutual coupling

inductors are connected in series (Fig. 13-6), the total inductance L_T is the sum of the individual inductances, or

Series:
$$L_T = L_1 + L_2 + L_3 + \cdots + L_n \tag{13-9}$$

If two series-connected coils are spaced close together so that their magnetic field lines interlink, their mutual inductance will have an effect on the circuit. In that case the total inductance is

$$L_T = L_1 + L_2 \pm 2L_M \tag{13-10}$$

where L_M is the mutual inductance between the coils. The plus (+) sign in Eq. (13-10) is used if the coils are arranged in *series-aiding* form, while the minus (−) sign is used if the coils are connected in *series-opposing* form. Series aiding means that the common current produces the same direction of magnetic field for the two coils. The series-opposing connection results in opposite fields.

Three different arrangements for coils L_1 and L_2 are shown both pictorially and schematically in Fig. 13-7. In Fig. 13-7a, the coils are spaced too far apart to interact electromagnetically. There is no mutual inductance, so L_M is zero. The total inductance is $L_T = L_1 + L_2$. In Fig. 13-7b, the coils are spaced close together and have windings in the same direction, as indicated by the dots. The coils are series-aiding, so $L_T = L_1 + L_2 + 2L_M$. In Fig. 13-7c, the coil windings are in the opposite direction, so the coils are series-opposing, and $L_T = L_1 + L_2 - 2L_M$.

The large dots above the coil (Fig. 13-7b and c) are used to indicate the polarity of the windings without having to show the actual physical construction. Coils with dots at the same end (Fig. 13-7b) have the same polarity or same direction of winding. When current enters the dotted ends for L_1 and L_2, their fields are aiding and L_M has the same sense as L.

If inductors are spaced sufficiently far apart so that their mutual inductance is negligible ($L_M = 0$), the rules for combining inductors in parallel are the same as for resistors. If a number of inductors are connected in parallel (Fig. 13-8), their total inductance L_T is

Parallel:
$$\frac{1}{L_T} = \frac{1}{L_1} + \frac{1}{L_2} + \frac{1}{L_3} + \cdots + \frac{1}{L_n} \tag{13-11}$$

The total inductance of two coils connected in parallel is

Parallel:
$$L_T = \frac{L_1 L_2}{L_1 + L_2} \tag{13-12}$$

All inductances must be given in the same units. The shortcuts for calculating parallel R can be used with parallel L. For example, if two 8-mH inductors are in parallel, the total inductance is $L_T = L/n = 8/2 = 4$ mH.

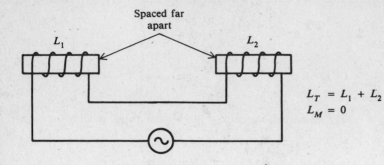
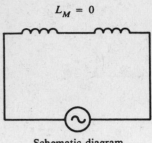

$$L_T = L_1 + L_2$$
$$L_M = 0$$

Schematic diagram

(a) No mutual inductance

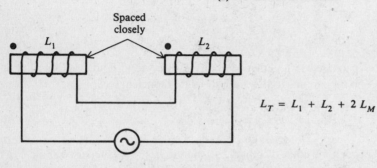
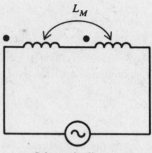

$$L_T = L_1 + L_2 + 2 L_M$$

Schematic diagram

(b) Series-aiding

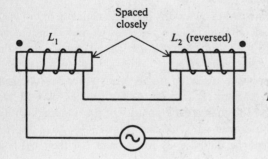
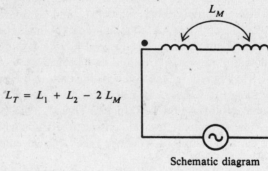

$$L_T = L_1 + L_2 - 2 L_M$$

Schematic diagram

(c) Series-opposing

Fig. 13-7 L_1 and L_2 in series with mutual coupling L_M

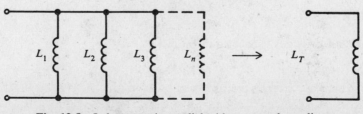

Fig. 13-8   Inductances in parallel without mutual coupling

Example 13.9 A 10- and a 12-H choke used to limit current in a circuit are connected in series. Initially they are spaced far apart. What is the total inductance?

$$L_T = L_1 + L_2 \tag{13-9}$$

$$= 10 + 12 = 22\ \text{H} \quad Ans.$$

Example 13.10 The two chokes of Example 13.9 are moved close together so that they are coupled by a mutual inductance of 7 H. What are the total inductances if (a) the coils are wound in the same direction and (b) the coils are wound in opposing directions?

(a) Series-aiding:

$$L_T = L_1 + L_2 + 2L_M \tag{13-10}$$

$$= 10 + 12 + 2(7) = 22 + 14 = 36\ \text{H} \quad Ans.$$

(b) Series-opposing:

$$L_T = L_1 + L_2 - 2L_M \tag{13-10}$$

$$= 10 + 12 - 2(7) = 22 - 14 = 8\ \text{H} \quad Ans.$$

Example 13.11 What is the total inductance of two parallel inductors with values of 8 and 12 H?

$$L_T = \frac{L_1 L_2}{L_1 + L_2} \tag{13-12}$$

$$= \frac{8(12)}{8 + 12} = \frac{96}{20} = 4.8\ \text{H} \quad Ans.$$

Example 13.12 A 6-H inductor and a 22-H inductor are connected in series and plugged into a 120-V ac 60-Hz outlet. Assume that their resistance is negligible and that they have no mutual inductance. What is their inductive reactance and what current will they draw?

$$L_T = L_1 + L_2 = 6 + 22 = 28\ \text{H}$$

$$X_L = 6.28 f L_T \tag{13-3}$$

$$= 6.28(60)(28) = 10\,550\ \Omega \quad Ans.$$

$$I_L = \frac{V_L}{X_L} = \frac{120}{10\,550} = 0.0114\ \text{A} \quad \text{or} \quad 11.4\ \text{mA} \quad Ans. \tag{13-6}$$

INDUCTIVE CIRCUITS

Inductance Only

If an ac voltage v is applied across a circuit having *only* inductance (Fig. 13-9a), the resulting ac current through the inductance, i_L, will lag the voltage across the inductance, v_L, by 90° (Fig. 13-9b and c). Voltages v and v_L are the same because the total applied voltage is dropped only across the inductance. Both i_L and v_L are sine waves with the same frequency. Lowercase letters such as i and v indicate instantaneous values; capital letters such as I and V show dc or ac rms values.

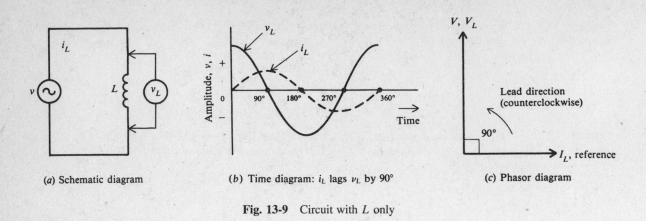

(a) Schematic diagram   (b) Time diagram: i_L lags v_L by 90°   (c) Phasor diagram

Fig. 13-9   Circuit with L only

RL in Series

When a coil has series resistance (Fig. 13-10a), the rms current I is limited by both X_L and R. I is the same in X_L and R since they are in series. The voltage drop across R is $V_R = IR$, and the voltage drop across X_L is $V_L = IX_L$. The current I through X_L must lag V_L by 90° because this is the phase angle between current through an inductance and its self-induced voltage (Fig. 13-10b). The current I through R and its IR voltage drop are in phase so the phase angle is 0° (Fig. 13-10b).

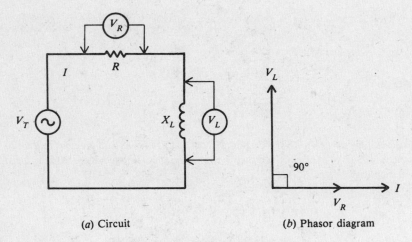

(a) Circuit   (b) Phasor diagram

Fig. 13-10 R and X_L in series

To combine two waveforms out of phase, we add their equivalent phasors. The method is to add the tail of one phasor to the arrowhead of the other, using the angle to show their relative phase. The sum of the phasors is a resultant phasor from the start of one phasor to the end of the other phasor. Since the V_R and V_L phasors form a right angle, the resultant phasor is the hypotenuse of a right triangle (Fig. 13-11). From the geometry of a right triangle, the Pythagorean theorem states that the hypotenuse is equal to the square root of the sum of the squares of the sides. Therefore, the resultant is

$$V_T = \sqrt{V_R^2 + V_L^2} \qquad (13\text{-}13)$$

where the total voltage V_T is the phasor sum of the two voltages V_R and V_L that are 90° out of phase. All the voltages must be in the same units—rms values, peak values, or instantaneous values. For example, when V_T is an rms value, V_R and V_L are also rms values. Most of the ac calculations will be made in rms units.

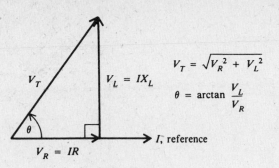

Fig. 13-11   Voltage-phasor triangle

The phase angle θ between V_T and V_R (Fig. 13-11) is

$$\tan \theta = \frac{V_L}{V_R}$$

$$\theta = \arctan \frac{V_L}{V_R} \tag{13-14}$$

Since V_R is in phase with I, θ is also the phase angle between V_T and I where I lags V_T.

Example 13.13   A *RL* series ac circuit has a current of 1 A peak with $R = 50\,\Omega$ and $X_L = 50\,\Omega$ (Fig. 13-12*a*). Calculate V_R, V_L, V_T, and θ. Draw the phasor diagram of V_T and I. Draw also the time diagram of i, v_R, v_L, and v_T.

$$V_R = IR = 1(50) = 50 \text{ V peak} \qquad Ans.$$

$$V_L = IX_L = 1(50) = 50 \text{ V peak} \qquad Ans.$$

Then (see Fig. 13-12*b*) $V_T = \sqrt{V_R^2 + V_L^2}$ \hfill (13-13)

$$= \sqrt{50^2 + 50^2} = \sqrt{2500 + 2500} = \sqrt{5000} = 70.7 \text{ V peak} \qquad Ans.$$

$$\theta = \arctan \frac{V_L}{V_R} = \arctan \frac{50}{50} = \arctan 1 = 45° \qquad Ans.$$

In a series circuit, since I is the same in R and X_L, it is convenient to show I as the reference phasor at 0°. The phasor diagram is shown as Fig. 13-12*c* and the time diagram as Fig. 13-12*d*.

Impedance in series RL.   The resultant of the phasor addition of R and X_L is called *impedance*. The symbol for impedance is Z. Impedance is the total opposition to the flow of current, expressed in ohms. The impedance triangle (Fig. 13-13) corresponds to the voltage triangle (Fig. 13-11), but the common factor I cancels. The equations for impedance and phase angle are derived as follows:

$$V_T^2 = V_R^2 + V_L^2$$

$$(IZ)^2 = (IR)^2 + (IX_L)^2$$

$$Z^2 = R^2 + X_L^2$$

$$Z = \sqrt{R^2 + X_L^2} \tag{13-15}$$

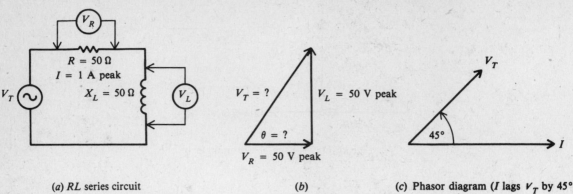

(a) RL series circuit   (b)   (c) Phasor diagram (I lags V_T by 45°)

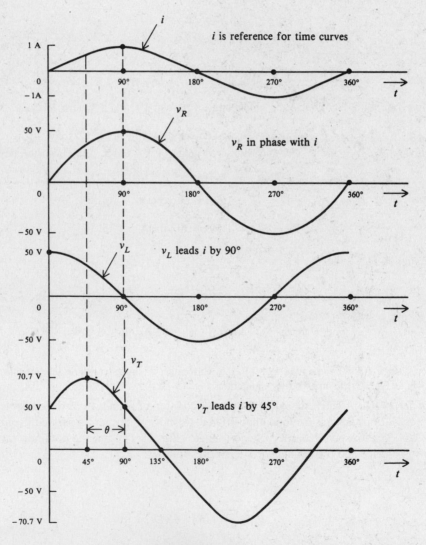

(d)   Time diagram of RL series circuit

Fig. 13-12

$$\tan \theta = \frac{X_L}{R}$$

$$\theta = \arctan \frac{X_L}{R} \qquad (13\text{-}16)$$

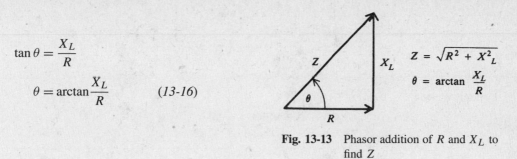

$$Z = \sqrt{R^2 + X_L^2}$$

$$\theta = \arctan \frac{X_L}{R}$$

Fig. 13-13 Phasor addition of R and X_L to find Z

Example 13.14   If a 50-Ω R and a 70-Ω X_L are in series with 120 V applied (Fig. 13-14a), find the following: Z, θ, I, V_R, and V_L. What is the phase angle of V_L, V_R, and V_T with respect to I? Prove that the sum of the series voltage drops equals the applied voltage V_T.

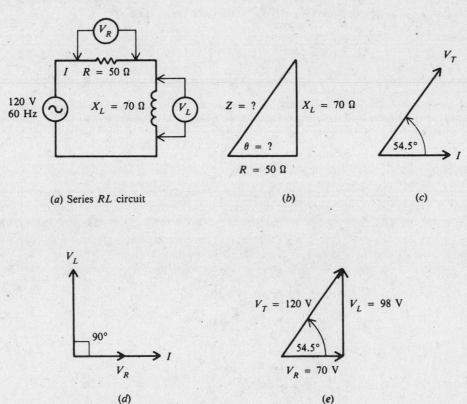

(a) Series RL circuit   (b)   (c)

(d)   (e)

Fig. 13-14

Step 1.   Find Z and θ (see Fig. 13-14b).

$$Z = \sqrt{R^2 + X_L^2} \qquad (13\text{-}15)$$

$$= \sqrt{50^2 + 70^2} = \sqrt{2500 + 4900} = \sqrt{7400} = 86 \ \Omega \qquad Ans.$$

$$\theta = \arctan \frac{X_L}{R} \qquad\qquad (13\text{-}16)$$

$$= \arctan \frac{70}{50} = \arctan 1.40 = 54.5° \qquad Ans.$$

V_T leads I by 54.5° (see Fig. 13-14c).

Step 2.   Find I, V_R, and V_L.

$$I = \frac{V_T}{Z} = \frac{120}{86} = 1.40 \text{ A} \qquad Ans.$$

$$V_R = IR = 1.40(50) = 70.0 \text{ V} \qquad Ans.$$

$$V_L = IX_L = 1.40(70) = 98.0 \text{ V} \qquad Ans.$$

I and V_R are in phase. V_L leads I by 90° (see Fig. 13-14d).

Step 3.   Show that V_T is the phasor sum of V_R and V_L (see Fig. 13-14e).

$$V_T = \sqrt{V_R^2 + V_L^2} \qquad\qquad (13\text{-}13)$$

$$= \sqrt{(70.0)^2 + (98.0)^2} = \sqrt{14\,504} \approx 120 \text{ V} \qquad Ans.$$

(The answer is not exactly 120 V because of rounding off I.) Therefore, the sum of the voltage drops equals the applied voltage.

RL in Parallel

For parallel circuits with R and X_L (Fig. 13-15a), the same applied voltage V_T is across R and X_L since both are in parallel with V_T. There is no phase difference between these voltages. Therefore, V_T will be used as the reference phasor. The resistive branch current $I_R = V_T/R$ is in phase with V_T. The inductive branch current $I_L = V_T/X_L$ lags V_T by 90° (Fig. 13-15b) because the current in an inductance lags the voltage across it by 90°. The phasor sum of I_R and I_L equals the total line current I_T (Fig. 13-15c), or

$$I_T = \sqrt{I_R^2 + I_L^2} \qquad\qquad (13\text{-}17)$$

$$\tan \theta = -\frac{I_L}{I_R}$$

$$\theta = \arctan\left(-\frac{I_L}{I_R}\right) \qquad\qquad (13\text{-}18)$$

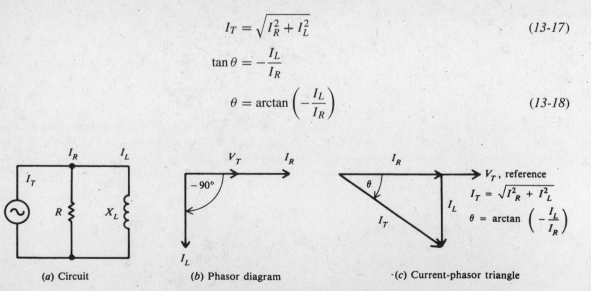

(a) Circuit   (b) Phasor diagram   (c) Current-phasor triangle

Fig. 13-15 R and X_L in parallel

Example 13.15 A RL parallel circuit has 100-V peak applied across $R = 20\ \Omega$ and $X_L = 20\ \Omega$ (Fig. 13-16a). Find I_R, I_L, I_T, and θ. (See Fig. 13-16b.) Draw the phasor and time diagrams of v_T, i_R, i_L, and i_T.

$$I_R = \frac{V_T}{R} = \frac{100}{20} = 5\text{ A peak} \qquad Ans.$$

$$I_L = \frac{V_T}{X_L} = \frac{100}{20} = 5\text{ A peak} \qquad Ans.$$

$$I_T = \sqrt{I_R^2 + I_L^2} \qquad\qquad\qquad (13\text{-}17)$$

$$= \sqrt{5^2 + 5^2} = \sqrt{50} = 7.07\text{ A peak} \qquad Ans.$$

$$\theta = \arctan\left(-\frac{I_L}{I_R}\right) \qquad\qquad\qquad (13\text{-}18)$$

$$= \arctan\left(-\frac{5}{5}\right) = \arctan(-1) = -45° \qquad Ans.$$

Since V_T is the same throughout the parallel circuit, V_T is shown as the reference phasor at $0°$. I_T lags V_T by $45°$. (See Fig. 13-16c.) For the time diagram, see Fig. 13-16d.

Impedance in parallel RL. For the general case of calculating the total impedance Z_T of R and X_L in parallel, assume any number for the applied voltage V_T because in the calculation of Z_T in terms of the branch currents the value of V_T cancels. A convenient value to assume for V_T is the value of either R or X_L, whichever is the higher number. This is only one method among others for calculating Z_T.

Example 13.16 What is the impedance Z_T of a 200-Ω R in parallel with a 400-Ω X_L? Assume 400 V for the applied voltage V_T.

$$I_R = \frac{V_T}{R} = \frac{400}{200} = 2\text{ A}$$

$$I_L = \frac{V_T}{X_L} = \frac{400}{400} = 1\text{ A}$$

$$I_T = \sqrt{I_R^2 + I_L^2} = \sqrt{4+1} = \sqrt{5} = 2.24\text{ A}$$

$$Z_T = \frac{V_T}{I_T} = \frac{400}{2.24} = 178.6\ \Omega \qquad Ans.$$

The combined impedance of a 200-Ω R in parallel with a 400-Ω X_L is equal to 178.6 Ω regardless of the value of the applied voltage. The combined impedance must be less than the *lowest* number of ohms in the parallel branches. The total impedance of a parallel RL circuit does not equal that of a series RL circuit; that is $Z_T \neq \sqrt{R^2 + X_L^2}$, because the resistance and inductive reactance combine to present a different load condition to the voltage source.

Q OF A COIL

The quality or merit Q of a coil is indicated by the equation

$$Q = \frac{X_L}{R_i} = \frac{6.28fL}{R_i} \qquad\qquad\qquad (13\text{-}19)$$

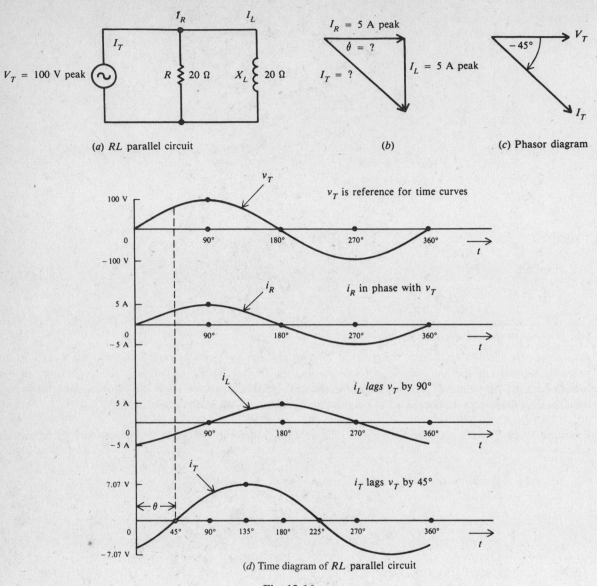

(a) RL parallel circuit   (b)   (c) Phasor diagram

(d) Time diagram of RL parallel circuit

Fig. 13-16

where R_i is the internal resistance of the coil equal to the resistance of the wire in the coil (Fig. 13-17). Q is a numerical value without any units since the ohms cancel in the ratio of reactance to resistance. If the Q of a coil is 200, it means that the X_L of the coil is 200 times more than its R_i.

The Q of a coil may range in value from less than 10 for a low-Q coil up to 1000 for a very high Q coil. Radio frequency (RF) coils have a Q of about 30–300.

As an example, a coil with an X_L of 300 Ω and a R_i of 3 Ω has a Q of $300/3 = 100$.

POWER IN *RL* CIRCUITS

In an ac circuit with inductive reactance, the line current I lags the applied voltage V. The *real power P* is equal to the voltage multiplied by only that portion of the line current which is in phase with the voltage.

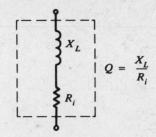

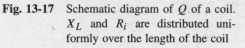

$$Q = \frac{X_L}{R_i}$$

Fig. 13-17   Schematic diagram of Q of a coil. X_L and R_i are distributed uniformly over the length of the coil

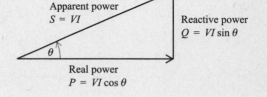

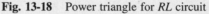

Fig. 13-18   Power triangle for *RL* circuit

Therefore,

$$\text{Real power } P = V(I\cos\theta) = VI\cos\theta \qquad (13\text{-}20)$$

where θ is the phase angle between V and I, and $\cos\theta$ is the *power factor* (PF) of the circuit. Also,

$$\text{Real power } P = I^2 R \qquad (13\text{-}21)$$

where R is the total resistive component of the circuit.

Reactive power Q in voltamperes reactive (VAR), is expressed as follows:

$$\text{Reactive power } Q = VI\sin\theta \qquad (13\text{-}22)$$

Apparent power S is the product of $V \times I$. The unit is volt amperes (VA). In formula form,

$$\text{Apparent power } S = VI \qquad (13\text{-}23)$$

In all the power formulas, the V and I are in rms values. The relationships of real, reactive, and apparent power can be illustrated by the phasor diagram of power (Fig. 13-18). Reactive power Q is inductive and shown above the horizontal axis.

Example 13.17   The ac circuit (Fig. 13-19a) has 2A through a 173-Ω R in series with an X_L of 100 Ω. Find the power factor, the applied voltage V, real power P, reactive power Q, and apparent power S.

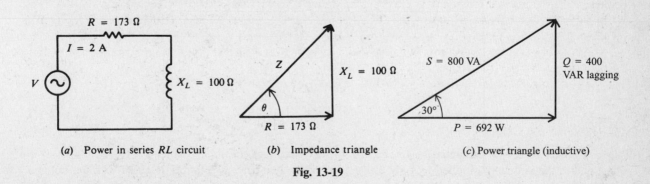

(a) Power in series *RL* circuit   (b) Impedance triangle   (c) Power triangle (inductive)

Fig. 13-19

Step 1. Find the phase angle θ, $\cos\theta$, and impedance Z by the impedance triangle (Fig. 13-19b).

$$\theta = \arctan\frac{X_L}{R} = \arctan\frac{100}{173} = \arctan 0.578 = 30° \qquad Ans.$$

$$PF = \cos\theta = \cos 30° = 0.866 \qquad Ans.$$

$$Z = \frac{R}{\cos\theta} = \frac{173}{\cos 30°} = 200 \ \Omega \qquad Ans.$$

An alternative method to find Z is by using $Z = \sqrt{R^2 + X_L^2}$.

Step 2. Find V.

$$V = IZ = 2(200) = 400 \ \text{V} \qquad Ans.$$

Step 3. Find P.

$$P = I^2 R \qquad\qquad\qquad (13\text{-}21)$$

$$= 2^2(173) = 692 \ \text{W} \qquad Ans.$$

or

$$P = VI\cos\theta \qquad\qquad\qquad (13\text{-}20)$$

$$= 400(2)(\cos 30°) = 692 \ \text{W}$$

In both P calculations, the real power is the same because this is the amount of power supplied by the voltage source and dissipated in the resistance. The inductive reactance merely transforms power back to the circuit. Either formula for P can be used, depending on which is more convenient.

Step 4. Find Q and S.

$$Q = VI\sin\theta \qquad\qquad\qquad (13\text{-}22)$$

$$= 400(2)(\sin 30°) = 400 \ \text{VAR lagging} \qquad Ans.$$

In an inductive circuit, reactive power is lagging because I lags V.

$$S = VI \qquad\qquad\qquad (13\text{-}23)$$

$$= 400(2) = 800 \ \text{VA} \qquad Ans.$$

See Fig. 13-19c.

Table 13-1 summarizes the relationships of current, voltage, impedance, and phase angle in RL circuits.

Table 13-1 Summary Table for Series and Parallel RL Circuits

X_L and R in Series	X_L and R in Parallel
I the same in X_L and R	V_T the same across X_L and R
$V_T = \sqrt{V_R^2 + V_L^2}$	$I_T = \sqrt{I_R^2 + I_L^2}$
$Z = \sqrt{R^2 + X_L^2} = \dfrac{V_T}{I}$	$Z_T = \dfrac{V_T}{I_T}$
V_R lags V_L by 90°	I_L lags I_R by 90°
$\theta = \arctan\dfrac{X_L}{R}$	$\theta = \arctan\left(-\dfrac{I_L}{I_R}\right)$

Solved Problems

13.1   A steady current of 20 mA is passed through a coil with an inductance of 100 mH. What is the voltage induced by the coil?

If the circuit is dc, the rate of change of current $\Delta i/\Delta t = 0$. So

$$v_L = L\frac{\Delta i}{\Delta t} \qquad (13\text{-}2)$$

$$= L(0) = 0\,\text{V} \qquad Ans.$$

A voltage can be induced only when a coil is carrying a *changing* current.

13.2   Current through a coil increases to 20 A in 1/1000 s. If its inductance is 100 μH, what is the induced voltage at that instant?

$$v_L = L\frac{\Delta i}{\Delta t} \qquad (13\text{-}2)$$

$$= \left(100 \times 10^{-6}\right)\left(\frac{20}{1/1000}\right) = \left(10^{-4}\right)\left(2 \times 10^{4}\right) = 2\,\text{V} \qquad Ans.$$

13.3   A 120-Hz 20-mA ac current is present in a 10-H inductor. What is the reactance of the inductor and the voltage drop across the inductor?

$$X_L = 6.28fL \qquad (13\text{-}3)$$

$$= 6.28(120)(10) = 7536\,\Omega \qquad Ans.$$

$$V_L = I_L X_L$$

$$= (20 \times 10^{-3})(7536) = 150.7\,\text{V} \qquad Ans.$$

13.4   In Problem 13.3, what are the maximum and average values of the voltage developed across the inductor?

In a reactive ac circuit, the same relations exist for the various values of voltage such as rms, peak, average, and instantaneous. The rms value is implied when no statement is made otherwise.

$$V_M = 1.414\,\text{V}$$

$$V_{L,M} = 1.414V_L = 1.414(150.7) = 213.1\,\text{V} \qquad Ans.$$

$$V_{\text{av}} = 0.91\,\text{V}$$

$$V_{L,\text{av}} = 0.91V_L = 0.91(150.7) = 137.1\,\text{V} \qquad Ans.$$

13.5   A 225-μH choke coil of negligible resistance is to limit the current through it to 25 mA when 40 V are impressed across it. What is the frequency of the current?

$$X_L = \frac{V_L}{I_L} \qquad (13\text{-}7)$$

$$= \frac{40}{25 \times 10^{-3}} = 1600\,\Omega$$

$$f = \frac{X_L}{6.28L} \tag{13-5}$$

$$= \frac{1600}{6.28(255 \times 10^{-6})} = 10^6 = 1 \text{ MHz or 1000 kHz} \quad Ans.$$

13.6 A simple high-pass filter (Fig. 13-20) is one in which high-frequency waves pass through the capacitor C to the output and low-frequency waves pass through the inductor L. What is the reactance of the 15-mH coil to (a) a 2000-Hz (low-frequency) current and (b) a 400-kHz (high-frequency) current?

(a) $$X_L = 6.28 fL \tag{13-3}$$

$$= 6.28(2 \times 10^3)(15 \times 10^{-3}) = 188.4 \, \Omega \quad Ans.$$

(b) $$X_L = 6.28(400 \times 10^3)(15 \times 10^{-3}) = 37\,680 \, \Omega \quad Ans.$$

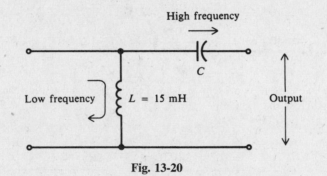

Fig. 13-20

13.7 What is the total inductance of the circuit shown in Fig. 13-21a?

Step 1. Reduce the parallel inductors to their equivalents.

$$L_4 = \frac{L_2 L_3}{L_2 + L_3} = \frac{3(6)}{3+6} = \frac{18}{9} = 2 \text{ mH}$$

See Fig. 13-21b.

Step 2. Add the series inductors.

$$L_T = L_1 + L_4 = 10 + 2 = 12 \text{ mH} \quad Ans.$$

See Fig. 13-21c.

13.8 If a frequency of 2 MHz is applied to the circuit of Fig. 13-21, what is the reactance of the circuit?

$$X_L = 6.28 f L_T = 6.28(2 \times 10^6)(12 \times 10^{-3}) = 150.72 \times 10^3 = 150\,720 \, \Omega \quad Ans.$$

13.9 With two coils L_1 and L_2 as wound (Fig. 13-22), find the total inductance.

Since the windings are wound in the same direction relative to the dots, L_1 and L_2 are series-aiding. Then, using Eq. (13-10),

$$L_T = L_1 + L_2 + 2L_M = 9 + 13 + 6 = 28 \text{ H} \quad Ans.$$

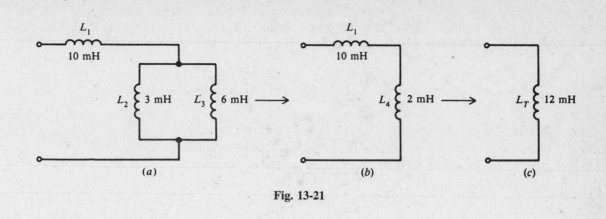

Fig. 13-21

13.10   The windings of L_2 (Fig. 13-22) are now wound in reverse (Fig. 13-23). What is the total inductance now?

Because the windings are wound in opposite directions relative to the dots, L_1 and L_2 are series-opposing. Then

$$L_T = L_1 + L_2 - 2L_M = 9 + 13 - 6 = 16 \text{ H} \qquad Ans.$$

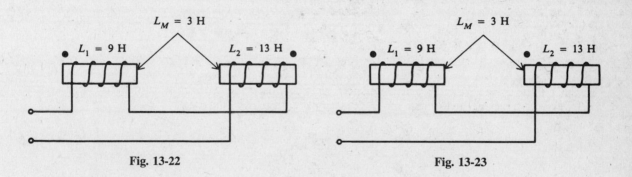

Fig. 13-22   Fig. 13-23

13.11   A 20-H coil is connected across a 110-V 60-Hz power line. If the coil has zero resistance, find the current and power drawn. Draw the phasor diagram.

$$X_L = 6.28fL \qquad (13\text{-}3)$$

$$= 6.28(60)(20) = 7536 \ \Omega$$

$$I_L = \frac{V_L}{X_L} \qquad (13\text{-}6)$$

$$= \frac{110}{7536} = 14.6 \text{ mA} \quad Ans.$$

$$P = VI \cos\theta \qquad (13\text{-}20)$$

$$= 110\left(14.6 \times 10^{-3}\right)(\cos 90^\circ) = 110\left(14.6 \times 10^{-3}\right)(0) = 0 \text{ W} \qquad Ans.$$

In a purely reactive circuit ($R = 0$), real power is zero because no energy is dissipated. Also $P = I^2 R = I^2(0) = 0$. In the phasor diagram, I_L lags V_L by 90°.

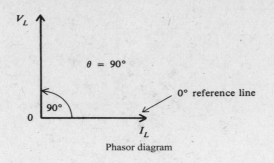

Phasor diagram

13.12   A tuning coil has an inductance of 39.8 μH and an internal resistance of 20 Ω. Find its impedance to a frequency of 100 kHz and the current through the coil if the voltage drop is 80 V across the entire coil. Also find the resistive drop and the inductive drop of the coil, and draw the phasor diagram.

A coil with R_i and X_L is treated as a series RL circuit.

Step 1.   Find X_L and then Z, θ.

$$X_L = 6.28fL \tag{13-3}$$

$$= 6.28(10^5)(39.8 \times 10^{-6}) = 25 \ \Omega$$

$$Z = \sqrt{R_i^2 + X_L^2} \tag{13-15}$$

$$= \sqrt{20^2 + 25^2} = 32 \ \Omega \qquad Ans.$$

$$\theta = \arctan \frac{X_L}{R_i} \tag{13-16}$$

$$= \arctan \frac{25}{20} = \arctan 1.25 = 51.3°$$

Step 2.   Find I.

$$I = \frac{V}{Z} = \frac{80}{32} = 2.5 \ \text{A} \qquad Ans.$$

Step 3.   Find V_R, V_L, and check θ.

$$V_R = IR_i = 2.5(20) = 50 \ \text{V} \qquad Ans.$$

$$V_L = IX_L = 2.5(25) = 62.5 \ \text{V} \qquad Ans.$$

Also

$$\theta = \arctan \frac{V_L}{V_R} \tag{13-14}$$

$$= \arctan \frac{62.5}{50} = 1.25 = 51.3° \qquad Check$$

Step 4.   Draw phasor diagram.

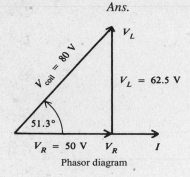

Phasor diagram

13.13   A choke is an inductance coil with a very low resistance. The ac voltage drop across R is therefore very low. That is, practically all the ac voltage drop is across L. For this to occur, X_L is taken as 10 or more times the series R. Find the minimum inductance required for a choke with a resistance of $100\,\Omega$ when the frequency of the circuit is (a) 5 kHz, (b) 5 MHz, and (c) 50 MHz. If the applied voltage V_T is 200 V, (d) what is the voltage across the choke and the resistance?

$$X_L = 10R = 10(100) = 1000\,\Omega$$

(a) $$L = \frac{X_L}{6.28f} \qquad (13\text{-}4)$$

$$= \frac{1000}{6.28(5 \times 10^3)} = 32\,\text{mH} \qquad Ans.$$

(b) $$L = \frac{1000}{6.28(5 \times 10^6)} = 32\,\mu\text{H} \qquad Ans.$$

(c) $$L = \frac{1000}{6.28(5 \times 10^7)} = 3.2\,\mu\text{H} \qquad Ans.$$

(d) $$Z = \sqrt{R_i^2 + X_L^2} \qquad (13\text{-}15)$$

$$= \sqrt{100^2 + 1000^2} = 1005\,\Omega$$

$$I = \frac{V_T}{Z} = \frac{200}{1005} = 0.199\,\text{A}$$

$$V_R = IR = 0.199(100) = 19.9\,\text{V} \qquad Ans.$$

$$V_L = IX_L = 0.199(1000) = 199\,\text{V} \qquad Ans.$$

Note that V_L is practically all the applied voltage and V_R is small by comparison.

13.14   The purpose of a high-pass filter circuit (Fig. 13-24) is to permit high frequencies to pass on to the load but to prevent the passing of low frequencies. Find the branch currents, the total current, and the percentage of the total current passing through the resistor for (a) a 1.5-kHz (low) audio-frequency signal and (b) a 1-MHz (high) radio-frequency signal.

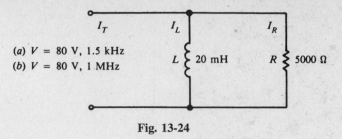

Fig. 13-24

(a)   **Step 1.**   Find X_L at $f = 1.5\,\text{kHz}$.

$$X_L = 6.28fL = 6.28(1.5 \times 10^3)(20 \times 10^{-3}) = 188.4\ \Omega$$

Step 2.   Find branch currents I_L and I_R.

$$I_L = \frac{V}{X_L} = \frac{80}{188.4} = 0.425\ \text{A} \qquad Ans.$$

$$I_R = \frac{V}{R} = \frac{80}{5000} = 0.016\ \text{A} \qquad Ans.$$

Step 3.   Find total current I_T.

$$I_T = \sqrt{I_R^2 + I_L^2} \tag{13-17}$$

$$= \sqrt{(0.425)^2 + (0.016)^2} = \sqrt{0.1809} = 0.425\ \text{A} \qquad Ans.$$

Since $X_L \ll R$, the current is mostly inductive.

Step 4.   Find I_R as a percentage of I_T.

$$\frac{I_R}{I_T} \times 100 = \frac{0.016}{0.425}100 = 0.038(100) = 3.8\% \qquad Ans.$$

Therefore, only 3.8 percent of the 1.5-kHz audio signal passes through the resistor.

(b)   **Step 1.**   Find X_L now at $f = 1\,\text{MHz}$.

$$X_L = 6.28fL = 6.28(1 \times 10^6)(20 \times 10^{-3}) = 125.6\,\text{k}\Omega$$

Step 2.   Find I_L and I_R.

$$I_L = \frac{V}{X_L} = \frac{80}{125.6 \times 10^3} = 0.637\ \text{mA} \qquad Ans.$$

$$I_R = 16\ \text{mA} \qquad Ans.$$

I_R remains the same as in part (a).

Step 3.   Find I_T.

$$I_T = \sqrt{I_R^2 + I_L^2} = \sqrt{16^2 + (0.637)^2} = \sqrt{256.41} = 16.01 \text{ mA} \qquad Ans.$$

Since $R \ll X_L$, the current is mostly resistive.

Step 4.   Find I_R as a percentage of I_T.

$$\frac{I_R}{I_T} \times 100 = \frac{16}{16.01} 100 = 0.999(100) = 99.9\% \qquad Ans.$$

Thus, theoretically 100 percent of the 1-MHz radio signal passes through the resistor.

It is clear that the circuit is an excellent high-pass filter by passing almost 100 percent of the high radio frequency to the load and only 3.8 percent of the low audio frequency to the load.

13.15   If the Q of a coil is greater than 5, its internal resistance R_i may be disregarded so that $Z = X_L$. If the Q is smaller than 5, then the resistance must be added to the reactance to obtain the impedance by the formula $Z = \sqrt{R_i^2 + X_L^2}$. A coil has $R_i = 5 \,\Omega$ and $X_L = 30 \,\Omega$ at a certain frequency. Find the Q and Z of the coil.

$$Q = \frac{X_L}{R_i} \qquad \qquad (13\text{-}19)$$

$$= \frac{30}{5} = 6 \qquad Ans.$$

Since $Q > 5$, the resistance may be disregarded, so the impedance is equal to the inductive reactance.

$$Z = X_L = 30 \,\Omega \qquad Ans.$$

We can determine the percent error by finding Z with R included and comparing values.

$$Z = \sqrt{R_i^2 + X_L^2} = \sqrt{5^2 + 30^2} = 30.4 \,\Omega$$

With R not included, $Z = 30 \,\Omega$, as found above. The error is $30.4 - 30 = 0.4 \,\Omega$. Therefore, the percent error is

$$\frac{0.4}{30.4} 100 = 1.3\%$$

The error is well within the range of human error in taking measurements and therefore is negligible.

13.16   What is the inductance of a coil whose resistance is $100 \,\Omega$ if it draws 0.55 A from a 110-V, 60-Hz power line?

Step 1.   Find Z.

$$Z = \frac{V}{I} = \frac{100}{0.55} = 200 \,\Omega$$

Step 2. Find X_L.

$$Z = \sqrt{R_i^2 + X_L^2}$$

$$Z^2 = R_i^2 + X_L^2$$

$$X_L = \sqrt{Z^2 - R_i^2} = \sqrt{200^2 - 100^2} = \sqrt{3 \times 10^4} = 173 \,\Omega$$

Step 3. Find L.

$$L = \frac{X_L}{6.28 f} = \frac{173}{6.28 \times 60} = 0.459 \,\text{H} \qquad Ans.$$

13.17   A 500-Ω R is in parallel with 300-Ω X_L (Fig. 13-25). Find I_T, θ, and Z_T.

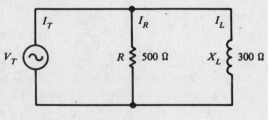

Fig. 13-25

Assume $V_T = 500$ V. (Refer to discussion on total impedance, p. 287.) Then

$$I_R = \frac{V_T}{R} = \frac{500}{500} = 1\,\text{A}$$

$$I_L = \frac{V_T}{X_L} = \frac{500}{300} = 1.67\,\text{A}$$

$$I_T = \sqrt{I_R^2 + I_L^2} = \sqrt{1^2 + (1.67)^2} = 1.95\,\text{A} \qquad Ans.$$

$$\theta = \arctan\left(-\frac{I_L}{I_R}\right) = \arctan(-1.67) = -59.1° \qquad Ans.$$

$$Z_T = \frac{V_T}{I_T} = \frac{500}{1.95} = 256.4 \,\Omega \qquad Ans.$$

13.18   The frequency in Problem 13.17 is increased by a factor of 2. Now find I_T, θ, and Z_T.

Since X_L is directly proportional to f,

$$X_L = 300(2) = 600 \,\Omega$$

Assume $V_T = 600$ V. Then

$$I_R = \frac{V_T}{R} = \frac{600}{500} = 1.2 \text{ A}$$

$$I_L = \frac{V_T}{X_L} = \frac{600}{600} = 1 \text{ A}$$

$$I_T = \sqrt{I_R^2 + I_L^2} = \sqrt{(1.2)^2 + 1^2} = 1.56 \text{ A} \qquad Ans.$$

$$\theta = \arctan\left(-\frac{I_L}{I_R}\right) = \arctan\left(-\frac{1}{1.2}\right) = \arctan(-0.83) = -39.8° \qquad Ans.$$

$$Z_T = \frac{V_T}{I_T} = \frac{600}{1.56} = 384.6 \,\Omega \qquad Ans.$$

Increasing the frequency in an RL parallel circuit decreases θ, since more X_L means less I_L.

13.19 Show that the real power $P = (V_M I_M/2)\cos\theta$.

$$P = VI\cos\theta \tag{13-20}$$

The effective or rms value of voltage (or current) is its maximum value divided by $\sqrt{2}$. So substitute

$$V = \frac{V_M}{\sqrt{2}} \quad \text{and} \quad I = \frac{I_M}{\sqrt{2}}$$

into Eq. (13-20) to obtain

$$P = \left(\frac{V_M}{\sqrt{2}} \frac{I_M}{\sqrt{2}}\right)\cos\theta = \frac{V_M I_M}{2}\cos\theta \qquad Ans.$$

13.20 An induction motor operating at a power factor of 0.8 draws 1056 W from a 110-V ac line. What is the current?

Given PF $= \cos\theta = 0.80$, $V = 110$ V, and $P = 1056$ W.

$$P = VI\cos\theta \tag{13-20}$$

from which

$$I = \frac{P}{V\cos\theta} = \frac{1056}{110(0.8)} = 12 \text{ A} \qquad Ans.$$

Supplementary Problems

13.21 If the rate of change of current in a coil is large, the voltage induced is high. Compare the induced voltages of a coil with an inductance of 10 mH when the rate of change of current is 2000 A/s and when the rate is 5 times faster at 10 000 A/s.
Ans.   When $\Delta i/\Delta t = 2000$ A/s, $v_L = 20$ V. When $\Delta i/\Delta t = 10\,000$ A/s, $v_L = 100$ V

13.22 How fast must current change in a 100-μH coil so that a voltage of 3 V is induced?
Ans. $\Delta i / \Delta t = 30\,000$ A/s

13.23 At a particular instant the current changes at 1000 A/s. If 1.5 V is induced, what is the inductance of the coil? *Ans.* $L = 1.5$ mH

13.24 Find the inductive reactance of a 0.5-H choke coil at (*a*) 200 Hz, (*b*) 2000 Hz, (*c*) 20 kHz, and (*d*) 2 MHz. *Ans.* (*a*) $X_L = 628\ \Omega$; (*b*) $X_L = 6280\ \Omega$; (*c*) $X_L = 62\,800\ \Omega$; (*d*) $X_L = 6280$ kΩ

13.25 A choke coil in a FM receiver has an inductance of 20 μH. What is its reactance at 10 MHz?
Ans. $X_L = 1256\ \Omega$

13.26 A 2-mH coil in a tuning circuit is resonant at 460 kHz. What is its inductive reactance at this frequency?
Ans. $X_L = 5778\ \Omega$

13.27 A transmitter tuning coil must have a reactance of 95.6 Ω at 3.9 MHz. Find the inductance of the coil.
Ans. $L = 3.9\ \mu$H

13.28 A 25-H choke coil in a filter circuit of a power supply operates at 60 Hz. Find (*a*) its inductive reactance, (*b*) the coil current flowing if the voltage across the coil is 105 V, and (*c*) the rms, peak, and average values of this current.
Ans. (*a*) $X_L = 9420\ \Omega$; (*b*) $I_L = 11.1$ mA; (*c*) $I_L = 11.1$ mA; $I_{L,M} = 15.7$ mA; $I_{L,\text{av}} = 10.1$ mA

13.29 A choke coil with no resistance acts as a current limiter to 25 mA when 40 V is applied across it at a frequency of 500 kHz. What is its inductance? *Ans.* $L = 0.51$ mH

13.30 Two 2-H coils are connected in series so that the mutual inductance between them is 0.2 H. Find the total inductance when they are connected to be (*a*) series-aiding and (*b*) series-opposing.
Ans. (*a*) $L_T = 4.4$ H; (*b*) $L_T = 3.6$ H

13.31 A number of coils are connected together to form an inductance network. Group *A* is made up of three 12-H chokes in parallel; group *B* of a 3- and a 5-H choke in parallel; and group *C* of a 4- and 6-H choke in parallel. Groups *A*, *B*, and *C* are then connected in series. Find the equivalent inductance of (*a*) group *A*, (*b*) group *B*, and (*c*) group *C*; and (*d*) find the total inductance of the network.
Ans. (*a*) 4 H; (*b*) 1.88 H; (*c*) 2.4 H; (*d*) 8.3 H

13.32 Find the total inductance of the circuits (Fig. 13-26). The coils are spaced far apart so that mutual inductance is negligible.
Ans. (*a*) $L_T = 15$ mH; (*b*) $L_T = 4$ mH; (*c*) $L_T = 1.48$ H; (*d*) $L_T = 5.1$ H

13.33 Find the total inductance of the circuits (Fig. 13-27). The coils are spaced sufficiently close so that mutual inductance is present.
Ans. (*a*) $L_T = 11$ H; (*b*) $L_T = 8$ H; (*c*) $L_T = 14$ H; (*d*) $L_T = 8$ H

13.34 A resistance of 12 Ω is connected in series with a coil whose inductive reactance is 5 Ω. If the impressed ac voltage is 104 V, find the impedance, the line current, the voltage drop across the resistor and coil, the phase angle, and the power. Draw the voltage-phasor diagram.
Ans. $Z = 13\ \Omega$; $I = 8$ A; $V_R = 96$ V; $V_L = 40$ V; $\theta = 22.6°$, I lags V_T; $P = 768$ W; phasor diagram: see Fig. 13-28.

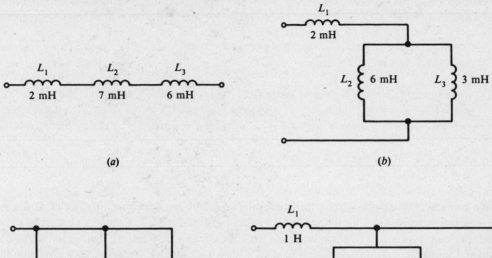

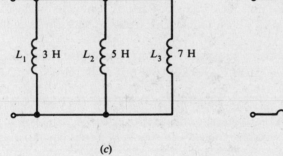

(a)

(b)

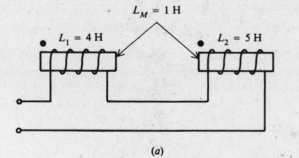

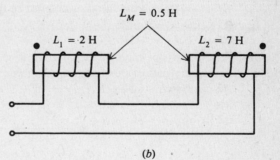

(c)

(d)

Fig. 13-26

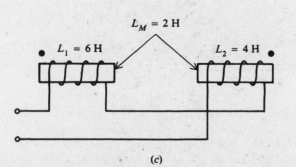

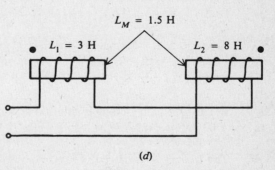

(a)

(b)

(c)

(d)

Fig. 13-27

13.35 A lightning protector circuit contains 55.7-mH coil in series with a 6-Ω resistor. What current will flow when it is tested with a 110-V 60-Hz voltage? How much power will the lightning protector consume?   *Ans.* $I = 5.05$ A ($Z = 21.8$ Ω); $P = 153$ W

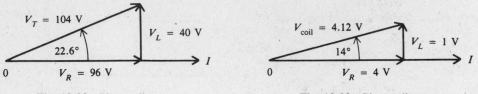

Fig. 13-28   Phasor diagram   Fig. 13-29   Phasor diagram

13.36 The coil of a telephone relay has a resistance of 400 Ω and an inductance of 16 mH. If the relay is operated at a frequency of 1 kHz, find the impedance of the coil and the voltage that must be impressed across the coil in order to operate the relay at its rated current of 10 mA. Draw the voltage-phasor diagram.   *Ans.* $Z = 412.4$ Ω; $V_{\text{coil}} = 4.12$ V; phasor diagram: see Fig. 13-29

13.37 A 60-V source at 1.5 kHz is impressed across a loudspeaker of 5000 Ω and 2.12 H inductance. Find the current and power drawn.   *Ans.* $I = 2.9$ mA ($Z = 20\,600$ Ω); $P = 42.1$ mW

13.38 What is the inductive reactance of a single-phase motor if the line voltage is 220 V, the line current is 15 A, and the resistance of the motor coils is 10 Ω? Also what is the angle of lag, the power factor, and the power consumed by the motor? (Treat this problem as a simple *RL* series circuit.)
Ans. $X_L = 10.7$ Ω; $\theta = 47°$; PF $= 0.682$; $P = 2250$ W

13.39 What is the minimum inductance for a RF choke in series with a resistance of 50 Ω at a radio frequency of 1000 kHz, if the resistive drop is to be considered negligible? If the applied voltage is 100 V, what is the voltage across the resistance? (See Problem 13-13.)
Ans. $L = 0.08$ mH; $V_R = 9.95$ V

13.40 A 20-Ω resistor and a 15-Ω inductive reactance are placed in parallel across a 120-V ac line. Find the branch currents, the total current, the impedance, and the power drawn; and draw the phasor diagram.
Ans. $I_R = 6$ A; $I_L = 8$ A; $I_T = 10$ A; $Z_T = 12$ Ω; $P = 720$ W; phasor diagram: see Fig. 13-30.

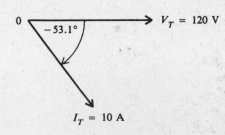

Fig. 13-30   Phasor diagram

13.41 A 100-Ω *R* is in parallel with a 100-Ω X_L. If $V_T = 100$ V, calculate I_T, θ, and Z_T.
Ans. $I_T = 1.41$ A; $\theta = -45°$; $Z_T = 70.7$ Ω

13.42 The frequency is halved in Problem 13.41. Find I_T, θ, and Z_T. Compare the difference in values.
Ans. $I_T = 2.24$ A; $\theta = -63.4°$; $Z_T = 44.6$ Ω. Reducing the frequency in an R_L parallel circuit increases θ because less X_L means more I_L. With less X_L, Z_T is less.

13.43   A 50-Ω R and a 120-Ω X_L are connected in parallel across a 120-V ac line. Find the (a) branch currents, (b) total current, (c) impedance, (d) power drawn, and (e) draw the phasor diagram.
   Ans. (a) $I_R = 2.4$ A; $I_L = 1$ A;   (b) $I_T = 2.6$ A;   (c) $Z_T = 46.2\,\Omega$;   (d) $P = 288$ W;   (e) phasor diagram: see Fig. 13-31

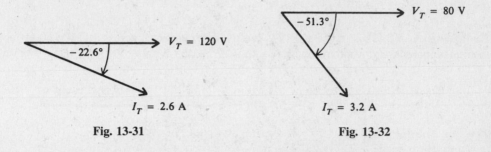

Fig. 13-31	Fig. 13-32

13.44   A 40-Ω resistor and a 10-mH coil are in parallel across an 80-V 500-Hz ac line. Find the (a) branch currents, (b) total current, (c) impedance, (d) power drawn; and (e) draw the phasor diagram.
   Ans. (a) $I_R = 2$ A; $I_L = 2.5$ A;   (b) $I_T = 3.2$ A;   (c) $Z_T = 25\,\Omega$;   (d) $P = 160$ W;   (e) phasor diagram: see Fig. 13-32.

13.45   For the high-pass filter circuit (Fig. 13-33), find the (a) branch currents, (b) total current, and (c) percentage of total current that passes through the resistor for the case of audio frequency (AF) at 1 kHz and for the case of radio frequency (RF) at 2 MHz. (Calculate current values to three significant figures.)
   Ans. AF case: (a) $I_L = 0.797$ A; $I_R = 0.0333$ A; (b) $I_T = 0.798$ A;   (c) 4.2%. RF case: (a) $I_L = 0.398$ mA; $I_R = 33.3$ mA;   (b) $I_T = 33.3$ mA;   (c) 100%

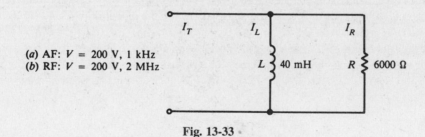

(a) AF: $V = 200$ V, 1 kHz
(b) RF: $V = 200$ V, 2 MHz

Fig. 13-33

13.46   A 120-V 60-Hz power line is connected across a 12-H choke coil whose resistance is 500 Ω. Find (a) the inductive reactance, (b) the Q of the coil, (c) the impedance, and (d) the current.
   Ans. (a) $X_L = 4522\,\Omega$;   (b) $Q = 9.0$;   (c) $Z = 4522\,\Omega$;   (d) $I_L = 27$ mA

13.47   The primary of an audio-frequency transformer has a resistance of 100 Ω and an inductance of 25 mH. What are the inductive reactance and the impedance at 2 kHz?
   Ans. $X_L = 314\,\Omega$; $Z = 330\,\Omega$

13.48   Find the inductance of a coil whose resistance is 500 Ω if it draws 10 mA from a 110-V 60-Hz source.   *Ans.* $L = 29.2$ H ($X_L = 11\,000\,\Omega$)

13.49   A coil having a Q of 25 draws 20 mA when connected to a 12-V, 1-kHz power supply. What is its inductance?   *Ans.* $L = 95.5$ mH ($X_L = 600\,\Omega$)

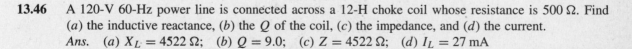

13.50 For the circuit (Fig. 13-34), find (*a*) the inductive reactance, (*b*) the impedance, (*c*) the rms current, and (*d*) the peak current.
Ans.   (*a*) $X_L = 37.7\,\Omega$;   (*b*) $Z = 41.0\,\Omega$ (circuit $R = 16\,\Omega$);   (*c*) $I = 5.85\,\text{A}$;   (*d*) $I_M = 8.27\,\text{A}$

13.51 An inductive load operating at a phase angle of 53° draws 1400 W from a 120-V line. Find the current drawn.   *Ans.* $I = 19.4\,\text{A}$

13.52 An inductance of 5 Ω resistance and 12 Ω reactance is connected across a 117-V 60-Hz ac line. Find the real, apparent, and reactive power.   *Ans.* $P = 405\,\text{W}$; $S = 1053\,\text{VA}$; $Q = 972\,\text{VAR}$ lagging

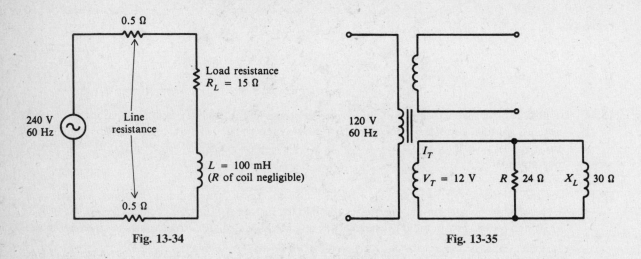

Fig. 13-34   Fig. 13-35

13.53 A toy electric train semaphore has a 24-Ω lamp in parallel with a solenoid coil of 30-Ω inductive reactance (Fig. 13-35). If it operates from the 12-V winding of a 60-Hz power transformer, find (*a*) the total current, (*b*) the impedance, (*c*) the phase angle, (*d*) the power drawn, and (*e*) the reactive power.
Ans.   (*a*) $I_T = 0.64\,\text{A}$;   (*b*) $Z_T = 18.8\,\Omega$;   (*c*) $\theta = -38.7°$;   (*d*) $P = 6\,\text{W}$;   (*e*) $Q = 4.8\,\text{VAR}$ lagging

Chapter 14

Capacitance, Capacitive Reactance, and Capacitive Circuits

CAPACITOR

A *capacitor* is an electrical device which consists of two conducting plates of metal separated by an insulating material called a *dielectric* (Fig. 14-1a). Schematic symbols shown (Fig. 14-1b and c) apply to all capacitors.

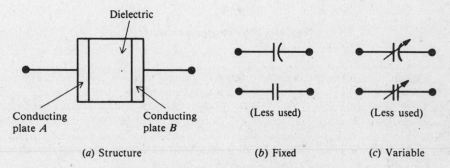

(a) Structure   (b) Fixed   (c) Variable

Fig. 14-1   Capacitor and schematic symbols

A capacitor stores electric charges in the dielectric. The two plates of the capacitor shown in Fig. 14-2a are electrically neutral since there are as many protons (positive charge) as electrons (negative charge) on each plate. Thus the capacitor has *no charge*. Now a battery is connected across the plates (Fig. 14-2b). When the switch is closed (Fig. 14-2c), the negative charge on plate A is attracted to the positive terminal of the battery, while the positive charge on plate B is attracted to the negative terminal of the battery. This movement of charges will continue until the difference in charge between plates A and B is equal to the electromotive force (voltage) of the battery. The capacitor is now *charged*. Since almost none of the charge can cross the space between plates, the capacitor will remain in this condition even if the battery is removed (Fig. 14-3a). However, if a conductor is placed across the plates (Fig. 14-3b), the electrons find a path back to plate A and the charges on each plate are again neutralized. The capacitor is now *discharged*.

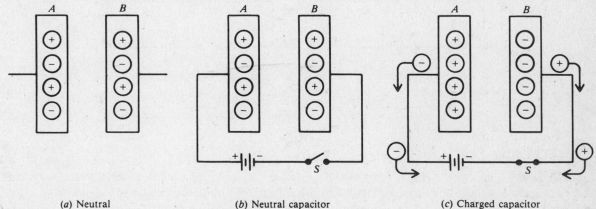

(a) Neutral   (b) Neutral capacitor   (c) Charged capacitor

Fig. 14-2   Charging a capacitor

305

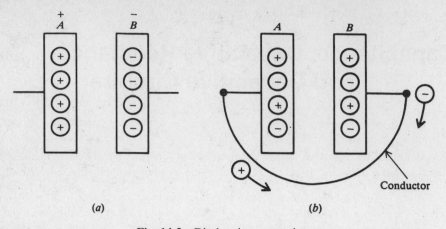

(a)   (b)

Fig. 14-3   Discharging a capacitor

Example 14.1   Explain the charging and discharging action of a simple capacitor circuit when switch 1 is closed with switch 2 open (Fig. 14-4a), and when switch 1 is open with switch 2 closed (Fig. 14-4b).

When switch $S1$ is closed and switch $S2$ is open (Fig. 14-4a), the battery voltage is applied across the two plates A and B. The capacitor charges to a voltage equal to that of the battery. Plate A is charged positively and plate B is charged negatively. When $S1$ is open and $S2$ is closed, the excess electrons on plate B will move through $S2$ to plate A (Fig. 14-4b). Now the capacitor acts as a voltage source with plate A the positive terminal and plate B the negative terminal. The motion of electrons off plate B reduces its negative charge, and their arrival at plate A reduces its positive charge. This motion of electrons continues until there is no charge on plate A or plate B and the voltage between the two plates is zero.

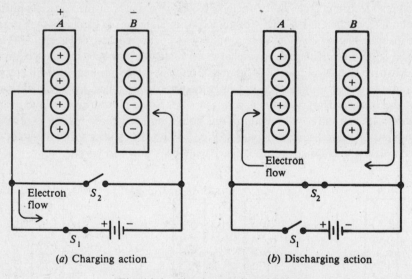

(a) Charging action   (b) Discharging action

Fig. 14-4   Simple capacitor circuit

CAPACITANCE

Electrically, *capacitance* is the ability to store an electric charge. Capacitance is equal to the amount of charge that can be stored in a capacitor divided by the voltage applied across the plates:

$$C = \frac{Q}{V}$$

$$(14\text{-}1)$$

where C = capacitance, F
        Q = amount of charge, C
        V = voltage, V

Equation (*14-1*) can be rewritten as follows:

$$Q = CV \qquad\qquad (14\text{-}2)$$

$$V = \frac{Q}{C} \qquad\qquad (14\text{-}3)$$

The unit of capacitance is the *farad* (F). The farad is that capacitance that will store one coulomb of charge in the dielectric when the voltage applied across the capacitor terminals is one volt.

The characteristic of a dielectric that describes its ability to store electric energy is called the *dielectric constant*. Air is used as a reference and is given a dielectric constant of 1. Some other dielectric materials are Teflon, paper, mica, Bakelite, or ceramic. Paper, for example, has an average dielectric constant of 4, meaning it can provide an electric flux density four times as great as that for air for the same applied voltage and equal physical size.

The capacitance of a capacitor depends on the area of the conductor plates, the separation between the plates, and the dielectric constant of the insulating material. For a capacitor with two parallel plates, the formula to find its capacitance is

$$C = k\frac{A}{d}(8.85 \times 10^{-12}) \qquad\qquad (14\text{-}4)$$

where C = capacitance, F
        k = dielectric constant of the insulating material
        A = area of the plate, m^2
        d = distance between the plates, m

The farad is too high a unit for most capacitors. Therefore, we conveniently use the microfarad (μF), which equals one-millionth farad (10^{-6} F), the nanofarad (nF), which equals one-billionth farad (10^{-9} F), and the picofarad (pF), which equals one-millionth microfarad (10^{-6} μF). Thus, 1 F = 10^6 μF = 10^9 nF = 10^{12} pF.

Example 14.2   What is the capacitance of a capacitor that stores 4 C of charge at 2 V?

$$C = \frac{Q}{V} \qquad\qquad (14\text{-}1)$$

$$= \frac{4}{2} = 2\,\text{F} \qquad Ans.$$

Example 14.3   What is the charge taken on by a 10-F capacitor at 3 V?

$$Q = CV \qquad\qquad (14\text{-}2)$$

$$= 10(3) = 30\,\text{C} \qquad Ans.$$

Example 14.4   What is the voltage across a 0.001-F capacitor that stores 2 C?

$$V = \frac{Q}{C} \qquad\qquad (14\text{-}3)$$

$$= \frac{2}{0.001} = 2000\,\text{V} \qquad Ans.$$

Example 14.5   The area of one plate of a two-plate mica capacitor is 0.0025 m^2 and the separation between plates is 0.02 m. If the dielectric constant of mica is 7, find the capacitance of the capacitor.

$$C = k\frac{A}{d}\left(8.85 \times 10^{-12}\right) \tag{14-4}$$

$$= 7\frac{0.0025}{0.02}\left(8.85 \times 10^{-12}\right) = 7.74 \times 10^{-12}\ \text{F} = 7.74\ \text{pF} \qquad Ans.$$

Example 14.6   If the area of each plate in Example 14.5 is increased five times, and neither the dielectric nor the spacing is changed for the capacitor, what is the new capacitance?

Since capacitance is proportional to area, increasing the area five times increases the capacitance five times so that

$$C = 5(7.74) = 38.7\ \text{pF} \qquad Ans.$$

TYPES OF CAPACITORS

Commercial capacitors are named according to their dielectric. Most common are air, mica, paper, and ceramic capacitors, plus the electrolytic type. These types are compared in Table 14-1. Most types of capacitors can be connected to an electric circuit without regard to polarity. But electrolytic capacitors and certain ceramic capacitors are marked to show which side must be connected to the more positive side of a circuit.

Table 14-1 Types of Capacitors

Dielectric	Construction	Capacitance Range
Air	Meshed plates	10–400 pF
Mica	Stacked sheets	10–5000 pF
Paper	Rolled foil	0.001–1 μF
Ceramic	Tubular	0.5–1600 pF
	Disk	0.002–0.1 μF
Electrolytic	Aluminum	5–1000 μF
	Tantalum	0.01–300 μF

CAPACITORS IN SERIES AND PARALLEL

When capacitors are connected in series (Fig. 14-5), the total capacitance C_T is

Series:
$$\frac{1}{C_T} = \frac{1}{C_1} + \frac{1}{C_2} + \frac{1}{C_3} + \cdots + \frac{1}{C_n} \tag{14-5}$$

The total capacitance of two capacitors in series is

Series:
$$C_T = \frac{C_1 C_2}{C_1 + C_2} \tag{14-6}$$

When n number of series capacitors have the same capacitance, $C_T = C/n$.

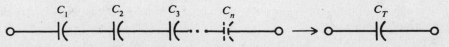

Fig. 14-5   Capacitances in series

When capacitors are connected in parallel (Fig. 14-6), the total capacitance C_T is the sum of the individual capacitances.

Parallel:
$$C_T = C_1 + C_2 + C_3 + \cdots + C_n \qquad (14\text{-}7)$$

There is a limit to the voltage that may be applied across any capacitor. If too high a voltage is applied, a current will be forced through the dielectric, sometimes burning a hole in it. The capacitor then will short-circuit and must be discarded. The maximum voltage that may be applied to a capacitor is called the *working voltage* and should not be exceeded.

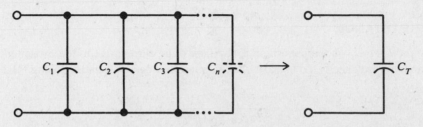

Fig. 14-6 Capacitances in parallel

Example 14.7 Find the total capacitance of a 3-μF, a 5-μF, and a 10-μF capacitor connected in series (Fig. 14-7). Write Eq. (*14-5*) for three capacitors in series.

$$\frac{1}{C_T} = \frac{1}{C_1} + \frac{1}{C_2} + \frac{1}{C_3} = \frac{1}{3} + \frac{1}{5} + \frac{1}{10} = \frac{19}{30}$$

$$C_T = \frac{30}{19} = 1.6\,\mu\text{F} \qquad Ans.$$

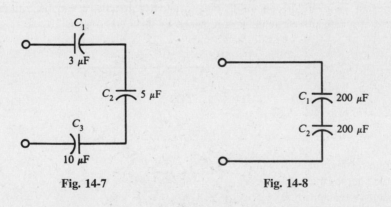

Fig. 14-7   **Fig. 14-8**

Example 14.8 What is the total capacitance and working voltage of a capacitor series combination if C_1 and C_2 are both 200-μF 150-V capacitors (Fig. 14-8)?

$$C_T = \frac{C}{n} = \frac{200}{2} = 100\,\mu\text{F} \qquad Ans.$$

The total voltage that may be applied across a group of capacitors in series is equal to the sum of the working voltages of the individual capacitors. Therefore,

$$\text{Working voltage} = 150 + 150 = 300\,\text{V} \qquad Ans.$$

Example 14.9 A capacitor in a radio receiver tuning circuit has a capacitance of 310 pF (Fig. 14-9). When the stage is aligned, a variable capacitor (called a *trimmer*) in parallel with it is adjusted to a capacitance of 50 pF. What is the total capacitance of the combination?

Write Eq. (*14-7*) for two capacitors in parallel.

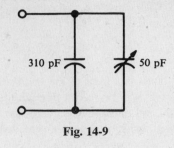

310 pF   50 pF

Fig. 14-9

$$C_T = C_1 + C_2 = 310 + 50 = 360 \text{ pF} \quad Ans.$$

CAPACITIVE REACTANCE

Capacitive reactance X_C is the opposition to the flow of ac current due to the capacitance in the circuit. The unit of capacitive reactance is the ohm. Capacitive reactance can be found by using the equation

$$X_C = \frac{1}{2\pi fC} = \frac{1}{6.28\,fC} = \frac{0.159}{fC} \tag{14-8}$$

where X_C = capacitive reactance, Ω
f = frequency, Hz
C = capacitance, F

If any two quantities in Eq. (*14-8*) are known, the third can be found.

$$C = \frac{0.159}{fX_C} \tag{14-9}$$

$$f = \frac{0.159}{CX_C} \tag{14-10}$$

Voltage and current in a circuit containing only capacitive reactance can be found using Ohm's law. However, in the case of a capacitive circuit, R is replaced by X_C.

$$V_C = I_C X_C \tag{14-11}$$

$$I_C = \frac{V_C}{X_C} \tag{14-12}$$

$$X_C = \frac{V_C}{I_C} \tag{14-13}$$

where I_C = current through the capacitor, A
V_C = voltage across the capacitor, V
X_C = capacitive reactance, Ω

Example 14.10 What is the capacitive reactance of a 0.001-F capacitor at 60 Hz (Fig. 14-10)?

$$X_C = \frac{0.159}{fC} \tag{14-8}$$

$$= \frac{0.159}{60(0.001)} = 2.65\ \Omega \quad Ans.$$

Example 14.11 A capacitor in a telephone circuit has a capacitance of 3 μF (Fig. 14-11). What current flows through it when 15 V at 800 Hz is impressed across it?

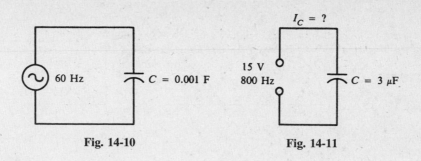

Fig. 14-10   **Fig. 14-11**

Find X_C and then I_C by Ohm's law.

$$X_C = \frac{0.159}{fC} \qquad (14\text{-}8)$$

$$= \frac{0.159}{800(3 \times 10^{-6})} = 66.25 \ \Omega$$

$$I_C = \frac{V_C}{X_C} \qquad (14\text{-}12)$$

$$= \frac{15}{66.25} = 0.226 = 226 \text{ mA} \qquad Ans.$$

Example 14.12   A 120-Hz 25-mA ac current flows in a circuit containing a 10 μF capacitor (Fig. 14-12). What is the voltage drop across the capacitor?

Find X_C and then V_C by Ohm's law.

$$X_C = \frac{0.159}{fC} \qquad (14.8)$$

$$= \frac{0.159}{120(10 \times 10^{-6})} = 132.5 \ \Omega$$

$$V_C = I_C X_C \qquad (14.11)$$

$$= (25 \times 10^{-3})(132.5) = 3.31 \text{ V} \qquad Ans.$$

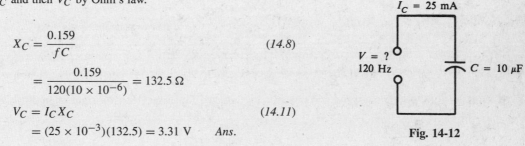

Fig. 14-12

CAPACITIVE CIRCUITS

Capacitance Only

If an ac voltage v is applied across a circuit having *only* capacitance (Fig. 14-13a), the resulting ac current through the capacitance, i_c, will lead the voltage across the capacitance, v_c, by 90° (Fig. 14-13b and c). (Quantities expressed as lowercase letters, i_c and v_c, indicate instantaneous values.) Voltages v and v_c are the same because they are parallel. Both i_c and v_c are sine waves with the same frequency. In series circuits, the current I_C is the horizontal phasor for reference (Fig. 14-13d) so the voltage V_C can be considered to lag I_C by 90°.

RC in Series

As with inductive circuits, the combination of resistance and capacitive reactance (Fig. 14-14a) is called *impedance*. In a series circuit containing R and X_C, the same current I flows in X_C and R. The voltage drop across R is $V_R = IR$, and the voltage drop across X_C is $V_C = IX_C$. The voltage across X_C lags the current through X_C by 90° (Fig. 14-14b). The voltage across R is in phase with I since resistance does not produce a phase shift (Fig. 14-14b).

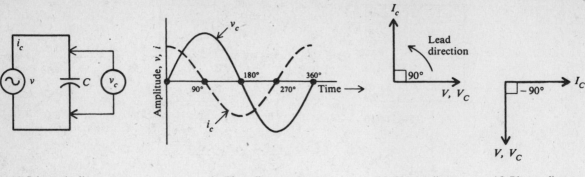

(a) Schematic diagram   (b) Time diagram, i_c (c) Phasor diagram,   (d) Phasor diagram,
                             leads v_c by 90° V reference I_C reference

Fig. 14-13   Circuit with C only

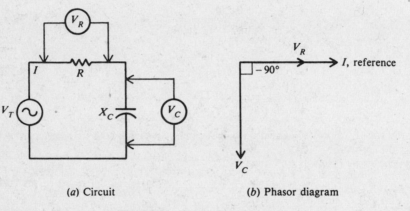

(a) Circuit   (b) Phasor diagram

Fig. 14-14 R and X_C in series

To find the total voltage V_T, we add phasors V_R and V_C. Since they form a right triangle (Fig. 14-15),

$$V_T = \sqrt{V_R^2 + V_C^2} \qquad (14\text{-}14)$$

Note that the IX_C phasor is downward, exactly opposite from an IX_L phasor (see Fig. 13-11), because of the opposite phase angle.

The phase angle θ between V_T and V_R (Fig. 14-15) is expressed according to the following equation:

$$\tan \theta = -\frac{V_C}{V_R}$$

$$\theta = \arctan \left(-\frac{V_C}{V_R} \right) \qquad (14\text{-}15)$$

Example 14.13   An RC series ac circuit has a current of 1 A peak with $R = 50\ \Omega$ and $X_C = 120\ \Omega$ (Fig. 14-16a). Calculate V_R, V_C, V_T, and θ. Draw the phasor diagram of V_C and I. Also draw the time diagram of i, v_R, v_C, and v_T.

$$V_R = IR = 1(50) = 50\ \text{V peak} \qquad Ans.$$

$$V_C = IX_C = 1(120) = 120\ \text{V peak} \qquad Ans.$$

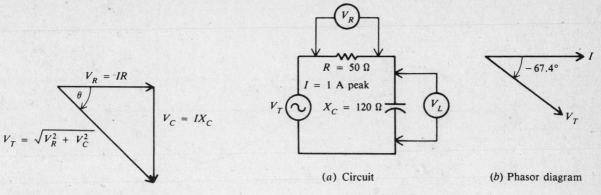

Fig. 14-15   Voltage-phasor triangle

(a) Circuit

(b) Phasor diagram

Fig. 14-16 a, b

Then
$$V_T = \sqrt{V_R^2 + V_C^2} \tag{14-14}$$

$$= \sqrt{50^2 + 120^2}$$

$$= \sqrt{2500 + 14\,400} = \sqrt{16\,900} = 130 \text{ V peak} \quad Ans.$$

$$\theta = \arctan\left(-\frac{V_C}{V_R}\right) \tag{14-15}$$

$$= \arctan\left(-\frac{120}{50}\right) = \arctan(-2.4) = -67.4° \quad Ans.$$

In a series circuit since I is the same in R and X_C, I is shown as the reference phasor at 0° (Fig. 14-16b). I leads V_T by 67.4° or, equivalently, V_T lags I by 67.4°. For the time diagram, see Fig. 14-16c.

Impedance in series RC.   The voltage triangle (Fig. 14-15) corresponds to the impedance triangle (Fig. 14-17) because the common factor I in V_C and V_R cancels.

$$V_C = IX_C$$

$$V_R = IR$$

$$\tan\theta = -\frac{IX_C}{IR} = -\frac{X_C}{R}$$

Impedance Z is equal to the phasor sum for R and X_C.

$$Z = \sqrt{R^2 + X_C^2} \tag{14-16}$$

The phase angle θ is

$$\theta = \arctan\left(-\frac{X_C}{R}\right) \tag{14-17}$$

Example 14.14   A 40-Ω X_C and a 30-Ω R are in series across a 120-V source (Fig. 14-18a). Calculate Z, I, and θ. Draw the phasor diagram.

$$Z = \sqrt{R^2 + X_C^2} \tag{14-16}$$

$$= \sqrt{30^2 + 40^2} = \sqrt{2500} = 50 \ \Omega \quad Ans.$$

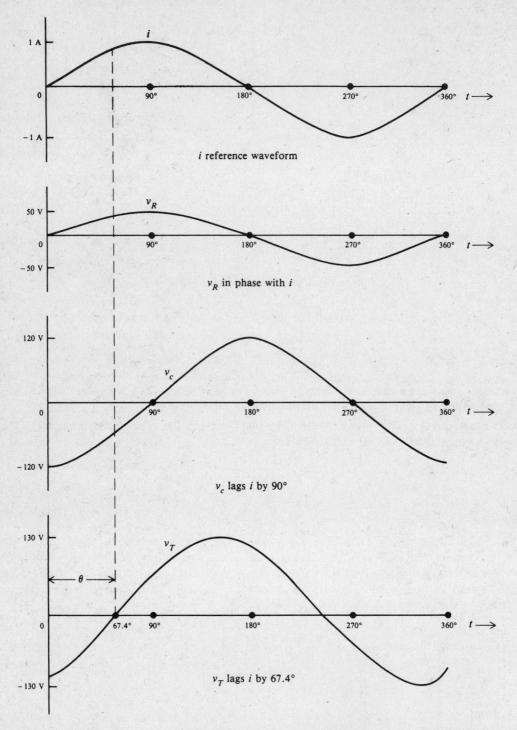

(c) Time diagram of *RC* series circuit

Fig. 14-16c

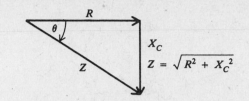

Fig. 14-17   Series RC impedance triangle

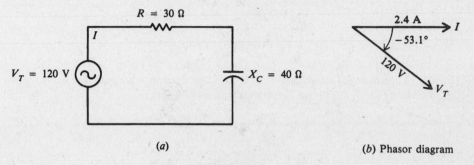

(a)

(b) Phasor diagram

Fig. 14-18

By Ohm's law,

$$I = \frac{V_T}{Z} = \frac{120}{50} = 2.4 \, \text{A} \qquad Ans.$$

$$\theta = \arctan\left(-\frac{X_C}{R}\right) \qquad\qquad (14\text{-}17)$$

$$= \arctan\left(-\frac{40}{30}\right) = \arctan(-1.33) = -53.1° \qquad Ans.$$

For the phasor diagram, see Fig. 14-18b.

RC in Parallel

In the RC parallel circuit (Fig. 14-19a), the voltage is the same across the source, R, and X_C since they are all in parallel. Each branch has its individual current. The resistive branch current $I_R = V_T/R$ is in phase with V_T. The capacitive branch current $I_C = V_T/X_C$ leads V_T by 90° (Fig. 14-19b). The phasor diagram has the source voltage V_T as the reference phasor because it is the same throughout the circuit. The total line current I_T equals the phasor sum of I_R and I_C (Fig. 14-19c).

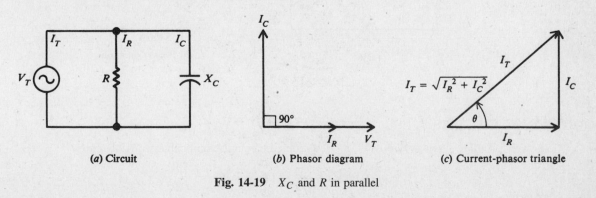

(a) Circuit

(b) Phasor diagram

(c) Current-phasor triangle

Fig. 14-19 X_C and R in parallel

$$I_T = \sqrt{I_R^2 + I_C^2} \tag{14-18}$$

$$\tan \theta = \frac{I_C}{I_R}$$

$$\theta = \arctan \frac{I_C}{I_R} \tag{14-19}$$

Impedance in parallel RC.   The impedance of a parallel circuit equals the total voltage V_T divided by the total current I_T.

$$Z_T = \frac{V_T}{I_T} \tag{14-20}$$

Example 14.15   A 15-Ω resistor and a capacitor of 20 Ω capacitive reactance are placed in parallel across a 120-V ac line (Fig. 14-20a). Calculate I_R, I_C, I_T, θ, and Z_T. Draw the phasor diagram.

$$I_R = \frac{V_T}{R} = \frac{120}{15} = 8\ \text{A} \qquad Ans.$$

$$I_C = \frac{V_T}{X_C} = \frac{120}{20} = 6\ \text{A} \qquad Ans.$$

$$I_T = \sqrt{I_R^2 + I_C^2} \tag{14-18}$$

$$= \sqrt{8^2 + 6^2} = \sqrt{100} = 10\ \text{A} \qquad Ans.$$

$$\theta = \arctan \frac{I_C}{I_R} \tag{14-19}$$

$$= \arctan \frac{6}{8} = \arctan 0.75 = 36.9^\circ \qquad Ans.$$

$$Z_T = \frac{V_T}{I_T} \tag{14-20}$$

$$= \frac{120}{10} = 12\ \Omega \qquad Ans.$$

For the phasor diagram, see Fig. 14-20b.

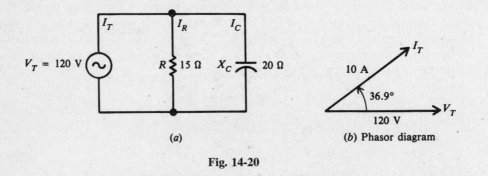

(a)

(b) Phasor diagram

Fig. 14-20

POWER IN RC CIRCUITS

The power formulas given previously for RL circuits are equally applicable to RC circuits.

$$\text{Real power } P = VI \cos\theta,\ \text{W} \tag{14-21}$$

or
$$P = I^2 R = \frac{V^2}{R}, \text{W} \qquad (14\text{-}22)$$

$$\text{Reactive power } Q = VI \sin\theta, \text{VAR} \qquad (14\text{-}23)$$

$$\text{Apparent power } S = VI, \text{VA} \qquad (14\text{-}24)$$

Capacitance, like inductance, consumes no power. The only part of the circuit consuming power is the resistance. Reactive power θ in an RC circuit is capacitive and shown *below* the horizontal axis.

Table 14-2 summarizes the relationships of current, voltage, impedance, and phase angle in RC circuits.

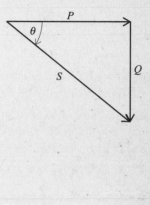

Table 14-2 Summary Table for Series and Parallel RC Circuits

X_C and R in Series	X_C and R in Parallel
I the same in X_C and R	V_T the same across X_C and R
$V_T = \sqrt{V_R^2 + V_C^2}$	$I_T = \sqrt{I_R^2 + I_C^2}$
$Z = \sqrt{R^2 + X_C^2} = \dfrac{V_T}{I}$	$Z_T = \dfrac{V_T}{I_T}$
V_C lags V_R by 90°	I_C leads I_R by 90°
$\theta = \arctan\left(-\dfrac{X_C}{R}\right)$	$\theta = \arctan\dfrac{I_C}{I_R}$

Solved Problems

14.1 What is the total capacitance of three capacitors connected in parallel if their values are 0.15 μF, 50 V; 0.015 μF, 100 V; and 0.003 μF, 150 V (Fig. 14-21)? What would be the working voltage of this combination?

Write Eq. (*14-5*) for three capacitors in parallel.

$$C_T = C_1 + C_2 + C_3 = 0.15 + 0.015 + 0.003 = 0.168 \, \mu\text{F} \qquad Ans.$$

The working voltage of a group of parallel capacitors is only as high as the lowest working voltage. Therefore, the working voltage of this combination is only 50 V.

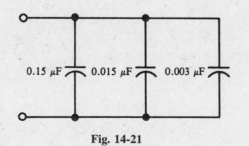

0.15 μF   0.015 μF   0.003 μF

Fig. 14-21

14.2 A technician has the following capacitors available: 300 pF, 75 V; 250 pF, 50 V; 200 pF, 50 V; 150 pF, 75 V; and 50 pF, 75 V. Which of these should be connected in parallel to form a combination with a capacitance of 500 pF and 75 V working voltage?

Capacitors with voltage ratings less than 75 V must not be used because of possible short-circuit damage. The remaining capacitors with 75-V ratings are 300, 150, and 50 pF, the sum of which is 500 pF. Therefore, the safe parallel combination is as shown in Fig. 14-22, where $C_T = 300 + 150 + 50 = 500$ pF.

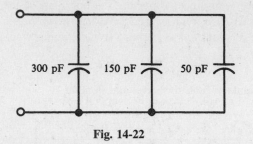

Fig. 14-22

14.3   What is the range of total capacitances available in an oscillator circuit that uses a variable tuning capacitor of 35- to 300-pF range in series with a fixed capacitor of 250 pF (Fig. 14-23)?

At the *low* point in the range of total capacitance, we have 35 pF in series with 250 pF.

$$C_{T1} = \frac{C_1 C_2}{C_1 + C_2} = \frac{35(250)}{35 + 250} = \frac{8750}{285} = 30.7 \text{ pF} \qquad Ans.$$

At the *high* point in the range, we have 300 pF in series with 250 pF.

$$C_{T2} = \frac{300(250)}{300 + 250} = \frac{75\,000}{550} = 136.4 \text{ pF} \qquad Ans.$$

Therefore, the range is from 30.7 to 136.4 pF.

14.4   What capacitance must be added in parallel with a 550-pF capacitor in order to get a total capacitance of 750 pF (Fig. 14-24)?

Write Eq. (*14-7*) for two capacitors in parallel.

$$C_T = C_1 + C_2$$
$$750 = 550 + C_2$$
$$C_2 = 750 - 550 = 200 \text{ pF} \qquad Ans.$$

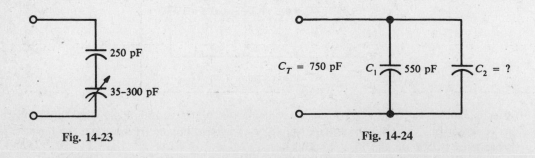

Fig. 14-23   **Fig. 14-24**

14.5 Find the total capacitance of the capacitive networks shown in Figs. 14-25, 14-26a, and 14-27a.

(a)   See Fig. 14-25. Simple series combination:

$$C_T = \frac{C_1 C_2}{C_1 + C_2} = \frac{3(2)}{3+2} = \frac{6}{5} = 1.2\ \mu\text{F} \qquad \textit{Ans.}$$

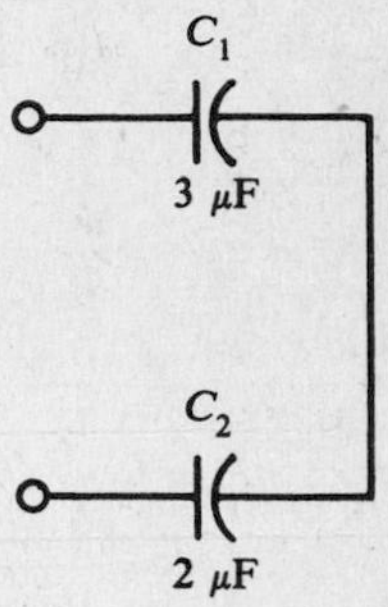

(b)   See Fig. 14-26a, b, and c. Series–parallel combination:

Parallel: $C_a = C_2 + C_3 = 0.1 + 0.2 = 0.3\ \mu\text{F}$

Series: $C_T = \dfrac{C_1 C_a}{C_1 + C_a} = \dfrac{0.3(0.3)}{0.3 + 0.3} = \dfrac{0.09}{0.6} = 0.15\ \mu\text{F} \qquad \textit{Ans.}$

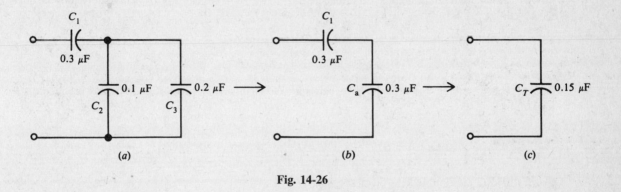

Fig. 14-25

(a)

(b)

(c)

Fig. 14-26

(c)   See Fig. 14-27a, b, and c. Parallel–series combination:

Series: $C_a = \dfrac{C_3 C_4}{C_3 + C_4} = \dfrac{4(5)}{4+5} = \dfrac{20}{9} = 2.22\ \text{pF}$

Parallel: $C_T = C_1 + C_2 + C_a = 2 + 3 + 2.22 = 7.22\ \text{pF} \qquad \textit{Ans.}$

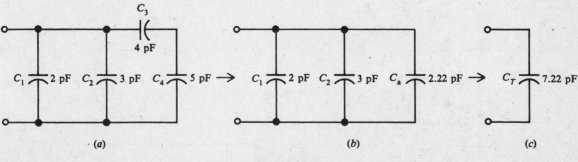

(a)

(b)

(c)

Fig. 14-27

14.6 Find the total capacitance of the series circuit and the capacitive reactance of the group of capacitors when used in a 60-Hz circuit (Fig. 14-28).

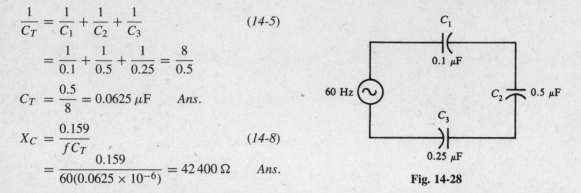

$$\frac{1}{C_T} = \frac{1}{C_1} + \frac{1}{C_2} + \frac{1}{C_3} \qquad (14\text{-}5)$$

$$= \frac{1}{0.1} + \frac{1}{0.5} + \frac{1}{0.25} = \frac{8}{0.5}$$

$$C_T = \frac{0.5}{8} = 0.0625\ \mu\text{F} \qquad Ans.$$

$$X_C = \frac{0.159}{fC_T} \qquad (14\text{-}8)$$

$$= \frac{0.159}{60(0.0625 \times 10^{-6})} = 42\,400\ \Omega \qquad Ans.$$

Fig. 14-28

14.7 A capacitor draws 6 mA when connected across a 110-V 60-Hz line. What will be the current drawn if both the frequency and capacitance are doubled?

We look at two relationships. First, because $I_C = V_C/X_C$, we can say I_C is inversely proportional to X_C, or

$$I_C \propto \frac{1}{X_C}$$

Second, because $X_C = 0.159/fC$, we can say X_C is inversely proportional to the product of f and C, or

$$X_C \propto \frac{1}{fC}$$

So if f and C are doubled, X_C is decreased by 1/4.

$$X_C \propto \frac{1}{(2f)(2C)} = \frac{1}{4fC}$$

And when X_C is decreased by 1/4, I_C is increased four times. Therefore

$$I_C = 4(6) = 24\ \text{mA} \qquad Ans.$$

14.8 A 20-μF capacitor in an audio amplifier circuit produces a voltage drop of 5 V at 1 kHz. Find the current passed by the capacitor.

Find X_C and then I_C.

$$X_C = \frac{0.159}{fC} \qquad (14\text{-}8)$$

$$= \frac{0.159}{(1 \times 10^3)(20 \times 10^{-6})} = 7.95\ \Omega$$

$$I_C = \frac{V_C}{X_C} \qquad (14\text{-}12)$$

$$= \frac{5}{7.95} = 0.629\ \text{A} \qquad Ans.$$

14.9 Calculate the value of the bypass capacitor in an audio circuit if it is to have a reactance of 800 Ω at 10 kHz.

$$C = \frac{0.159}{fX_C} \tag{14-9}$$

$$= \frac{0.159}{(10 \times 10^3)(800)} = 0.02 \,\mu\text{F} \quad Ans.$$

14.10 A capacitor is inserted in a circuit to obtain a leading current of 5 A. If the voltage is 110 V, 60 Hz, what is the capacitance?

Find X_C and then C.

$$X_C = \frac{V_C}{I_C} \tag{14-13}$$

$$= \frac{110}{5} = 22 \,\Omega$$

$$C = \frac{0.159}{fX_C} \tag{14-9}$$

$$= \frac{0.159}{60(22)} = 121 \times 10^{-6} = 121 \,\mu\text{F} \quad Ans.$$

14.11 A capacitance of 20 pF draws 10 mA when connected across a 95-V source. Find the frequency of the ac voltage.

Find X_C and then f.

$$X_C = \frac{V_C}{I_C} \tag{14-13}$$

$$= \frac{95}{10 \times 10^{-3}} = 9500 \,\Omega$$

$$f = \frac{0.159}{CX_C} \tag{14-10}$$

$$= \frac{0.159}{(20 \times 10^{-12})(9500)} = 838 \,\text{kHz} \quad Ans.$$

14.12 Find the impedance of an RC combination when the coupling capacitor is 0.01 μF, the audio frequency is 1 kHz, and the resistance of the circuit is 3 kΩ (Fig. 14-29). A coupling capacitor, because it provides more reactance at lower frequencies, results in less ac voltage across R and more across C.

Find X_C and then Z.

$$X_C = \frac{0.159}{fC} \tag{14-8}$$

$$= \frac{0.159}{(1 \times 10^3)(0.01 \times 10^{-6})} = 15\,900 \,\Omega$$

$$Z = \sqrt{R^2 + X_C^2} \tag{14-16}$$

$$= \sqrt{3000^2 + 15\,900^2} = 16\,180 \,\Omega \quad Ans.$$

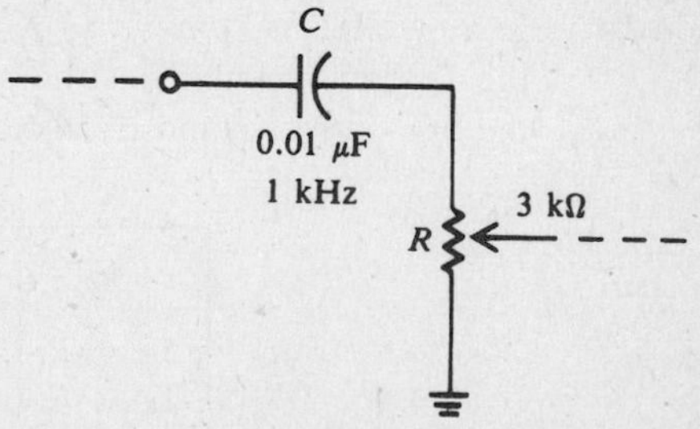

Fig. 14-29

14.13   In a series RC circuit, the *higher* the X_C compared with R, the more capacitive is the circuit. With higher X_C there is more voltage drop across the capacitive reactance, and the phase angle increases toward $-90°$. To illustrate, find the indicated quantity.

	Case	R, Ω	X_C, Ω	Z, Ω	θ	Nature of Circuit
(a)	$X_C = R$	10	10	?	?	?
(b)	$X_C < R$	10	1	?	?	?
(c)	$X_C > R$	1	10	?	?	?

(a)
$$Z = \sqrt{R^2 + X_C^2} \qquad\qquad (14\text{-}16)$$
$$= \sqrt{10^2 + 10^2} = 14.1 \ \Omega \qquad Ans.$$
$$\theta = \arctan\left(-\frac{X_C}{R}\right) \qquad\qquad (14\text{-}17)$$
$$= \arctan\frac{-10}{10} = \arctan(-1) = -45° \qquad Ans.$$

The circuit is capacitive.

(b)
$$Z = \sqrt{10^2 + 1^2} = 10.0 \ \Omega \qquad Ans.$$
$$\theta = \arctan(-0.1) = -5.7° \qquad Ans.$$

The circuit is only slightly capacitive.

(c)
$$Z = \sqrt{1^2 + 10^2} = 10.0 \ \Omega \qquad Ans.$$
$$\theta = \arctan(-10) = -84.3° \qquad Ans.$$

The circuit is almost entirely capacitive. Recall that if $R = 0$ (pure capacitive circuit), $Z = X_C = 10 \ \Omega$ at $\theta = -90°$. The complete table is as follows:

	Case	R, Ω	X_C, Ω	Z, Ω	θ	Nature of Circuit
(a)	$X_C = R$	10	10	14.1	$-45°$	Capacitive
(b)	$X_C < R$	10	1	10.0	$-5.7°$	Slightly capacitive
(c)	$X_C > R$	1	10	10.0	$-84.3°$	Very capacitive

14.14   In a parallel RC circuit, as X_C becomes *smaller* compared with R, practically all the line current is the I_C component. Thus the parallel circuit is capacitive. The phase angle approaches 90° because the line current is mostly capacitive. To illustrate, find the indicated quantity. Assume $V_T = 10$ V.

	Case	R, Ω	X_C, Ω	I_R, A	I_C, A	I_T, A	θ	Nature of Circuit
(a)	$X_C = R$	10	10	?	?	?	?	?
(b)	$X_C > R$	1	10	?	?	?	?	?
(c)	$X_C < R$	10	1	?	?	?	?	?

(a)

$$I_R = \frac{V_T}{R} = \frac{10}{10} = 1 \text{ A} \qquad Ans.$$

$$I_C = \frac{V_T}{X_C} = \frac{10}{10} = 1 \text{ A} \qquad Ans.$$

$$I_T = \sqrt{I_R^2 + I_C^2} \tag{14-18}$$

$$= \sqrt{1^2 + 1^2} = 1.41 \text{ A} \qquad Ans.$$

$$\theta = \arctan \frac{I_C}{I_R} \tag{14-19}$$

$$= \arctan 1 = 45° \qquad Ans.$$

The circuit is capacitive.

(b)

$$I_R = \frac{10}{1} = 10 \text{ A} \qquad Ans.$$

$$I_C = \frac{10}{10} = 1 \text{ A} \qquad Ans.$$

$$I_T = \sqrt{10^2 + 1^2} = 10.0 \text{ A} \qquad Ans.$$

$$\theta = \arctan \frac{1}{10} = 5.7° \qquad Ans.$$

The circuit is only slightly capacitive.

(c)

$$I_R = \frac{10}{10} = 1 \text{ A} \qquad Ans.$$

$$I_C = \frac{10}{1} = 10 \text{ A} \qquad Ans.$$

$$I_T = \sqrt{1^2 + 10^2} = 10.0 \text{ A} \qquad Ans.$$

$$\theta = \arctan 10 = 84.3° \qquad Ans.$$

The circuit is almost entirely capacitive.

Recall that if $R = 0$ (pure capacitive circuit), $I_T = I_C = 10$ A at $\theta = 90°$. The complete table is as follows:

Case	R, Ω	X_C, Ω	I_R, A	I_C, A	I_T, A	θ	Nature of Circuit
(a)	$X_C = R$	10	10	1	1	1.4	45°	Capacitive
(b)	$X_C > R$	1	10	10	1	10.0	5.7°	Slightly capacitive
(c)	$X_C < R$	10	1	1	10	10.0	84.3°	Very capacitive

14.15   A voltage of 10 V at a frequency of 20 kHz is applied across a 1-μF capacitor. Find the current and the real power used. Draw the phasor diagram.

Find X_C and then I_C.

$$X_C = \frac{0.159}{fC} \qquad (14\text{-}8)$$

$$= \frac{0.159}{(20 \times 10^3)(1 \times 10^{-6})} = 7.95 \ \Omega$$

$$I_C = \frac{V_C}{X_C} \qquad (14\text{-}12)$$

$$= \frac{10}{7.95} = 1.26 \text{ A} \qquad Ans.$$

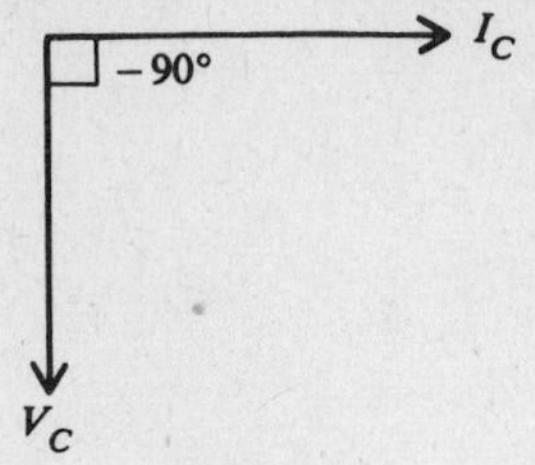

Now find P.

Fig. 14-30
Phasor diagram

$$P = VI \cos\theta \qquad (14\text{-}21)$$

$$V = V_C \quad \text{and} \quad I = I_C \quad \text{so}$$

$$P = 10(1.26)(\cos 90°) = 10(1.26)(0) = 0 \text{ W} \qquad Ans.$$

No net power is consumed in the circuit when there is no resistance. For the phasor diagram, see Fig. 14-30. I_C leads V_C by 90°.

14.16 A capacitance of 3.53 μF and a resistance of 40 Ω are connected in series across a 110-V 1.5-kHz ac source (Fig. 14-31). Find X_C, Z, θ, I, V_R, V_C, and P. Draw the phasor diagram.

Step 1. Find X_C.

$$X_C = \frac{0.159}{fC} = \frac{0.159}{(1.5 \times 10^3)(3.53 \times 10^{-6})} = 30 \ \Omega \qquad Ans.$$

Step 2. Find Z and θ.

$$Z = \sqrt{R^2 + X_C^2} = \sqrt{40^2 + 30^2} = 50 \ \Omega \qquad Ans.$$

$$\theta = \arctan\left(-\frac{X_C}{R}\right) = \arctan\left(-\frac{30}{40}\right) = \arctan(-0.75) = -36.9° \qquad Ans.$$

Step 3. Find I.

$$I = \frac{V_T}{Z} = \frac{110}{50} = 2.2 \text{ A} \qquad Ans.$$

Step 4. Find V_R and V_C. By Ohm's law,

$$V_R = IR = 2.2(40) = 88 \text{ V} \qquad Ans.$$

$$V_C = IX_C = 2.2(30) = 66 \text{ V} \qquad Ans.$$

Step 5. Find P.

$$P = I^2 R = (2.2)^2(40) = 193.6 \text{ W} \qquad Ans.$$

Step 6. Draw the phasor diagram (see Fig. 14-31b). I leads V_T by 36.9°.

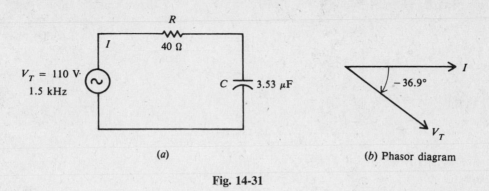

(a)

(b) Phasor diagram

Fig. 14-31

14.17 The purpose of a low-pass filter circuit (Fig. 14-32) is to permit low frequencies to pass on to the load but to prevent the passing of high frequencies. Find the branch currents, total current, phase angle, and the percentage of the total current passing through the resistor for (a) a 1.5-kHz (low) audio-frequency signal and (b) a 1-MHz (high) radio-frequency signal.

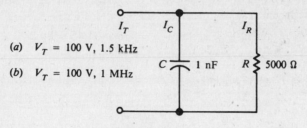

(a) $V_T = 100$ V, 1.5 kHz

(b) $V_T = 100$ V, 1 MHz

Fig. 14-32

(a)   **Step 1.**   Find X_C at $f = 1.5$ kHz.

$$X_C = \frac{0.159}{fC} = \frac{0.159}{(1.5 \times 10^3)(1 \times 10^{-9})} = 106 \text{ k}\Omega$$

Step 2.   Find branch currents I_C and I_R.

$$I_C = \frac{V_T}{X_C} = \frac{100}{106 \times 10^3} = 0.94 \text{ mA} \qquad Ans.$$

$$I_R = \frac{V_T}{R} = \frac{100}{5 \times 10^3} = 20 \text{ mA} \qquad Ans.$$

Step 3.   Find I_T and θ.

$$I_T = \sqrt{I_R^2 + I_C^2} = \sqrt{20^2 + (0.94)^2} = \sqrt{400 + 0.88} = \sqrt{400.88} = 20.02 \text{ mA} \qquad Ans.$$

$$\theta = \arctan \frac{I_C}{I_R} = \arctan \frac{0.94}{20} = \arctan 0.047 = 2.7° \qquad Ans.$$

Step 4.   Find I_R as a percentage of I_T.

$$\frac{I_R}{I_T} \times 100 = \frac{20}{20.02} 100 = 0.999(100) = 99.9\% \qquad Ans.$$

Thus, practically all the 1.5 kHz audio-signal current passes through the resistor.

(b) **Step 1.**   Find X_C now at $f = 1$ MHz.

$$X_C = \frac{0.159}{fC} = \frac{0.159}{(1 \times 10^6)(1 \times 10^{-9})} = 159\,\Omega$$

Step 2.   Find I_C and I_R.

$$I_C = \frac{V_T}{X_C} = \frac{100}{159} = 0.629 = 629 \text{ mA} \qquad Ans.$$

$$I_R = 20 \text{ mA} \qquad Ans.$$

I_R is the same as in part (a).

Step 3.   Find I_T and θ.

$$I_T = \sqrt{I_R^2 + I_C^2} = \sqrt{20^2 + 629^2} = \sqrt{400 + 395\,641} = \sqrt{396\,041} = 629.3 \text{ mA} \qquad Ans.$$

$$\theta = \arctan \frac{I_C}{I_R} = \arctan \frac{629}{20} = \arctan 31.45 = 88.2° \qquad Ans.$$

Step 4.   Find I_R as a percentage of I_T.

$$\frac{I_R}{I_T} \times 100 = \frac{20}{629.3} 100 = 0.332(100) = 3.2\% \qquad Ans.$$

Thus, only 3.2 percent of the 1-MHz radio signal current passes through the resistor.

It is clear that the circuit is an excellent low-pass filter by passing practically all the low-frequency (1.5-kHz) current through the resistor, but very little of the high-frequency (1-MHz) current through it.

Supplementary Problems

14.18   What is the capacitance of a capacitor that stores 10.35 C at 3 V?   *Ans.* $C = 3.45$ F

14.19   What charge is taken on by a 0.5-F capacitor connected across a 50-V source?   *Ans.* $Q = 25$ C

14.20   Find the capacitance of a capacitor that has one plate area of 0.5 m², a distance between plates of 0.01 m, and a dielectric of paper with dielectric constant of 3.5.   *Ans.* $C = 1549$ pF

14.21   What is the reactance of a 500-pF capacitor at (a) 40 kHz, (b) 100 kHz, and (c) 1200 kHz?
Ans.   (a) $X_C = 7950\,\Omega$;   (b) $X_C = 3180\,\Omega$;   (c) $X_C = 265\,\Omega$

14.22   Two capacitors in parallel are connected across a 120-V line. The first capacitor has a charge of 0.00006 C and the second capacitor has a charge of 0.000048 C. What is the capacitance of each capacitor and what is the total capacitance?   *Ans.* $C_1 = 0.5\,\mu$F; $C_2 = 0.4\,\mu$F; $C_T = 0.9\,\mu$F

14.23 Find the indicated quantity.

	C, F	Q, C	V, V
(a)	?	11	110
(b)	0.3	?	220
(c)	0.2	50	?

Ans.

	C, F	Q, C	V, V
(a)	0.1	....	....
(b)	....	66	....
(c)	....	....	250

14.24 The capacitance of a known parallel-plate capacitor with air as a dielectric is $0.248\ \mu$F. What is the capacitance if (a) Teflon with a dielectric constant of 2.1 replaces air; (b) the area of one plate is reduced by one-half; (c) the separation distance is increased by a factor of $1\frac{1}{2}$; and (d) rubber with a dielectric constant of 3 replaces air, the area of a plate is increased by $1\frac{1}{4}$, and the separation distance is reduced to three-fourths its original value?
   Ans.   (a) $C = 0.521\ \mu$F;   (b) $C = 0.124\ \mu$F;   (c) $C = 0.165\ \mu$F;   (d) $C = 1.24\ \mu$F

14.25 What is the reactance of an oscillator capacitor of 400 pF to a frequency of 630 kHz?
   Ans. $X_C = 631\ \Omega$

14.26 Find the total capacitance of the capacitive networks shown (Fig. 14-33).
   Ans.   (a) $C_T = 0.40\ \mu$F;   (b) $C_T = 0.065\ \mu$F;   (c) $C_T = 0.04\ \mu$F;   (d) $C_T = 0.05\ \mu$F;
   (e) $C_T = 0.84\ \mu$F

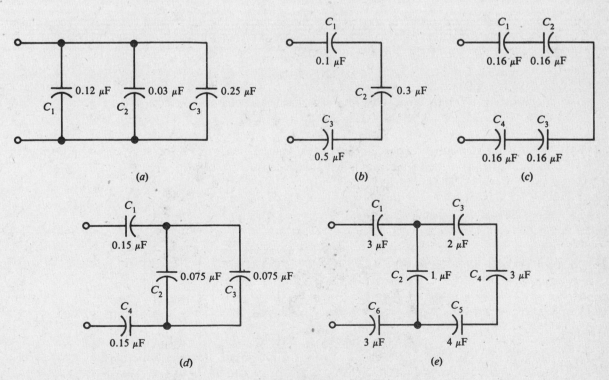

Fig. 14-33

14.27 A 10-V 1-MHz signal appears across a 1200-pF capacitor. Find the current passed by the capacitor.
Ans. $I_C = 75.2$ mA; ($X_C = 133\ \Omega$)

14.28 A filter circuit consists of an inductor and two capacitors (Fig. 14-34). Its purpose is to smooth the power-supply voltages so that a pure direct current is delivered to the load. If the reactance of C_1 is 175 Ω at a frequency of 60 Hz, what is its capacitance?   *Ans.* $C_1 = 15.1\ \mu$F

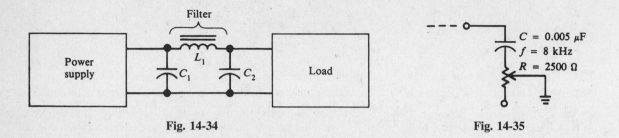

Fig. 14-34   Fig. 14-35

14.29 Find the impedance of the output of the tone control circuit (Fig. 14-35).
Ans. $Z = 4700\ \Omega$; ($X_C = 3980\ \Omega$)

14.30 What is the total capacitance of each of the following three capacitors connected in parallel: 300-pF 100-V, 0.001-μF 150 V, and 0.003-μF 50-V? What is the working voltage of the parallel combination?   *Ans.* $C_T = 4300$ pF or 0.0043 μF, 50 V

14.31 What capacitance must be added in parallel to a 200-pF capacitor in order to obtain a total capacitance of 1100 pF?   *Ans.* 900 pF

14.32 Two capacitors are placed in series across the secondary line of a transformer to reduce voltage peaks. What is the total capacitance and what is the working voltage of the pair of 0.008-μF 650-V capacitors?   *Ans.* $C_T = 0.004\ \mu$F, 1300 V

14.33 A Colpitts oscillator has two capacitors in series, $C_1 = 300$ pF and $C_2 = 300$ pF, and oscillates at a frequency of 100 kHz. What is the total capacitance and what is the reactance?
Ans. $C_T = 150$ pF; $X_C = 10\,600\ \Omega$

14.34 What is the capacitive reactance between two transmission wires if the stray capacitance between them is 10 pF and one wire carries a radio frequency of 1200 kHz?
Ans. $X_C = 13\,250\ \Omega$ or 13.25 kΩ

14.35 Find the indicated quantity.

	X_C, Ω	f	C	I_C	V_C, V
(a)	?	120 Hz	10 μF	25 mA	?
(b)	?	4.2 MHz	?	160 mA	400
(c)	200	600 kHz	?	?	10
(d)	?	800 Hz	2 μF	?	20
(e)	1000	500 Hz	?	22 mA	?
(f)	?	?	30 pF	20 mA	106
(g)	?	?	0.01 μF	4 A	3

Ans.

	X_C, Ω	f	C	I_C	V_C, V
(a)	133	....	....	....	3.32
(b)	2500	....	15.2 pF	....	....
(c)	....	....	1325 pF	50 mA	....
(d)	99.4	....	....	0.2 A	....
(e)	....	....	0.318 μF	....	22
(f)	5300	1 MHz	....	....	....
(g)	0.75	21.25 MHz	....	....	....

14.36   Find the impedance of a capacitor if its reactance is 40 Ω and its resistance is 20 Ω.
Ans. $Z = 44.7\,\Omega$

14.37   Find the impedance of a capacitive circuit to a 1.5-kHz audio signal if the resistance is 2000 Ω and the capacitance is 0.02 μF.   Ans. $Z = 5670\,\Omega$

14.38   What is the impedance of a capacitive circuit to 20-kHz frequency if its resistance is 400 Ω and its capacitance is 0.032 μF?   Ans. $Z = 471\,\Omega$

14.39   A 120-V 60-Hz ac voltage is impressed across a series circuit of a 10-Ω resistor and a capacitor whose reactance is 15 Ω. Find the impedance, phase angle, line current, voltage drop across the resistor and the capacitor, and power. Draw the phasor diagram.
Ans. $Z = 18\,\Omega$; $\theta = 56.3°$, I leading V_T; $I = 6.67$ A; $V_R = 66.7$ V; $V_C = 100$ V; $P = 444$ W. For the phasor diagram, see Fig. 14-36.

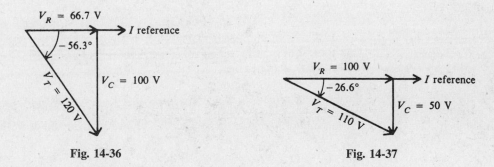

Fig. 14-36   Fig. 14-37

14.40   A 110-V 200-Hz ac voltage is applied across a series circuit of a 100-Ω resistor and a 15.9-μF capacitor. Find Z, θ, I, V_R, V_C, and P. Draw the phasor diagram.
Ans. $Z = 122\,\Omega$; $\theta = 26.6°$, I leading V_T; $I = 0.982$ A; $V_R = 100$ V; $V_C = 50$ V; $P = 100$ W. For the phasor diagram, see Fig. 14-37.

14.41   A 5-kΩ resistor and an unknown capacitor are placed in series across a 60-Hz line. If the voltage across the resistor is 30 V and the voltage across the capacitor is 60 V, find the impressed voltage, the current in the resistor, the capacitive reactance, and the capacitance of the capacitor.
Ans. $V_T = 67.1$ V; $I = I_R = 6$ mA; $X_C = 10$ kΩ; $C = 0.265\,\mu$F

14.42   A circuit consisting of a 30-μF capacitance in series with a rheostat is connected across a 120-V 60-Hz line. What must the value of the resistance be in order to permit a current of 1 A to flow? (*Hint*: Solve the impedance triangle in Fig. 14-38 for R.)   Ans. $R = 81.2\,\Omega$

14.43   In a resistance-coupled stage (Fig. 14-39), the voltage drop across points A and B is 14.14 V. If the frequency of the current is 1 kHz, find the voltage across the resistor. Draw the phasor diagram.
Ans. $V_R = 10$ V; for the phasor diagram, see Fig. 14-40.

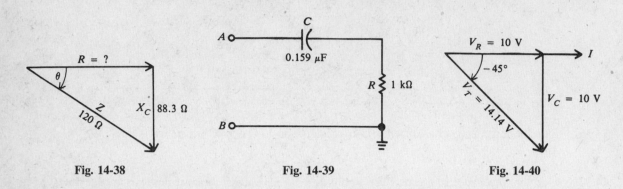

Fig. 14-38   Fig. 14-39   Fig. 14-40

14.44   A 15-Ω resistor and an 8-Ω capacitive reactance are placed in parallel across a 120-V ac line. Find the phasor branch currents, total current and phase angle, impedance, and power drawn; and draw the phasor diagram.
Ans. $I_R = 8$ A; $I_C = 15$ A; $I_T = 17$ A, $\theta = 61.9°$, I_T leads V_T; $Z_T = 7.1\ \Omega$; $P = 960$ W; for the phasor diagram, see Fig. 14-41

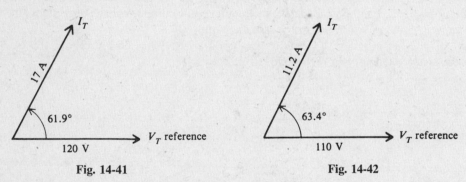

Fig. 14-41   Fig. 14-42

14.45   A 20-Ω resistor and a 7.95-μF capacitor are connected in parallel across a 100-V 2-kHz source. Find the branch currents, total current and phase angle, impedance, and power drawn; and draw the phasor diagram.
Ans. $I_R = 5$ A; $I_C = 10$ A; $I_T = 11.2$ A, $\theta = 63.4°$, I_T leads V_T; $Z_T = 8.9\ \Omega$; $P = 500$ W; for the phasor diagram, see Fig. 14-42

14.46   For the low-pass filter circuit (Fig. 14-43), find I_C, I_R, I_T, and the percentage of I_T that passes through the resistor for an audio frequency at 1 kHz and for a radio frequency at 2 MHz.
Ans.   AF case: $I_C = 0.5$ mA; $I_R = 19.9$ mA; $I_T = 19.9$ mA; 100 percent; RF case: $I_C = 1$ A; $I_R = 19.9$ mA; $I_T = 1.002$ A; 2 percent

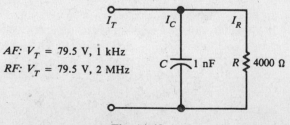

Fig. 14-43

14.47   Find the indicated values for an *RC* parallel circuit.

	V_T, V	R	X_C	I_R	I_C	I_T	Z_T	$\theta°$	P, W
(a)	120	120 Ω	60 Ω	?	?	?	?	?	?
(b)	8	4 kΩ	8 kΩ	?	?	?	?	?	?
(c)	20	40 Ω	40 Ω	?	?	?	?	?	?

Ans.

	V_T, V	R	X_C	I_R	I_C	I_T	Z_T	$\theta°$	P, W
(a)	....	....	....	1 A	2 A	2.24 A	54.5 Ω	63.4	120
(b)	....	....	....	2 mA	1 mA	2.24 mA	3.57 kΩ	26.6	0.016
(c)	....	....	....	0.5 A	0.5 A	0.707 A	28.2 Ω	45	10

Chapter 15

Single-Phase Circuits

THE GENERAL *RLC* CIRCUIT

The preceding two chapters have explained how a combination of inductance and resistance and then capacitance and resistance behave in a series circuit and in a parallel circuit. We saw how the *RL* and *RC* combination affect the current, voltage, power, power factor, and phase angle of a circuit. In this chapter, all three fundamental circuit parameters—inductance, capacitance, and resistance—are combined and their effect on circuit values studied.

RLC IN SERIES

Current in a series circuit containing resistance, inductive reactance, and capacitive reactance (Fig. 15-1*a*) is determined by the total impedance of the combination. The current *I* is the same in *R*, X_L, and X_C since they are in series. The voltage drop across each element is found by Ohm's law:

$$V_R = IR \qquad V_L = IX_L \qquad V_C = IX_C$$

where V_R = voltage drop across the resistance, V
   V_L = voltage drop across the inductance, V
   V_C = voltage drop across the capacitance, V

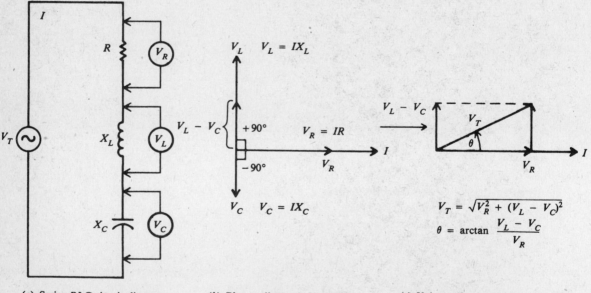

(*a*) Series *RLC* circuit diagram   (*b*) Phasor diagram, $V_L > V_C$ (*c*) Voltage-phasor triangle, $V_L > V_C$

Fig. 15-1   *R*, X_L, and X_C in series; $X_L > X_C$ for inductive circuit

The voltage drop across the resistance is in phase with the current through the resistance (Fig. 15-1*b*). The voltage across the inductance leads the current through the inductance by 90°. The voltage across the capacitance lags the current through the capacitance by 90° (Fig. 15-1*b*). Since V_L and V_C are exactly 180° out of phase and acting in exactly opposite directions, they are added algebraically. When X_L is greater than X_C, the circuit is *inductive*, V_L is greater than V_C, and *I* lags V_T (Fig. 15-1*c*).

When X_C is greater than X_L, the circuit is *capacitive*. Now V_C is greater than V_L so that I leads V_T (Fig. 15-2).

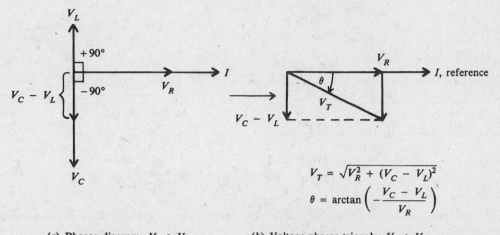

$$V_T = \sqrt{V_R^2 + (V_C - V_L)^2}$$

$$\theta = \arctan\left(-\frac{V_C - V_L}{V_R}\right)$$

(*a*) Phasor diagram, $V_C > V_L$ (*b*) Voltage-phasor triangle, $V_C > V_L$

Fig. 15-2 R, X_L, and X_C in series; $X_C > X_L$ for capacitive circuit

When $X_L > X_C$, the voltage-phasor diagram (Fig. 15-1*c*) shows that the total voltage V_T and phase angle are as follows:

$$V_T = \sqrt{V_R^2 + (V_L - V_C)^2} \tag{15-1}$$

$$\theta = \arctan\frac{V_L - V_C}{V_R} \tag{15-2}$$

When $X_C > X_L$ (Fig. 15-2*b*),

$$V_T = \sqrt{V_R^2 + (V_C - V_L)^2} \tag{15-3}$$

$$\theta = \arctan\left(-\frac{V_C - V_L}{V_R}\right) \tag{15-4}$$

where V_T = applied voltage, V
            V_R = voltage drop across the resistance, V
            V_L = voltage drop across the inductance, V
            V_C = voltage drop across the capacitance, V
            θ = phase angle between V_T and I, degrees

Example 15.1   In an *RLC* series ac circuit (Fig. 15-3*a*), find the applied voltage and phase angle. Draw the voltage-phasor diagram.

By Ohm's law,

$$V_R = IR = 2(4) = 8\text{ V} \qquad V_L = IX_L = 2(19.5) = 39\text{ V} \qquad V_C = IX_C = 2(12) = 24\text{ V}$$

With $V_L > V_C$,

$$V_T = \sqrt{V_R^2 + (V_L - V_C)^2} = \sqrt{8^2 + (39 - 24)^2} = \sqrt{8^2 + 15^2} = 17\text{ V} \qquad Ans. \tag{15-1}$$

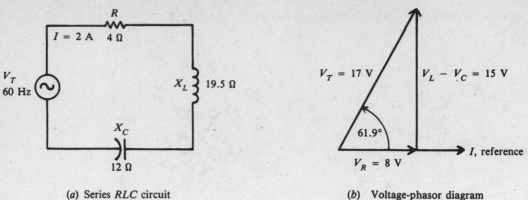

| (a) Series *RLC* circuit | (b) Voltage-phasor diagram |

Fig. 15-3

$$\theta = \arctan \frac{V_L - V_C}{V_R} \tag{15-2}$$

$$= \arctan \frac{39 - 24}{8} = \arctan \frac{15}{8} = \arctan 1.88 = 61.9° \quad I \text{ lags } V_T \quad Ans.$$

For the phasor diagram, see Fig. 15-3*b*.

Impedance in Series *RLC*

Impedance Z is equal to the phasor sum of R, X_L, and X_C. In Fig. 15-4*a*:

When $X_L > X_C$, $$Z = \sqrt{R^2 + (X_L - X_C)^2} \tag{15-5}$$

In Fig. 15-4*b*:

When $X_C > X_L$, $$Z = \sqrt{R^2 + (X_C - X_L)^2} \tag{15-6}$$

It is convenient to define net reactance X as

$$X = X_L - X_C \tag{15-7}$$

Then, from Eqs. (*15-5*) and (*15-6*),

$$Z = \sqrt{R^2 + X^2} \tag{15-8}$$

for both inductive and capacitive *RLC* series circuits (Fig. 15-4).

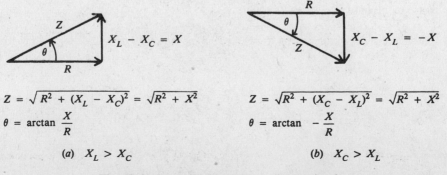

$$Z = \sqrt{R^2 + (X_L - X_C)^2} = \sqrt{R^2 + X^2} \qquad Z = \sqrt{R^2 + (X_C - X_L)^2} = \sqrt{R^2 + X^2}$$

$$\theta = \arctan \frac{X}{R} \qquad\qquad\qquad \theta = \arctan -\frac{X}{R}$$

| (a) $X_L > X_C$ | (b) $X_C > X_L$ |

Fig. 15-4   Series *RLC* impedance-phasor triangle

Example 15.2   Find the impedance of the series *RLC* circuit in Example 15.1.

$$Z = \sqrt{R^2 + X^2} \qquad\qquad (15\text{-}8)$$

$$X = X_L - X_C = 19.5 - 12 = 7.5 \ \Omega$$

Then

$$Z = \sqrt{4^2 + (7.5)^2} = 8.5 \ \Omega \qquad Ans.$$

Or, more simply, by Ohm's law,

$$Z = \frac{V_T}{I} = \frac{17}{2} = 8.5 \ \Omega \qquad Ans.$$

Example 15.3   When inductive reactance X_L and capacitive reactance X_C are exactly equal in a series *RLC* circuit, a condition called *series resonance* exists. If in a series circuit $R = 4 \ \Omega$ and $X_L = X_C = 19.5 \ \Omega$, find Z and V_T.

$$Z = \sqrt{R^2 + X^2} \qquad\qquad (15\text{-}8)$$

$$X = X_L - X_C = 19.5 - 19.5 = 0$$

Then

$$Z = \sqrt{R^2} = R = 4 \ \Omega \qquad Ans.$$

So

$$V_T = IZ = IR = 2(4) = 8 \ \text{V} \qquad Ans.$$

Note that in series resonance, the impedance of the circuit is equal to the resistance of the circuit. This is the minimum value of impedance for the circuit. Therefore at resonance, the highest current will flow.

RLC IN PARALLEL

A three-branch parallel ac circuit (Fig. 15-5*a*) has resistance in one branch, inductance in the second branch, and capacitance in the third branch. The voltage is the same across each parallel branch, so $V_T = V_R = V_L = V_C$. The applied voltage V_T is used as the reference line to measure phase angle θ. The total current I_T is the phasor sum of I_R, I_L, and I_C. The current in the resistance I_R is in phase with the applied voltage V_T (Fig. 15-5*b*). The current in the inductance I_L lags the voltage V_T by 90°. The current in the capacitor I_C leads the voltage V_T by 90°. I_L and I_C are exactly 180° out of phase and thus acting in opposite

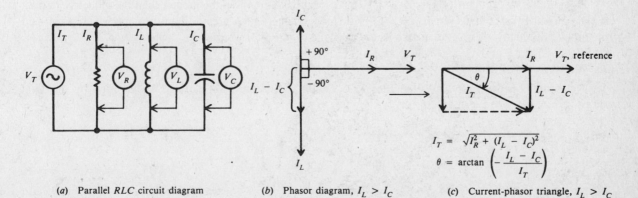

(a) Parallel *RLC* circuit diagram   (b) Phasor diagram, $I_L > I_C$ (c) Current-phasor triangle, $I_L > I_C$

Fig. 15-5 R, X_L, and X_C in parallel; $I_L > I_C$ for inductive circuit

directions (Fig. 15-5*b*). When $I_L > I_C$, I_T lags V_T (Fig. 15-5*c*) so the parallel *RLC* circuit is considered *inductive*.

If $I_C > I_L$, the current relationships and phasor triangle (Fig. 15-6) show that I_T now leads V_T so this type of parallel *RLC* circuit is considered *capacitive*.

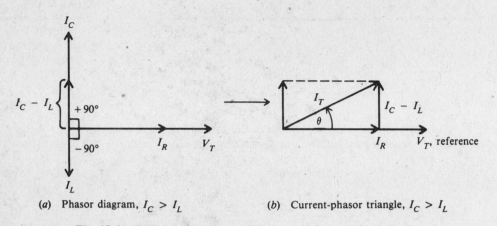

(*a*) Phasor diagram, $I_C > I_L$ (*b*) Current-phasor triangle, $I_C > I_L$

Fig. 15-6 R, X_L, and X_C in parallel; $I_C > I_L$ for capacitive circuit

When $I_L > I_C$, the circuit is inductive and

$$I_T = \sqrt{I_R^2 + (I_L - I_C)^2} \tag{15-9}$$

$$\theta = \arctan\left(-\frac{I_L - I_C}{I_R}\right) \tag{15-10}$$

and when $I_C > I_L$, the circuit is capacitive and

$$I_T = \sqrt{I_R^2 + (I_C - I_L)^2} \tag{15-11}$$

$$\theta = \arctan\frac{I_C - I_L}{I_R} \tag{15-12}$$

In a parallel *RLC* circuit, when $X_L > X_C$, the capacitive current will be greater than the inductive current and the circuit is capacitive. When $X_C > X_L$, the inductive current is greater than the capacitive current and the circuit is inductive. These relationships are opposite to those for a series *RLC* circuit.

Impedance in Parallel *RLC*

The total impedance Z_T of a parallel *RLC* circuit equals the total voltage V_T divided by the total current I_T.

$$Z_T = \frac{V_T}{I_T} \tag{15-13}$$

Example 15.4   A 400-Ω resistor, a 50-Ω inductive reactance, and a 40-Ω capacitive reactance are placed in parallel across a 120-V ac line (Fig. 15-7*a*). Find the phasor branch currents, total current, phase angle, and impedance. Draw the phasor diagram.

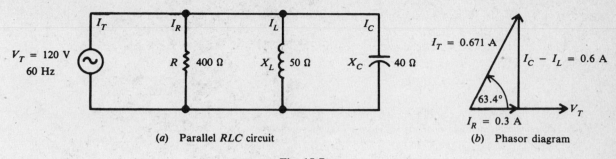

(a) Parallel *RLC* circuit   (b) Phasor diagram

Fig. 15-7

Step 1.   Find I_R, I_L, and I_C.

$$I_R = \frac{V_T}{R} = \frac{120}{400} = 0.3 \text{ A} \qquad Ans.$$

$$I_L = \frac{V_T}{X_L} = \frac{120}{50} = 2.4 \text{ A} \qquad Ans.$$

$$I_C = \frac{V_T}{X_C} = \frac{120}{40} = 3.0 \text{ A} \qquad Ans.$$

Step 2.   Find I_T and θ. Since $X_L > X_C$ (50 Ω > 40 Ω) or $I_C > I_L$ (3.0 A > 2.4 A), the circuit is capacitive.

$$I_T = \sqrt{I_R^2 + (I_C - I_L)^2} \qquad\qquad (15\text{-}11)$$

$$= \sqrt{(0.3)^2 + (3.0 - 2.4)^2} = \sqrt{(0.3)^2 + (0.6)^2} = 0.671 \text{ A} \qquad Ans.$$

$$\theta = \arctan \frac{I_C - I_L}{I_R} \qquad\qquad (15\text{-}12)$$

$$= \arctan \frac{3.0 - 2.4}{0.3} = \arctan \frac{0.6}{0.3} = \arctan 2 = 63.4° \quad I_T \text{ leads } V_T \qquad Ans.$$

Step 3.   Find Z_T, using Eq. (*15-13*).

$$Z_T = \frac{V_T}{I_T} = \frac{120}{0.671} = 179 \text{ Ω} \qquad Ans.$$

Step 4.   Draw the phasor diagram (Fig. 15-7*b*).

Example 15.5   A parallel *RLC* circuit in which $X_L = X_C$ is said to be in *parallel resonance*. Since X_L and X_C are dependent upon the values of L, C, and frequency f, parallel resonance (that is, $X_L = X_C$) can be obtained by choosing the proper values of L and C for a given frequency. If the values of L and C are already given, then the frequency can be varied until $X_L = X_C$. If in Example 15.4, $X_L = X_C = 40$ Ω, find the values of the same quantities as in that example.

Step 1.   Find I_R, I_L, and I_C.

$$I_R = \frac{V_T}{R} = \frac{120}{400} = 0.3 \text{ A} \qquad Ans.$$

$$I_L = I_C = \frac{V_T}{X_L} = \frac{120}{40} = 3 \text{ A} \qquad Ans.$$

Step 2.   Find I_T and θ. Since I_L and I_C are equal and opposite in phase,

$$I_T = \sqrt{I_R^2 + (I_C - I_L)^2} \qquad\qquad\qquad (15\text{-}11)$$

$$= \sqrt{I_R^2} = I_R = 0.3\,\text{A} \qquad Ans.$$

Note that only 0.3 A comes from the source though the reactive branch currents are each 3 A. At resonance these currents are equal and opposite so that they cancel each other.

$$\theta = \arctan\frac{I_L - I_C}{I_T} \qquad\qquad\qquad (15\text{-}12)$$

$$= \arctan\frac{0}{I_T} = \arctan 0 = 0° \qquad I_T \text{ in phase with } V_T \qquad Ans.$$

Step 3.   Find Z_T, using Eq. (*15-13*).

$$Z_T = \frac{V_T}{I_T} = \frac{120}{0.3} = 400\,\Omega \qquad Ans.$$

Step 4.   Draw the phasor diagram.

Note that a parallel resonant circuit reduces to a resistive circuit ($\theta = 0°$). Because I_L and I_C cancel each other, the current I_T is a minimum so impedance Z_T is a maximum.

RL AND RC BRANCHES IN PARALLEL

Total current I_T for a circuit containing parallel branches of RL and RC (Fig. 15-8) is the phasor sum of the branch currents I_1 and I_2. A convenient way to find I_T is to (1) add algebraically the horizontal components of I_1 and I_2 with respect to the phasor reference V_T, (2) add algebraically the vertical components of I_1 and I_2, and (3) form a right triangle with these two sums as legs and calculate the value of the hypotenuse (I_T) and its angle to the horizontal.

Example 15.6   An ac circuit has an RL branch parallel to an RC branch (Fig. 15-9a). Find the total current, phase angle, and impedance of this circuit.

Step 1.   For the RL branch, find Z_1, θ_1, and I_1.

$$Z_1 = \sqrt{R_1^2 + X_1^2} = \sqrt{6^2 + 8^2} = 10\,\Omega \qquad \theta_1 = \arctan\frac{X_1}{R_1} = \arctan\frac{8}{6} = 53.1°$$

$$I_1 = \frac{V_T}{Z_1} = \frac{60}{10} = 6\,\text{A}$$

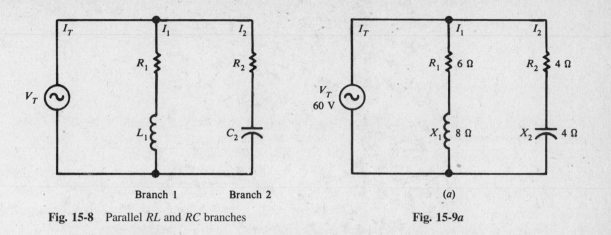

Branch 1   Branch 2   (a)

Fig. 15-8 Parallel *RL* and *RC* branches   **Fig. 15-9a**

I_1 lags V_T in the *RL* branch (inductive circuit) by 53.1°. Resolve I_1 into its horizontal and vertical components with respect to V_T (Fig. 15-9b).

Horizontal component: $I_1 \cos \theta_1 = 6 \cos (-53.1°) = 3.6 \text{ A}$

Vertical component: $I_1 \sin \theta_1 = 6 \sin (-53.1°) = -4.8 \text{ A}$

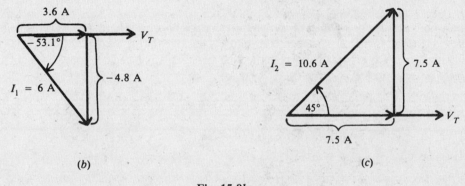

(b)   (c)

Fig. 15-9b, c

Step 2.   For the *RC* branch, find Z_2, θ_2, and I_2.

$$Z_2 = \sqrt{R_2^2 + X_2^2} = \sqrt{4^2 + 4^2} = 5.66 \ \Omega \qquad \theta_2 = \arctan\left(-\frac{X_2}{R_2}\right) = \arctan -\frac{4}{4} = \arctan -1 = -45°$$

$$I_2 = \frac{V_T}{Z_2} = \frac{60}{5.66} = 10.6 \text{ A}$$

I_2 leads V_T in the *RC* branch (capacitive circuit) by 45°. Resolve I_2 into its horizontal and vertical components with respect to V_T (Fig. 15-9c).

Horizontal component: $I_2 \cos \theta_2 = 10.6 \cos 45° = 7.5 \text{ A}$

Vertical component: $I_2 \sin \theta_2 = 10.6 \sin 45° = 7.5 \text{ A}$

Step 3.   Find I_T.

I_T is the phasor sum of I_1 and I_2. Add the horizontal components of I_1 and I_2.

$$3.6 + 7.5 = 11.1 \text{ A}$$

Add the vertical components of I_1 and I_2.

$$-4.8 + 7.5 = 2.7\,\text{A}$$

The resultant phasor is I_T (Fig. 15-9d).

$$I_T = \sqrt{(11.1)^2 + (2.7)^2} = \sqrt{130.5} = 11.4\,\text{A} \qquad Ans.$$

$$\theta = \arctan\frac{2.7}{11.1} = \arctan 0.243 = 13.7° \qquad Ans.$$

$$Z_T = \frac{V_T}{I_T} = \frac{60}{11.4} = 5.26\,\Omega \qquad Ans.$$

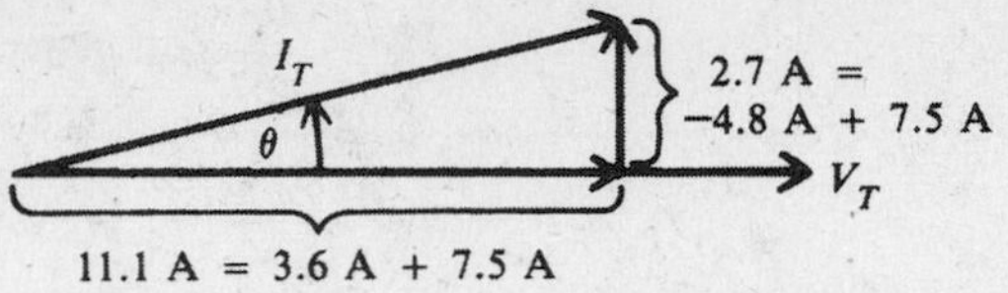

Note that I_T leads V_T (Fig. 15-9d) so that the circuit is leading and thus capacitive.

Fig. 15-9d

POWER AND POWER FACTOR

The instantaneous power p is the product of the current i and the voltage v at that instant of time t.

$$P = vi \qquad\qquad (15\text{-}14)$$

When v and i are both positive or both negative, their product p is positive. Therefore, power is being expended throughout the cycle (Fig. 15-10). If v is negative while i is positive during any part of the cycle (Fig. 15-11), or if i is negative while v is positive, their product will be negative. This "negative power" is not available for work; it is power returned to the line.

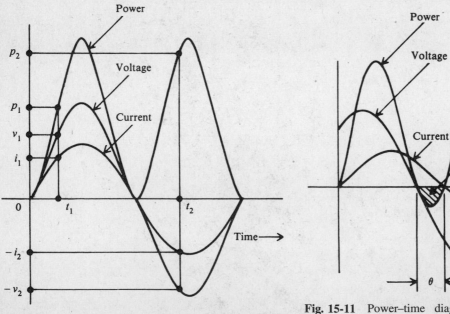

Fig. 15-10 Power–time diagram when voltage and current are in phase

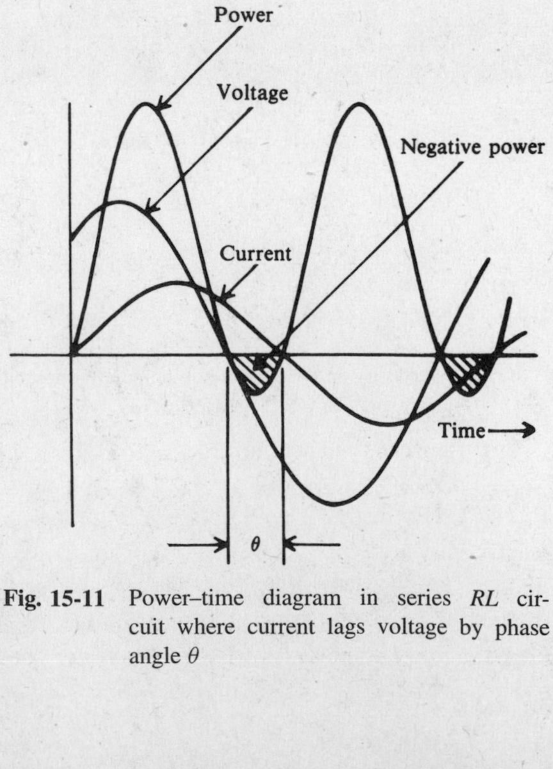

Fig. 15-11 Power–time diagram in series RL circuit where current lags voltage by phase angle θ

The product of the voltage across the resistance and the current through the resistance is always positive and is called *real power*. Real power can be considered as resistive power that is dissipated as heat. Since the voltage across a reactance is always 90° out of phase with the current through the reactance, the product $p_x = v_x i_x$ is always negative. This product is called *reactive power* and is due to the reactance of a circuit. Similarly, the product of the line voltage and the line current is known as *apparent power*.

Real power, reactive power, and apparent power can be represented by a right triangle (Fig. 15-12a). From this triangle the power formulas are:

$$\text{Real power } P = V_R I_R = VI \cos\theta, \text{ W} \qquad (15\text{-}15)$$

or

$$P = I^2 R, \text{ W} \qquad (15\text{-}16)$$

$$P = \frac{V^2}{R}, \text{ W} \qquad (15\text{-}17)$$

$$\text{Reactive power } Q = V_X I_X = VI \sin\theta, \text{ VAR} \qquad (15\text{-}18)$$

$$\text{Apparent power } S = VI, \text{ VA} \qquad (15\text{-}19)$$

In a circuit where the net reactance is inductive, Q is *lagging* and shown above the horizontal axis (Fig. 15-12b); when the net reactance is capacitive, Q is *leading* and shown below the horizontal axis (Fig. 15-12c).

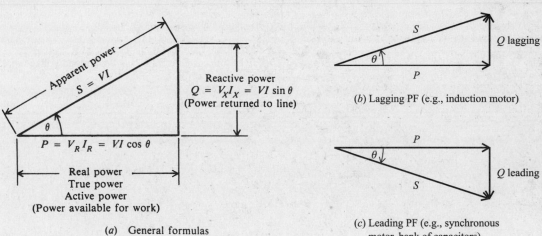

(a)   General formulas

(b) Lagging PF (e.g., induction motor)

(c) Leading PF (e.g., synchronous motor, bank of capacitors)

Fig. 15-12   Power triangle

The ratio of real power to apparent power, called the *power factor* (PF), is

$$\text{PF} = \frac{\text{real power}}{\text{apparent power}} = \frac{V_R I_R}{VI} = \frac{VI \cos\theta}{VI} = \cos\theta \qquad (15\text{-}20)$$

Also from Eq. (*15-15*),

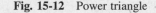

$$\text{PF} = \cos\theta = \frac{P}{VI} \qquad (15\text{-}21)$$

The $\cos\theta$ of a circuit is the power factor, PF, of the circuit. The power factor determines what portion of the apparent power is real power and can vary from 1 when the phase angle θ is 0°, to 0 when θ is 90°. When $\theta = 0°$, $P = VI$, the formula for voltage and current of a circuit in phase. When $\theta = 90°$, $P = VI \times 0 = 0$, indicating that no power is being expended or consumed.

A circuit in which the current lags the voltage (i.e., an inductive circuit) is said to have a *lagging* PF (Fig. 15-12*b*); a circuit in which the current leads the voltage (i.e., a capacitive circuit) is said to have a *leading* PF (Fig. 15-12*c*).

Power factor is expressed as a decimal or as a percentage. A power factor of 0.7 is the same as a power factor of 70 percent. At unity (PF = 1 or 100 percent), the current and voltage are in phase. A 70 percent PF means that the device uses only 70 percent of the voltampere input. It is desirable to design circuits that have a high PF since such circuits make the most efficient use of the current delivered to the load.

When we state that a motor draws 10 kVA (1 kVA = 1000 VA) from a power line, we recognize that this is the apparent power taken by the motor. Kilovoltamperes always refers to the apparent power. Similarly, when we say a motor draws 10 kW, we mean that the real power taken by the motor is 10 kW.

Example 15.7   A current of 7 A lags voltage of 220 V by 30°. What is the PF and real power taken by the load?

$$\text{PF} = \cos\theta \tag{15-20}$$

$$= \cos 30° = 0.866 \quad Ans.$$

$$P = VI\cos\theta \tag{15-15}$$

$$= 220(7)(0.866) = 1334\,\text{W} \quad Ans.$$

Example 15.8   A motor rated at 240 V, 8 A draws 1536 W at full load. What is its PF?
Use Eq. (*15-21*).

$$\text{PF} = \frac{P}{VI} = \frac{1536}{240(8)} = 0.8 \text{ or } 80\% \quad Ans.$$

Example 15.9   In the *RLC* series ac circuit shown in Fig. 15-3*a*, the line current of 2 A lags the applied voltage of 17 V by 61.9°. Find PF, *P*, *Q*, and *S*. Draw the power triangle.

$$\text{PF} = \cos\theta \tag{15-20}$$

$$= \cos 61.9° = 0.471 \text{ or } 47.1\% \text{ lagging} \quad Ans.$$

$$P = VI\cos\theta \tag{15-15}$$

$$= 17(2)(0.471) = 16\,\text{W} \quad Ans.$$

or

$$P = I^2 R \tag{15-16}$$

$$= 2^2(4) = 16\,\text{W} \quad Ans.$$

$$Q = VI\sin\theta \tag{15-18}$$

$$= 17(2)(\sin 61.9°) = 17(2)(0.882) = 30\,\text{VAR lagging} \quad Ans.$$

$$S = VI \tag{15-19}$$

$$= 17(2) = 34\,\text{VA} \quad Ans.$$

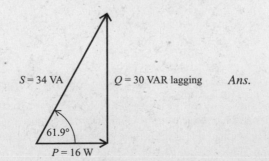

Power Factor Correction

In order to make the most efficient use of the current delivered to a load, we desire a high PF or a PF that approaches unity. A low PF is generally due to the large inductive loads, such as induction motors, which take a lagging current. In order to correct this low PF, it is necessary to bring the current as closely in phase with the voltage as possible. That is, the phase angle θ is made as small as possible. This is usually done by placing a capacitive load, which produces a leading current, in parallel with the inductive load.

Example 15.10   With the use of a phasor diagram show how the PF produced by an inductive motor can be corrected to unity.

We show an induction motor circuit (Fig. 15-13a) and its current-phasor diagram (Fig. 15-13b). Current I lags the voltage V by an angle θ where PF $= \cos\theta = 0.7$. We desire to raise the PF to 1.0. We do this by connecting a capacitor across the motor (Fig. 15-14a). If the capacitor current is equal to the inductive current, the two cancel one another (Fig. 15-14b). The line current I_1 is now less than its original value and in phase with V so that PF $= \cos 0° = 1$. Notice that the current through the motor I remains unchanged. The reactive part of the current to the motor is being supplied by the capacitor. The line now has only to supply the component of current for the resistive part of the motor load.

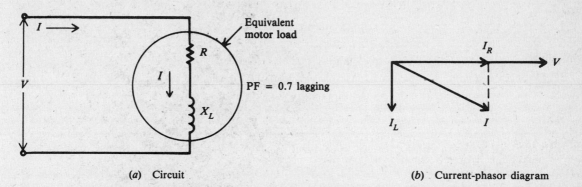

(a) Circuit   (b) Current-phasor diagram

Fig. 15-13   An induction motor represented by a series RL circuit

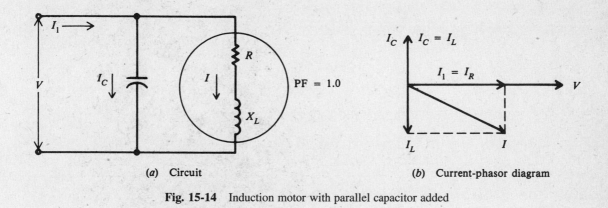

(a) Circuit   (b) Current-phasor diagram

Fig. 15-14   Induction motor with parallel capacitor added

Example 15.11   An induction motor draws 1.5 kW and 7.5 A from a 220-V 60-Hz line. What must be the capacitance of a capacitor in parallel in order to raise the total PF to unity (Fig. 15-15a)?

Step 1.   Find phase angle θ_M and then the reactive power Q_M of the motor load.

$$P_M = V_M I_M \cos\theta_M \qquad\qquad (15\text{-}15)$$

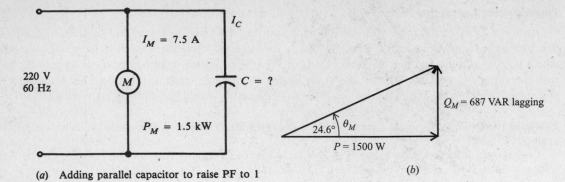

(a)   Adding parallel capacitor to raise PF to 1   (b)

Fig. 15-15

from which

$$\cos \theta_M = \frac{P_M}{V_M I_M} = \frac{1500}{220(7.5)} = 0.909$$

$$\theta_M = \arccos 0.909 = 24.6°$$

From the motor power triangle (Fig. 15-15b),

$$Q_M = 1500 \tan 24.6° = 687 \text{ VAR lagging}$$

Step 2.   Find the current I_C drawn by the capacitor. For the current to have PF = 1, the capacitor must have a $Q_C = 687$ VAR leading to balance the $Q_M = 687$ VAR lagging. Since reactive power in a pure capacitor is also its apparent power,

$$Q_C = S_C = V_C I_C \qquad\qquad\qquad (15\text{-}18)$$

$$I_C = \frac{S_C}{V_C} = \frac{687}{220} = 3.12 \text{ A}$$

Step 3.   Find the reactance of the capacitor.

$$X_C = \frac{V_C}{I_C} = \frac{220}{3.12} = 70.5 \ \Omega$$

Step 4.   Find the capacitance of the capacitor, using Eq. (14-9).

$$C = \frac{0.159}{f X_C} = \frac{0.159}{60(70.5)} = 37.6 \times 10^{-6} = 37.6 \ \mu\text{F} \qquad Ans.$$

Example 15.12   An induction motor takes 15 kVA at 440 V and 75 percent PF lagging. What must be the PF of a 10-kVA capacitive load connected in parallel in order to raise the total PF to unity?

Step 1.   Find the reactive power of the induction motor, Q_M.

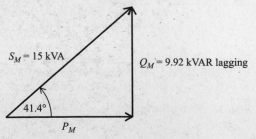

$$\text{PF}_M = \cos \theta = 0.75$$

$$\theta = \arccos 0.75 = 41.4°$$

$$Q_M = VI \sin \theta = 15 \sin 41.4° = 9.92 \text{ kVAR lagging}$$

Induction motor power triangle

Step 2. Find the phase angle and then the PF for the capacitive load. To have a circuit with PF = 1, the total reactive power must equal zero. Since the motor uses 9.92 kVAR *lagging*, the *leading* PF load must also use 9.92 kVAR. From the power triangle for leading load,

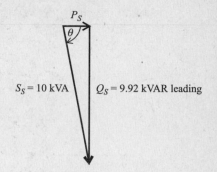

$S_S = 10$ kVA $Q_S = 9.92$ kVAR leading

Leading load power triangle

$$\sin\theta = \frac{Q_S}{S_S} = \frac{9.92}{10} = 0.992$$

$$\theta = \arcsin 0.992 = 82.7°$$

$$\text{PF} = \cos\theta = \cos 82.7° = 0.127 = 12.7\% \text{ leading} \qquad Ans.$$

Solved Problems

15.1   For the *RLC* series circuit (Fig. 15-16*a*), find X_L, X_C, Z, I, V_R, V_L, V_C, P, and PF. Draw the voltage-phasor diagram.

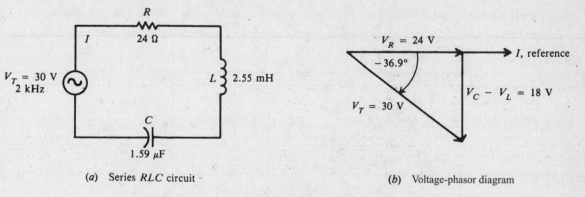

(*a*)   Series *RLC* circuit

$V_R = 24$ V

$-36.9°$

$V_T = 30$ V

I, reference

$V_C - V_L = 18$ V

(*b*)   Voltage-phasor diagram

Fig. 15-16

Step 1.   Find X_L, X_C, and then Z.

$$X_L = 6.28\,fL \qquad\qquad (13\text{-}4)$$

$$= 6.28(2 \times 10^3)(2.55 \times 10^{-3}) = 32\ \Omega \qquad Ans.$$

$$X_C = \frac{0.159}{fC} \qquad\qquad (14\text{-}8)$$

$$= \frac{0.159}{(2 \times 10^3)(1.59 \times 10^{-6})} = 50\ \Omega \qquad Ans.$$

$$X = X_L - X_C = 32 - 50 = -18\ \Omega$$

Minus indicates the circuit is capacitive ($X_C > X_L$).

$$Z = \sqrt{R^2 + X^2} = \sqrt{24^2 + (-18)^2} = 30\ \Omega \qquad Ans.$$

$$\theta = \arctan\frac{X}{R} = \arctan\left(-\frac{18}{24}\right) = -36.9° \qquad Ans.$$

Step 2. Find I, V_R, V_L, and V_C by Ohm's law.

$$I = \frac{V_T}{Z} = \frac{30}{30} = 1\,\text{A} \qquad Ans.$$

$$V_R = IR = 1(24) = 24\,\text{V} \qquad Ans.$$

$$V_L = IX_L = 1(32) = 32\,\text{V} \qquad Ans.$$

$$V_C = IX_C = 1(50) = 50\,\text{V} \qquad Ans.$$

Step 3. Find P and PF.

$$P = I^2 R = 1^2(24) = 24\,\text{W} \qquad Ans.$$

$$\text{PF} = \cos\theta = \cos(-36.9°) = 0.8\,\text{leading} \qquad Ans.$$

Step 4. Draw the voltage-phasor diagram (see Fig. 15-16*b*).

15.2 The output signal of a rectifier is 200 V at 120 Hz. It is fed to a filter circuit consisting of a 30-H filter choke coil and a 20-μF capacitor (Fig. 15-17). How much 120-Hz voltage appears across the capacitor? (The purpose of the filter is to reduce considerably the voltage across the capacitor.)

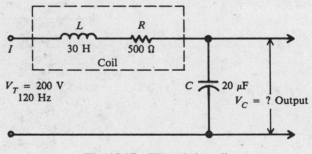

Fig. 15-17 Filter choke coil

Step 1. Find X_L, X_C, and Z.

$$X_L = 6.28\,fL = 6.28(120)(30) = 22\,600\,\Omega$$

$$X_C = \frac{0.159}{fC} = \frac{0.159}{120(20 \times 10^{-6})} = 66\,\Omega$$

$$X = X_L - X_C = 22\,600 - 66 = 22\,534\,\Omega = 22\,500\,\Omega \text{ (rounded to three significant figures)}$$

$$Z = \sqrt{R^2 + X^2}$$

Since $X \gg R$,

$$Z \approx X = 22\,500\,\Omega$$

Step 2. Find I.

$$I = \frac{V_T}{Z} = \frac{200}{22\,500} = 8.89\,\text{mA}$$

Step 3. Find V_C.

$$V_C = IX_C = (8.89 \times 10^{-3})(66) = 0.587 \text{ V} \qquad Ans.$$

This is a satisfactory filter since only 0.587 V out of a total of 200 V ac gets through to the output circuit.

15.3 Find the impedance and current of an *RLC* series circuit containing a number of series resistances and reactances (Fig. 15-18).

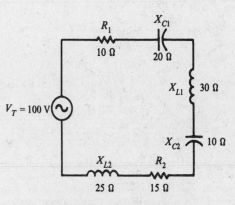

Fig. 15-18 Series *RLC* circuit

Add the values of similar circuit elements.

Resistance: $\qquad\qquad\qquad R_T = R_1 + R_2 = 10 + 15 = 25 \ \Omega$

Capacitance: $\qquad\qquad X_{C,T} = X_{C1} + X_{C2} = 20 + 10 = 30 \ \Omega$

Inductance: $\qquad\qquad X_{L,T} = X_{L1} + X_{L2} = 30 + 25 = 55 \ \Omega$

Net reactance: $\qquad\quad X_T = X_{L,T} - X_{C,T} = 55 - 30 = 25 \ \Omega$

$$Z = \sqrt{R_T^2 + X_T^2} = \sqrt{25^2 + 25^2} = 35.4 \ \Omega \qquad Ans.$$

$$I = \frac{V_T}{Z} = \frac{100}{35.4} = 2.82 \text{ A} \qquad Ans.$$

15.4 A 30-Ω resistor, a 40-Ω inductive reactance, and a 60-Ω capacitive reactance are connected in parallel across a 120-V 60-Hz ac line (Fig. 15-19a). Find I_T, θ, Z_T, and P. Is the circuit inductive or capacitive? Draw the current-phasor diagram.

Step 1. Find I_T and θ.

$$I_R = \frac{V_T}{R} = \frac{120}{30} = 4 \text{ A} \qquad I_L = \frac{V_T}{X_L} = \frac{120}{40} = 3 \text{ A} \qquad I_C = \frac{V_T}{X_C} = \frac{120}{60} = 2 \text{ A}$$

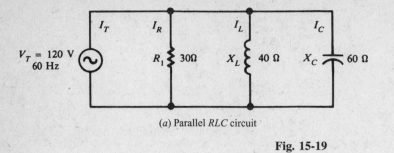
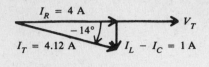

(a) Parallel RLC circuit   (b) Current-phasor diagram

Fig. 15-19

Since $I_L > I_C$, the circuit is inductive.

$$I_T = \sqrt{I_R^2 + (I_L - I_C)^2} \qquad\qquad (15\text{-}9)$$

$$= \sqrt{4^2 + 1^2} = 4.12\ \text{A} \qquad Ans.$$

$$\theta = \arctan\left(-\frac{I_L - I_C}{I_R}\right) \qquad\qquad (15\text{-}10)$$

$$= \arctan\left(-\frac{1}{4}\right) = -14° \qquad Ans.$$

Step 2.   Find Z_T.

$$Z_T = \frac{V_T}{I_T} = \frac{120}{4.12} = 29.1\ \text{A} \qquad Ans.$$

Step 3.   Find P.

$$P = V_T I_T \cos\theta = 120(4.12)(\cos 14°) = 480\ \text{W} \qquad Ans.$$

or $\qquad\qquad\qquad P = I_R^2 R = 4^2(30) = 480\ \text{W} \qquad Ans.$

Step 4.   Draw the current-phasor diagram (see Fig. 15-19b). I_T lags V_T by 14°.

15.5   An ac circuit can have a number of parallel resistances and reactances (Fig. 15-20). This circuit has more components than Fig. 15-19a (Problem 15.4). Verify that it has the same value of I_T, Z_T, and θ.

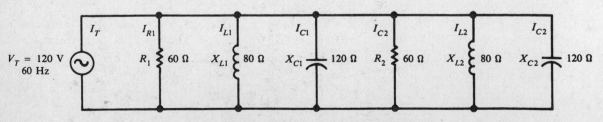

Fig. 15-20   Six-branch parallel RLC circuit

$$I_{R1} = \frac{120}{60} = 2\,\text{A} \qquad I_{L1} = \frac{120}{80} = 1.5\,\text{A} \qquad I_{C1} = \frac{120}{120} = 1\,\text{A}$$

$$I_{R2} = \frac{120}{60} = 2\,\text{A} \qquad I_{L2} = \frac{120}{80} = 1.5\,\text{A} \qquad I_{C2} = \frac{120}{120} = 1\,\text{A}$$

$$I_{R,T} = 2 + 2 = 4\,\text{A} \qquad I_{L,T} = 1.5 + 1.5 = 3\,\text{A} \qquad I_{C,T} = 1 + 1 = 2\,\text{A}$$

The total resistance branch current is 4 A; total inductive branch current, 3 A; and total capacitive branch current, 2 A. These current values are the same, respectively, in Fig. 15-19a. Therefore, I_T, Z_T, and θ have the same values for both circuits.

15.6   An induction motor of 6-Ω resistance and 8-Ω inductive reactance is in parallel with a synchronous motor of 8-Ω resistance and 15-Ω capacitive reactance (Fig. 15-21a). Find the total current drain from a 150-V 60-Hz source, phase angle, total impedance, power factor, and power drawn by the circuit. Find the series impedance, current, and phase angle for each branch.

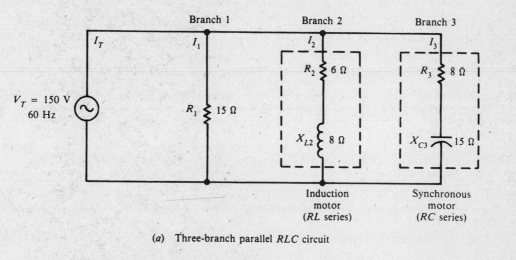

(a)   Three-branch parallel *RLC* circuit

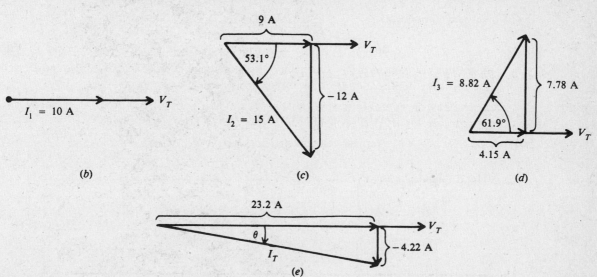

(b)   (c)   (d)

(e)

Fig. 15-21

Step 1. Find Z_1, I_1, and θ_1, for branch 1.

$$Z_1 = R_1 = 15 \, \Omega \qquad I_1 = \frac{V_T}{R_1} = \frac{150}{15} = 10 \, \text{A}$$

$$\theta_1 = 0° \qquad \text{since } I_1 \text{ is in phase with } V_T$$

Calculate horizontal and vertical components of current with respect to V_T (Fig. 15-21b).

Horizontal component: 10 A

Vertical component: 0 A

Step 2. Find Z_2, I_2, and θ_2 for branch 2.

$$Z_2 = \sqrt{R_2^2 + X_{L2}^2} = \sqrt{6^2 + 8^2} = 10 \, \Omega \qquad I_2 = \frac{V_T}{Z_2} = \frac{150}{10} = 15 \, \text{A}$$

$$\theta_2 = \arctan \frac{X_{L2}}{R_2} = \arctan \frac{8}{6} = 53.1°$$

I_2 lags V_T in RL series branch by 53.1° (Fig. 15-21c).

Horizontal component: $\qquad\qquad 15\cos(-53.1°) = 9\,\text{A}$

Vertical component: $\qquad\qquad 15\sin(-53.1°) = -12\,\text{A}$

Step 3. Find Z_3, I_3, and θ_3 for branch 3.

$$Z_3 = \sqrt{R_3^2 + X_{C3}^2} = \sqrt{8^2 + 15^2} = 17 \, \Omega \qquad I_3 = \frac{V_T}{Z_3} = \frac{150}{17} = 8.82 \, \text{A}$$

$$\theta_3 = \arctan\left(-\frac{X_{C3}}{R_3}\right) = \arctan\left(-\frac{15}{8}\right) = -61.9°$$

I_3 leads V_T in RC series branch by 61.9° (Fig. 15-21d).

Horizontal component: $\qquad\qquad 8.82\cos 61.9° = 4.15\,\text{A}$

Vertical component: $\qquad\qquad 8.82\sin 61.9° = 7.78\,\text{A}$

Step 4. Find L_T, θ, Z_T, PF, and P of the circuit. I_T is the phasor sum of I_1, I_2, and I_3. The horizontal component of I_T is the sum of the horizontal components of I_1, I_2, and I_3, and the vertical component of I_T is the sum of their vertical components.

$$\text{Horizontal component of } I_T = 10 + 9 + 4.15 = 23.2 \, \text{A}$$

$$\text{Vertical component of } I_T = 0 - 12 + 7.78 = -4.22 \, \text{A}$$

The resultant phasor is I_T (Fig. 15-21e).

$$I_T = \sqrt{(23.2)^2 + (-4.22)^2} = 23.6 \, \text{A} \qquad Ans.$$

$$\theta = \arctan\left(-\frac{4.22}{23.2}\right) = -10.3° \qquad Ans.$$

$$Z_T = \frac{V_T}{I_T} = \frac{150}{23.6} = 6.36 \, \Omega \qquad Ans.$$

$$PF = \cos\theta = \cos(-10.3°) = 0.984 \text{ lagging} \qquad (I_T \text{ lags } V_T) \qquad Ans.$$

$$P = V_T I_T \cos\theta = 150(23.6)(0.984) = 3480 \text{ W} \qquad Ans.$$

15.7   What value of resistance dissipates 800 W of ac power with 5 A rms current?

Use
$$P = I^2 R \qquad\qquad (15\text{-}16)$$

from which
$$R = \frac{P}{I^2} = \frac{800}{5^2} = \frac{800}{25} = 32\,\Omega \qquad Ans.$$

15.8   An ac motor operating at 75 percent PF draws 8 A from a 110-V ac line. Find the apparent and real power.

Apparent power: $S = VI$ $(15\text{-}19)$
$$= 110(8) = 880 \text{ VA} \qquad Ans.$$
Real (true) power: $P = VI \cos\theta$ $(15\text{-}15)$
$$PF = \cos\theta = 0.75$$
$$P = 110(8)(0.75) = 660 \text{ W} \qquad Ans.$$

15.9   A motor operating at 85 percent PF draws 300 W from a 120-V line. What is the current drawn?

$$P = VI \cos\theta \qquad\qquad (15\text{-}15)$$

from which
$$I = \frac{P}{V \cos\theta} = \frac{300}{120(0.85)} = 2.94 \text{ A} \qquad Ans.$$

15.10   A 10-kVA induction motor operating at 80 percent lagging PF and a 5-kVA synchronous motor operating at 70 percent leading PF are connected in parallel across a 220-V 60-Hz power line (Fig. 15-22a). Find the total real power P_T, total reactive power Q_T, total power factor $(PF)_T$, total apparent power S_T, and total current I_T.

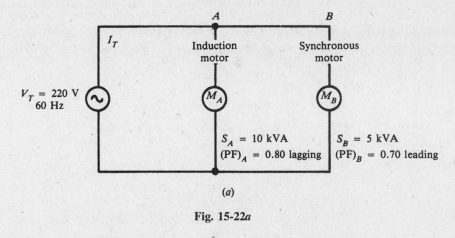

(a)

Fig. 15-22a

The approach to this problem with two motors in parallel is first to solve the power triangle for the induction motor, then the power triangle for the synchronous motor, and finally to combine by phasor addition the components of the two power triangles to arrive at the power triangle of the combined motors.

Step 1.   Find the power relationships of the induction motor.

$$S_A = VI = 10 \text{ kVA, given}$$
$$P_A = VI \cos\theta_A = 10(0.80) = 8 \text{ kW}$$

$\theta_A = \arccos 0.8 = 36.9°$

$Q_A = VI \sin \theta_A = 10(\sin 36.9°) = 10(0.6) = 6$ kVAR lagging because circuit is inductive

The induction motor power triangle is as shown in Fig. 15-22b.

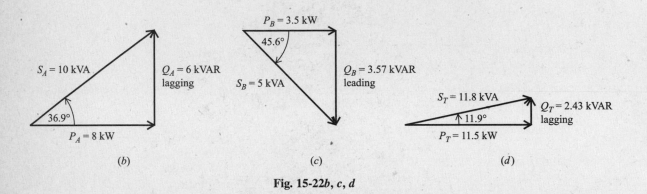

Fig. 15-22b, c, d

Step 2. Find power relationships of the synchronous motor.

$$S_B = VI = 5 \text{ kVA, given}$$

$$P_B = VI \cos \theta_B = 5(0.70) = 3.5 \text{ kW}$$

$$\theta_B = \arccos 0.7 = 45.6°$$

$Q_B = VI \sin \theta_B = 5(\sin 45.6°) = 5(0.71) = 3.57$ kVAR leading because circuit is capacitive

The synchronous motor power triangle is as shown in Fig. 15.22c.

Step 3. Find power relations of combined motors. If the power drawn by one branch is not in phase with the power drawn by another branch, power must be added by phasor addition. Therefore, P_T and Q_T are the phasor sums of P_A, P_B and Q_A, Q_B, respectively. S_T is the resultant phasor of P_T and Q_T.

$$P_T = P_A + P_B = 8 + 3.5 = 11.5 \text{ kW} \qquad Ans.$$

$$Q_T = Q_A - Q_B = 6 - 3.57 \qquad (Q_A > Q_B)$$

$$= 2.43 \text{ kVAR lagging} \qquad Ans.$$

$$\theta_T = \arctan \frac{Q_T}{P_T} = \arctan \frac{2.43}{11.5} = \arctan 0.211 = 11.9°$$

$$(\text{PF})_T = \cos \theta_T = \cos 11.9° = 0.979 = 97.9\% \text{ lagging} \qquad Ans.$$

$$S_T = \sqrt{P_T^2 + Q_T^2} = \sqrt{(11.5)^2 + (2.43)^2} = 11.8 \text{ kVA} \qquad Ans.$$

The combined motors' power triangle is as shown in Fig. 15-22d. Notice that the addition of the synchronous motor has raised the PF from 0.80 to 0.979.

15.11 An induction motor takes 7.2 kW at 80 percent PF lagging from a 220-V 60-Hz power line (Fig. 15-23). Find the capacitance of a capacitor placed across the motor terminals in order to increase the PF to 1.

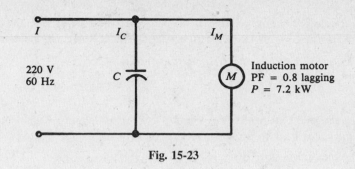

Fig. 15-23

Step 1. Find S and Q of the induction motor.

$$S = \frac{P}{\text{PF}} = \frac{7.2}{0.8} = 9 \text{ kVA} \quad \text{PF} = \cos\theta = 0.8 \quad \theta = \arccos 0.8 = 36.9°$$

$$Q = P \tan\theta = 7.2 \tan 36.9° = 5.4 \text{ kVAR lagging}$$

Step 2. Find the current drawn by the capacitor I_C. To balance 5.4 kVAR lagging, it is necessary that the capacitor take 5.4 kVAR leading. Since reactive power in a pure capacitor is also its apparent power,

$$S_C = V_C I_C = 5.4 \text{ kVA}$$

$$I_C = \frac{S_C}{V_C} = \frac{5400}{220} = 24.6 \text{ A}$$

Step 3. Find the reactance of the capacitor.

$$X_C = \frac{V_C}{I_C} = \frac{220}{24.6} = 8.94 \text{ }\Omega$$

Step 4. Find the capacitance of the capacitor.

$$C = \frac{0.159}{f X_C} = \frac{0.159}{60(8.94)} = 296 \times 10^{-6} = 296 \text{ }\mu\text{F} \qquad Ans.$$

15.12 An inductive load draws 5 kW at 60 percent PF lagging from a 220-V line. Find the kilovoltampere rating of the capacitor needed to raise the total PF unity.

Step 1. Find the power relations of the inductive load (see Fig. 15-24).

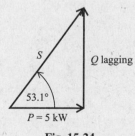

Fig. 15-24

$$S = \frac{P}{\text{PF}} = \frac{5}{0.6} = 8.33 \text{ kVA} \qquad \text{PF} = \cos\theta = 0.6 \qquad \theta = \arccos 0.6 = 53.1°$$

$$Q = 8.33 \sin 53.1° = 6.66 \text{ kVAR lagging}$$

Step 2. Find the kilovoltampere rating of the capacitor. In order to raise the PF to 1, the capacitor must provide 6.66 kVAR of reactive power leading. Since the apparent power in a capacitor is equal to its reactive power,

$$S_C = Q_C = 6.66 \text{ kVA} \qquad Ans.$$

15.13 A plant load draws 2000 kVA from a 240-V line at a PF of 0.7 lagging. Calculate the kilovoltamperage required of a capacitor bank in parallel with the plant for the overall PF to be 0.9 lagging.

We use the method of power triangles.

$$\text{PF} = \cos\theta = 0.7 \qquad \theta = \arccos 0.7 = 45.6°$$

Real power of the plant load:

$$P = VI\cos\theta = 2000(0.7) = 1400 \text{ kW}$$

Reactive power of the plant load:

$$Q = VI\sin\theta = 2000\sin 45.6° = 1430 \text{ kVAR lagging}$$

Therefore, the power triangle of the original plant is as shown in Fig. 15-25a.

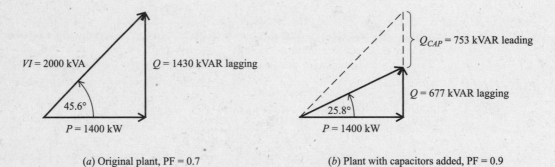

(a) Original plant, PF = 0.7   (b) Plant with capacitors added, PF = 0.9

Fig. 15-25   Method of power triangles

We add a parallel capacitor bank to improve the power factor of the plant to 0.9 lagging. The total real power remains the same. We calculate the new Q of the plant.

$$(\text{PF})_1 = \cos\theta = 0.9 \qquad \theta_1 = \arccos 0.9 = 25.8° \qquad Q_1 = 1400\tan 25.8° = 677 \text{ kVAR lagging}$$

The total reactive power of a circuit is equal to the algebraic sum of the reactive powers of each branch. Therefore,

$$Q \text{ of the capacitor bank} = \text{original } Q \text{ of the plant} - \text{ new } Q \text{ of the plant}$$
$$= 1430 - 677 = 753 \text{ kVAR leading}$$

Thus S of the capacitor bank = 753 kVA   *Ans.*

The power triangle with the capacitor bank added (Fig. 15-25b) shows how the leading Q of the capacitor reduces the overall plant Q to 677 kVAR lagging to produce an improved PF of 0.9.

15.14   A series RL combination in an ac circuit has $R = 10\ \Omega$ and $X_L = 12\ \Omega$. A capacitor is connected across the combination (Fig. 15-26a). What should be the reactance of the capacitor if the circuit is to have a PF of unity?

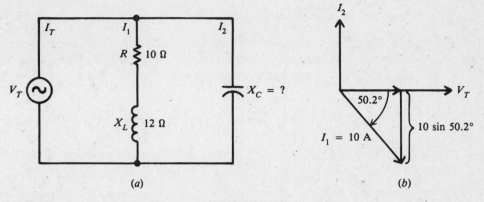

Fig. 15-26

Step 1.   Find I_1.

$$Z_1 = \sqrt{R^2 + X_L^2} = \sqrt{10^2 + 12^2} = 15.6\ \Omega \qquad \theta = \arctan\frac{X_L}{R} = \arctan\frac{12}{10} = 50.2°$$

Assume $V_T = 156$ V. This choice is arbitrary but is conveniently used because the impedance of the RL branch is 15.6 Ω. Then

$$I_1 = \frac{V_T}{Z_1} = \frac{156}{15.6} = 10\ \text{A}$$

The phasor diagram is as shown in Fig. 15-26b.

Step 2.   Find I_2. For PF = 1, I_T must be in phase with V_T. I_T is the phasor sum of I_1 and I_2. The current in the capacitor, I_2 (leading V_T by 90°), must cancel the vertical component of I_1 in order for I_T to be in phase with V_T (Fig. 15-26b). Therefore,

$$I_2 = 10\sin 50.2° = 7.68\ \text{A}$$

Step 3.   Find X_C.

$$X_C = \frac{V_T}{I_2} = \frac{156}{7.68} = 20.3\ \Omega \qquad Ans.$$

Supplementary Problems

15.15   In a series circuit, $R = 12\ \Omega$, $X_L = 7\ \Omega$, and $X_C = 2\ \Omega$. Find the impedance and phase angle of the circuit and the line current when the ac voltage is 110 V. Also find all voltage drops and draw the voltage-phasor diagram.
Ans. $Z = 13\ \Omega$, $\theta = 22.6°$; $I = 8.46$ A (lagging); $V_R = 101.5$ V; $V_L = 59.2$ V; $V_C = 16.9$ V; phasor diagram: Fig. 15-27.

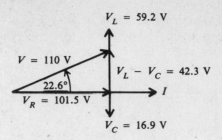

Fig. 15-27   Phasor diagram

15.16   In a series circuit, $R = 6\,\Omega$, $X_L = 4\,\Omega$, and $X_C = 12\,\Omega$. Find Z, θ, I, V_R, V_L, V_C, and P when the line voltage is 115 V.
  Ans. $Z = 10\,\Omega$; $\theta = -53.1°$; $I = 11.5\,\text{A}$ (leading); $V_R = 69\,\text{V}$; $V_L = 46\,\text{V}$; $V_C = 138\,\text{V}$; $P = 794\,\text{W}$

15.17   A 130-V 200-Hz power supply is connected across a 10-kΩ resistor, a 0.05-μF capacitor, and a 10-H coil in series. Find X_L, X_C, Z, θ, I, and P.
  Ans. $X_L = 12\,560\,\Omega$; $X_C = 15\,900\,\Omega$; $Z = 10\,540\,\Omega$; $\theta = -18.5°$; $I = 12.3\,\text{mA}$ (leading); $P = 1.51\,\text{W}$

15.18   A rectifier delivers 250 V at 120 Hz to a filter circuit consisting of a filter choke coil having 25-H inductance and 400-Ω resistance, and a 25-μF capacitor (Fig. 15-28). Find the reactance of the coil, the reactance of the capacitor, the impedance of the circuit, the current, and the amount of the 120-Hz voltage that will appear across the capacitor.
  Ans. $X_L = 18\,840\,\Omega$; $X_C = 53\,\Omega$; $Z = 18\,790\,\Omega$. For practical purposes the circuit is predominantly inductive ($Z \approx X_L$). $I = 13.3\,\text{mA}$; $V_C = 0.7\,\text{V}$. Note that only 0.7 V out of a total of 250 V of the 120-Hz ac reaches the output.

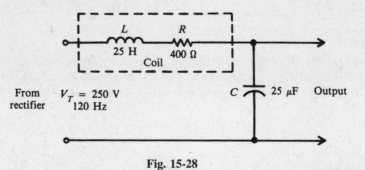

Fig. 15-28

15.19   A series circuit has $R = 300\,\Omega$, $X_{C1} = 300\,\Omega$, $X_{C2} = 500\,\Omega$, $X_{L1} = 400\,\Omega$, and $X_{L2} = 800\,\Omega$, all in series with an applied voltage V_T of 400 V. Calculate Z, I, and θ.
  Ans. $Z = 500\,\Omega$; $I = 0.8\,\text{A}$ (lagging); $\theta = 53.1°$

15.20   A 10-H coil and a 0.75-μF capacitor are in series with a variable resistor. What must be the value of the resistance in order to draw 0.4 A from a 120-V 60-Hz line?
  Ans. $R = 186\,\Omega$; $(X_L - X_C = 235\,\Omega)$

15.21   A series resonant circuit ($X_L = X_C$) has a 0.1-H inductance, a 1.013-μF capacitor, and a 5-Ω resistor connected across a 50-V 500-Hz supply line. Find the inductive and capacitive reactance, impedance, phase angle, current, and voltage across each part of the circuit. Draw the phasor diagram.

Ans. $X_L = 314\,\Omega$; $X_C = 314\,\Omega$; $Z = 5\,\Omega$; $\theta = 0°$; $I = 10\,\text{A}$; $V_R = 50\,\text{V}$; $V_L = 3140\,\text{V}$; $V_C = 3140\,\text{V}$; phasor diagram: Fig. 15-29.

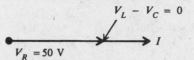

Fig. 15-29   Phasor diagram

15.22   Fill in the indicated values for a series *RLC* circuit.

	V_T, V	$\theta°$	I, A	R, Ω	X_L, Ω	X_C, Ω	Z, Ω	V_R, V	V_L, V	V_C, V	P, W	Circuit Type
(a)	?	?	1	3	8	4	?	?	?	?	?	?
(b)	104	?	?	12	2	7	?	?	?	?	?	?
(c)	110	?	?	22	18	18	?	?	?	?	?	?
(d)	?	45°	?	15	30	?	?	30	?	?	?	?
(e)	14.1	?	0.1	?	150	250	?	?	?	?	?	?

Ans.

	V_T, V	$\theta°$	I, A	R, Ω	X_L, Ω	X_C, Ω	Z, Ω	V_R, V	V_L, V	V_C, V	P, W	Circuit Type
(a)	5	53.1°	....	....	....	....	5	3	8	4	3	Inductive
(b)	....	−22.6°	8	....	....	....	13	96	16	56	768	Capacitive
(c)	....	0	5	....	....	....	22	110	90	90	550	Resonant
(d)	42.4	....	2	....	....	15	21.2	....	60	30	60	Inductive
(e)	....	−45°	....	99.4	....	....	141	10	15	25	1	Capacitive

15.23   A 30-Ω resistor, a 15-Ω inductive reactance, and a 12-Ω capacitive reactance are connected in parallel across a 120-V 60-Hz line. Find (*a*) the phasor branch currents, (*b*) total current and phase angle, (*c*) impedance, and (*d*) power drawn by the circuit; and (*e*) draw the current-phasor diagram.
Ans.   (*a*) $I_R = 4\,\text{A}$; $I_L = 8\,\text{A}$; $I_C = 10\,\text{A}$;   (*b*) $I_T = 4.47\,\text{A}$ (leading); $\theta = 26.6°$;   (*c*) $Z_T = 26.8\,\Omega$;   (*d*) $P = 480\,\text{W}$;   (*e*) phasor diagram: Fig. 15-30

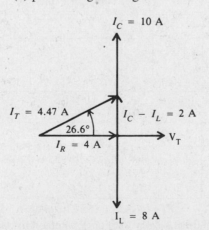

Fig. 15-30   Phasor diagram

15.24   A 100-Ω resistor, a 3-mH coil, and a 0.05-μF capacitor are in parallel across a 200-V 10-kHz ac source. Find (*a*) the reactance of the coil and capacitor, (*b*) phasor current drawn by each branch, (*c*) total current, (*d*) impedance and phase angle, and (*e*) power drawn by the circuit; and (*f*) draw the phasor diagram.

Ans.   (*a*) $X_L = 188\ \Omega$; $X_C = 318\ \Omega$;   (*b*) $I_R = 2$ A; $I_L = 1.06$ A; $I_C = 0.63$ A;   (*c*) $I_T = 2.05$ A (lagging);   (*d*) $Z_T = 97.6\ \Omega$; $\theta = -12.1°$;   (*e*) $P = 400$ W;   (*f*) phasor diagram: Fig. 15-31

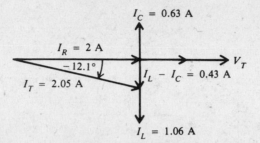

Fig. 15-31   Phasor diagram

15.25   With 420 mV applied, an ac circuit has the following parallel branches: $R_1 = 100\ \Omega$; $R_2 = 175\ \Omega$; $X_{L1} = 60\ \Omega$; $X_{L2} = 420\ \Omega$; $X_C = 70\ \Omega$. Calculate I_T, Z_T, and θ.

Ans. $I_T = 6.9$ mA (lagging); $Z_T = 60.9\ \Omega$; $\theta = -16.9°$

15.26   Repeat Problem 15.23 but substitute a 15-Ω capacitive reactance for the given 12-Ω capacitive reactance. Because $X_L = X_C = 15\ \Omega$, we have now a parallel resonant circuit. A parallel resonant circuit has a maximum impedance and a minimum current at the resonant frequency.

Ans.   (*a*) $I_R = 4$ A; $I_L = 8$ A; $I_C = 8$ A;   (*b*) $I_T = 4$ A (compare with Problem 15.23); $\theta = 0°$; (*c*) $Z_T = 30\ \Omega$ (compare with Problem 15.23);   (*d*) $P = 480$ W;   (*e*) phasor diagram: Fig. 15-32

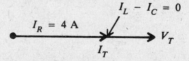

Fig. 15-32   Phasor diagram

15.27   Find the indicated values for an *RLC* parallel circuit.

	V, V	R, Ω	X_L, Ω	X_C, Ω	I_R	I_L	I_C	I_T	Z_T	θ	P, W	Circuit Type
(*a*)	110	27.5	22	55	?	?	?	?	?	?	?	?
(*b*)	90	45	40	30	?	?	?	?	?	?	?	?
(*c*)	90	45	40	40	?	?	?	?	?	?	?	?

Ans.

	V, V	R, Ω	X_L, Ω	X_C, Ω	I_R, A	I_L, A	I_C, A	I_T, A	Z_T, Ω	$\theta°$	P, W	Circuit Type
(*a*)	....	....	....	....	4	5	2	5	22	$-36.9°$	440	Inductive
(*b*)	....	....	....	....	2	2.25	3	2.14	42.1	$20.6°$	180	Capacitive
(*c*)	....	....	....	....	2	2.25	2.25	2	45	0	180	Resonant

15.28   Find I_T, θ, PF, and Z_T of an ac circuit with an *RL* branch in parallel to an *RC* branch (Fig. 15-33).

Ans. $I_T = 7.82$ A; $\theta = -33.5°$; PF = 0.834 lagging; $Z = 3.07\ \Omega$

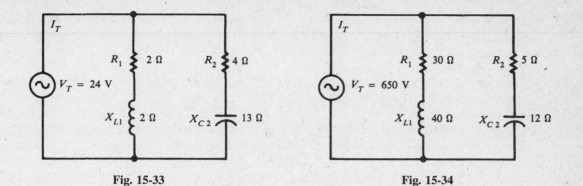

Fig. 15-33   **Fig. 15-34**

15.29   For the circuit shown, calculate I_T, θ, Z_T, PF, and P (Fig. 15-34).
  Ans. $I_T = 44.8$ A; $\theta = 53°$; $Z_T = 14.5$ Ω; PF $= 0.602$ leading; $P = 17\,530$ W

15.30   With 420 mV applied, an ac circuit has the five following parallel branches: $R_1 = 100$ Ω; $R_2 = 175$ Ω; $X_{L1} = 60$ Ω; $X_{L2} = 420$ Ω; and $X_C = 70$ Ω. Find I_T, θ, and Z_T.
  Ans. $I_T = 6.90$ mA (lagging); $\theta = -16.9°$; $Z_T = 60.9$ Ω

15.31   Find the power factor of a washing-machine motor if it draws 4 A and 420 W from a 110-V ac line.
  Ans.   PF $= 0.955$, or 95.5%

15.32   Find the PF of a refrigerator motor if it draws 300 W and 3.5 A from a 120-V ac line.
  Ans.   PF $= 71.4\%$

15.33   The lights and motors in a shop draw 20 kW of power. The PF of the entire load is 60 percent. Find the number of kilovoltamperes of power delivered to the load.   *Ans.* $S = 33.3$ kVA

15.34   A 50-V 60-Hz power supply is connected across an *RLC* series ac circuit with $R = 3$ Ω, $X_L = 6$ Ω, and $X_C = 2$ Ω. Find the apparent power, real power, reactive power, and power factor; and draw the power triangle.
  Ans. $S = 500$ VA; $P = 300$ W; $Q = 400$ VAR lagging; PF $= 0.6$; power triangle: Fig. 15-35

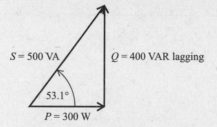

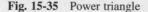

Fig. 15-35   Power triangle

15.35   A current of 8 A lags a voltage of 250 V by 30°. What is the power factor and the real power taken by the load?   *Ans.*   PF $= 0.866$ lagging; $P = 1732$ W

15.36   A motor operating at an 85 percent PF draws 300 W from a 120-V line. What is the current drawn?
  Ans. $I = 2.94$ A

15.37   A 220-V line delivers 15 kVA to a load at 80 percent PF lagging. Find the PF of a 12-kVA synchronous motor in parallel to raise the PF to 100 percent.
  Ans.   PF $= 66.1\%$ leading

15.38 A bank of motors draws 30 kW at 75 percent PF lagging from a 440-V 60-Hz line. Find the kilovolt-amperage and the capacitance of a capacitor placed across the motor terminals if it is to raise the total PF to 1.0.   *Ans.* $S = 26.5$ kVA; $C = 363$ μF

15.39 A motor draws 2 kW and 10 A from a 220-V 60-Hz line. Find the voltamperage and the capacitance of a capacitor in parallel that will raise the total PF to 1.   *Ans.* $S = 916$ VA; $C = 50$ μF

15.40 A 220-V 20-A induction motor draws 3 kW of power. A 4-kVA capacitive load is placed in parallel to adjust the PF to unity. What must be the PF of the capacitive load?   *Ans.*   PF $= 59.3\%$ leading

15.41 A plant load draws 2000 kVa from a 240-V line at a PF of 0.7 lagging. Find the kilovoltamperage required of a capacitor bank in parallel with the plant for the overall PF to be 0.9 leading.
Ans.   2107 kVA

15.42 A series RL branch in an ac circuit has $R = 8$ Ω and $X_L = 10$ Ω. A capacitor is connected in parallel across the branch. What should be the reactance of the capacitor if the unit is to have PF of unity?
Ans. $X_C = 16.4$ Ω

15.43 Find I_T, θ, Z_T, PF, and P for the circuit shown (Fig. 15-36).
Ans. $I_T = 4.49$ A; $\theta = -20.9°$; $Z_T = 22.3$ Ω; PF $= 0.934$ lagging; $P = 419$ W

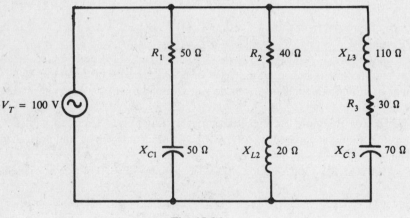

Fig. 15-36

Chapter 16

Alternating-Current Generators and Motors

ALTERNATORS

Alternating-current generators are also called *alternators*. Almost all electric power for homes and industry is supplied by alternators in power plants. A simple alternator consists of (1) a strong, constant magnetic field; (2) conductors that rotate across the magnetic field; and (3) some means of making a continuous connection to the conductors as they are rotating (Fig. 16-1). The magnetic field is produced by current flowing through the stationary, or stator, field coil. Field-coil excitation is supplied by a battery or other dc source. The armature, or rotor, rotates within the magnetic field. For a single turn of wire around the rotor, each is connected to a separate slip ring, which is insulated from the shaft. Each time the rotor turn makes one complete revolution, one complete cycle of alternating current is developed (Fig. 16-2). A practical alternator has several hundred turns wound into the slots of the rotor. Two brushes are spring-held against the slip rings to provide the continuous connection between the alternating current induced in the rotor or armature coil and the external circuits.

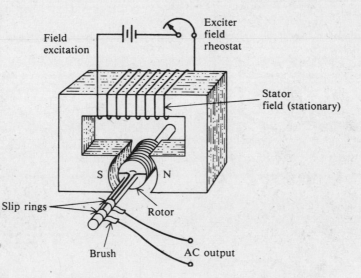

Fig. 16-1 A simple alternator having a stationary field and a rotating armature

The small ac generator usually has a stationary field and a rotating armature (Fig. 16-1). One disadvantage is that the slip-ring and brush contacts are in series with the load. If these parts become worn or dirty, current flow may be interrupted. However, if the dc field excitation is connected to the rotor, the formerly stationary coils will have alternating current induced into them (Fig. 16-3). A load can be connected across these armature coils without the necessity of any moving contacts in the circuit. Field excitation is fed to the rotating field through the slip rings and brushes. Another advantage of this rotating-field and stationary-armature generator is the greater ease of insulating stator fields compared with insulating rotating field coils. Since voltages as high as 18 000–24 000 V are commonly generated, this high voltage need not be brought out through slip rings and brushes but can be brought directly to the switch gear through insulated leads from the stationary armature.

The amount of generated voltage of an ac generator depends on the field strength and speed of the rotor. Since most generators are operated at constant speed, the amount of emf depends on field excitation.

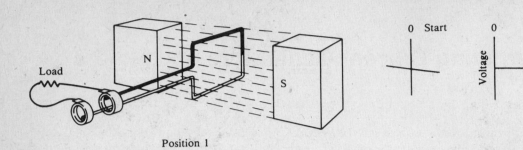

Position 1

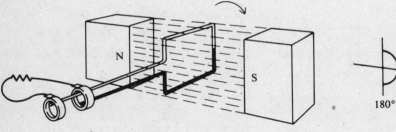

Position 2

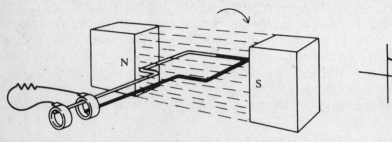

Position 3

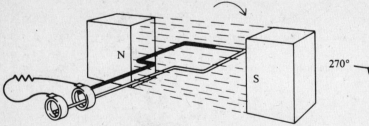

Position 4

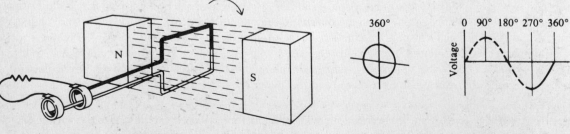

Position 5

Fig. 16-2 Generating 1 cycle of ac voltage with a single-coil alternator

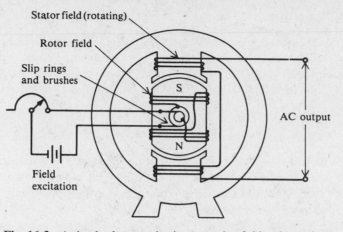

Fig. 16-3 A simple alternator having a rotating field and a stationary armature

The frequency of the generated emf depends on the number of field poles and on the speed at which the
generator is operated, or

$$f = \frac{pn}{120} \tag{16-1}$$

where f = frequency of generated voltage, Hz
         p = total number of poles
         n = rotor speed, revolutions per minute (rpm)

Regulation of an ac generator is the percentage rise in terminal voltage as load is reduced from the rated
full-load current to zero, with the speed and excitation being constant, or

$$\text{Voltage regulation} = \frac{\text{no-load voltage} - \text{full-load voltage}}{\text{full-load voltage}} \tag{16-2}$$

Example 16.1   What is the frequency of a four-pole alternator operating at a speed of 1500 rpm?

$$f = \frac{pn}{120} \tag{16-1}$$

$$= \frac{4(1500)}{120} = 50 \, \text{Hz} \qquad Ans.$$

Example 16.2   An alternator is operating at 120 V with no load. A load is now applied to the generator. The voltage
output drops (the field current remains the same) to 110 V. What is the regulation?

$$\text{Voltage regulation} = \frac{\text{no-load voltage} - \text{full-load voltage}}{\text{full-load voltage}} \tag{16-2}$$

$$= \frac{120 - 110}{110} = \frac{10}{110} = 0.091 = 9.1\% \qquad Ans.$$

When the output voltage is not steady, there would be a constant flickering of electric lights, and TV sets would not operate
properly. Automatic voltage-regulator devices are used to make up for the drop in output voltage by increasing the field
current. Voltage regulation is usually a function external to the alternator.

PARALLELING GENERATORS

Most power plants have several ac generators operating in parallel in order to increase the power available. Before two generators may be paralleled, their terminal voltages must be equal, their voltages must be in phase, and their frequencies must be equal. When these conditions are met, the two generators are operating in *synchronism*. The operation of getting the generators into synchronism is called *synchronizing*.

RATINGS

Nameplate data for a typical ac generator (Fig. 16-4) include manufacturer's name, serial, and type number; speed (rpm), number of poles, frequency of output, number of phases, and maximum supply voltage; capacity rating in kilovoltamperes and kilowatts at a specified power factor and maximum output voltage; armature and field current per phase; and maximum temperature rise.

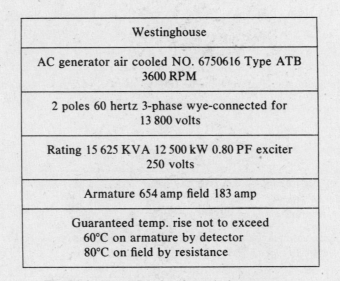

Westinghouse
AC generator air cooled NO. 6750616 Type ATB 3600 RPM
2 poles 60 hertz 3-phase wye-connected for 13 800 volts
Rating 15 625 KVA 12 500 kW 0.80 PF exciter 250 volts
Armature 654 amp field 183 amp
Guaranteed temp. rise not to exceed 60°C on armature by detector 80°C on field by resistance

Fig. 16-4   Nameplate data for typical ac generator

LOSSES AND EFFICIENCY

Losses of an ac generator are similar to those of a dc generator and include armature copper loss, field-excitation copper loss, and mechanical losses.

Efficiency (Eff) is the ratio of the useful power output to the total power input:

$$\text{Eff} = \frac{\text{output}}{\text{input}} \qquad (16\text{-}3)$$

Example 16.3   A 2-hp motor running at rated output acts as the prime mover for an alternator that has a load demand of 1.1 kW. What is the efficiency of the alternator in percent? Neglect field excitation.

$$\text{Input power} = 2\,\text{hp} \times \frac{746\,\text{W}}{\text{hp}} = 1492\,\text{W} \qquad \text{Output power} = 1.1\,\text{kW} = 1100\,\text{W}$$

$$\text{Eff} = \frac{\text{output}}{\text{input}} = \frac{1100}{1492} = 0.737 = 73.7\% \qquad Ans.$$

Since the prime mover is supplying 1492 W but the alternator is delivering 1100 W to the load, there must be 392 W of loss in the alternator.

POLYPHASE INDUCTION MOTORS

Principle of Operation

The induction motor is the most commonly used type of ac motor because of its simple, rugged construction and good operating characteristics. It consists of two parts: the stator (stationary part) and the rotor (rotating part). The stator is connected to the ac supply. The rotor is not connected electrically to the ac supply. The most important type of polyphase induction motor is the three-phase motor. (Three-phase machines have three windings and deliver an output between several pairs of wires.) When the stator winding is energized from a three-phase supply, a rotating magnetic field is created. As the field sweeps across the rotor conductors, an emf is induced in these conductors which causes current to flow in them. The rotor conductors carrying current in the stator field thus have a torque exerted upon them that spins the rotor.

Squirrel-Cage Motor and Wound-Rotor Motor

Three-phase induction motors are classified into two types: squirrel-cage (Fig. 16-5) and wound-rotor motors (Fig. 16-6). Both motors have the same stator construction, but differ in rotor construction. The stator core is built of slotted sheet-steel laminations. Windings are spaced in the stator slots to form the three separate sets of poles.

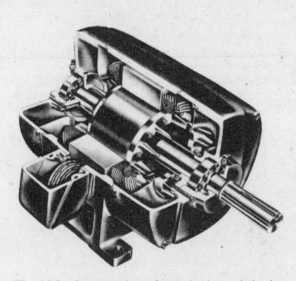

Fig. 16-5 Cutaway view of a squirrel-cage induction motor (*Courtesy of General Electric Company; from E. C. Lister, Electric Circuits and Machines, 5th ed., McGraw-Hill, New York, 1975, p. 247*)

The rotor of a squirrel-cage motor has a laminated core, with conductors placed parallel to the shaft and embedded in slots around the perimeter of the core. The rotor conductors are not insulated from the core. At each end of the rotor, the rotor conductors are all short-circuited by continuous end rings. If the laminations were not present, the rotor conductors and their end rings would resemble a revolving squirrel cage (Fig. 16-7).

The rotor of a wound-rotor motor is wound with an insulated winding similar to the stator winding. The rotor phase windings are brought out to the three slip rings mounted on the motor shaft (Fig. 16-6). The rotor winding is not connected to the supply. The slip rings and brushes merely provide a means of connecting an external rheostat into the rotor circuit. The purpose of the rheostat is to control the speed of the motor.

Fig. 16-6 Cutaway view of a wound-rotor induction motor (*Courtesy of General Electric Company; from Lister, p. 248*)

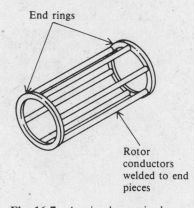

Fig. 16-7 A simple squirrel-cage rotor with rotor conductors welded to end rings on a shaft

Speed and Slip

The speed of the rotating magnetic field is called the *synchronous* speed of the motor.

$$n = \frac{120f}{p} \qquad (16\text{-}4)$$

where n = speed of rotating magnetic field, rpm
            f = frequency of rotor current, Hz
            p = total number of poles

It is noted that the same relation exists between the frequency, number of poles, and synchronous speed of a motor [Eq. (*16-4*)] as exists between the frequency, number of poles, and speed of rotation of an ac generator [Eq. (*16-1*)].

An induction motor cannot run at synchronous speed since then the rotor would be standing still with respect to the rotating field and no emf would be induced in the rotor. The rotor speed must be slightly less than synchronous speed in order that current be induced in the rotor to permit rotor rotation. The difference between rotor speed and synchronous speed is called *slip* and is expressed as a percent of synchronous speed.

$$\text{Percent } S = \frac{N_S - N_R}{N_S} 100 \qquad (16\text{-}5)$$

where $S = $ slip
       $N_S = $ synchronous speed, rpm
       $N_R = $ rotor speed, rpm

Example 16.4   A four-pole 60-Hz squirrel-cage motor has a full-load speed of 1754 rpm. What is the percent slip at full load?

$$\text{Synchronous speed } N_S = \frac{120 f}{p} \qquad (16\text{-}4)$$

$$= \frac{120(60)}{4} = 1800 \text{ rpm}$$

$$\text{Slip} = N_S - N_R = 1800 - 1754 = 46 \text{ rpm}$$

$$\text{Percent } S = \frac{N_S - N_R}{N_S} 100 \qquad (16\text{-}5)$$

$$= \frac{46}{1800} 100 = 2.6\% \qquad Ans.$$

Rotor Frequency

For any value of slip, the rotor frequency is equal to the stator frequency times the percent slip, or

$$f_R = S f_S \qquad (16\text{-}6)$$

where $f_R = $ rotor frequency, Hz
       $S = $ percent slip (written as a decimal)
       $f_S = $ stator frequency, Hz

Example 16.5   At a slip of 2.6 percent for the induction motor in Example 16.4, what is the rotor frequency?

$$f_S = 60 \text{ Hz} \qquad \text{given}$$

$$f_R = S f_S \qquad (16\text{-}6)$$

$$= 0.026(60) = 1.56 \text{ Hz} \qquad Ans.$$

Torque

The torque of an induction motor depends on the strength of the interacting rotor and stator fields and the phase relations between them.

$$T = k \phi I_R \cos \theta_R \qquad (16\text{-}7)$$

where $T = $ torque, lb · ft
        $k = $ constant
        $\phi = $ rotating stator flux, lines of flux
       $I_R = $ rotor current, A
   $\cos \theta_R = $ rotor power factor

Throughout the normal range of operation, k, ϕ, and $\cos\theta_R$ are nearly constant so that T is directly proportional to I_R. Rotor current I_R in turn increases in almost direct proportion to the motor slip. Variation of torque with slip (Fig. 16-8) shows that as slip increases from zero to about 10 percent, the torque linearly increases with the slip. As load and slip are increased beyond rated or full-load torque, the torque reaches a maximum value at about 25 percent slip. This maximum value of torque is called the *breakdown* torque of the motor. If the load is further increased beyond the breakdown point, the motor will quickly come to a stop. For typical squirrel-cage motors, the breakdown torque varies from 20 to 300 percent of full-load torque. The starting torque is the value at 100 percent slip (rotor speed is zero) and is normally 150–200 percent of full-load rating. As the rotor accelerates, the torque increases to its maximum value and then decreases to the value required to carry the load on the motor at a constant speed.

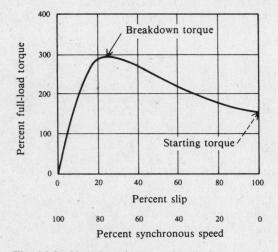

Fig. 16-8 Variation of torque with slip for a typical squirrel-cage motor

SYNCHRONOUS MOTORS

Like induction motors, synchronous motors have stator windings that produce a rotating magnetic field. But unlike the induction motor, the rotor circuit of a synchronous motor is excited by a dc source. The rotor locks into step with the rotating magnetic field and rotates at the same speed, as given by Eq. (*16-4*). If the rotor is pulled out of step with the rotating stator field, no torque is developed and the motor stops. Since a synchronous motor develops torque only when running at synchronous speed, it is not self-starting and hence needs some device to bring the rotor to synchronous speed.

Example 16.6 What is the slip of a synchronous motor?
Since the synchronous speed is equal to the rotor speed, $N_S = N_R$,

$$\text{Percent } S = \frac{N_S - N_R}{N_S}100 \qquad\qquad (16\text{-}5)$$

$$= \frac{0}{N_S}100 = 0\% \qquad Ans.$$

An ac electric clock uses a synchronous motor to maintain the correct time (as long as the frequency of the ac supply remains constant).

Starting Synchronous Motors

A synchronous motor may be started rotating with a dc motor on a common shaft. After the motor is brought to synchronous speed, alternating current is applied to the stator windings. The dc starting motor now acts as a dc generator, which supplies dc field excitation for the rotor. The load then can be coupled to the motor. More often synchronous motors are started by means of a squirrel-cage winding embedded in the face of the rotor poles. The motor is then started as an induction motor and is brought to about 95 percent of synchronous speed. At the proper time, direct current is applied and the motor pulls into synchronism. The amount of torque needed to pull the motor into synchronism is called the *pull-in torque*.

Effect of Loading Synchronous Motors

In the synchronous motor, the rotor is locked into step magnetically with the rotating magnetic field and must continue to rotate at synchronous speed for all loads. At no load the center lines of a pole of the rotating magnetic field and a dc field pole coincide (Fig. 16-9*a*). When load is added to the motor, there is a backward shift of the rotor pole relative to the stator pole (Fig. 16-9*b*). There is no change in speed. The angular displacement between the rotor and stator poles is called the torque angle α.

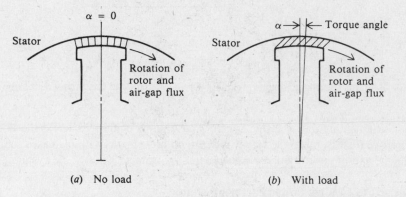

Fig. 16-9 Relative positions of a stator pole and a dc field pole

When a synchronous motor operates at no load (torque angle practically 0°), the counter emf V_g is equal to the applied or terminal voltage V_t (neglecting motor losses) (Fig. 16-10*a*). With increasing loads and torque angles, the phase position of V_g changes with respect to V_t which allows more stator current to flow to carry additional load (Fig. 16-10*b*). V_t and V_g are no longer in direct opposition. Their resultant voltage V_r causes a current I to flow in the stator windings. I lags V_r by nearly 90° because of the high inductance of the stator windings. θ is the phase angle between V_t and I. An increase in load results in a larger torque angle, which increases V_r and I (Fig. 16-10*c*).

If the mechanical load is too high, the rotor is pulled out of synchronism and comes to a stop. The maximum value of torque that a motor can develop without losing its synchronism is called its *pull-out torque*. If the synchronous motor has a squirrel-cage winding, it will continue to operate as an induction motor.

Ratings and Efficiency

Synchronous-motor nameplate data include the same items found on ac generator nameplates, with the horsepower rating replacing the kilovoltampere rating.

The efficiency of synchronous motors is generally higher than that of induction motors of the same horsepower and speed rating. The losses are the same as those of synchronous generators.

Synchronous motors are used for constant-speed power applications in sizes above 20 hp. A common application is driving air or gas compressors.

Power Factor Correction with Synchronous Motors

An outstanding advantage of a synchronous motor is that it operates at unity or leading power factor (PF). By varying the strength of the dc field, the overall power factor of a synchronous motor can be adjusted over a considerable range. Thus the motor can be made to appear as a leading load across the line. If an electrical

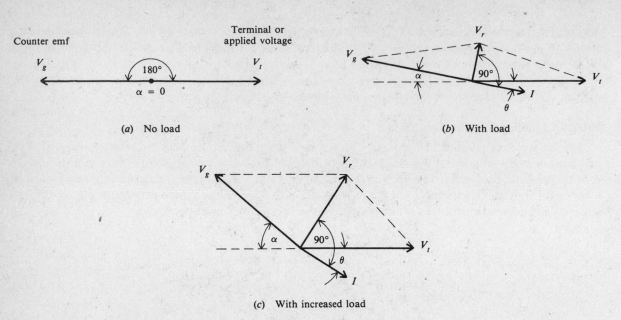

Fig. 16-10 Phasor diagrams for a synchronous motor for three different load conditions with same dc field excitation

system is operating at a lagging power factor, synchronous motors connected across the line and adjusted for leading PF can improve (that is, raise) the system PF. Any improvement in PF increases supply capacity to the load, raises efficiency, and in general improves the operating characteristics of the system.

Field Excitation Used to Change Power Factor of Motor

For a constant mechanical load, the PF of a synchronous motor may be varied from a leading value to a lagging value by adjusting its dc field excitation (Fig. 16-11). Field excitation is adjusted so that PF = 1 (Fig. 16-11a). At the same load, when the field excitation is increased, the counter emf V_g increases. This results in a change in phase between stator current I and terminal voltage V_t so that the motor operates at a leading PF (Fig. 16-11b). If the field excitation is reduced below the value represented (Fig. 16-11a), the motor operates at

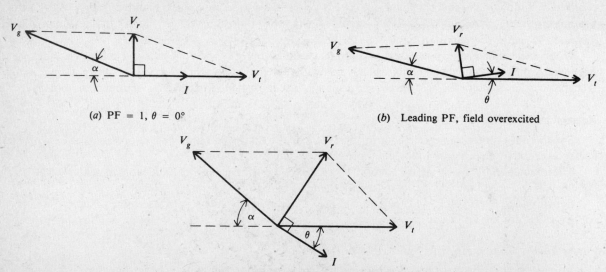

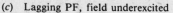

Fig. 16-11 Phasor diagrams for a synchronous motor with a constant load with different amounts of field excitation

a lagging PF (Fig. 16-11c). An example of a *V* curve for a synchronous motor, obtained from a manufacturer, shows how stator current varies at a constant load with rotor field excitation (Fig. 16-12). Power factor may also be read when the field current is varied.

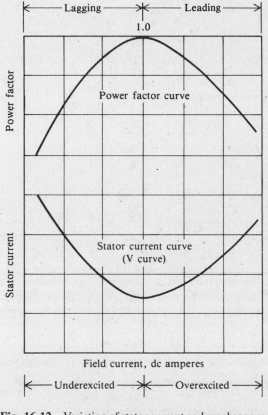

Fig. 16-12   Variation of stator current and synchronous motor PF with varying dc field excitation at a constant load

Example 16.7   The load of an industrial plant is 400 kVA at a PF of 75 percent lagging. What must be the PF of the added 100-kW load of a synchronous motor if it improved the overall plant PF to 100 percent?

For PF = 1, the net reactive power of the plant must be equal to zero.

Step 1.   Find the initial reactive power of the plant (Fig. 16-13a and b).

$$PF = \cos\theta = 0.75 \qquad \text{given}$$

$$\theta = \arccos 0.75 = 41.4° \qquad P = S\cos\theta = 400(0.75) = 300 \text{ kW}$$

$$Q = S\sin\theta = 400\sin 41.4° = 264.5 \text{ kVAR lagging}$$

Step 2.   Find the PF of the synchronous motor load (Fig. 16-13c). For a net PF = 1, the reactive power of the motor must equal the initial reactive power of the plant in the opposite direction. The Q of the plant (Step 1) is 264.5 kVAR lagging. So the Q_L of the added load must be 264.5 kVAR leading.

$$\theta_L = \arctan \frac{264.5}{100} = \arctan 2.64 = 69.3°$$

$$PF = \cos\theta_L = \cos 69.3° = 0.353 = 35.3\% \text{ leading} \qquad Ans.$$

The resultant power triangle (Fig. 16-13d) shows a plant load of 400 kW (300 kW + 100 kW) at a PF of unity.

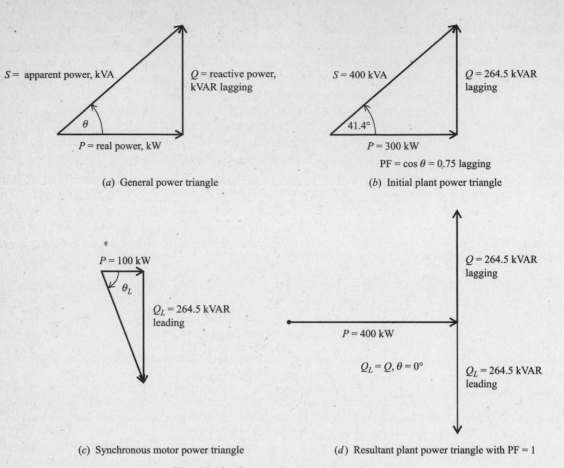

(a) General power triangle

(b) Initial plant power triangle

(c) Synchronous motor power triangle

(d) Resultant plant power triangle with PF = 1

Fig. 16-13   Power-triangle relationships

SINGLE-PHASE MOTORS

Single-phase motors are so called because their field windings are connected directly to a single-phase source. Single-phase motors are classified as commutator, induction, or synchronous motors according to the method used to start them, as follows:

1. Commutator motor

    (a) AC series motor

    (b) Repulsion motor

2. Induction motor

    (a) Split-phase motors

        (1) Capacitor-start motor

        (2) Capacitor motor

    (b) Repulsion-start induction motor

    (c) Shaded-pole motor

3. Synchronous motor

Commutator Motor

AC series motor.   When an ordinary dc series motor is connected to an ac supply, the current drawn by the motor is low due to the high series-field impedance. The result is low running torque. To reduce the field reactance to a minimum, ac series motors are built with as few turns as possible. Armature reaction is overcome by using compensating windings in the pole pieces.

The operating characteristics are similar to those of dc series motors. The speed increases to a high value with a decrease in load. The torque is high for high armature currents so that the motor has a good starting torque. AC series motors operate more efficiently at low frequencies. Some of the larger sizes used in railroad engines operate at 25 Hz or less. However, fractional horsepower sizes are designed to operate at 50 or 60 Hz.

Repulsion motor.   The repulsion motor has an armature and commutator similar to that of a dc motor. However, the brushes are not connected to the supply but are short-circuited (Fig. 16-14). The stator windings produce a current in the rotor windings by induction. This current produces magnetic poles in the rotor. The orientation of these poles is dependent on the position of the brushes. The interaction of the rotor field with the stator field creates the motor torque. The repulsion motor has a high starting torque and a high speed at light loads. It is used where heavy starting loads are expected.

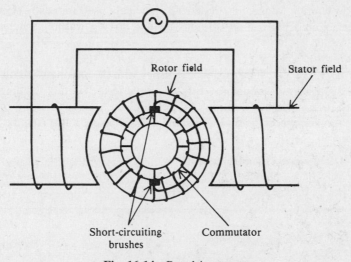

Fig. 16-14   Repulsion motor

Induction Motor

A single-phase induction motor is not self-starting. The magnetic field set up in the stator by the ac power supply stays lined up in one direction. This magnetic field, though stationary, pulsates as the voltage sine wave does. This pulsating field induces a voltage in the rotor windings, but the rotor field can only line up with the stator field. With these two fields in direct line, no torque is developed. It is necessary then to turn the rotor by some auxiliary device. Once the rotor is rotating with sufficient speed, the interaction between the rotor and stator fields will maintain rotation. The rotor will continue to increase in speed, trying to lock into synchronous speed. It finally will reach an equilibrium speed equal to the synchronous speed minus slip.

Split-phase motor.   If two stator windings of unequal impedance are spaced 90 electrical degrees apart but connected in parallel to a single-phase source, the field produced will appear to rotate. This is the principle of *phase splitting*.

In the split-phase motor, the starting winding has a higher resistance and lower reactance than the main winding (Fig. 16-15a). When the same voltage V_t is applied to both windings, the current in the main winding I_m lags behind the current in the starting winding I_s (Fig. 16-15b). The angle ϕ between the main and starting windings is enough phase difference to provide a weak rotating magnetic field to produce a starting torque.

When the motor reaches a predetermined speed, usually 70–80 percent of synchronous speed, a centrifugal switch mounted on the motor shaft opens, thereby disconnecting the starting winding.

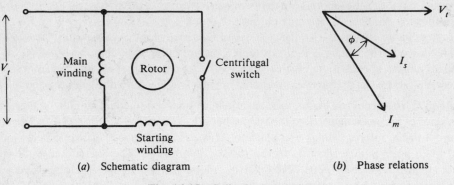

(a) Schematic diagram   (b) Phase relations

Fig. 16-15 Split-phase motor

Because it has a low starting torque, this motor type is widely used for easily started loads. It is seldom used in sizes larger than $\frac{1}{3}$ hp. Common applications include driving washing machines and woodworking tools.

Capacitor-Start Motor.   By placing a capacitor in series with the starting winding of a split-phase motor (Fig. 16-15a), starting characteristics are improved. The current in the starting winding may be made to lead the voltage (Fig. 16-16). ϕ may be made nearly 90°, resulting in a higher starting torque. This motor also uses a centrifugal switch to disconnect the starting winding. Thus the capacitor is in the circuit only during the starting period.

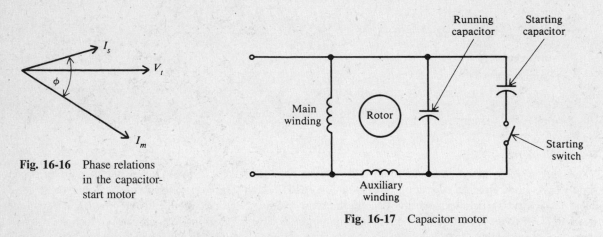

Fig. 16-16 Phase relations in the capacitor-start motor

Fig. 16-17 Capacitor motor

Capacitor Motor.   The capacitor motor operates with an auxiliary winding and series capacitor permanently connected to the line (Fig. 16-17). The capacitance in series may be of one value for starting and another value for running. As the motor approaches synchronous speed, the centrifugal switch disconnects one section of the capacitor.

Repulsion-start induction motor.   Like a dc motor, the rotor of the repulsion-start induction motor has windings connected to a commutator. Starting brushes make contact with the commutator so the motor starts as a repulsion motor. As the motor nears full speed, a centrifugal device short-circuits all the commutator segments so that it operates as an induction motor. This type of motor is made in sizes ranging from $\frac{1}{2}$ to 15 hp and is used in applications requiring a high starting torque.

Shaded-pole motor.   A shaded-pole is produced by a short-circuited coil wound around a part of each pole of a motor. The coil is usually a single band or strap of copper. The effect of the coil is to produce a small sweeping motion of the field flux from one side of the pole piece to the other as the field pulsates (Fig. 16-18). This slight shift in the magnetic field produces a small starting torque. Thus shaded-pole motors are self-starting. As the field in the pole piece increases, a current is induced in the shading coil. This current causes a magnetic field that opposes the main field. The main field therefore will concentrate on the opposite side of the pole pieces (Fig. 16-18*a*). As the field begins to decrease, the shading-coil field will aid the main field. The concentration of flux then moves to the other edge of the pole piece (Fig. 16-18*b*). This method of motor starting is used in very small motors, up to about $\frac{1}{25}$ hp, for driving small fans, small appliances, and clocks.

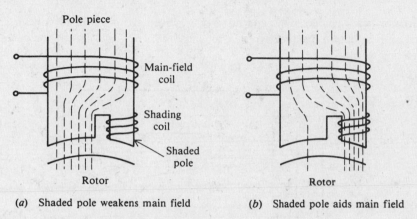

(*a*) Shaded pole weakens main field   (*b*) Shaded pole aids main field

Fig. 16-18   Action of the magnetic field in a shaded-pole motor

Synchronous Motor

Several types exist to drive electric clocks, phonograph turntables, and other devices requiring precise rotation. One type is called the Warren synchronous motor. It starts by the use of shading coils in the pole piece. The motor is brought up to synchronous speed from the effects of eddy currents flowing in the rotor iron and of hysteresis. Its principal use is in clocks and other timing devices.

Example 16.8   List the type of field excitation (dc or ac) and whether the field is usually the stator or the rotor for each of the following: alternator, polyphase induction, synchronous motor.

The list is shown in the table below.

| | Field Excitation | |
Device	Input Field	Output Field
Alternator		
Rotating armature	DC (stator)	Rotor (ac output)
Stationary armature	DC (rotor)	Stator (ac output)
Polyphase induction motor	AC (stator)	
Synchronous motor	DC (rotor)	
	AC (stator)	

Example 16.9   Fill in the appropriate word to complete each of the following sentences.

(*a*)   The magnetic field of a single-phase motor does not appear to _____ .

(*b*)   Repulsion motors have _____ starting torques.

(c) Induction motors are classified by different _____ methods.

(d) _____ must exist so that the stator and rotor fields are not exactly lined up.

(e) Split-phase motors have _____ separate windings.

(a) rotate (a three-phase field does appear to rotate),   (b) high,   (c) starting,   (d) Slip,   (e) two   *Ans.*

Solved Problems

16.1 An alternator has a characteristic curve showing percentage of terminal voltage and percentage of full-load ampere output for three types of loading (Fig. 16-19). Calculate the percent regulation for the three types of loading.

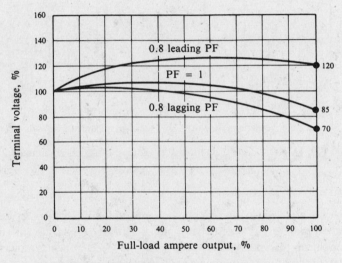

Fig. 16-19   Effect of power factor on alternator output

When 100 percent rated current is delivered by the alternator, the full-load (FL) voltages are 85, 70, and 120 percent of the no-load (NL) values for PF = 1, 0.8 lagging, and 0.8 leading, respectively (Fig. 16-19).

$$\text{Voltage regulation} = \frac{\text{NL} - \text{FL}}{\text{FL}} \qquad (16\text{-}2)$$

When PF = 1:

$$\text{Voltage regulation} = \frac{100 - 85}{85} = 0.176 = 17.6\% \qquad \textit{Ans.}$$

PF = 0.8 lagging:

$$\text{Voltage regulation} = \frac{100 - 70}{70} = 0.429 = 42.9\% \qquad \textit{Ans.}$$

PF = 0.8 leading:

$$\text{Voltage regulation} = \frac{100 - 120}{120} = -0.167 = -16.7\% \qquad \textit{Ans.}$$

The negative regulation indicates that the full-load voltage is greater than the no-load voltage.

16.2 Draw phasor diagrams of an ac generator operating when PF = 1.0, 0.8 lagging, and 0.8 leading.

Let IR and IX_L be voltage drops due to resistance and inductive reactance in the armature winding, respectively.

$$V_g = \text{generated emf} \qquad V_t = \text{terminal voltage} \qquad I = \text{armature current}$$

V_g is the phasor sum of V_t, the IR drop which is in phase with I, and the IX_L drop which leads I by 90° (Fig. 16-20). V_g is not constant but varies with the amount of load and PF of the load. At lagging PF, V_g is lowered. The lower the PF in the lagging direction, the less V_g. At leading PF, V_g increases with load.

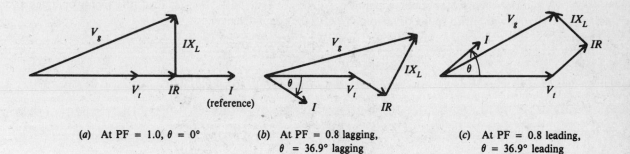

(a) At PF = 1.0, $\theta = 0°$ (b) At PF = 0.8 lagging,   (c) At PF = 0.8 leading,
                                            θ = 36.9° lagging θ = 36.9° leading

Fig. 16-20   Phasor diagrams of ac generator operating at three different load factors

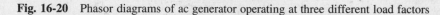

16.3 A diesel-driven 60-Hz synchronous generator produces 60 Hz when operated at 200 rpm. How many poles does it have?

$$f = \frac{pn}{120} \tag{16-1}$$

from which
$$p = \frac{120f}{n} = \frac{120(60)}{200} = 36 \text{ poles} \qquad Ans.$$

16.4 At what speed will a two-pole 25-Hz synchronous generator produce 25 Hz?

$$f = \frac{pn}{120} \tag{16-1}$$

from which
$$n = \frac{120f}{p} = \frac{120(25)}{2} = 1500 \text{ rpm} \qquad Ans.$$

16.5 An alternator has a voltage regulation of 10.0 percent. If the full-load voltage is 220 V, what is the no-load voltage?

$$\text{Voltage regulation} = \frac{NL - FL}{FL} \tag{16-2}$$

from which $NL = FL(\text{voltage regulation} + 1) = 220(0.10 + 1) = 242 \text{ V} \qquad Ans.$

16.6 A four-pole 60-Hz induction motor has a full-load slip of 5 percent. What is the full-load rotor speed?

$$N_S = \frac{120f}{p} \tag{16-4}$$

$$= \frac{120(60)}{4} = 1800 \text{ rpm}$$

$$S = \frac{N_S - N_R}{N_S} \tag{16-5}$$

$$SN_S = N_S - N_R$$

$$N_R = N_S - SN_S = N_S(1 - S)$$

Then $N_R = 1800(1 - 0.05) = 1800(0.95) = 1710 \text{ rpm}$ *Ans.*

16.7 A squirrel-cage motor stator winding is wound for four poles. At full load, the motor operates at 170 rpm with a slip speed of 60 rpm. What is the supply frequency?

$$\text{Slip} = N_S - N_R$$

$$N_S = N_R + \text{Slip} = 1740 + 60 = 1800 \text{ rpm}$$

$$N_S = \frac{120f}{p} \tag{16-4}$$

from which $f = \dfrac{pN_S}{120} = \dfrac{4(1800)}{120} = 60 \text{ Hz}$ *Ans.*

16.8 What is the rotor frequency of an eight-pole 60-Hz squirrel-cage motor operating at 850 rpm?

$$N_S = \frac{120f}{p} = \frac{120(60)}{8} = 900 \text{ rpm}$$

$$\text{Slip} = N_S - N_R = 900 - 850 = 50 \text{ rpm}$$

$$S = \frac{N_S - N_R}{N_S} \tag{16-5}$$

$$= \frac{50}{900} = 0.056$$

$$f_R = Sf_S \tag{16-6}$$

$$= 0.056(60) = 3.33 \text{ Hz}$$ *Ans.*

This means that a rotor conductor will have induced in it an emf with a frequency of 3.33 Hz.

16.9 How much larger is the rotor reactance of a squirrel-cage motor at start-up (with the rotor at a standstill) than it is when the motor operates at 4 percent slip?

$$\text{Rotor reactance } X_R = 2\pi f_R L_R$$

With L_R constant, $X_R \propto f_R$ so
rotor reactance is directly proportional to rotor frequency.

At start-up, the speed of the motor $N_R = 0$ so slip $= 1.00$. During motor operation slip $= 0.04$ (given), so

$$f_{R1} = S_1 f_S \tag{16-6}$$

$$f_{R2} = S_2 f_S$$

Dividing,

$$\frac{f_{R1}}{f_{R2}} = \frac{S_1}{S_2}$$

$$f_{R1} = \frac{S_1}{S_2} \times f_{R2} = \frac{1.00}{0.04} \times f_{R2} = 25 f_{R2}$$

Since $X_R \propto f_R$, the rotor reactance at start is 25 times greater than that at 4 percent slip.   *Ans.*

16.10 A motor-generator set used for frequency conversion has a 10-pole 25-Hz synchronous motor and a direct-connected 24-pole synchronous generator. What is the generator frequency?

Synchronous motor:

$$N_S = \frac{120 f}{p} = \frac{120(25)}{10} = 300 \text{ rpm}$$

Synchronous generator:

$$f = \frac{p N_S}{120} = \frac{24(300)}{120} = 60 \text{ Hz}   \text{Ans.}$$

16.11 The load of an industrial plant is 400 kVA at a PF of 75 percent lagging. An additional motor load of 100 kW is needed. Find the new kilovoltampere load and the PF of the load, if the motor to be added is (*a*) an induction motor with a PF of 90 percent lagging, and (*b*) a synchronous motor with a PF of 80 percent leading.

The solution is simplified by drawing and solving a series of power triangles.

Step 1. Set up a power triangle for current industrial load (IL) (Fig. 16-21*a*).

$$P_{\text{IL}} = 400 \cos \theta = 400(0.75) = 300 \text{ kW}$$

$$Q_{\text{IL}} = 400 \sin \theta = 400 \sin 41.4° = 264.5 \text{ kVAR lagging}$$

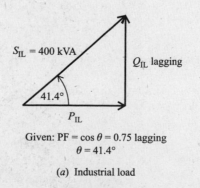

Given: PF = $\cos \theta$ = 0.75 lagging
θ = 41.4°

(*a*) Industrial load

Fig. 16-21*a*

Step 2. Add induction motor (IM) to industrial load (Fig. 16-21*b*).

$$Q_{IM} = 100 \tan 25.8° = 48.3 \text{ kVAR lagging}$$

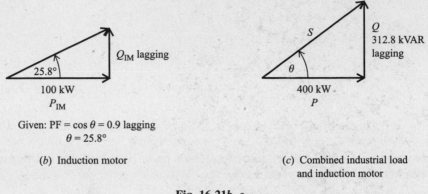

Given: PF = $\cos \theta$ = 0.9 lagging
$\theta = 25.8°$

(*b*) Induction motor

(*c*) Combined industrial load
and induction motor

Fig. 16-21*b, c*

The resultant power triangle is as shown in Fig. 16-21*c*.

$$P = P_{IL} + P_{IM} = 300 + 100 = 400 \text{ kW}$$

$$Q = Q_{IL} + Q_{IM} = 264.5 + 48.3 = 312.8 \text{ kVAR lagging}$$

$$\theta = \arctan \frac{Q}{P} = \arctan \frac{312.8}{400} = 38°$$

(*a*) $$PF = \cos \theta = \cos 38° = 0.788 = 78.8\% \text{ lagging}   \quad Ans.$$

$$S = \frac{P}{\cos \theta} = \frac{400}{\cos 38°} = 508 \text{ kVA (3 significant figures)}   \quad Ans.$$

Step 3. Add synchronous motor (SM) to industrial load (Fig. 16-21*d*).

$$Q_{SM} = 100 \tan 36.9° = 75.1 \text{ kVAR leading}$$

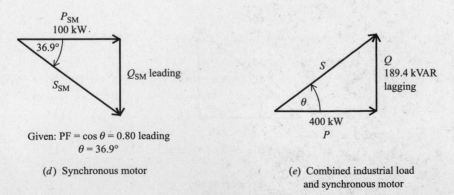

Given: PF = $\cos \theta$ = 0.80 leading
$\theta = 36.9°$

(*d*) Synchronous motor

(*e*) Combined industrial load
and synchronous motor

Fig. 16-21*d, e*

The resultant power triangle is as shown in Fig. 16-21e.

$$P = P_{IL} + P_{SM} = 300 + 100 = 400 \text{ kW}$$

$$Q = Q_{IL} - Q_{SM} = 264.5 - 75.1 = 189.4 \text{ kVAR lagging}$$

$$\theta = \arctan \frac{189.4}{400} = 25.3°$$

(b) $$PF = \cos \theta = \cos 25.3° = 0.904 = 90.4\% \text{ lagging}   \textit{Ans.}$$

$$S = \frac{400}{\cos 25.3°} = 442 \text{ kVA}   \textit{Ans.}$$

16.12   A 220-V 50-A induction motor draws 10 kW of power (Fig. 16-22a). An 8-kVA synchronous motor is placed in parallel with it in order to adjust the PF to unity. What must be the PF of the synchronous motor?

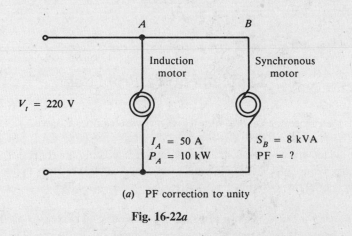

(a)   PF correction to unity

Fig. 16-22a

Step 1.   Set up a power triangle for the induction motor (Fig. 16-22b).

$$S_A = V_t I_A = 220(50) = 11\,000 \text{ VA} = 11 \text{ kVA}$$

$$\theta = \arccos \frac{P_A}{S_A} = \arccos \frac{10}{11} = 24.6°$$

$$Q_A = S_A \sin \theta = 11 \sin 24.6° = 4.58 \text{ kVAR lagging}$$

Step 2.   Set up a power triangle for the synchronous motor (Fig. 16-22c). For the load PF = 1, the net number of kilovoltamperes reactive must be 0. Therefore, the reactive power of the synchronous motor is

$$Q_B = 4.58 \text{ kVAR leading}   \theta = \arcsin \frac{Q_B}{S_B} = \arcsin \frac{4.58}{8} = 34.9°$$

$$PF = \cos \theta = \cos 34.9° = 0.820 = 82.0\% \text{ leading}   \textit{Ans.}$$

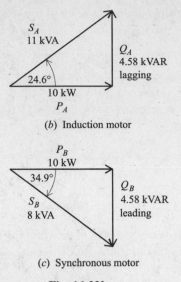

(b) Induction motor

(c) Synchronous motor

Fig. 16-22b, c

Supplementary Problems

16.13   A 60-Hz alternator operates at 900 rpm. How many poles does it have?   *Ans.*   Eight poles

16.14   How many cycles are generated in 1 revolution of a 12-pole ac generator? How many revolutions per second (rps) must it make to generate a frequency of 60 Hz? How many revolutions per minute? *Ans.*   6 cycles; 10 rps; 600 rpm

16.15   (*a*) At what speed must a six-pole synchronous generator be driven to produce 25 Hz? (*b*) At what speed must a four-pole 60-Hz synchronous generator be driven to produce 60 Hz? *Ans.*   (*a*) 500 rpm;   (*b*) 1800 rpm

16.16   To produce 50 Hz with a two-pole rotating-coil alternator, what must be the prime-mover rpm? *Ans.*   3000 rpm

16.17   Find the regulation of an ac generator that has a full-load voltage of 2600 V and a no-load voltage of 3310 V at a PF of 80 percent lagging. Will the percent regulation at a PF of unity be higher than, lower than, or the same as at a PF of 80 percent lagging?   *Ans.*   27.3%; lower than

16.18   When the load is light, is it more efficient to use one alternator operating at its rated capacity or to share the load between two alternators operating at less than their rated capacity? *Ans.*   It is more efficient to use one alternator at rated capacity.

16.19   An alternator has a voltage regulation of 10.0 percent. If the no-load voltage is 220 V, what is the full-load voltage?   *Ans.*   220 V

16.20   A fully loaded 10-hp electric motor is driving a 120-V ac output alternator delivering 6.5 kW to a remote lighting system. If the transmission-line losses are 300 W, what is the approximate loss in the alternator? What is the efficiency of the alternator?   *Ans.*   600 W; Eff = 91.2%

16.21   A 440-V alternator operating from a 30-hp prime mover turning at full capacity produces 20 kW into a load. Find the efficiency of the alternator.   *Ans.*   Eff = 89.4%

16.22   A 5-kW alternator is known to be 92 percent efficient when it is at full load. What is the power requirement in horsepower for the prime mover?   *Ans.*   7.3 hp

16.23   Find the synchronous speed of a 60-Hz motor that has an eight-pole stator winding.
Ans.   900 rpm

16.24   Make a table showing the synchronous speeds of 2-, 4-, 6-, 8-, and 12-pole induction motors for frequencies of 25, 50, and 60 Hz.

Ans.

	n, rpm		
p	$f = 25$ Hz	$f = 50$ Hz	$f = 60$ Hz
2	1500	3000	3600
4	750	1500	1800
6	500	1000	1200
8	375	750	900
12	250	500	600

16.25   A six-pole 60-Hz induction motor has a full-load slip of 4 percent. Find the rotor speed at full load.
Ans.   1152 rpm

16.26   What is the rotor frequency of a six-pole 60-Hz squirrel-cage motor operating at 1130 rpm?
Ans.   3.5 Hz

16.27   The three-phase induction motors driving an aircraft carrier have stators that may be connected to either 22 or 44 poles. The frequency of the supply may be varied from 20 to 65 Hz. What are the maximum and minimum speeds available from the motors?
Ans.   Maximum speed, 354.3 rpm; minimum speed, 54.5 rpm

16.28   How much greater is the rotor reactance of an induction motor at start-up than it is when operating at 5 percent slip?   *Ans.*   20 times greater

16.29   What is the speed and speed regulation of a 30-pole 60-Hz 440-V synchronous motor?
Ans.   240 rpm; 0.0%

16.30   The propulsion motors used in a naval vessel are rated 5900 hp, three-phase, 2400 V, 62.5 Hz, and 139 rpm. How many poles do they have? The speed of these motors may be changed by varying the supply frequency between 16 and 62.5 Hz. What are the maximum and minimum speeds?
Ans.   54 poles; maximum speed, 139 rpm; minimum speed, 35.6 rpm

16.31   A transmission line delivers a load of 7500 kVA at a PF of 70 percent lagging. If a synchronous condenser is to be located at the end of the line to improve the load power factor to 100 percent, how many kilovoltamperes must it draw from the line? Assume the condenser is 100 percent reactive.
Ans.   5360 kVA

16.32   A 440-V line delivers 15 kVA to a load at a PF of 75 percent lagging. To what PF should a 10-kVA synchronous motor be adjusted in order to raise the PF to unity when connected in parallel?
Ans.   PF = 12.6% leading

16.33   A 220-V 20-A induction motor draws 3 kW of power. A 4-kVA synchronous motor is placed in parallel to adjust the PF to unity. What must be the PF of the synchronous motor?
Ans.   PF = 59.3% leading

16.34   A 30-kW induction motor operates at a PF of 80 percent lagging. In parallel with it is a 50-kW induction motor operating at a PF of 90 percent lagging. (*a*) Find the new kilovoltampere load and PF of the load. (*b*) Find the PF adjustment which must be made on a 20-kW synchronous motor in parallel with the two induction motors in order to raise the PF of the line to unity.
Ans.   (*a*) 92.6 kVA, PF = 86.3% lagging;   (*b*) PF = 39.4% leading

16.35   A synchronous motor which has an input of 500 kW is added to a system which has an existing load of 800 kW at a PF of 80 percent lagging. What will be the new system kilowatt load, kilovoltampere load, and PF if the new motor is operated at a PF of (*a*) 85 percent lagging, (*b*) 100 percent, and (*c*) 85 percent leading?
Ans.   (*a*) 1300 kW, 1590 kVA, 81.9% lagging;   (*b*) 1300 kW, 1430 kW, 90.8% lagging;   (*c*) 1300 kW, 1320 kVA, 97.6% lagging

16.36   When is a synchronous motor said to be (*a*) overexcited, (*b*) underexcited?
Ans.   (*a*) Operates at leading PF (field excitation greater than that for PF = 1);   (*b*) Operates at lagging PF (field excitation less than that for PF = 1). (See Fig. 16-12.)

16.37   For a constant field excitation, what is the effect on a synchronous motor with a lagging PF if the load is increased?   *Ans.*   Phase angle increases in lagging direction so PF becomes less.

16.38   What is meant by (*a*) pull-in torque, and (*b*) pull-out torque for a synchronous motor?
Ans.   (*a*) Torque value to pull the motor into synchronous speed;   (*b*) Maximum torque developed without losing synchronous speed (stalling).

16.39   Why is a centrifugal switch used in a split-phase motor?
Ans.   Starting winding is designed only to help develop starting torque. Once the motor approaches running speed, it is no longer needed. Starting windings are usually wound with lighter-gauge wire, which could overheat and burn out if not disconnected.

16.40   How does a shaded-pole motor create a rotating magnetic field?
Ans.   Short-circuited coil on one edge of pole piece produces a field that first weakens and then aids the main field.

Chapter 17

Complex Numbers and Complex Impedance
for Series AC Circuits

INTRODUCTION

It is important to understand complex numbers because impedance, voltage, and current of ac circuits are best expressed in terms of complex numbers. Circuit calculations are simplified when using complex numbers. In previous chapters, ac circuits were analyzed without applying complex numbers.

DEFINITION OF A COMPLEX NUMBER

A *complex number* z has the form $x + jy$ where x and y are real numbers and j is the unit imaginary number. It is conventional to use a bold face letter symbol for all complex numbers. In complex number $x + jy$, the first term x is called the *real part* and the second term jy is called the *imaginary part*.

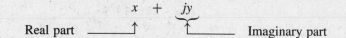

$$x \quad + \quad jy$$

Real part ⎯⎯⎯⎯ Imaginary part

Complex numbers can be represented by perpendicular axes, one axis representing the real part and the other axis the imaginary part (Fig. 17-1).

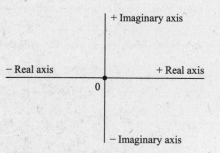

Fig. 17-1 Real and imaginary axes

Eight complex numbers, z_1, through z_8, are plotted in Fig. 17-2.

Note that for $z_1 = 4$, so $y = 0$ and hence z_1 is a real number 4 and corresponds to a point on the real axis. Note that for $z_4 = j6$, $x = 0$, so that z_4 is a pure imaginary number $j6$ and corresponds to a point on the j axis. Thus, complex numbers include all real and all pure imaginary numbers.

OPERATOR j

The unit imaginary number j is known as operator j. When operator j is multiplied by a real number a, $j \times a$ means a $90°$ change of a in a counterclockwise direction (Fig. 17-3a). When we multiply j twice, $j \times j \times a = j^2 \times a = -a$, the result is $180°$ counterclockwise change in direction shown in (b). When we multiply j three times, $j \times j \times j \times a = j^3 \times a = j(j^2) \times a = -j \times a$, the change in direction is $270°$ shown in (c). And when j is multiplied four times, $j \times j \times j \times j \times a = j^4 \times a = (j^2)(j^2) \times a = (-1)(-1) \times a = a$, we go back full circle as shown in (d).

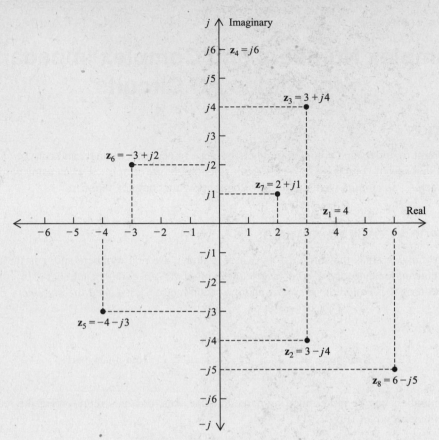

Fig. 17-2   Plots of complex numbers

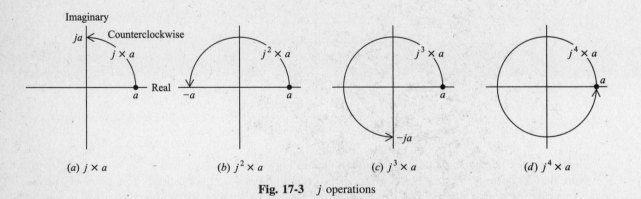

Fig. 17-3 j operations

Mathematically, j is defined as $j = \sqrt{-1}$

$$j^2 = j \times j = -1$$

$$j^3 = j^2 \times j = -j$$

$$j^4 = j^2 \times j^2 = (-1) \times (-1) = 1$$

$$j^5 = j^4 \times j = 1 \times j = j$$

and so on. [i is used outside electrical engineering to represent j.] Figure 17-4 illustrates the operator j principle when $a = 3$ for $j3$, -3, $-j3$, and 3.

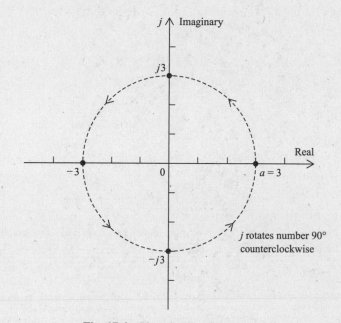

Fig. 17-4   Plot of points when $a = 3$

RECTANGULAR AND POLAR FORMS OF COMPLEX NUMBERS

Consider the complex number

$$\mathbf{z} = x + jy \qquad (17\text{-}1)$$

The graph of $\mathbf{z}$ is shown in Fig. 17-5. The quadrature (90-degree) components of $\mathbf{z}$ are given by the numbers x and y. Since y is multiplied by j, it lies on the imaginary axis. The form $x + jy$ is called the *rectangular form*. Another way to indicate a complex number is the *polar form* expressed as

$$\mathbf{z} = z \,\underline{/\theta} \qquad (17\text{-}2)$$

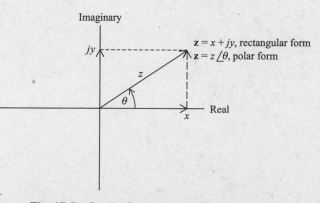

Fig. 17-5   Graph of complex number $\mathbf{z} = x + jy$

where z = magnitude of $\mathbf{z}$
        θ = direction of $\mathbf{z}$ with respect to the positive real axis

To convert from rectangular to polar form, refer to Fig. 17-5. From trigonometry

$$x = z \cos \theta$$

$$y = z \sin \theta$$

Substituting $\mathbf{z} = x + jy = z(\cos \theta + j \sin \theta)$

$$z = \sqrt{x^2 + y^2}$$

$$\theta = \arctan(y/x)$$

Substituting z and θ values into Eq. (17-2),

$$\mathbf{z} = \sqrt{x^2 + y^2} \; \underline{/\arctan(y/x)} \qquad\qquad (17\text{-}3)$$

Example 17.1   Convert the polar form $\mathbf{z} = 10\,\underline{/30°}$ into the rectangular form and show graph.
   Write the rectangular form:

$$\mathbf{z} = x + jy \qquad\qquad (17\text{-}1)$$

Find x:

$$x = z \cos \theta = 10 \cos 30° = 10(0.866) = 8.66$$

Find y:

$$y = z \sin \theta = 10 \sin 30° = 10(0.500) = 5$$

Therefore, $\mathbf{z} = 8.66 + j5$ *Ans.*

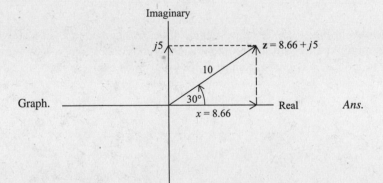

Graph.   *Ans.*

Example 17.2   Convert the rectangular form $\mathbf{z} = 8.66 + j5$ to the polar form.
   Write the polar form:

$$\mathbf{z} = z\,\underline{/\theta} \qquad\qquad (17\text{-}2)$$

Find z:

$$z = \sqrt{x^2 + y^2} = \sqrt{(8.66)^2 + (5)^2} = \sqrt{100} = 10$$

Find θ:

$$\theta = \arctan \frac{y}{x} = \arctan \frac{5}{8.66} = \arctan 0.577 = 30°$$

Therefore, $\mathbf{z} = 10 \underline{/30°}$ *Ans.*

Many scientific calculators have keys that can convert between rectangular and polar forms. Some have the ability to work with complex numbers without conversion. Refer to your calculator manual for the particular steps used. If your calculator does not have a conversion feature, the following formulas can be used:

polar-to-rectangular, $z \underline{/\theta} = z \cos \theta + jz \sin \theta$

rectangular-to-polar, $x + jy = \sqrt{x^2 + y^2} \underline{/\arctan(y/x)}$

OPERATIONS WITH COMPLEX NUMBERS

As with ordinary numbers, complex numbers can be added, subtracted, multiplied, and divided.

Addition

Complex numbers may be added when they appear in rectangular form. To add two or more complex numbers, add the reals, add the imaginaries, and then add the result. For example,

$$(2 + j4) + (3 - j1) = \underbrace{(2 + 3)}_{\substack{\text{sum} \\ \text{reals}}} + \underbrace{j(4 - 1)}_{\substack{\text{sum} \\ \text{imaginaries}}} = \underbrace{5 + j3}_{\substack{\text{sum} \\ \text{result}}}$$

Addition of complex numbers can be done graphically (Fig. 17-6).

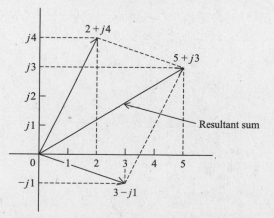

Fig. 17-6   Adding complex numbers graphically

Complete the parallelogram formed by sides $2 + j4$ and $3 - j1$. The diagonal $5 + j3$ is the resultant sum.
If complex numbers are given in the polar form, convert them first into rectangular form and then add.

Subtraction

To subtract one complex number from another, subtract the reals, subtract the imaginaries, and then add the result. For example,

$$(2 + j4) - (3 - j1) = \underbrace{(2 - 3)}_{\substack{\text{difference} \\ \text{reals}}} + \underbrace{j(4 + 1)}_{\substack{\text{difference} \\ \text{imaginaries}}} = \underbrace{-1 + j5}_{\substack{\text{add} \\ \text{result}}}$$

Subtraction of complex numbers also can be done graphically (Fig. 17-7).

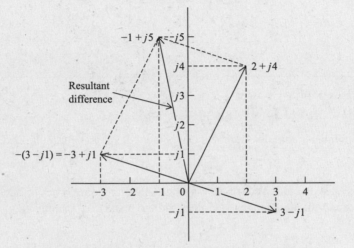

Fig. 17-7   Subtracting complex numbers graphically

Multiplication

Complex numbers may be multiplied in either the rectangular or polar form. Given two numbers in rectangular form, follow the rule of algebra for multiplying two terms. For example,

$$(2 + j4)(3 - j1) = 2(3) + (j4)(3) + 2(-j1) + (j4)(-j1)$$

$$= 6 + j12 - j2 + 4$$

$$= (6 + 4) + j(12 - 2)$$

$$= 10 + j10$$

When complex numbers are given in polar form, we have

$$(z_1 \underline{/\theta_1})(z_2 \underline{/\theta_2}) = z_1 z_2 \underline{/\theta_1 + \theta_2} \tag{17-4}$$

Let us verify the result of the previous example by using Eq. (17-4).

First, convert to polar form:

$$2 + j4 = 4.48 \underline{/63.4°}$$

$$3 - j1 = 3.16 \underline{/-18.4°}$$

Then multiply the magnitudes and add the angles algebraically:

$$\left(4.48 \underline{/63.4°}\right)\left(3.16 \underline{/-18.4°}\right) = (4.48)(3.16) \underline{/63.4° - 18.4°}$$
$$= 14.16 \underline{/45°}$$

Finally, convert product to rectangular form:

$$x = 14.16 \cos 45° = 14.16(0.707) = 10$$
$$y = 14.16 \sin 45° = 14.16(0.707) = 10$$

Therefore,

$$(2 + j4)(3 - j1) = 14.16 \underline{/45°} = 10 + j10 \qquad Check \quad Ans.$$

The Conjugate Complex Number

The *conjugate* of a complex number is obtained when the sign of the imaginary part of the number is changed. For example, $3 + j2$ has the conjugate $3 - j2$. These two complex numbers are referred to as *conjugate pair*. The conjugate of $\mathbf{z}$ is indicated by $\mathbf{z}^*$.

When we multiply $\mathbf{zz}^*$, we find

$$\mathbf{zz}^* = (3 + j2)\,(3 - j2) = 3^2 + j6 - j6 + 2^2$$
$$= 9 + 4 = 13, \text{a real number}$$

We shall use this product property in the division of complex numbers.

Division

Complex numbers may be divided in either the rectangular or polar form. For example,

$$\frac{8 - j4}{2 + j1} = ?$$

To eliminate the imaginary part from the denominator, we multiply both numerator and denominator by the conjugate of the denominator.

$$\frac{8 - j4}{2 + j1} \times \frac{2 - j1}{2 - j1} = \frac{16 - j8 - j8 + j^2 4}{4 + 1} = \frac{12 - j16}{5} = \frac{12}{5} - j\frac{16}{5}$$
$$= 2.4 - j3.2$$

This process of converting the denominator to a real number without any j term is called *rationalization*.

When complex numbers are given in polar form, we may perform division by using the formula,

$$\frac{z_1 \underline{/\theta_1}}{z_2 \underline{/\theta_2}} = \frac{z_1}{z_2} \underline{/\theta_1 - \theta_2} \tag{17-5}$$

Let us verify the result of the previous example by using Eq. (*17-5*).

First, convert to polar form:

$$8 - j4 = 8.94 \underline{/-26.5°}$$

$$2 + j1 = 2.24 \underline{/26.5°}$$

Then divide the magnitudes and subtract the angles algebraically:

$$\frac{8.94 \underline{/-26.5°}}{2.24 \underline{/26.5°}} = 3.99 \underline{/-53°}$$

Finally convert polar to rectangular form:

$$x = 3.99 \cos(-53°) = 3.99(0.602) = 2.4$$

$$y = 3.99 \sin(-53°) = 3.99(-0.799) = -3.2$$

Therefore,

$$\frac{8 - j4}{2 + j1} = 3.99 \underline{/-53°} = 2.4 - j3.2 \qquad \textit{Check} \quad \textit{Ans.}$$

COMPLEX IMPEDANCE IN SERIES

Consider the impedance **Z** of a circuit as a phasor quantity with magnitude and direction. A series *RL* circuit is shown in Fig. 17-8*a*.

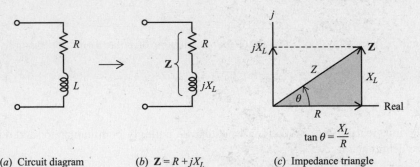

(*a*) Circuit diagram   (*b*) $\mathbf{Z} = R + jX_L$ (*c*) Impedance triangle

Fig. 17-8   Series *RL* circuit

The impedance of a series *RL* circuit is

$$\mathbf{Z} = R + jX_L \qquad (17\text{-}6)$$

where $\mathbf{Z}$ = complex impedance of the circuit, Ω
R = resistance of the circuit, Ω
X_L = inductive reactance of the circuit, Ω $(X_L = 2\pi fL)$

Equation (*17-6*) defines impedance **Z** as the phasor sum of the real quantity R and the imaginary quantity jX_L, as shown in Fig. 17-8*c*. Inductive reactance is indicated as X_L in (*b*).

In polar form, the impedance of a series RL circuit is

$$\mathbf{Z} = Z \underline{/\theta} \qquad (17\text{-}7)$$

where $Z = \sqrt{R^2 + X_L^2}$,   magnitude of the impedance

$\theta = \arctan \dfrac{X_L}{X_R}$,   phase angle with respect to the positive real axis

θ is the phase angle between the input voltage and its resulting current in the circuit. Although impedance $\mathbf{Z}$ does not vary sinusoidally, it is considered as a phasor because it determines the phase angle between voltage and current.

A series RC circuit is shown in Fig. 17-9a.

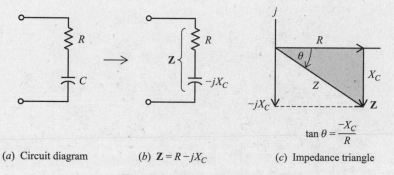

(a) Circuit diagram   (b) $\mathbf{Z} = R - jX_C$ (c) Impedance triangle

Fig. 17-9   Series RC circuit

The impedance of a series RC circuit is

$$\mathbf{Z} = R - jX_C \qquad (17\text{-}8)$$

where $\mathbf{Z}$ = impedance, Ω

R = resistance, Ω

X_C = capacitive reactance, Ω ($X_C = 1/2\pi f C$)

Equation ($17\text{-}8$) shows impedance as the phasor sum of R and $-jX_C$, as indicated in the impedance triangle of Fig. 17-9c. Capacitive reactance is shown as $-X_C$ in (b).

In polar form, the impedance of a series RC circuit is

$$\mathbf{Z} = Z \underline{/\theta} \qquad (17\text{-}7)$$

where $Z = \sqrt{R^2 + X_C^2}$,   magnitude of the impedance

$\theta = \arctan(-X_C/R)$,   phase angle with respect to the positive real axis

We may generalize for a circuit which contains R, L, and C in series. The impedance of a series RLC circuit is

$$\mathbf{Z} = R + jX \qquad (17\text{-}9)$$

where $\mathbf{Z}$ = impedance, Ω

R = resistance, Ω

$X = X_L - X_C$ = net reactance, Ω

When $X_L > X_C$, X is *positive* so that X is *inductive*; and when $X_C > X_L$, X is *negative* so that X is *capacitive*.

In polar form, the impedance of a series RLC circuit is

$$\mathbf{Z} = Z \underline{/\theta} \qquad (17\text{-}7)$$

$$Z = \sqrt{R^2 + X^2}, \quad \text{magnitude of the impedance, } \Omega$$

$$\theta = \arctan\left(\frac{X}{R}\right), \quad \text{phase angle with respect to the positive real axis}$$

A series RLC circuit is shown in Fig. 17-10a and impedance triangles if $X_L > X_C$ (inductive) shown in (c) or $X_C > X_L$ (capacitive) shown in (d).

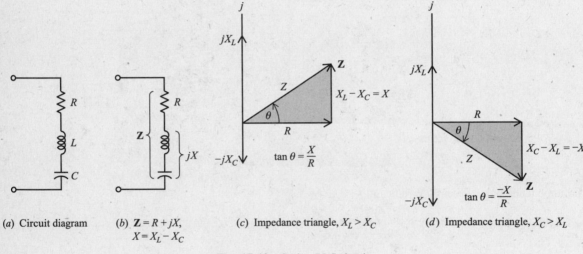

(a) Circuit diagram   (b) $\mathbf{Z} = R + jX$,   (c) Impedance triangle, $X_L > X_C$ (d) Impedance triangle, $X_C > X_L$
                              $X = X_L - X_C$

Fig. 17-10 Series RLC circuit

Example 17.3   For a series RL circuit with $R = 5\ \Omega$ and $X_L = 10\ \Omega$ (Fig. 17-11), find the complex impedance $\mathbf{Z}$ in rectangular and polar form. Draw the impedance triangle.

Label inductive reactance $j10$.

$$\mathbf{Z} = R + jX_L \qquad (17\text{-}6)$$

$$\mathbf{Z} = 5 + j10\ \Omega \qquad Ans.$$

To convert to polar form, write

$$\mathbf{Z} = Z \underline{/\theta} \qquad (17\text{-}7)$$

where

$$Z = \sqrt{R^2 + X_L^2} = \sqrt{(5)^2 + (10)^2} = \sqrt{125} = 11.2\ \Omega$$

$$\theta = \arctan\left(\frac{X_L}{R}\right) = \arctan\left(\frac{10}{5}\right) = \arctan 2$$

$$= 63.4°$$

Then

$$\mathbf{Z} = 11.2 \underline{/63.4°}\ \Omega \qquad Ans.$$

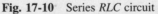

Fig. 17-11

Impedance triangle:

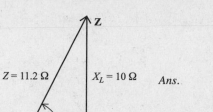

Impedance triangle: $Z = 11.2 \, \Omega$ $X_L = 10 \, \Omega$ *Ans.*

$63.4°$

$R = 5 \, \Omega$

Example 17.4 In a series RC circuit, $R = 15 \, \Omega$ and $X_C = 15 \, \Omega$ (Fig. 17-12). Find $\mathbf{Z}$ in rectangular and polar form. Draw the impedance triangle.

Label capacitive reactance $-j10$.

$$\mathbf{Z} = R - jX_C \qquad (17\text{-}8)$$

$$\mathbf{Z} = 15 - j15 \, \Omega \qquad Ans.$$

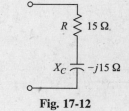

R $15 \, \Omega$

X_C $-j15 \, \Omega$

Fig. 17-12

To convert to polar form, write

$$\mathbf{Z} = Z \, \underline{/\theta} \qquad (17\text{-}7)$$

where $Z = \sqrt{R^2 + X_C^2} = \sqrt{(15)^2 + (15)^2} = \sqrt{450} = 21.2 \, \Omega$

$$\theta = \arctan\left(\frac{-X_C}{R}\right) = \arctan\left(\frac{-15}{15}\right) = \arctan(-1) = -45°$$

Then $\mathbf{Z} = 21.2 \, \underline{/-45°} \, \Omega \qquad Ans.$

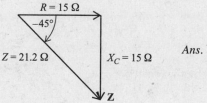

$R = 15 \, \Omega$

$-45°$

Impedance triangle: $Z = 21.2 \, \Omega$ $X_C = 15 \, \Omega$ *Ans.*

$\mathbf{Z}$

Example 17.5 Find the complex impedance $\mathbf{Z}$ in rectangular and polar form (Fig. 17-13), and show the impedance triangle.

Label X_L as $j8$ and X_C as $-j4$.

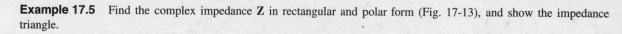

$$\mathbf{Z} = R + jX \qquad (17\text{-}9)$$

$$X = X_L - X_C = 8 - 4 = 4$$

so

$$\mathbf{Z} = 3 + j4 \qquad Ans.$$

R $3 \, \Omega$

$X_L > X_C$

X_L $j8 \, \Omega$

X_C $-j4 \, \Omega$

or directly from Fig. 17-13, write, $\mathbf{Z} = 3 + j8 - j4 = 3 + j4 \, \Omega \qquad Ans.$

$$Z = \sqrt{R^2 + X^2} = \sqrt{(3)^2 + (4)^2} = \sqrt{25} = 5 \, \Omega$$

Fig. 17-13

$$\theta = \arctan\left(\frac{X}{R}\right) = \arctan\frac{4}{3} = \arctan 1.33 = 53.1°$$

Then $\mathbf{Z} = 5 \, \underline{/53.1°} \, \Omega \qquad Ans.$

Impedance triangle:

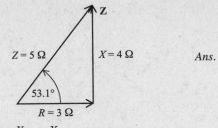

Ans.

Note: If $X_L = j4$ and $X_C = -j8$ where now $X_C > X_L$,

$$\mathbf{Z} = R + jX = 3 + j4 - j8 = 3 - j4$$

$$Z = \sqrt{R^2 + X^2} = \sqrt{(3)^2 + (-4)^2} = \sqrt{25} = 5\,\Omega$$

$$\theta = \arctan\left(\frac{-4}{3}\right) = \arctan -1.33 = -53.1°$$

so

$$\mathbf{Z} = 5 \underline{/-53.1°}$$

and the impedance triangle is

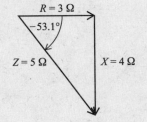

Solved Problems

17.1   Plot the following complex numbers: $\mathbf{z}_1 = -j3$, $\mathbf{z}_2 = 2 - j2$, $\mathbf{z}_3 = 1$, $\mathbf{z}_4 = -3 - j2$, $\mathbf{z}_5 = 2 + j3$, $\mathbf{z}_6 = -2 + j2$.

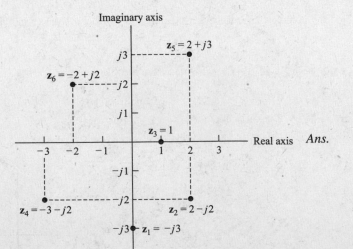

17.2   Convert from rectangular form $x + jy$ to polar form $z \underline{/\theta}$. Show the plot of $\mathbf{z}$.

    (a) $4 + j4$ (b) $4 + j3$ (c) $R + jX$ (d) $5 - j3$

(a)
$$z = \sqrt{x^2 + y^2} = \sqrt{(4)^2 + (4)^2} = \sqrt{32} = 5.67$$
$$\theta = \arctan \frac{y}{x} = \arctan \frac{4}{4} = \arctan 1 = 45°$$
$$\mathbf{z} = z \underline{/\theta} = 5.67 \underline{/45°} \qquad Ans.$$

Ans.

(b)
$$z = \sqrt{x^2 + y^2} = \sqrt{(4)^2 + (3)^2} = \sqrt{25} = 5$$
$$\theta = \arctan \frac{y}{x} = \arctan \frac{3}{4} = \arctan 0.75 = 36.9°$$
$$\mathbf{z} = z \underline{/\theta} = 5 \underline{/36.9°} \qquad Ans.$$

Ans.

(c)
$$z = \sqrt{R^2 + X^2}$$
$$\theta = \arctan (X/R)$$
$$\mathbf{z} = z \underline{/\theta} = \sqrt{R^2 + X^2} \underline{/\arctan(X/R)} \qquad Ans.$$

Ans.

(d)
$$z = \sqrt{x^2 + y^2} = \sqrt{(5)^2 + (-3)^2} = \sqrt{34} = 5.83$$
$$\theta = \arctan \frac{y}{x} = \arctan \frac{-3}{5} = \arctan -0.6 = -31°$$
$$\mathbf{z} = z \underline{/\theta} = 5.83 \underline{/-31°} \qquad Ans.$$

Ans.

17.3   Convert from polar to rectangular form. Show the plot of $\mathbf{z}$.

    (a) $100 \underline{/35°}$ (b) $20 \underline{/-30°}$ (c) $8 \underline{/45°}$ (d) $12 \underline{/240°}$

(a)
$$\mathbf{z} = x + jy$$
$$x = 100 \cos 35° = 100(0.819) = 81.9$$
$$y = 100 \sin 35° = 100(0.574) = 57.4$$
$$\therefore \quad \mathbf{z} = 81.9 + j57.4 \qquad Ans.$$

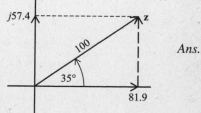

Ans.

(b)

$$\mathbf{z} = x + j4$$

$$x = 20\cos(-30°) = 20(0.866) = 17.3$$

$$y = 20\sin(-30°) = 20(-0.500) = -10.0$$

$$\therefore \quad \mathbf{z} = 17.3 - j10.0 \quad \text{Ans.}$$

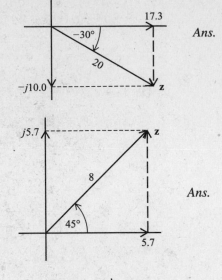

Ans.

(c)

$$\mathbf{z} = x + jy$$

$$x = 8\cos 45° = 8(0.707) = 5.7$$

$$y = 8\sin 45° = 8(0.707) = 5.7$$

$$\therefore \quad \mathbf{z} = 5.7 + j5.7 \quad \text{Ans.}$$

Ans.

(d)

$$\mathbf{z} = x + jy$$

$$x = 12\cos 240° = 12(-0.500) = -6.0$$

$$y = 12\sin 240° = 12(-0.866) = -10.4$$

$$\therefore \quad \mathbf{z} = -6.0 - j10.4 \quad \text{Ans.}$$

Ans.

17.4   Find the sum of complex numbers $5 + j6$ and $1 - j3$. Also find the sum graphically.

Add the reals and imaginaries.

$$(5 + j6) + (1 - j3) = (5 + 1) + j(6 - 3)$$

$$= 6 + j3 \quad \text{Ans.}$$

Plot the point $(5 + j6)$. Draw a straight line from the origin to that point. Follow the same procedure to draw $(1 - j3)$. The two lines are the sides of a parallelogram. Draw the dotted lines to complete the parallelogram. Its diagonal is the resultant sum $(6 + j3)$.

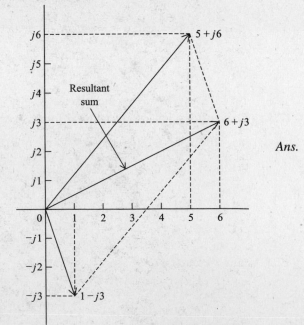

Ans.

17.5 What is the difference between $(5 + j6)$ and $(1 - j3)$? Also find the difference graphically.

Subtract the reals and imaginaries.

$$(5 + j6) - (1 - j3) = (5 - 1) + j(6 + 3) = 4 + j9 \qquad \textit{Ans.}$$

Plot points $(5 + j6)$ and $(1 - j3)$. Show $(1 - j3)$ in the opposite or 180° direction, which becomes $-(1 - j3) = -1 + j3$. Draw straight lines from the origin to points $(5 + j6)$ and $(-1 + j3)$. Draw dotted lines to complete the parallelogram. Its diagonal is the resultant difference $(4 + j9)$.

17.6 Find the product of complex numbers $3 + j5$ and $4 - j6$.

Use algebraic multiplication of two terms.

$$(3 + j5)(4 - j6) = 3(4) + 3(-j6) + j5(4) + j5(-j6)$$
$$= 12 - j18 + j20 - j^2 30$$
$$= (12 + 30) + j(-18 + 20)$$
$$= 42 + j2 \qquad \textit{Ans.}$$

17.7 Find the quotient of $6 + j2$ divided by $3 - j4$.

$$\frac{6 + j2}{3 - j4} = ?$$

To clear the denominator of the imaginary, multiply both numerator and denominator by the conjugate of the denominator.

$$\frac{6+j2}{3-j4} \times \frac{3+j4}{3+j4} = \frac{18+j24+j6+j^2 8}{9+16}$$

$$= \frac{18-8+j30}{25} = \frac{10+j30}{25}$$

$$= \frac{10}{25} + j\frac{30}{25}$$

$$= 0.4 + j1.2 \qquad Ans.$$

17.8   Multiply the polar form of complex numbers $10 \underline{/20°}$ and $7 \underline{/25°}$.

Use formula $(z_1 \underline{/\theta_1})(z_2 \underline{/\theta_2}) = z_1 z_2 \underline{/\theta_1 + \theta_2}$ (17-4)

Then, $(10 \underline{/20°})(7 \underline{/25°}) = (10)(7) \underline{/20° + 25°}$

$$= 70 \underline{/45°} \qquad Ans.$$

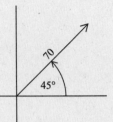

17.9   Divide $10 \underline{/20°}$ by $7 \underline{/25°}$.

Use formula $\dfrac{z_1 \underline{/\theta_1}}{z_2 \underline{/\theta_2}} = \dfrac{z_1}{z_2} \underline{/\theta_1 - \theta_2}$ (17-5)

Then, $\dfrac{10 \underline{/20°}}{7 \underline{/25°}} = \dfrac{10}{7} \underline{/20° - 25°} = 1.43 \underline{/-5°} \qquad Ans.$

17.10   Evaluate $\begin{vmatrix} 4-j3 & -j3 \\ -j4 & 5+j6 \end{vmatrix}$

    The value of this second-order determinant equals the product of the elements on the principal diagonal minus the product of the elements on the other diagonal, the same as for a determinant with real elements. [Refer to Eq. (8-1).]

$$\begin{vmatrix} 4-j3 & -j3 \\ -j4 & 5+j6 \end{vmatrix} = (4-j3)(5+j6) - (-j3)(-j4)$$

$$= (20 - j15 + j24 - j^2 18) - (j^2 12)$$

$$= 20 + j9 + 18 + 12$$

$$= 50 + j9 \qquad Ans.$$

17.11   Perform the following operations:

(a) $\frac{1}{j5}$ so that the denominator is a real number

(b) $(6+j2)(3-j5)(2-j3)$ in polar and rectangular form

(a)   Multiply numerator and denominator by $j1$:

$$\frac{1}{j5} \times \frac{j1}{j1} = \frac{j1}{5j^2} = \frac{j}{-5} = -0.2j \qquad Ans.$$

(b)   Convert each complex number to polar form and then multiply:

$$6 + j2 = 6.32 \,\underline{/18.4^\circ}$$
$$3 - j5 = 5.83 \,\underline{/-59.0^\circ}$$
$$(2 - j3) = 3.61 \,\underline{/-56.3^\circ}$$

$$(6 + j2)(3 - j5)(2 - j3) = \left(6.32 \,\underline{/18.4^\circ}\right)\left(5.83 \,\underline{/-59.0^\circ}\right)\left(3.61 \,\underline{/-56.3^\circ}\right)$$

$$= (6.32)(5.83)(3.61) \,\underline{/18.4^\circ - 59.0^\circ - 56.3^\circ}$$

$$= 133 \,\underline{/-96.9^\circ}, \text{polar form} \qquad Ans.$$

Convert polar form to rectangular form:

By calculator,

$$133 \,\underline{/-96.9^\circ} = -16 - j132 \qquad Ans.$$

Or by trigonometry,

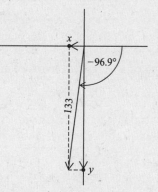

$$x = 133 \sin(-96.9^\circ) = 133(-0.12) = -16$$
$$y = 133 \cos(-96.9^\circ) = 133(-0.99) = -132$$

Then, $x + jy = -16 - j132$, rectangular form   Ans.

17.12   If $\mathbf{z} = z \,\underline{/\theta}$, show that its conjugate $\mathbf{z}^* = z \,\underline{/-\theta}$.

Write $\mathbf{z} = z \,\underline{/\theta} = x + jy$ where $z = \sqrt{x^2 + y^2}$ and $\theta = \arctan\left(\dfrac{y}{x}\right)$.

For conjugate $\mathbf{z}^* = x - jy$, by definition

Magnitude of $\mathbf{z}^* = z = \sqrt{x^2 + (-y)^2} = \sqrt{x^2 + y^2}$, same magnitude as $\mathbf{z}$

Angle of $\mathbf{z}^* = \arctan \dfrac{-y}{x} = -\theta$, negative angle of $\mathbf{z}$

$\therefore$ $\mathbf{z}^* = z \,\underline{/-\theta}$ $\quad$ *Ans.*

Plotting $\mathbf{z}$ and $\mathbf{z}^*$ makes this relationship clear. $\mathbf{z}$ and $\mathbf{z}^*$ are symmetrical with respect to the real axis.

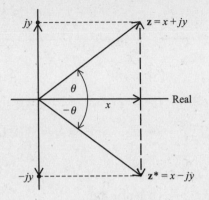

17.13 An interesting result is when the inductive reactance equals the capacitive reactance, $X_L = X_C$, in a series ac current.

$$\mathbf{Z} = R + jX \qquad (17\text{-}9)$$
$$= R + j(X_L - X_C)$$
$$= R + j0 \qquad (1)$$

The impedance of the circuit is equal to its resistance and thus has its lowest value. Such a circuit is called a *series resonant circuit*.

Find the impedance of a series *RLC* circuit when $R = 10\ \Omega$ and $X_L = X_C = 20\ \Omega$.

By Eq. (1), $\mathbf{Z} = R + j0 = 10 \,\underline{/0°}$ $\quad$ *Ans.*

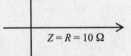

17.14 Prepare a summary table of complex impedance in series circuit with

(a) $R = 5\ \Omega$

(b) $X_L = 10\ \Omega$

(c) $X_C = 10\ \Omega$

(d) $R = 5\ \Omega,\ X_L = 15\ \Omega$

(e) $R = 15\ \Omega,\ X_C = 10\ \Omega$

(f) $R = 3\ \Omega,\ X_L = 8\ \Omega,\ X_C = 5\ \Omega$

For each part show the impedance schematic, rectangular form, polar form, and impedance triangle. It is instructive to look at the table and compare the different impedance expressions and triangles.

Table 17-1 Summary Table of Complex Impedance

Schematic	Rectangular Form $\mathbf{Z} = R + jX$	Polar Form $\mathbf{Z} = Z\,\underline{/\theta}$	Impedance Triangle
(a) R 5 Ω	$\mathbf{Z} = 5 + 0j\ \Omega$ Pure resistance	$\mathbf{Z} = 5\ \underline{/0°}\ \Omega$	
(b) X_L $j10$ Ω	$\mathbf{Z} = 0 + j10\ \Omega$ Pure inductive reactance	$\mathbf{Z} = 10\ \underline{/90°}\ \Omega$	
(c) X_C $-j10$ Ω	$\mathbf{Z} = 0 - j10\ \Omega$ Pure capacitive reactance	$\mathbf{Z} = 10\ \underline{/-90°}\ \Omega$	
(d) R 5 Ω, X_L $j15$ Ω	$\mathbf{Z} = 5 + j15\ \Omega$ RL series	$\mathbf{Z} = 15.8\ \underline{/71.6°}\ \Omega$	
(e) R 15 Ω, X_C $-j10$ Ω	$\mathbf{Z} = 15 + j10\ \Omega$ RC series	$\mathbf{Z} = 18.0\ \underline{/-33.7°}\ \Omega$	
(f) 3 Ω R, X_L $+j8$ Ω, X_C $-j5$ Ω	$\mathbf{Z} = 3 + j8 - j5$ $= 3 + j3\ \Omega$ RLC series	$\mathbf{Z} = 4.2\ \underline{/45°}\ \Omega$	

Supplementary Problems

17.15   Plot the following complex numbers.

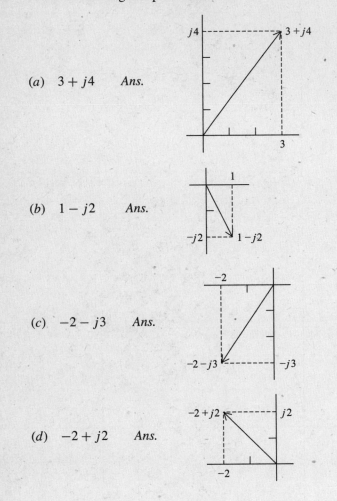

    (a) $3 + j4$ *Ans.*

    (b) $1 - j2$ *Ans.*

    (c) $-2 - j3$ *Ans.*

    (d) $-2 + j2$ *Ans.*

17.16   Write the conjugate pair to Problem 17.15 and plot it.

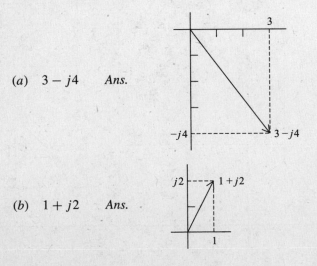

    (a) $3 - j4$ *Ans.*

    (b) $1 + j2$ *Ans.*

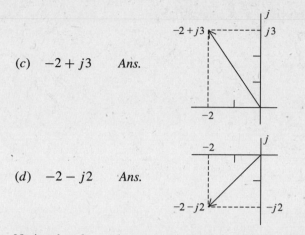

(c) $-2 + j3$ *Ans.*

(d) $-2 - j2$ *Ans.*

Notice that the conjugate is the reflection of the original number with respect to the real axis.

17.17   Plot the following complex numbers: $z_1 = 3 + j2$, $z_2 = 1 - j3$, $z_3 = 5$, $z_4 = -j2$, $z_5 = -1 + j2$, and $z_6 = -3 - j3$.

Ans.

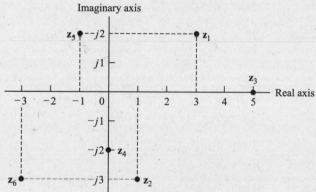

17.18   Evaluate and plot the following complex numbers: $z_1 = j5$, $z_2 = j^2 3$, $z_3 = j^4 2$, $z_4 = -j^2 4$, and $z_5 = -j^5 3$.

Ans.

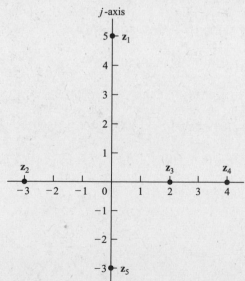

17.19   Convert each of the following complex numbers to polar form:

    (a) $40 + j30$ *Ans.* $50\underline{/36.9°}$

    (b) $4 - j4$ *Ans.* $5.66\underline{/-45°}$

    (c) $3 + j4$ *Ans.* $5\underline{/53.1°}$

    (d) $10 - j8$ *Ans.* $12.8\underline{/-38.7°}$

    (e) $20 + j5$ *Ans.* $20.6\underline{/14.0°}$

    (f) $-30 - j30$ *Ans.* $42.4\underline{/-135°}$

    (g) $5 - j15$ *Ans.* $15.8\underline{/-71.6°}$

    (h) $5 + j8$ *Ans.* $9.4\underline{/58.0°}$

    (i) $-10 + j20$ *Ans.* $22.4\underline{/116.6°}$

    (j) $R - jX$ *Ans.* $\sqrt{R^2 + X^2}\underline{/\arctan \frac{-X}{R}}$

17.20   Convert each of the following to rectangular form:

    (a) $15\underline{/30°}$ *Ans.* $13.0 + j7.5$

    (b) $15\underline{/-30°}$ *Ans.* $13.0 - j7.5$

    (c) $50\underline{/53.1°}$ *Ans.* $30 + j40$

    (d) $30\underline{/180°}$ *Ans.* $-30 + j0$

    (e) $100\underline{/-120°}$ *Ans.* $-50.0 - j86.6$

    (f) $50\underline{/90°}$ *Ans.* $j50$

    (g) $8\underline{/40°}$ *Ans.* $6.13 + j5.14$

    (h) $100\underline{/35°}$ *Ans.* $81.9 + j57.4$

    (i) $12\underline{/250°}$ *Ans.* $-4.1 - j11.3$

    (j) $Z\underline{/\theta°}$ *Ans.* $Z\cos\theta + jZ\sin\theta$

Perform the indicated operations:

17.21 $(4.1 + j1.2) + (3.6 - j0.8)$ *Ans.* $7.7 + j0.4$

17.22 $(50 - j50) + (100 + j50)$ *Ans.* 150

17.23 $(550 - j200) - (430 + j215)$ *Ans.* $120 - j415$

17.24 $(700 + j1000) - (-700 + j500)$ *Ans.* $1400 + j500$

17.25 $(3 + j5) + (12 - j3) - (6 + j10)$ *Ans.* $9 - j8$

17.26 $50\underline{/32°} + 20\underline{/18°}$ *Ans.* $61.4 + j32.7$

17.27 $45\underline{/45°} - 200\underline{/-35°}$ *Ans.* $-132 + j147$

17.28 $(-10 + j4)(6 + j2)$ *Ans.* $-68 + j4$

17.29 $(5 + j)(6 + j4)$ *Ans.* $26 + j26$

17.30 $(4 + j2)/(3 + j4)$ *Ans.* $0.8 - j0.4$

17.31 $(5 - j8)/(4 + j4)$ *Ans.* $-0.375 - j1.625$

17.32 $6\ \underline{/30°} \times 2\ \underline{/22°}$ *Ans.* $12\ \underline{/52°}$

17.33 $1.6\ \underline{/62°} \times 3.4\ \underline{/-30°}$ *Ans.* $5.44\ \underline{/32°}$

17.34 $324\ \underline{/40°}/10\ \underline{/20°}$ *Ans.* $32.4\ \underline{/20°}$

17.35 $25\ \underline{/15°}/2\ \underline{/-15°}$ *Ans.* $12.5\ \underline{/30°}$

17.36   If $\mathbf{Z} = 3 + j2$, find $\mathbf{Z}^*$, $\mathbf{ZZ}^*$, $(\mathbf{Z} + \mathbf{Z}^*)$, and $(\mathbf{Z} - \mathbf{Z}^*)$.

    Ans. $\mathbf{Z}^* = 3 - j2$, complex number (conjugate)

           $\mathbf{ZZ}^* = 13$, real number

      $\mathbf{Z} + \mathbf{Z}^* = 6$, real number

      $\mathbf{Z} - \mathbf{Z}^* = j4$, imaginary number

17.37   Find graphically:

    (*a*) $(3 + j2) + (2 - j4)$

    (*b*) $(3 + j2) - (2 - j4)$

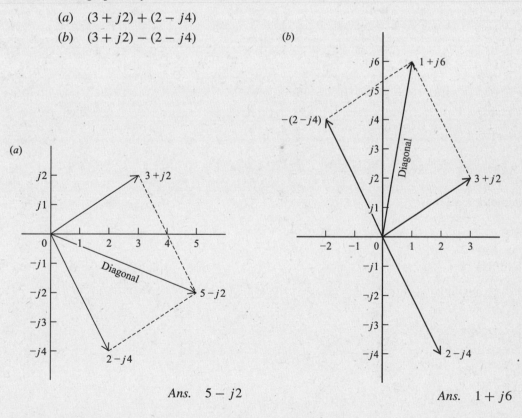

          Ans. $5 - j2$ *Ans.* $1 + j6$

Find the complex impedance $\mathbf{Z}$ of the following ac series circuits in rectangular and polar forms.

17.38 $R = 10\ \Omega$ *Ans.* $\mathbf{Z} = 10 + j5 = 11.2\ \underline{/26.6°}\ \Omega$
       $X_L = 5\ \Omega$

17.39 $R = 12\ \Omega$ *Ans.* $\mathbf{Z} = 12 + j5 = 13\underline{/22.6°}\ \Omega$
$X_L = 5\ \Omega$

17.40 $R = 100\ \Omega$ *Ans.* $\mathbf{Z} = 100 - j250 = 269\underline{/-68.2°}\ \Omega$
$X_C = 250\ \Omega$

17.41 $R = 50\ \Omega$ *Ans.* $\mathbf{Z} = 50 - j70 = 86\underline{/-54.5°}\ \Omega$
$X_C = 70\ \Omega$

17.42 $R = 10\ \Omega$ *Ans.* $\mathbf{Z} = 10 + j4 = 10.8\underline{/21.8°}\ \Omega$
$X_L = 12\ \Omega$
$X_C = 8\ \Omega$

17.43 $R = 10\ \Omega$ *Ans.* $\mathbf{Z} = 10 + j0 = 10\underline{/0°}\ \Omega$
$X_L = 40\ \Omega$
$X_C = 40\ \Omega$

17.44 $R = 20\ \Omega$ *Ans.* $\mathbf{Z} = 20 - j5 = 20.6\underline{/-14.0°}\ \Omega$
$X_L = 7\ \Omega$
$X_C = 12\ \Omega$

17.45 $R = 20\ \Omega$ *Ans.* $\mathbf{Z} = 20 + j0 = 20\underline{/0°}\ \Omega$
$X_L = 7\ \Omega$
$X_C = 7\ \Omega$

17.46 $R = 12.6\ \Omega$ *Ans.* $\mathbf{Z} = 12.6 + j9.2 = 15.6\underline{/36.1°}\ \Omega$
$X_L = 15.4\ \Omega$
$X_C = 6.2\ \Omega$

17.47 $R = 76.5\ \Omega$ *Ans.* $\mathbf{Z} = 76.5 - j33.4 = 83.5\underline{/-23.6°}\ \Omega$
$X_L = 13.2\ \Omega$
$X_C = 46.6\ \Omega$

17.48   Evaluate

(a) $\begin{vmatrix} 3 - j4 & 2 + j4 \\ 2 - j4 & 3 + j4 \end{vmatrix}$ *Ans.* 5

Note that by multiplying two sets of conjugate pairs and then subtracting them results in a real number.

(b) $\begin{vmatrix} 1 - j2 & 3 + j4 \\ 5 - j1 & 6 \end{vmatrix}$ *Ans.* $-13 - j29$

Chapter 18

AC Circuit Analysis with Complex Numbers

PHASORS

A *phasor* is a complex number associated with a phase-shifted alternating voltage or current. If the phasor is in polar form, the *magnitude* is the effective (rms) value of the voltage or current and its angle is the *phase angle* of its phase-shifted alternating voltage or current. (See section on phasors, Chapter 12, Principles of Alternating Current.)

TWO-TERMINAL NETWORK

For a two-terminal circuit with an input phasor voltage **V** and an input phasor current **I** (Fig. 18-1), the impedance **Z** of the circuit is defined as the ratio of **V** to **I**.

$$\mathbf{Z} = \frac{\mathbf{V}}{\mathbf{I}} \qquad (18\text{-}1)$$

Then

$$\mathbf{I} = \frac{\mathbf{V}}{\mathbf{Z}} \qquad (18\text{-}2)$$

and

$$\mathbf{V} = \mathbf{I}\mathbf{Z} \qquad (18\text{-}3)$$

Bold letters are used to show phasor quantities.

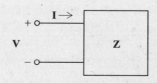

Fig. 18-1   Two-terminal network

Equation (*18-3*) is sometimes called "Ohm's law for alternating current." Voltage, current, and impedance quantities are complex numbers.

SERIES AC CIRCUIT

Figure 18-2 shows a series circuit with one voltage source **V** and three impedances, $\mathbf{Z}_1$, $\mathbf{Z}_2$, and $\mathbf{Z}_3$.

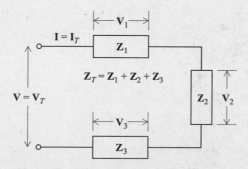

Fig. 18-2   General form of dc series circuit

$\mathbf{I}$ causes a voltage drop across each impedance: $\mathbf{V}_1$ across $\mathbf{Z}_1$, $\mathbf{V}_2$ across $\mathbf{Z}_2$, and $\mathbf{V}_3$ across $\mathbf{Z}_3$.

Kirchhoff's voltage law states that the sum of the voltage rises is equal to the sum of the voltage drops around any closed path. Applying this law to the series circuit (Fig. 18-2), we have

$$\mathbf{V} = \mathbf{V}_1 + \mathbf{V}_2 + \mathbf{V}_3 \tag{18-4}$$

Also

$$\mathbf{V} = \mathbf{IZ} \tag{18-3}$$

Then

$$\mathbf{V}_1 = \mathbf{IZ}_1, \quad \mathbf{V}_2 = \mathbf{IZ}_2, \quad \text{and} \quad \mathbf{V}_3 = \mathbf{IZ}_3$$

So

$$\mathbf{V} = \mathbf{IZ}_1 + \mathbf{IZ}_2 + \mathbf{IZ}_3$$

$$= \mathbf{I}(\mathbf{Z}_1 + \mathbf{Z}_2 + \mathbf{Z}_3)$$

We can rewrite

$$\mathbf{V} = \mathbf{V}_T = \mathbf{I}_T \mathbf{Z}_T \tag{18-5}$$

where $\mathbf{V}_T$ is total voltage, $\mathbf{I}_T$ is total current, and $\mathbf{Z}_T$ is total impedance

where

$$\mathbf{Z}_T = \mathbf{Z}_1 + \mathbf{Z}_2 + \mathbf{Z}_3 \tag{18-6}$$

For multiple n impedances in series,

$$\mathbf{Z}_T = \mathbf{Z}_1 + \mathbf{Z}_2 + \mathbf{Z}_3 + \cdots + \mathbf{Z}_n \tag{18-7}$$

Example 18.1   In the series ac circuit (Fig. 18-3a), find $\mathbf{Z}_T$ and $\mathbf{I}_T$, and draw the phasor diagram. Also verify KVL by showing that the sum of the voltage drops is equal to the input voltage.

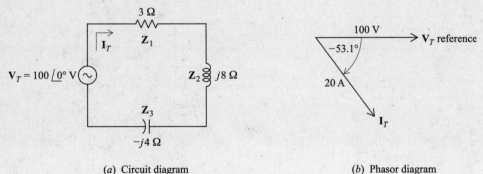

(a) Circuit diagram   (b) Phasor diagram

Fig. 18-3   *RLC* series circuit

Step 1.   Find $\mathbf{Z}_T$.

$$\mathbf{Z}_T = \mathbf{Z}_1 + \mathbf{Z}_2 + \mathbf{Z}_3$$

$$= 3 + j8 - j4$$

$$\mathbf{Z}_T = 3 + j4 = 5 \underline{/53.1°}\ \Omega \qquad Ans.$$

Step 2.   Find $\mathbf{I}_T$.

$$\mathbf{I}_T = \frac{\mathbf{V}_T}{\mathbf{Z}_T} \tag{18-2}$$

$$\mathbf{I}_T = \frac{100\ \underline{/0°}}{5\ \underline{/53.1°}} = 20\ \underline{/-53.1°}\ \text{A} \qquad Ans.$$

Step 3. Draw phasor diagram.
        See Fig. 18-3(*b*).

Step 4. Verify KVL.
        Write the impedances:

$$\mathbf{Z}_1 = 3 + 0j = 3 \;\underline{/0°}\; \Omega$$

$$\mathbf{Z}_2 = 0 + j8 = 8 \;\underline{/90°}\; \Omega$$

$$\mathbf{Z}_3 = 0 - j4 = 4 \;\underline{/-90°}\; \Omega$$

Find the individual voltage drops and then add them:

$$\mathbf{V}_1 = \mathbf{I}_T\mathbf{Z}_1 = (20\;\underline{/-53.1°})\,(3\;\underline{/0°}) = 60\;\underline{/-53.1°} = 36 - j48 \text{ V}$$

$$\mathbf{V}_2 = \mathbf{I}_T\mathbf{Z}_2 = (20\;\underline{/-53.1°})\,(8\;\underline{/90°}) = 160\;\underline{/36.9°} = 128 + j96 \text{ V}$$

$$\mathbf{V}_3 = \mathbf{I}_T\mathbf{Z}_3 = (20\;\underline{/-53.1°})\,(4\;\underline{/-90°}) = 80\;\underline{/-143.1°} = -64 - j48 \text{ V}$$

$$\mathbf{V}_T = \mathbf{V}_1 + \mathbf{V}_2 + \mathbf{V}_3 \qquad\qquad (18\text{-}4)$$

$$= (36 - j48) + (128 + j96) + (-64 - j48)$$

$$= (36 + 128 - 64) - j(48 - 96 + 48)$$

$$\mathbf{V}_T = 100 - j0 = 100\;\underline{/0°}\; \text{V}, \quad \text{which agrees with the given input voltage.} \qquad \textit{Ans.}$$

PARALLEL AC CIRCUIT

A single voltage source is applied to an ac parallel circuit with three impedances (Fig. 18-4). We may apply Kirchhoff's current law that the sum of the currents entering a junction, say at *A*, is equal to the sum of the currents leaving a junction, so that

$$\mathbf{I}_T = \mathbf{I}_1 + \mathbf{I}_2 + \mathbf{I}_3 \qquad\qquad (18\text{-}7)$$

where

$$\mathbf{I}_1 = \frac{\mathbf{V}_T}{\mathbf{Z}_1}, \quad \mathbf{I}_2 = \frac{\mathbf{V}_T}{\mathbf{Z}_2}, \quad \text{and} \quad \mathbf{I}_3 = \frac{\mathbf{V}_T}{\mathbf{Z}_3} \qquad\qquad (18\text{-}2)$$

Then substituting,

$$\mathbf{I}_T = \frac{\mathbf{V}_T}{\mathbf{Z}_1} + \frac{\mathbf{V}_T}{\mathbf{Z}_2} + \frac{\mathbf{V}_T}{\mathbf{Z}_3}$$

and factoring,

$$\mathbf{I}_T = \mathbf{V}_T \left(\frac{1}{\mathbf{Z}_1} + \frac{1}{\mathbf{Z}_2} + \frac{1}{\mathbf{Z}_3} \right) = \frac{\mathbf{V}_T}{\mathbf{Z}_T}$$

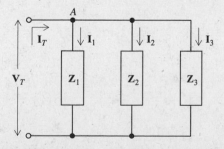

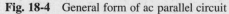

Fig. 18-4   General form of ac parallel circuit

where the total or equivalent impedance for *three* parallel impedances is

$$\frac{1}{\mathbf{Z}_T} = \frac{1}{\mathbf{Z}_1} + \frac{1}{\mathbf{Z}_2} + \frac{1}{\mathbf{Z}_3} \qquad (18\text{-}8)$$

which can also be written as

$$\mathbf{Z}_T = \frac{1}{\dfrac{1}{\mathbf{Z}_1} + \dfrac{1}{\mathbf{Z}_2} + \dfrac{1}{\mathbf{Z}_3}} \qquad (18\text{-}8a)$$

For *two* parallel impedances,

$$\frac{1}{\mathbf{Z}_T} = \frac{1}{\mathbf{Z}_1} + \frac{1}{\mathbf{Z}_2}$$

$$\mathbf{Z}_T = \frac{\mathbf{Z}_1 \mathbf{Z}_2}{\mathbf{Z}_1 + \mathbf{Z}_2} \qquad (18\text{-}9)$$

Example 18.2   In the parallel circuit (Fig. 18-5) find $\mathbf{I}_T$ and $\mathbf{Z}_T$. Also draw the phasor diagram.

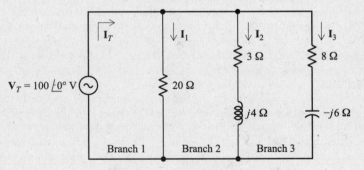

Fig. 18-5 *RLC* parallel circuit

Step 1.   Write the impedances of each branch.

Branch 1: $\mathbf{Z}_1 = 20 + j0 = 20\ \underline{/0^\circ}\ \Omega$

Branch 2: $\mathbf{Z}_2 = 3 + j4 = 5\ \underline{/53.1^\circ}\ \Omega$

Branch 3: $\mathbf{Z}_3 = 8 - j6 = 10\ \underline{/-36.9^\circ}\ \Omega$

Step 2.   Find the branch currents.

$$\mathbf{I}_1 = \frac{\mathbf{V}_T}{\mathbf{Z}_1} = \frac{100\ \underline{/0^\circ}}{20\ \underline{/0^\circ}} = 5\ \underline{/0^\circ} = 5 + j0\ \text{A}$$

$$\mathbf{I}_2 = \frac{\mathbf{V}_T}{\mathbf{Z}_2} = \frac{100\ \underline{/0^\circ}}{5\ \underline{/53.1^\circ}} = 20\ \underline{/-53.1^\circ} = 12 - j16\ \text{A}$$

$$\mathbf{I}_3 = \frac{\mathbf{V}_T}{\mathbf{Z}_3} = \frac{100\ \underline{/0^\circ}}{10\ \underline{/-36.9^\circ}} = 10\ \underline{/36.9^\circ} = 8 + j6\ \text{A}$$

Step 3.   Find $\mathbf{I}_T$.

$$\mathbf{I}_T = \mathbf{I}_1 + \mathbf{I}_2 + \mathbf{I}_3 \qquad (18\text{-}7)$$

Then substituting, $\mathbf{I}_T = (5 + j0) + (12 - j16) + (8 + j6)$

$$= (5 + 12 + 8) + j(0 - 16 + 6)$$

$$= 25 - j10 = 26.9 \,\underline{/-21.8°}\, \text{A} \qquad Ans.$$

Step 4.   Find $\mathbf{Z}_T$.

$$\mathbf{Z}_T = \frac{\mathbf{V}_T}{\mathbf{I}_T} \qquad\qquad\qquad\qquad (18\text{-}1)$$

$$\mathbf{Z}_T = \frac{100 \,\underline{/0°}}{26.9 \,\underline{/-21.8°}} = 3.72 \,\underline{/21.8°} = 3.45 + j1.38 \ \Omega \qquad Ans.$$

Step 5.   Draw the phasor diagram.

Ans.

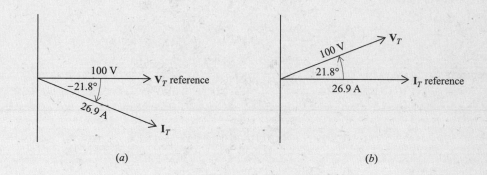

(a) (b)

It is convenient to use $\mathbf{V}_T$ as the reference line because its *given* phase angle is $0°$, as shown in (a). We also could show the phasor diagram with $\mathbf{I}_T$ as the reference line shown in (b). In both phasor diagrams, $\mathbf{I}_T$ lags $\mathbf{V}_T$ by $21.8°$.

Generally in drawing the phasor diagram for series circuits, we use *current* as the reference because current is the *same* in all parts of the circuit. In parallel circuits the current may be different in each part, but the *voltage* is the *same* for every branch. Thus, the reference line in parallel circuits is often chosen as the voltage.

Example 18.3   Find the input impedance at the terminals for two complex impedances in parallel (Fig. 18-6).

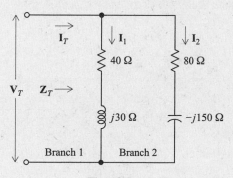

Fig. 18-6

Step 1.   Find the impedance of each branch.

Branch 1: $\mathbf{Z}_1 = 40 + j30 = 50 \,\underline{/36.9°}\ \Omega$

Branch 2: $\mathbf{Z}_2 = 80 - j150 = 170 \,\underline{/-61.9°}\ \Omega$

Step 2.   Find $\mathbf{Z}_T$.

$$\mathbf{Z}_T = \frac{\mathbf{Z}_1 \mathbf{Z}_2}{\mathbf{Z}_1 + \mathbf{Z}_2} \tag{18-9}$$

Substituting,

$$\mathbf{Z}_T = \frac{(50\ \underline{/36.9°})(170\ \underline{/-61.9°})}{(40 + j30) + (80 - j150)} = \frac{8500\ \underline{/-25.0°}}{120 - j120}$$

$$= \frac{8500\ \underline{/-25.0°}}{170\ \underline{/-45°}} = 50.0\ \underline{/20.0°} = 47.0 + j17.1\ \Omega \qquad Ans.$$

Another method for finding $\mathbf{Z}_T$ is to assume a convenient input voltage $\mathbf{V}_T$ at the terminals and solve for $\mathbf{I}_1$ and $\mathbf{I}_2$. Here is an example showing $\mathbf{V}_T$ cancelling out in the equation for $\mathbf{Z}_T$, thus allowing you to chose a convenient value for $\mathbf{V}_T$.

$$\mathbf{I}_1 = \frac{\mathbf{V}_T}{\mathbf{Z}_1} = \frac{\mathbf{V}_T}{50\ \underline{/36.9°}} = (0.02\ \underline{/-36.9°})\mathbf{V}_T = (0.016 - j0.012)\mathbf{V}_T$$

$$\mathbf{I}_2 = \frac{\mathbf{V}_T}{\mathbf{Z}_2} = \frac{\mathbf{V}_T}{170\ \underline{/-61.9°}} = (0.00588\ \underline{/61.9°})\mathbf{V}_T = (0.00277 + j0.00518)\mathbf{V}_T$$

$$\mathbf{I}_T = \mathbf{I}_1 + \mathbf{I}_2 = (0.01877 - j0.00682)\mathbf{V}_T = 0.0200\ \underline{/-20.0°}\ \mathbf{V}_T$$

Then

$$\mathbf{Z}_T = \frac{\mathbf{V}_T}{\mathbf{I}_T} = \frac{\mathbf{V}_T}{0.0200\ \underline{/-20.0°}\ \mathbf{V}_T} = 50.0\ \underline{/20.0°}\ \Omega \qquad Ans.$$

SERIES–PARALLEL AC CIRCUIT

We will illustrate the solution of a series–parallel circuit by presenting an example.

Example 18.4   In the series–parallel circuit (Fig. 18-7), find $\mathbf{Z}_T$ and $\mathbf{I}_T$.

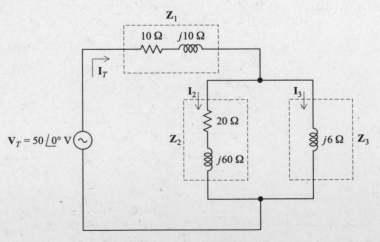

Fig. 18-7   Series–parallel circuit

Step 1.   Write the impedances.

$$\mathbf{Z}_1 = 10 + j10 = 14.1\ \underline{/45°}\ \Omega$$

$$\mathbf{Z}_2 = 20 + j60 = 63.2\ \underline{/71.6°}\ \Omega$$

$$\mathbf{Z}_3 = 0 + j6 = 6\ \underline{/90°}\ \Omega$$

Step 2.   Combine the impedances.
$\mathbf{Z}_2$ and $\mathbf{Z}_3$ are in parallel so their equivalent impedance is

$$(18 - 9) \qquad \mathbf{Z}_a = \frac{\mathbf{Z}_2\mathbf{Z}_3}{\mathbf{Z}_2 + \mathbf{Z}_3} = \frac{(63.2\ \underline{/71.6°})\,6\ \underline{/90°}}{(20 + j60) + j6} = \frac{379.2\ \underline{/161.6°}}{20 + j66}$$

$$\mathbf{Z}_a = \frac{379.2\ \underline{/161.6°}}{69.0\ \underline{/73.1°}} = 5.50\ \underline{/88.5°} = 0.144 + j5.50\ \Omega$$

$\mathbf{Z}_a$ is in series with $\mathbf{Z}_1$ so

$$\mathbf{Z}_T = \mathbf{Z}_1 + \mathbf{Z}_a = (10 + j10) + (0.144 + j5.50)$$

$$= 10.14 + j15.50 = 18.5\ \underline{/56.8°}\ \Omega \qquad Ans.$$

Step 3.   Find $\mathbf{I}_T$.

$$\mathbf{I}_T = \frac{\mathbf{V}_T}{\mathbf{Z}_T} = \frac{50\ \underline{/0°}}{18.5\ \underline{/56.8°}} = 2.7\ \underline{/-56.8°}\ A \qquad Ans.$$

The circuit is inductive with input current lagging input voltage by 56.8°.

Summary Table 18-1 for ac circuits shows the relationships between R, X_L, X_C, and $\mathbf{Z}$.

Table 18-1 Summary Table for AC Circuit Relationships

	Resistance R, Ω	Inductive reactance X_L, Ω	Capacitive reactance X_C, Ω	Impedance $\mathbf{Z}$, Ω
Definition	Opposition to ac due to resistance	Opposition to ac due to inductance	Opposition to ac due to capacitance	Opposition to ac due to combined resistance and reactance
Phase angle	I_R in phase with V_R	I_L lags V_L by 90°	I_C leads V_C by 90°	$Z = \sqrt{R^2 + X^2}$ $X = X_L - X_C$ $\tan\theta_Z = X/R$ in series $\tan\theta_I = \pm I_X/I_R$ in parallel
Single or combined impedance rectangular: polar:	$R + 0j$ $R\ \underline{/0°}$	$0 + jX_L$ $X_L\ \underline{/90°}$	$0 - jX_C$ $X_C\ \underline{/-90°}$	$R + jX$ $Z\ \underline{/\theta}$
Kirchhoff's law for voltage	$\mathbf{V} = \mathbf{I}R$	$\mathbf{V} = \mathbf{I}(jX_L)$ $= \mathbf{I}X_L\ \underline{/90°}$	$\mathbf{V} = \mathbf{I}(-jX_C)$ $= \mathbf{I}X_L\ \underline{/-90°}$	$\mathbf{V} = \mathbf{I}\mathbf{Z}$

COMPLEX POWER

Introduction

AC power has been discussed without use of complex numbers in Chapters 13–15. Power formulas from these chapters are repeated for quick reference.

$$\text{Real power } P = VI\cos\theta, \text{W} \qquad (13\text{-}20) \quad (14\text{-}21) \quad (15\text{-}15)$$

$$P = I^2 R, \text{W} \qquad (13\text{-}21) \quad (14\text{-}22) \quad (15\text{-}16)$$

$$\text{Reactive power } Q = VI\sin\theta, \text{VAR} \qquad (13\text{-}22) \quad (14\text{-}23) \quad (15\text{-}18)$$

$$\text{Apparent power } S = VI, \text{VA} \qquad (13\text{-}23) \quad (14\text{-}24) \quad (15\text{-}19)$$

$$\text{Power factor PF} = \cos\theta \qquad (15\text{-}20)$$

V is input voltage, I is input current, and θ is the phase angle between V and I.

One use for complex power is to obtain the total complex power of several loads in parallel energized by the same source. The total complex power is the sum of the individual complex powers regardless of how the loads are connected. Therefore, total real power is the sum of the individual real powers, and total reactive power is the sum of the individual reactive powers. The same is not true to obtain total apparent power. Another use for complex power is in power factor correction.

Complex Power Formulas

By definition, the complex power formula is

$$\mathbf{S} = P + jQ \qquad (18\text{-}10)$$

having components P and Q (Fig. 18-8)

where $\mathbf{S}$ = complex power in voltamperes, VA
$\quad\quad\quad$ S = magnitude of $\mathbf{S}$ = VI = apparent power in voltamperes, VA
$\quad\quad\quad$ P = real power in watts, W
$\quad\quad\quad$ Q = reactive power in voltamperes reactive, VAR; also referred to as *vars* for industrial applications.

We see from Fig. 18-8 that

$$\mathbf{S} = S\,\underline{/\theta} = VI\,\underline{/\theta} \qquad (18\text{-}11)$$

$$P = S\cos\theta = VI\cos\theta \qquad (18\text{-}12)$$

$$Q = S\sin\theta = VI\sin\theta \qquad (18\text{-}13)$$

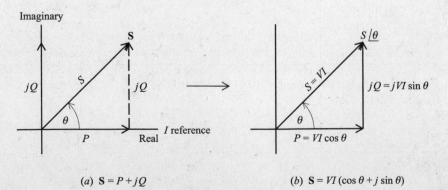

(a) $\mathbf{S} = P + jQ$ $\qquad\qquad$ (b) $\mathbf{S} = VI(\cos\theta + j\sin\theta)$

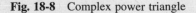

Fig. 18-8   Complex power triangle

Complex power **S** is a complex quantity with magnitude S equal to the product of the input voltage V and the input current I, both in rms or effective values, and with a phase angle θ between V and I.

Other formulas that can be used to determine components of the complex power triangle (Fig. 18-8), using $V = IZ$, are

$$S = VI = (IZ)I = I^2 Z \tag{18-14}$$

$$P = VI \cos\theta = (IZ)I \cos\theta = I^2(Z\cos\theta) = I^2 R \tag{18-15}$$

$$Q = VI \sin\theta = (IZ)I \sin\theta = I^2(Z\sin\theta) = I^2 X \tag{18-16}$$

V, I, and Z represent the *magnitude* of voltage, current, and impendence, respectively. The magnitude of these quantities can be shown also as $V = |\mathbf{V}|$, $I = |\mathbf{I}|$, and $Z = |\mathbf{Z}|$ by placing bars by the phasor quantity.

A third formula for complex power is

$$\mathbf{S} = \mathbf{VI}^* \tag{18-17}$$

where $\mathbf{I}^*$ is the conjugate of the input current phasor $\mathbf{I}$ and $\mathbf{V}$ is the input voltage phasor.

Note from Eq. *(18-15)* that $R = Z\cos\theta$ is the input resistance, the same as the real part of the input impedance. R is usually not the resistance of a physical resistor, but the real part of the input impedance and is usually dependent on inductive and capacitive reactances, as well as on resistances. It is important to note that **S** is a complex number, but does *not* represent a sinusoidally varying quantity.

Power Factor

The term "$\cos\theta$" is called the *power factor*, PF. The angle θ is called the *power factor angle*. θ is often also the impedance angle. The power factor of an inductive circuit is called a *lagging power factor*, and that of a capacitive circuit is called a *leading power factor*. If a circuit has only resistance, the PF $= 1$; if it has only reactance, PF $= 0$.

To deliver a large amount of power, a high PF, i.e., close to 1, is desirable.

We see from

$$P = VI \cos\theta \tag{18-12}$$

$$I = \frac{P}{V\cos\theta} = \frac{P}{V \times \mathrm{PF}}$$

that by having a smaller PF, the current I to the load becomes greater. Larger than necessary currents are undesirable due to the accompanying large IR voltage losses and $I^2 R$ power losses in power lines.

Reactive Power Q

Reactive power is often used for industrial power consideration. The sign convention for Q is positive $(+Q)$ for an inductive load shown *above* the real axis, and is minus $(-Q)$ for a capacitive load shown *below* the real axis; that is, $+Q$ *consumes* reactive power and $-Q$ *produces* reactive power. Reactance does not add to the real or effective power. Stored energy is being shuttled to and from the magnetic field of an inductance or the electric field of a capacitance.

Example 18.5 An ac voltage with a rms value of 115 V is applied to a load impedance of $R = 75\ \Omega$ and $X_L = 38\ \Omega$ (Fig. 18-9a). Find the value of real power P, reactive power Q, apparent power S, and complex power **S**. Show the phasor diagram and the power triangle.

Step 1. Find P.

It makes no difference what angle is assigned to **V**, so we conveniently assign $0°$ to voltage.

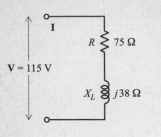

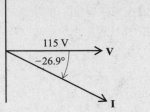

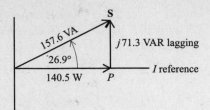

(a) Circuit diagram   (b) Phasor diagram   (c) Power triangle

Fig. 18-9   *RL* series circuit

$$\mathbf{V} = 115 \underline{/0°}\ \text{V}$$

$$\mathbf{Z} = 75 + j38 = 84.08 \underline{/26.9°}\ \Omega$$

Then
$$\mathbf{I} = \frac{\mathbf{V}}{\mathbf{Z}} = \frac{115 \underline{/0°}}{84.08 \underline{/26.9°}} = 1.37 \underline{/-26.9°}\ \text{A}$$

(18-12)
$$P = VI\cos\theta = (115)(1.37)(\cos 26.9°) = 140.5\ \text{W} \qquad Ans.$$

or
$$P = I^2 R = (1.37)^2(75) = 140.8 \approx 140.5\ \text{W} \qquad Ans.$$

Step 2.   Find Q.

(18-13)
$$Q = VI\sin\theta = (115)(1.37)(\sin 26.9°) = 71.3\ \text{VAR lagging} \qquad Ans.$$

since I lags **V**.

Step 3.   Find S.

$$S = VI = (115)(1.37) = 157.6\ \text{VA} \qquad Ans.$$

Step 4.   Write **S**.

(18-10)
$$\mathbf{S} = P + jQ = 140.5 + j71.3 = 157.6 \underline{/26.9°} \qquad Ans.$$

The phasor diagram and power triangle are shown in Fig. 18-9(b) and (c), respectively.

An alternate way of finding **S** without first solving for its components, *P* and *Q*, is to use the formula

$$\mathbf{S} = \mathbf{V}\mathbf{I}^* \qquad\qquad (18\text{-}17)$$

From Step 1, $\mathbf{I} = 1.37 \underline{/-26.9°}$

Then $\mathbf{I}^* = 1.37 \underline{/26.9°}$ by changing the sign of the phase angle of **I**

$$\mathbf{S} = \left(115 \underline{/0°}\right)\left(1.37 \underline{/26.9°}\right)$$

$$= 157.6 \underline{/26.9°}\ \text{VA} \qquad Ans.$$

from which $\mathbf{S} = \underbrace{140.5}_{P} + j\underbrace{71.3}_{Q}$ VA in rectangular form so

$$P = 140.5\ \text{W} \quad \text{and} \quad Q = 71.3\ \text{VAR lagging}.$$

Table 18-2 Summary of Complex Power Relationships

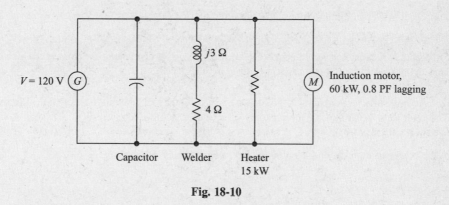

(a) Inductive, PF lagging
$\mathbf{S} = P + jQ$

(b) Capacitive, PF leading
$\mathbf{S} = P - jQ$

Complex power triangle:

	Complex power **S**, VA	Apparent power S, VA	Real power P, W	Reactive power Q, VAR	PF
Formulas:	$P + jQ$	VI	$VI \cos\theta$	$VI \sin\theta$	$\cos\theta$
	$S \underline{/\theta}$	$\sqrt{P^2 + Q^2}$	$I^2 R$	$I^2 X$	P/VI
	$\mathbf{VI^*}$				

Table 18-2 summarizes the quantities of complex power.

Example 18.6 A generator is to supply power to a welder, heater, and induction motor (Fig. 18-10). A capacitor is used to supply reactive power for the welder and motor in order that the net load on the generator will have unity power factor. Find the complex power and real or effective power that must be supplied by the generator, and the reactive power that must be supplied by the capacitor. If the capacitor were not used, what apparent power would have to be supplied by the generator?

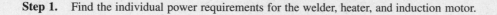

Fig. 18-10

Step 1.   Find the individual power requirements for the welder, heater, and induction motor.

Welder:

$$I = \frac{V}{Z} = \frac{120}{|4 + j3|} = \frac{120}{\sqrt{4^2 + 3^2}} = \frac{120}{5} = 24.0 \text{ A}$$

$$P = I^2 R = (24.0)^2(4) = 2304 \text{ W} \approx 2.3 \text{ kW}$$

$$Q = I^2 X_L = (24.0)^2(3) = 1728 \text{ VAR} \approx 1.7 \text{ kVAR lagging}$$

Heater: $P = 15 \text{ kW},\ \text{given} \quad Q = 0 \text{ VAR}$

Induction motor: $\text{PF} = \cos\theta = 0.8, \quad \theta = \arccos 0.8$

$$P = VI \cos\theta$$

$$VI = \frac{P}{\cos\theta} = \frac{60}{0.8} = 75 \text{ kVA}$$

$$Q = VI \sin\theta = 75 \sin(\arccos 0.8) = 75(0.6) = 45 \text{ kVAR lagging}$$

Step 2.   Find **S**, P, and Q.

S equals the sum of the individual complex power. Arrange a table for P and Q:

Item	P (kW)	Q (kVAR)
Welder	2.3	1.7
Heater	15	0
Motor	60	45
Total power requirements	77.3 kW	46.7 kVAR, lagging

$$\mathbf{S} = P + jQ \tag{18-10}$$

$$\mathbf{S} = 77.3 + j46.7 = 90.3\ \underline{/31.1°}\ \text{kVA} \qquad Ans.$$

$P = 77.3$ kW, real power from the generator.   *Ans.*

$Q = 46.7$ kVAR leading is the reactive power produced by the capacitor
to offset the inductive reactive power of 46.7 kVAR so that
the net reactive power $Q = 0$ and the PF $= 1$.   *Ans.*

Step 3.   Find apparent power of circuit without the capacitor.

We found $\mathbf{S} = 90.3\ \underline{/31.1°}\ \text{kVA} = VI\ \underline{/\theta}$, so that apparent power VI is 90.3 kVA.   *Ans.*

In practice, a capacitor would not be used because it would be cheaper to increase the size of the generator to supply 90.3 kVA than to buy a capacitor to supply 46.7 kVAR.

Example 18.7   Adding sufficient capacitance to increase power factor to 1 may not be economical in the circuit shown in Fig. 18-10, Example 18.6. What must be the capacitive reactance to achieve a power factor less than 1, say 0.90 lagging, and what is the new apparent power?

Step 1.   Draw the power triangle from the values solved in Example 18.6, where

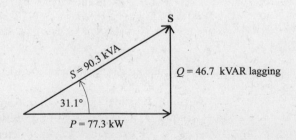

$$\mathbf{S} = 90.3\ \underline{/31.1°} = 77.3 + j46.7 \text{ kVA}$$

Without the capacitor, PF of the circuit $= \cos 31.1° = 0.856$ lagging.

Step 2. Draw the *desired* power triangle where PF = 0.90 lagging.

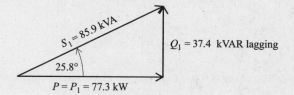

$$PF_1 = \cos\theta_1 = 0.90 \text{ lagging}$$

$$\theta_1 = \arccos 0.90 = 25.8°$$

$$Q_1 = 77.3 \tan 25.8° = 37.4 \text{ kVAR}$$

The new value of apparent power $S_1 = \dfrac{77.3}{\cos\theta_1} = \dfrac{77.3}{0.90}$

$$= 85.9 \text{ kVA} \qquad Ans.$$

Step 3. Draw the *combined* power triangle.

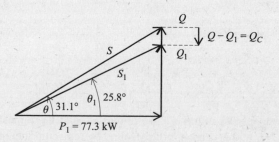

$Q - Q_1$ is the reduction of Q lagging by adding a capacitor.

The capacitive reactance Q_C required to improve the PF is

$$Q_c = Q - Q_1 = 46.7 - 37.4 = 9.3 \text{ kVAR leading} \qquad Ans.$$

Note that the addition of Q_1 has improved the PF from 0.856 to 0.90. The decrease of apparent power from 90.3 kVA to 85.9 kVA, 4.4 kVA, is 4.9%. The transformers, the distribution system, and the utility company attenuators are all noted in kVA or MVA. Thus an improvement in PF, with corresponding less kVA, releases some of the generation and transmission capability that can serve other customers.

DETERMINANT SOLUTION FOR AC CIRCUITS

When reactances are present in networks which cannot be resolved into simple series–parallel circuits, we can use the same determinant solution technique for ac networks as in dc networks.

Recall from Chapter 8 that in a two-mesh dc network, the *resistance* determinant Δ is

$$\Delta = \begin{vmatrix} R_{11} & -R_{12} \\ -R_{21} & R_{22} \end{vmatrix} \qquad (8\text{-}11)$$

and for a three-mesh dc network,

$$\Delta = \begin{vmatrix} R_{11} & -R_{12} & -R_{13} \\ -R_{21} & R_{22} & -R_{23} \\ -R_{31} & -R_{32} & R_{33} \end{vmatrix} \qquad (8\text{-}14)$$

For an ac network of two meshes and three meshes, we can similarly write the *impedance* determinant Δ.

$$\Delta = \begin{vmatrix} \mathbf{Z}_{11} & -\mathbf{Z}_{12} \\ -\mathbf{Z}_{21} & \mathbf{Z}_{22} \end{vmatrix} \qquad (18\text{-}18)$$

and

$$\Delta = \begin{vmatrix} \mathbf{Z}_{11} & -\mathbf{Z}_{12} & -\mathbf{Z}_{13} \\ -\mathbf{Z}_{21} & \mathbf{Z}_{22} & -\mathbf{Z}_{23} \\ -\mathbf{Z}_{31} & -\mathbf{Z}_{32} & \mathbf{Z}_{33} \end{vmatrix} \qquad (18\text{-}19)$$

To find the mesh current I, we solve

$$\mathbf{I}_1 = \frac{\mathbf{N}_{I_1}}{\Delta} \qquad (18\text{-}20)$$

where

$$\mathbf{N}_{I_1} = \begin{vmatrix} \mathbf{V}_1 & -\mathbf{Z}_{12} & -\mathbf{Z}_{13} \\ \mathbf{V}_2 & \mathbf{Z}_{22} & -\mathbf{Z}_{23} \\ \mathbf{V}_3 & -\mathbf{Z}_{32} & \mathbf{Z}_{33} \end{vmatrix} \qquad (18\text{-}21)$$

The 1st column of *net* voltages, $\mathbf{V}_1$ in mesh 1, $\mathbf{V}_2$ in mesh 2, and $\mathbf{V}_3$ in mesh 3, replaces the 1st column of the impedance determinant. $\mathbf{N}_{I_2}$ and $\mathbf{N}_{I_3}$ are similarly found by replacing the 2nd and 3nd column respectively by the net voltages.

Example 18.8   Find $\mathbf{I}_1$ by use of determinants for the circuit shown in Fig. 18-11.

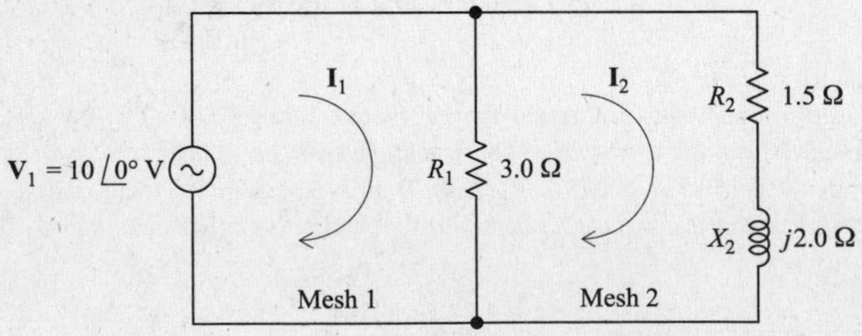

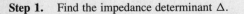

$\mathbf{V}_1 = 10\,\underline{/0°}\ \text{V}$ I_1 I_2 R_2 1.5 Ω

R_1 3.0 Ω X_2 $j2.0$ Ω

Mesh 1   Mesh 2

Fig. 18-11

Step 1.   Find the impedance determinant Δ.

$$\Delta = \begin{vmatrix} \mathbf{Z}_{11} & -\mathbf{Z}_{12} \\ -\mathbf{Z}_{21} & \mathbf{Z}_{22} \end{vmatrix} \qquad (18\text{-}18)$$

where $\mathbf{Z}_{11}$ = total impedance of mesh 1

$$= R_1 + 0j = 3.0 + 0j = 3.0 \; \underline{/0°} \; \Omega$$

$\mathbf{Z}_{22}$ = total impedance of mesh 2

$$= (R_1 + R_2) + jX_2 = 4.5 + j2.0 = 4.92 \; \underline{/24.0°} \; \Omega$$

$\mathbf{Z}_{12}$ = mutual impedance between mesh 1 and 2

$$= R_1 + j0 = 3.0 + j0 = 3 \; \underline{/0°} \; \Omega$$

$\mathbf{Z}_{21}$ = mutual impedance between mesh 2 and mesh 1

$$= R_1 + j0 = 3.0 + j0 = 3 \; \underline{/0°} \; \Omega$$

Substituting,

$$\Delta = \begin{vmatrix} 3.0 \; \underline{/0°} & -3.0 \; \underline{/0°} \\ -3.0 \; \underline{/0°} & 4.92 \; \underline{/24.0°} \end{vmatrix}$$

$$\Delta = \left(3.0 \; \underline{/0°}\right)\left(4.92 \; \underline{/24.0°}\right) - \left(-3.0 \; \underline{/0°}\right)\left(-3.0 \; \underline{/0°}\right)$$

$$\Delta = 14.76 \; \underline{/24.0°} - 9.0 \; \underline{/0°}$$

$$= (13.48 + j6.0) - 9.0 = 4.48 + j6.0 = 7.49 \; \underline{/53.3}$$

Step 2. Find $\mathbf{N}_{I_1}$.

$$\mathbf{N}_{I_1} = \begin{vmatrix} \mathbf{V}_1 & -\mathbf{Z}_{12} \\ \mathbf{V}_2 & \mathbf{Z}_{22} \end{vmatrix} \qquad (18\text{-}21)$$

$$= \begin{vmatrix} 10 \; \underline{/0°} & -3.0 \; \underline{/0°} \\ 0 \; \underline{/0°} & 4.92 \; \underline{/24.0°} \end{vmatrix}$$

$$= 49.2 \; \underline{/24.0°}$$

Step 3. Solve for $\mathbf{I}_1$.

$$\mathbf{I}_1 = \frac{\mathbf{N}_{I_1}}{\Delta} \qquad (18\text{-}20)$$

$$\mathbf{I}_1 = \frac{49.2 \; \underline{/24.0°}}{7.49 \; \underline{/53.3°}} = 6.57 \; \underline{/-29.3°} \; \text{A} \qquad Ans.$$

Note: This example illustrates the use of determinants.

$\mathbf{I}_1$ could be solved more simply by noting that

current through $R_1 = \mathbf{I}_{R_1} = \dfrac{\mathbf{V}_1}{R_1} = \dfrac{10 \; \underline{/0°}}{3} = 3.33 \; \underline{/0°} \; \text{A},$ and

current through $\mathbf{Z}_2 = \mathbf{I}_{Z_2} = \dfrac{\mathbf{V}_1}{\mathbf{Z}_2} = \dfrac{10 \; \underline{/0°}}{1.5 + j2.0} = \dfrac{10 \; \underline{/0°}}{2.5 \; \underline{/53.1°}} = 4 \; \underline{/-53.1°} \; \text{A}$

Then $\mathbf{I}_1 = \mathbf{I}_{R_1} + \mathbf{I}_{Z_2} = 3.33 \; \underline{/0°} + 4 \; \underline{/-53.1°} = 3.33 + 2.4 - j3.2$

$$= 5.73 - j3.20 = 6.57 \; \underline{/-29.2°} \; \text{A}, \text{ which agrees with the previous answer.}$$

Nevertheless, determinant solutions for current values are useful in a 2-mesh circuit, and particularly in a 3- or higher mesh circuit. See Solved Problem 18.15.

AC Δ-Y AND Y-Δ CONVERSIONS

Chapter 9 presents the Δ-Y and Y-Δ conversion formulas for resistances. The only difference for impedances is in the use of $\mathbf{Z}$'s instead of R's. Specifically for the Δ-Y arrangement shown in Fig. 18-12, the Δ-Y conversion formulas are

$$\mathbf{Z}_a = \frac{\mathbf{Z}_1\mathbf{Z}_3}{\mathbf{Z}_1 + \mathbf{Z}_2 + \mathbf{Z}_3} \qquad \text{(from Eq. 9-1)} \qquad (18\text{-}22)$$

$$\mathbf{Z}_b = \frac{\mathbf{Z}_1\mathbf{Z}_2}{\mathbf{Z}_1 + \mathbf{Z}_2 + \mathbf{Z}_3} \qquad \text{(from Eq. 9-2)} \qquad (18\text{-}23)$$

$$\mathbf{Z}_c = \frac{\mathbf{Z}_2\mathbf{Z}_3}{\mathbf{Z}_1 + \mathbf{Z}_2 + \mathbf{Z}_3} \qquad \text{(from Eq. 9-3)} \qquad (18\text{-}24)$$

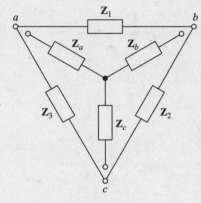

Fig. 18-12

and the Y-Δ conversion formulas are

$$\mathbf{Z}_1 = \frac{\mathbf{Z}_a\mathbf{Z}_b + \mathbf{Z}_b\mathbf{Z}_c + \mathbf{Z}_c\mathbf{Z}_a}{\mathbf{Z}_c} \qquad \text{(from Eq. 9-4)} \qquad (18\text{-}25)$$

$$\mathbf{Z}_2 = \frac{\mathbf{Z}_a\mathbf{Z}_b + \mathbf{Z}_b\mathbf{Z}_c + \mathbf{Z}_c\mathbf{Z}_a}{\mathbf{Z}_a} \qquad \text{(from Eq. 9-5)} \qquad (18\text{-}26)$$

$$\mathbf{Z}_3 = \frac{\mathbf{Z}_a\mathbf{Z}_b + \mathbf{Z}_b\mathbf{Z}_c + \mathbf{Z}_c\mathbf{Z}_a}{\mathbf{Z}_b} \qquad \text{(from Eq. 9-6)} \qquad (18\text{-}27)$$

Example 18.9   Using Δ-Y conversion, find $\mathbf{I}_T$ for the circuit shown in Fig. 18-13(a).

Step 1.   Convert Δ configuration at abc terminals to Y configuration (Fig. 18-13b). Use Eqs. (18-22)–(18-24). The denominator is the same for all the formulas.

$$\mathbf{Z}_D = \mathbf{Z}_1 + \mathbf{Z}_2 + \mathbf{Z}_3 = 3 + 4 - j4 = 8.062 \underline{/-29.7°}\ \Omega$$

$$(18\text{-}23) \qquad \mathbf{Z}_a = \frac{\mathbf{Z}_1\mathbf{Z}_3}{\mathbf{Z}_D} = \frac{(3)(-j4)}{\mathbf{Z}_D} = \frac{12\underline{/-90°}}{8.062\underline{/-29.7°}} = 1.49\underline{/-60.3°}\ \Omega$$

$$= 0.74 - j1.29\ \Omega$$

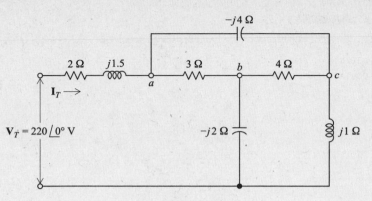

(a) Circuit

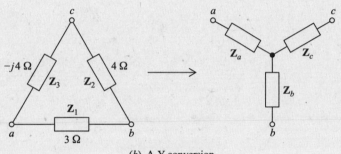

(b) Δ-Y conversion

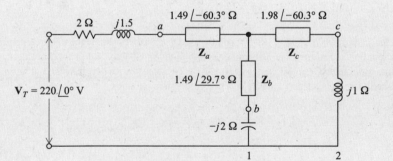

(c) Circuit with Δ-Y conversion

Fig. 18-13

(18-23)
$$\mathbf{Z}_b = \frac{\mathbf{Z}_1\mathbf{Z}_2}{\mathbf{Z}_D} = \frac{(3)(4)}{\mathbf{Z}_D} = \frac{12}{8.062\,\underline{/-29.7^\circ}} = 1.49\,\underline{/29.7^\circ}\ \Omega$$

$$= 1.29 + j0.74\ \Omega$$

(18-24)
$$\mathbf{Z}_c = \frac{\mathbf{Z}_2\mathbf{Z}_3}{\mathbf{Z}_D} = \frac{(4)(-j4)}{\mathbf{Z}_D} = \frac{16\,\underline{/-90^\circ}}{8.062\,\underline{/-29.7^\circ}} = 1.98\,\underline{/-60.3^\circ}\ \Omega$$

$$= 0.98 - j1.72\ \Omega$$

The circuit with the Δ-Y conversion is shown in Fig. 18-13(c).

Step 2.   Find $\mathbf{Z}_T$ by circuit reduction.

First, find the equivalent impedance of the two parallel branches:

Branch 1: $\mathbf{Z}_b - j2 = (1.29 + j0.74) - j2 = 1.29 - j1.26 = 1.80\ \underline{/-44.3^\circ}$

Branch 2: $\mathbf{Z}_c + j1 = 0.98 - j1.72 + j1 = 0.98 - j0.72 = 1.22\ \underline{/-36.3^\circ}$

$$\mathbf{Z}_{eq} = \frac{(1.80\ \underline{/-44.3^\circ})(1.22\ \underline{/-36.3^\circ})}{(1.29 - j1.26) + (0.98 - j0.72)} = \frac{2.20\ \underline{/-80.6^\circ}}{2.27 - j1.98}$$

$$= \frac{2.20\ \underline{/-80.6^\circ}}{3.01\ \underline{/-41.1^\circ}} = 0.73\ \underline{/-39.5^\circ} = 0.56 - j0.46\ \Omega$$

$\mathbf{Z}_{eq}$ is in series with $2 + j1.5$ and $\mathbf{Z}_a$.

Then $\mathbf{Z}_T = (2 + j1.5) + \mathbf{Z}_a + \mathbf{Z}_{eq}$

$$= (2 + j1.5) + (0.74 - j1.29) + (0.56 - j0.46)$$

$$\mathbf{Z}_T = 3.30 - j0.25 = 3.31\ \underline{/-4.3^\circ}\ \Omega \qquad Ans.$$

Step 3.   Find $\mathbf{I}_T$.

$$\mathbf{I}_T = \frac{\mathbf{V}_T}{\mathbf{Z}_T} = \frac{220\ \underline{/0^\circ}}{3.31\ \underline{/-4.3^\circ}} = 66.5\ \underline{/4.3^\circ}\ A \qquad Ans.$$

Example 18.10   An ac bridge circuit (Fig. 18-14) can be used to measure inductance or capacitance in the same way that a Wheatstone bridge can be used to measure resistance, as explained in Chapter 9. For measurement, two of the resistances are varied until the galvanometer G in the center arm reads zero when the switch is closed. The bridge is then balanced, and the unknown impedance $\mathbf{Z}_X$ can be found by the bridge balance equation:

$$\mathbf{Z}_X = \frac{\mathbf{Z}_1\mathbf{Z}_3}{\mathbf{Z}_2}$$

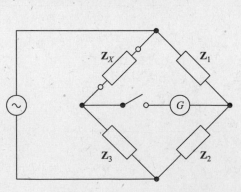

Fig. 18-14   Wheatstone bridge

This equation is the same as that for the Wheatstone bridge except for having $\mathbf{Z}$'s instead of R's. [Though no further problems will be offered, the reason for presenting the Δ-Y and Y-Δ conversion formulas is to inform the student that complex number techniques can be applied to solve for networks with Δ or Y configurations.]

Solved Problems

18.1   The resistance of a coil is 1.5 Ω and its inductive reactance is 2 Ω (Fig. 18-15). When the current is 4 A, find the voltage **V** and draw the phasor diagram.

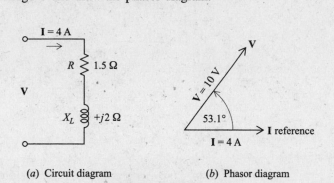

(a) Circuit diagram   (b) Phasor diagram

Fig. 18-15   *RL* series circuit

$$\mathbf{Z} = R + jX = 1.5 + j2 = 2.5\,\underline{/53.1°}\ \Omega$$

Choose **I** as the reference phasor at 0°.

(*18-3*) $\mathbf{V} = \mathbf{IZ} = \left(4\,\underline{/0°}\right)\left(2.5\,\underline{/53.1°}\right) = 10\,\underline{/53.1°}\ \text{V}$ *Ans.*

V leads **I** by 53.1° or equivalently **I** lags **V** by 53.1°. Phasor diagram is shown in (*b*).   *Ans.*

18.2   For *RLC* series circuit (Fig. 18-16), find the impedance $\mathbf{Z}_T$, current $\mathbf{I}_T$, and the voltage drops around the current. Draw the voltage phasor diagram. Check the solution by use of Kirchhoff's voltage law.

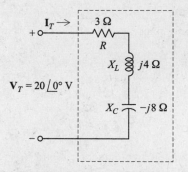

Fig. 18-16   *RLC* series circuit

Step 1.   Find $\mathbf{Z}_T$.

By inspection, $\mathbf{Z}_T = R + jX = R + j4 - j8 = 3 - j4 = 5\,\underline{/-53.1°}\ \Omega$ *Ans.*

Step 2.   Find $\mathbf{I}_T$.

(*18-2*) $\mathbf{I}_T = \dfrac{\mathbf{V}_T}{\mathbf{Z}_T} = \dfrac{20\,\underline{/0°}}{5\,\underline{/-53.1°}} = 4\,\underline{/53.1°}\ \text{A}$ *Ans.*

V_T is the reference phasor at 0° angle.

Step 3.   Find the voltage drops.

$$\mathbf{V}_R = \mathbf{I}_T R = \left(4 \underline{/53.1°}\right)\left(3 \underline{/0°}\right) = 12 \underline{/53.1°} \text{ V} \qquad Ans.$$

$$\mathbf{V}_L = \mathbf{I}_T X_L = \left(4 \underline{/53.1°}\right)\left(4 \underline{/90°}\right) = 16 \underline{/143.1°} \text{ V} \qquad Ans.$$

$$\mathbf{V}_C = \mathbf{I}_T X_C = \left(4 \underline{/53.1°}\right)\left(8 \underline{/-90°}\right) = 32 \underline{/-36.9°} \text{ V} \qquad Ans.$$

Step 4.   Draw the voltage phasor diagram.

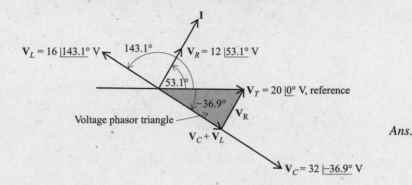

Ans.

Step 5.   Check solution.

Convert voltage drops from polar form to rectangular and then add.

$$\mathbf{V}_R = 12 \underline{/53.1°} = 7.2 + j9.6 \text{ V}$$

$$\mathbf{V}_L = 16 \underline{/143.1°} = -12.8 + j9.6 \text{ V}$$

$$\mathbf{V}_C = 32 \underline{/-36.9°} = 25.6 - j19.2 \text{ V}$$

KVL states that the applied voltage of a series circuit equals the total voltage drops.

$$\mathbf{V}_T = \mathbf{V}_R + \mathbf{V}_L + \mathbf{V}_C$$

$$= (7.2 + j9.6) + (-12.8 + j9.6) + (25.6 - j19.2)$$

$$= 20 + 0j = 20 \underline{/0°} \text{ V} \qquad Check$$

18.3   Figure 18-17 shows a parallel two-branch circuit. (*a*) Show that $\mathbf{I}_1 = (\mathbf{Z}_T/\mathbf{Z}_1)\mathbf{I}_T$ and $\mathbf{I}_2 = (\mathbf{Z}_T/\mathbf{Z}_2)\mathbf{I}_T$. These equations are *current division* formulas between two parallel branches. (*b*) Also show that $\mathbf{I}_1 = [\mathbf{Z}_2/(\mathbf{Z}_1 + \mathbf{Z}_3)]\mathbf{I}_T$ and $\mathbf{I}_2 = [\mathbf{Z}_1/(\mathbf{Z}_1 + \mathbf{Z}_2)]\mathbf{I}_T$.

(*a*)   Write: $\mathbf{V}_T = \mathbf{I}_T \mathbf{Z}_T = \mathbf{I}_1 \mathbf{Z}_1 = \mathbf{I}_2 \mathbf{Z}_2$, $\mathbf{V}_T$ is the common voltage

Solve for $\mathbf{I}_1$:
$$\mathbf{I}_1 = \frac{\mathbf{Z}_T}{\mathbf{Z}_1}\mathbf{I}_T \qquad Ans.$$

Solve for $\mathbf{I}_2$:
$$\mathbf{I}_2 = \frac{\mathbf{Z}_T}{\mathbf{Z}_2}\mathbf{I}_T \qquad Ans.$$

The general formula to find individual currents for the nth branch if we know total current $\mathbf{I}_T$ and equivalent or total impedance $\mathbf{Z}_T$ is $\mathbf{I}_n = (\mathbf{Z}_T/\mathbf{Z}_n)\mathbf{I}_T$. For example, if there were $n = 3$ parallel branches, the current in the 3rd branch would be $\mathbf{I}_3 = (\mathbf{Z}_T/\mathbf{Z}_3)\mathbf{I}_T$.

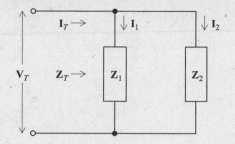

Fig. 18-17   Current division between 2 parallel branches

(*b*)   Write:

$$\mathbf{Z}_T = \frac{\mathbf{Z}_1 \mathbf{Z}_2}{\mathbf{Z}_1 + \mathbf{Z}_2} \qquad (18\text{-}9)$$

Substituting $\mathbf{Z}_T$, $\mathbf{I}_1 = \dfrac{\mathbf{Z}_T}{\mathbf{Z}_1} \mathbf{I}_T = \left(\dfrac{\mathbf{Z}_1 \mathbf{Z}_2}{\mathbf{Z}_1 + \mathbf{Z}_2} \right) \left(\dfrac{1}{\mathbf{Z}_1} \right) \mathbf{I}_T = \left(\dfrac{\mathbf{Z}_2}{\mathbf{Z}_1 + \mathbf{Z}_2} \right) \mathbf{I}_T$ *Ans.*

$$\mathbf{I}_2 = \frac{\mathbf{Z}_T}{\mathbf{Z}_2} \mathbf{I}_T = \left(\frac{\mathbf{Z}_1 \mathbf{Z}_2}{\mathbf{Z}_1 + \mathbf{Z}_2} \right) \left(\frac{1}{\mathbf{Z}_2} \right) \mathbf{I}_T = \left(\frac{\mathbf{Z}_1}{\mathbf{Z}_1 + \mathbf{Z}_2} \right) \mathbf{I}_T \qquad Ans.$$

18.4   Connect a resistor of 3.0 Ω in parallel with a coil having a resistance of 1.5 Ω and inductance of 2.0 Ω (Fig. 18-18). The applied voltage is $10 \underline{/0°}$ V. Find the total current and draw the current-phasor diagram.

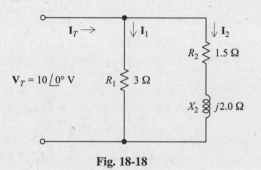

Fig. 18-18

Step 1.   Write the impedances of the two branches.

Branch 1: $\mathbf{Z}_1 = R_1 + 0j = 3.0 + 0j = 3 \underline{/0°}$ Ω

Branch 2: $\mathbf{Z}_2 = R_2 + jX_2 = 1.5 + j2.0 = 2.5 \underline{/53.1°}$ Ω

Step 2.   Find $\mathbf{I}_1$ and $\mathbf{I}_2$.

$$\mathbf{I}_1 = \frac{\mathbf{V}_T}{\mathbf{Z}_1} = \frac{10 \underline{/0°}}{3 \underline{/0°}} = 3.33 \underline{/0°} = 3.33 + j0 \text{ A}$$

$$\mathbf{I}_2 = \frac{\mathbf{V}_T}{\mathbf{Z}_2} = \frac{10 \underline{/0°}}{2.5 \underline{/53.1°}} = 4.00 \underline{/-53.1°} = 2.40 - j3.20 \text{ A}$$

Step 3.   Find $\mathbf{I}_T$ by adding $\mathbf{I}_1$ and $\mathbf{I}_2$.

$$\mathbf{I}_T = \mathbf{I}_1 + \mathbf{I}_2 = (3.33 + j0) + (2.40 - j3.20)$$

$$= 5.73 - j3.20 = 6.56 \underline{/-29.2°} \text{ A} \qquad Ans.$$

Step 4. Draw the current-phasor diagram.

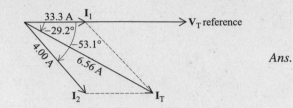

Ans.

18.5   In the series circuit (Fig. 18-19), the effective value of the current indicated by the ammeter is 10 A. What are the readings on a voltmeter placed across the entire circuit and then across each element?

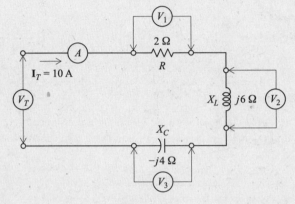

Fig. 18-19

Step 1. Find $\mathbf{Z}_T$.

$$\mathbf{Z}_T = 2 + j6 - j4 = 2 + j2 = 2.83 \underline{/45°} \ \Omega$$

Step 2. Find the voltmeter readings.

Since we are interested in finding the effective values of voltage, we need only multiply magnitudes of current and of impedance to find voltage. Then

$$V_T = I_T Z_T = (10)(2.83) = 28.3 \text{ V} \qquad Ans.$$

$$V_1 = I_T R = (10)(2) = 20 \text{ V} \qquad Ans.$$

$$V_2 = I_T X_L = (10)(6) = 60 \text{ V} \qquad Ans.$$

$$V_3 = I_T X_C = (10)(4) = 40 \text{ V} \qquad Ans.$$

18.6   In a two branch parallel circuit, the voltmeter reads 50 V across the 5 Ω-resistor (Fig. 18-20). What is the reading of the ammeter?

Step 1. Find magnitude of $\mathbf{I}_2$.

$$\mathbf{I}_2 = \frac{50}{5} = 10 \text{ A}$$

Assume $\mathbf{I}_2$ has a phase angle of $0°$. Then

$$\mathbf{I}_2 = 10 \underline{/0°} \text{ A}$$

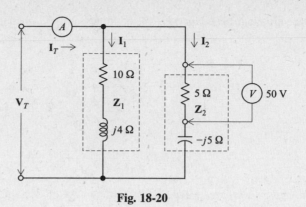

Fig. 18-20

Step 2.   Write $\mathbf{Z}_2$ and find $\mathbf{V}_T$.

$$\mathbf{Z}_2 = 5 - j5 = 7.07 \underline{/-45°}\ \Omega$$

$$\mathbf{V}_T = \mathbf{I}_2\mathbf{Z}_2 = \left(10\ \underline{/0°}\right)\left(7.07\ \underline{/-45°}\right)$$

$$= 70.7\ \underline{/-45°}\ \text{V}$$

Step 3.   Write $\mathbf{Z}_1$ and find $\mathbf{I}_1$.

$$\mathbf{Z}_1 = 10 + j4 = 10.8\ \underline{/21.8°}\ \Omega$$

$$\mathbf{I}_1 = \frac{\mathbf{V}_T}{\mathbf{Z}_1} = \frac{70.7\ \underline{/-45°}}{10.8\ \underline{/21.8°}} = 6.55\ \underline{/-66.8°} = 2.58 - j6.02\ \text{A}$$

Step 4.   Find $\mathbf{I}_T$.

$$\mathbf{I}_T = \mathbf{I}_1 + \mathbf{I}_2 = (2.58 - j6.02) + (10 + j0) = 12.58 - j6.02$$

$$= 13.9\ \underline{/-25.6}\ \text{A}$$

The ammeter reads 13.9 A.   *Ans.*

18.7   Find the input impedance of the series–parallel circuit (Fig. 18-21).

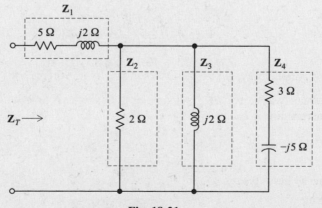

Fig. 18-21

Step 1. Find the equivalent impedance $\mathbf{Z}_A$ to the three parallel impedances, $\mathbf{Z}_2$, $\mathbf{Z}_3$, and $\mathbf{Z}_4$.

$$\frac{1}{\mathbf{Z}_A} = \frac{1}{\mathbf{Z}_2} + \frac{1}{\mathbf{Z}_3} + \frac{1}{\mathbf{Z}_4} \tag{18-8}$$

$$= \frac{\mathbf{Z}_3\mathbf{Z}_4 + \mathbf{Z}_2\mathbf{Z}_4 + \mathbf{Z}_2\mathbf{Z}_3}{\mathbf{Z}_2\mathbf{Z}_3\mathbf{Z}_4}$$

Then
$$\mathbf{Z}_A = \frac{\mathbf{Z}_2\mathbf{Z}_3\mathbf{Z}_4}{\mathbf{Z}_3\mathbf{Z}_4 + \mathbf{Z}_2\mathbf{Z}_4 + \mathbf{Z}_2\mathbf{Z}_3} \tag{18-8a}$$

$$\mathbf{Z}_2 = 2 + j0 = 2\ \underline{/0°}\ \Omega$$

$$\mathbf{Z}_3 = 0 + j2 = 2\ \underline{/90°}\ \Omega$$

$$\mathbf{Z}_4 = 3 - j5 = 5.83\ \underline{/-59.0°}\ \Omega$$

Substitute these impedances into Eq. (*18-8a*),

$$\mathbf{Z}_A = \frac{(2\ \underline{/0°})(2\ \underline{/90°})(5.83\ \underline{/-59.0°})}{(2\ \underline{/90°})(5.83\ \underline{/-59.0°}) + (2\ \underline{/0°})(5.83\ \underline{/-59.0°}) + (2\ \underline{/0°})(2\ \underline{/90°})}$$

$$= \frac{23.32\ \underline{/31.0°}}{11.66\ \underline{/31.0°} + 11.66\ \underline{/-59.0°} + 4\ \underline{/90°}}$$

$$= \frac{23.32\ \underline{/31.0°}}{(9.99 + j6.01) + (6.01 - j9.99) + (0 + j4)}$$

$$\mathbf{Z}_A = \frac{23.32\ \underline{/31.0°}}{16 + j0.02} = \frac{23.32\ \underline{/31.0°}}{16.00\ \underline{/0.07°}}$$

$$\mathbf{Z}_A = 1.46\ \underline{/30.9°} = 1.25 + j0.75\ \Omega$$

Step 2. Find $\mathbf{Z}_T$.

$\mathbf{Z}_1$ is in series with $\mathbf{Z}_A$.

$$\mathbf{Z}_T = \mathbf{Z}_1 + \mathbf{Z}_A$$

$$= (5 + j2) + (1.25 + j0.75)$$

$$\mathbf{Z}_T = 6.25 + j2.75 = 6.83\ \underline{/23.7°}\ \Omega \qquad Ans.$$

An alternate way to find $\mathbf{Z}_A$, and simpler in this case, is to write Eq. (*18-8*) and rationalize the denominator in each term.

$$\frac{1}{\mathbf{Z}_A} = \frac{1}{\mathbf{Z}_2} + \frac{1}{\mathbf{Z}_3} + \frac{1}{\mathbf{Z}_4}$$

Substitute:
$$\frac{1}{\mathbf{Z}_A} = \frac{1}{2} + \frac{1}{j2} + \frac{1}{3 - j5}$$

Rationalize:
$$\frac{1}{Z_A} = 0.5 + \left(\frac{1}{j2}\right)\left(\frac{j}{j}\right) + \left(\frac{1}{3-j5}\right)\left(\frac{3+j5}{3+j5}\right)$$

$$= 0.5 - j0.5 + \frac{3+j5}{34}$$

$$= 0.5 - j0.5 + 0.088 + j0.148$$

$$\frac{1}{Z_A} = 0.588 - j0.352 = 0.685 \underline{/-30.9°}$$

Take the reciprocal: $Z_A = \dfrac{1}{0.685 \underline{/-30.9°}} = 1.46 \underline{/30.9°} \; \Omega$ *Check*

When all three complex impedances have components of resistance and reactance, it is recommended that formula (*18-8a*) be used.

18.8   Determine the total impedance Z_T as seen from terminals A and B in the bridge circuit (Fig. 18-22).

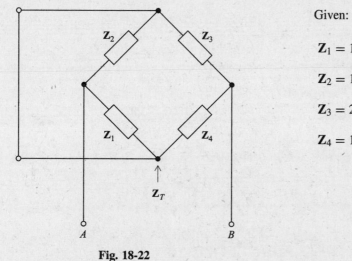

Given:

$$Z_1 = 100 \underline{/0°} = 100 + j0 \; \Omega$$

$$Z_2 = 150 \underline{/30°} = 130 + j75 \; \Omega$$

$$Z_3 = 250 \underline{/0°} = 250 + j0 \; \Omega$$

$$Z_4 = 100 \underline{/-30°} = 86.6 - j50 \; \Omega$$

Fig. 18-22

Step 1.   Write the formula for Z_T between terminals A and B. The parallel combination of Z_1 and Z_2 is in series with the parallel combination of Z_3 and Z_4. Therefore,

$$Z_T = \frac{Z_1 Z_2}{Z_1 + Z_2} + \frac{Z_3 Z_4}{Z_3 + Z_4}   \quad (1) \quad \text{from Eq. (18-9)}$$

Step 2.   Solve for Z_T by substituting the given impedance values into Eq. (*18-9*) (*1*) and simplifying.

$$Z_T = \frac{(100 \underline{/0°})(150 \underline{/30°})}{(100 + j0) + (130 + j75)} + \frac{(250 \underline{/0°})(100 \underline{/-30°})}{(250 + j0) + (86.6 - j50)}$$

$$= \frac{15\,000 \underline{/30°}}{230 + j75} + \frac{25\,000 \underline{/-30°}}{336.6 - j50} = \frac{15\,000 \underline{/30°}}{241.9 \underline{/18.1°}} + \frac{25\,000 \underline{/-30°}}{340.3 \underline{/-8.4°}}$$

$$= 62.0 \underline{/11.9°} + 73.5 \underline{/-21.6°} = 60.7 + j12.8 + (68.3 - j27.1)$$

$$Z_T = 129.0 - j14.3 = 129.8 \underline{/-6.3°} \; \Omega \quad \textit{Ans.}$$

18.9 The total current entering the parallel circuit is $\mathbf{I}_T = 20\ \underline{/45°}$ A (Fig. 18-23). Find the potential difference between points A and B.

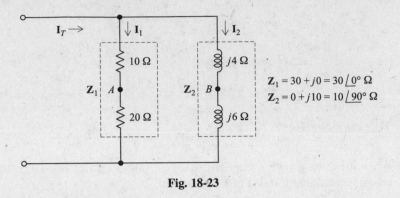

$\mathbf{Z}_1 = 30 + j0 = 30\ \underline{/0°}\ \Omega$
$\mathbf{Z}_2 = 0 + j10 = 10\ \underline{/90°}\ \Omega$

Fig. 18-23

Step 1. Find $\mathbf{I}_1$ and $\mathbf{I}_2$ by the parallel current division formula. See Solved Problem 18.3(b) for formulas.

$$\mathbf{I}_1 = \left(\frac{\mathbf{Z}_2}{\mathbf{Z}_1 + \mathbf{Z}_2}\right)\mathbf{I}_T$$

$$= \left(\frac{10\ \underline{/90°}}{30 + j10}\right)(20\ \underline{/45°})$$

$$= \frac{200\ \underline{/135°}}{30 + j10} = \frac{200\ \underline{/135°}}{31.6\ \underline{/18.4°}}$$

$$\mathbf{I}_1 = 6.33\ \underline{/116.6°}\ \text{A}$$

Similarly,
$$\mathbf{I}_2 = \left(\frac{\mathbf{Z}_1}{\mathbf{Z}_1 + \mathbf{Z}_2}\right)\mathbf{I}_T$$

$$= \left(\frac{30\ \underline{/0°}}{30 + j10}\right)(20\ \underline{/45°})$$

$$= \frac{600\ \underline{/45°}}{31.6\ \underline{/18.4°}}$$

$$\mathbf{I}_2 = 19.0\ \underline{/26.6°}\ \text{A}$$

Step 2. Calculate the voltage drop across the 20-Ω resistance and the $j6$-Ω reactance.

$$\mathbf{V}_{20\Omega} = \mathbf{I}_1(20\ \underline{/0°}) = (6.33\ \underline{/116.6°})(20\ \underline{/0°}) = 126.6\ \underline{/116.6°} = -56.7 + j113.2\ \text{V}$$

$$\mathbf{V}_{j6\Omega} = \mathbf{I}_2(6\ \underline{/90°}) = (19.0\ \underline{/26.6°})(6\ \underline{/90°}) = 114\ \underline{/116.6°} = -51.0 + j101.9\ \text{V}$$

Step 3. Find the voltage difference between A and B, $\mathbf{V}_{AB}$.

The sketch shows the correct polarity. So $\mathbf{V}_{AB}$ is the difference between $\mathbf{V}_{20\Omega}$ and $\mathbf{V}_{j6\Omega}$.

$$\mathbf{V}_{AB} = \mathbf{V}_{20\Omega} - \mathbf{V}_{j6\Omega} = (-56.7 + j113.2) - (-51.0 + j101.9)$$

$$\mathbf{V}_{AB} = -5.7 + j11.3 = 12.7 \underline{/116.8°}$$

A voltmeter placed between points A and B would read 12.7 V.   *Ans.*

18.10   An ac power line of 110 V operates a parallel bank of lamps and a small induction motor (Fig. 18-24). Find (a) the total effective power P_T, (b) the total apparent power S_T, (c) the total reactive power Q_T, (d) the total PF, and (e) the total current $\mathbf{I}_T$.

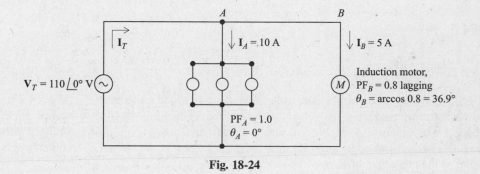

Fig. 18-24

Apparent power in branch A, S_A, is not in phase with the apparent power in branch B, S_B, since they have different power factors. Thus, we must find the complex power for branch A, $\mathbf{S}_A$, and for branch B, $\mathbf{S}_B$, and then add them to obtain total power of the circuit, $\mathbf{S}_T$.

Step 1.   Find $\mathbf{S}_A$.

$$S_A = V_A I_A = (110)(10) = 1100 \text{ VA}$$

$$P_A = S_A \cos\theta_A = S_A \cos 0° = (1100)(1) = 1100 \text{ W} \qquad (18\text{-}12)$$

$$Q_A = S_A \sin\theta_A = 1100 \sin 0° = 1100 \times 0 = 0 \text{ vars} \qquad (18\text{-}13)$$

$$\mathbf{S}_A = P_A + jQ_A = 1100 + j0 \qquad (18\text{-}10)$$

Step 2.   Find $\mathbf{S}_B$.

Similarly, $$S_B = V_B I_B = (110)(5) = 550 \text{ VA}$$

$$P_B = S_B \cos\theta_B = (550)(0.8) = 440 \text{ W}$$

$$Q_B = S_B \sin\theta_B = (550)(\sin 36.9°) = 330 \text{ vars, lagging}$$

$$\mathbf{S}_B = P_B + jQ_B = 440 + j330$$

Step 3.   Find $\mathbf{S}_T$.

$$\mathbf{S}_T = \mathbf{S}_A + \mathbf{S}_B$$

$$= (1100 + j0) + (440 + j330)$$

$$\mathbf{S}_T = \underbrace{1540}_{P_T} + \underbrace{j330}_{Q_T} = \underbrace{1575}_{S_T} \ \underbrace{\underline{/12.1°}}_{\text{PF}_T \text{ angle}}$$

Interpreting S_T as complex power, we have

(a) $P_T = 1540\,\text{W}$ *Ans.*

(b) $S_T = 1575\,\text{VA}$ *Ans.*

(c) $Q_T = 330\,\text{vars, lagging}$ *Ans.*

(d) $\text{PF}_T = \cos 12.1° = 0.978\,\text{lagging}$ *Ans.*

Step 4.   Find $\mathbf{I}_T$.

$$\mathbf{S}_T = \mathbf{V}_T \mathbf{I}_T$$

$$\mathbf{I}_T = \frac{\mathbf{S}_T}{\mathbf{V}_T}$$

$$= \frac{1575\,\underline{/12.1°}}{110\,\underline{/0°}} = 14.3\,\underline{/12.1°}\,\text{A}   \quad \textit{Ans.}$$

18.11   A generator is in series with a fixed impedance $\mathbf{Z}_G$ and a load impedance $\mathbf{Z}_L$ (Fig. 18-25). Maximum power is transferred from the generator to the load by making the load resistance R_L equal to the generator resistance R_G, $R_L = R_G$, and the load reactance X_L (not to be confused with inductive reactance) equal and opposite to the generator reactance X_G, $X_L = -X_G$. In other words, make $\mathbf{Z}_L$ equal to the conjugate of $\mathbf{Z}_G$,

$$\mathbf{Z}_L = \mathbf{Z}_G^*   \tag{1}$$

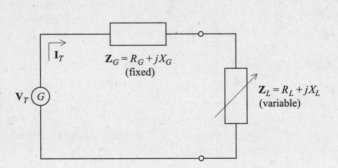

Fig. 18-25   Maximum power transfer

In an actual circuit, $\mathbf{Z}_G$ is the internal impedance of the generator plus the impedance of connecting lines and any other elements in the circuit. If $\mathbf{Z}_G$ were a pure resistance, $\mathbf{Z}_L$ would be an equal resistance in order to receive maximum power. If $\mathbf{Z}_G$ were inductive, $\mathbf{Z}_L$ would need to be capacitive. It is clear that a load can be adjusted to receive maximum power *only* if its resistance and reactance can both be varied independently. *Impedance matching* to obtain maximum power to the load is important in practically all communication engineering. Though the conjugate match is the ideal, in practice both generator impedance and load impedance are likely to be mainly resistive, and often it is adequate to make their impedance equal in magnitude,

$$|Z_L| = |Z_G| \tag{2}$$

If a 120-V generator has an internal resistance of $\mathbf{Z}_G = 4 + j3$, find (a) the load impedance $\mathbf{Z}_L$ for the maximum power to be transferred to the load, (b) the power P_L delivered to the load, (c) the power factor, and (d) the power efficiency of the circuit.

Step 1. Find $\mathbf{Z}_L$.

   (a)   For maximum transfer of power, $\mathbf{Z}_L = \mathbf{Z}_G^* = 4 - j3$ *Ans.*

Step 2. Find I_T.

$$\mathbf{Z}_T = \mathbf{Z}_G + \mathbf{Z}_L = (4 + j3) + (4 - j3)$$

$$= 8 + 0j = 8 \ \underline{/0°} \ \Omega$$

Let $\mathbf{V}_T = 120 \ \underline{/0°} \ \text{V}$ be the reference.

$$\mathbf{I}_T = \frac{\mathbf{V}_T}{\mathbf{Z}_T} = \frac{120 \ \underline{/0°}}{8 \ \underline{/0°}} = 15 \ \underline{/0°} \ \text{A}, \ \mathbf{V}_T \text{ and } \mathbf{I}_T \text{ are in phase at } 0°.$$

Step 3. Find P_L and PF.

   (b) $P_L = I_T^2 R_L = (15)^2 (4) = 900 \ \text{W}$ *Ans.*

   (c) $\text{PF} = \cos\theta = \cos 0° = 1$ *Ans.*

Step 4. Find the power efficiency.

$$\text{Power efficiency} = \frac{\text{Power output at load}}{\text{Power input}} = \frac{P_L}{P_{IN}}$$

$$P_{IN} = V_T I_T = (120)(15) = 1800 \ \text{W}   \text{\textit{Ans.}}$$

   (d)   Power efficiency $= \dfrac{900}{1800} = 0.50 = 50\%$ *Ans.*

A matched load is always 50% efficient, meaning that the load receives half the output from the generator source, the other half dissipated by the generator resistance. This also means that the terminal voltage drops to half when maximum load is applied. Neither of these conditions is tolerable on a power system.

18.12   Two motors on the same line (Fig. 18-26) use 33 kW at 0.96 PF leading. The motor in branch *B* draws 25 kW at 0.86 PF lagging. What is the real power, reactive power, apparent power, and PF of motor *A*?

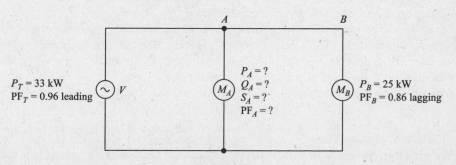

Fig. 18-26

Step 1. Find the power relations of the total circuit, $\mathbf{S}_T$.

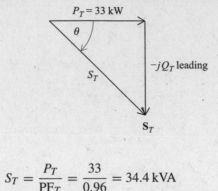

$$S_T = \frac{P_T}{\text{PF}_T} = \frac{33}{0.96} = 34.4\,\text{kVA}$$

$$\text{PF}_T = \cos\theta = 0.96$$

$$\theta = \arccos 0.96 = 16.3°$$

$$Q_T = S_T \sin\theta = 34.4 \sin 16.3° = (34.4)(0.281)$$

$$= 9.65\,\text{kvars leading}$$

Then $$\mathbf{S}_T = P_T - jQ_T = 33 - j9.65\,\text{kVA}$$

Step 2. Find the power relations of motor B, $\mathbf{S}_B$.

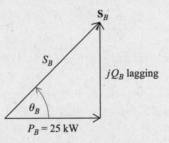

$$S_B = \frac{P_B}{\text{PF}_B} = \frac{25}{0.86} = 29.1\,\text{kVA}$$

$$\text{PF}_B = \cos\theta_B = 0.86$$

$$\theta_B = \arccos 0.86 = 30.7°$$

$$Q_B = S_B \sin\theta = 29.1 \sin 30.7° = (29.1)(0.511)$$

$$= 14.9\,\text{kvars lagging}$$

Then $$\mathbf{S}_B = 25 + j14.86\,\text{kVA}$$

Step 3. Find the power relations of motor A, $\mathbf{S}_A$.

$$\mathbf{S}_T = \mathbf{S}_A + \mathbf{S}_B, \quad \text{total complex power equals the sum of the complex power of branch } A \text{ and branch } B.$$

Then $$\mathbf{S}_A = \mathbf{S}_T - \mathbf{S}_B$$

Total circuit: $\mathbf{S}_T = 33 - j9.65$

Motor B: $\mathbf{S}_B = 25 + j14.86$

Motor A: $\mathbf{S}_A = \mathbf{S}_T - \mathbf{S}_B = \underbrace{8}_{P_A} - j\underbrace{24.51}_{Q_A} = \underbrace{25.8}_{VI_A} \underline{/-71.9°}\ \text{kVA}$

Thus, $P_A = 8\,\text{kW}$ *Ans.*

$\qquad\qquad\quad Q_A = 24.5\,\text{kvars leading}$ *Ans.*

$\qquad\qquad\quad S_A = VI_A = 25.8\,\text{kVA}$ *Ans.*

$\qquad\qquad\quad \text{PF}_A = \cos\theta_A = \cos(-71.9°) = 0.31\ \text{leading}$ *Ans.*

Note that total PF_T does not equal the sum of the individual branches, $\text{PF}_A + \text{PF}_B$. That is

$$\text{PF}_T \neq \text{PF}_A + \text{PF}_B$$

$$0.96\ \text{leading} \neq 0.31\ \text{leading} + 0.86\ \text{lagging}$$

18.13   Find currents $\mathbf{I}_1$ and $\mathbf{I}_2$ in a series–parallel circuit (Fig. 18-27).

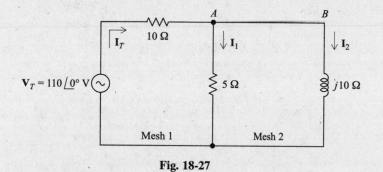

Fig. 18-27

Step 1.   Find total or equivalent impedance $\mathbf{Z}_T$.

10 Ω-resistor is in series with parallel impedances in branch A and branch B.

$$\mathbf{Z}_1 = 5 + j0 = 5\ \underline{/0°}$$

$$\mathbf{Z}_2 = 0 + j10 = 10\ \underline{/90°}$$

$$\mathbf{Z}_T = 10 + \frac{\mathbf{Z}_1\mathbf{Z}_2}{\mathbf{Z}_1 + \mathbf{Z}_2}$$

$$\frac{\mathbf{Z}_1\mathbf{Z}_2}{\mathbf{Z}_1 + \mathbf{Z}_2} = \frac{(5\ \underline{/0°})(10\ \underline{/90°})}{(5 + j0) + (0 + j10)} = \frac{50\ \underline{/90°}}{5 + j10} = \frac{50\ \underline{/90°}}{11.18\ \underline{/63.43°}}$$

$$= 4.47\ \underline{/26.57°} = 4.00 + j2.00$$

So $\mathbf{Z}_T = 10 + (4.00 + j2.00) = 14.00 + j2.00 = 14.14\ \underline{/8.13°}\ \Omega$

Step 2.   Solve for $\mathbf{I}_T$.

$$\mathbf{I}_T = \frac{\mathbf{V}_T}{\mathbf{Z}_T} = \frac{110\ \underline{/0°}}{14.14\ \underline{/8.13°}} = 7.78\ \underline{/-8.13°} = 7.70 - j1.10\,\text{A}$$

Step 3. Find nodal voltage at A, $\mathbf{V}_A$.

$$\mathbf{V}_A = \mathbf{V}_T - \mathbf{V}_{10\Omega}, \quad \text{the difference between the input voltage and the voltage drop across the 10-Ω series resistor.}$$

$$\mathbf{V}_{10\Omega} = \mathbf{I}_T R = (7.70 - j1.10)(10) = 77.0 - j11.0 \text{ V}$$

$$\mathbf{V}_A = (110 + j0) - (77.0 - j11.0) = 33 + j11 = 34.79 \underline{/18.43°} \text{ V}$$

Step 4. Solve for $\mathbf{I}_1$ and $\mathbf{I}_2$.

$$\mathbf{I}_1 = \frac{\mathbf{V}_A}{\mathbf{Z}_1} = \frac{34.79 \underline{/18.43°}}{5 \underline{/0°}} = 6.96 \underline{/18.4°} \text{ A} \qquad Ans.$$

$$\mathbf{I}_2 = \frac{\mathbf{V}_A}{\mathbf{Z}_2} = \frac{34.79 \underline{/18.43°}}{10 \underline{/90°}} = 3.48 \underline{/-71.6°} \text{ A} \qquad Ans.$$

18.14   Find $\mathbf{I}_2$ in Fig. 18-27 by the determinant method of solution.

Step 1. Find the impedance determinant Δ.

By inspection,

$$\mathbf{Z}_{11} = 10 + 5 = 15 + j0 = 15 \underline{/0°} \ \Omega$$

$$\mathbf{Z}_{12} = \mathbf{Z}_{21} = 5 + j0 = 5 \underline{/0°} \ \Omega$$

$$\mathbf{Z}_{22} = 5 + j10 = 11.18 \underline{/63.43°} \ \Omega$$

$$\Delta = \begin{vmatrix} \mathbf{Z}_{11} & -\mathbf{Z}_{12} \\ -\mathbf{Z}_{21} & \mathbf{Z}_{22} \end{vmatrix} = \begin{vmatrix} 15 \underline{/0°} & -5 \underline{/0°} \\ -5 \underline{/0°} & 11.18 \underline{/63.43°} \end{vmatrix} \qquad (18\text{-}18)$$

$$= (15 \underline{/0°})(11.18 \underline{/63.43°}) - (-5 \underline{/0°})(-5 \underline{/0°})$$

$$= 167.70 \underline{/63.43°} - 25 \underline{/0°}$$

$$= 75.01 + j149.99 - 25 = 50.01 + j149.99 = 158.11 \underline{/71.56°}$$

Step 2. Find $\mathbf{N}_{I_2}$ by replacing column 2 of Eq. *18-18* by the net mesh voltages, $\mathbf{V}_1$ and $\mathbf{V}_2$.

$$\mathbf{N}_{I_2} = \begin{vmatrix} \mathbf{Z}_{11} & \mathbf{V}_1 \\ -\mathbf{Z}_{12} & \mathbf{V}_2 \end{vmatrix} = \begin{vmatrix} 15 \underline{/0°} & 110 \underline{/0°} \\ -5 \underline{/0°} & 0 \end{vmatrix}$$

$$= (15 \underline{/0°})(0) - (110 \underline{/0°})(-5 \underline{/0°}) = 550 \underline{/0°}$$

Step 3. Solve for $\mathbf{I}_2$.

$$\mathbf{I}_2 = \frac{\mathbf{N}_{I_2}}{\Delta} = \frac{550 \underline{/0°}}{158.11 \underline{/71.56°}} = 3.48 \underline{/71.6°} \text{ A} \qquad Ans.$$

Value of $\mathbf{I}_2$ agrees with that found in Solved Problem 18.13.

18.15 Find by mesh analysis the total current $\mathbf{I}_T$ through the generator and the impedance seen by the generator if its voltage is 120 V (Fig. 18-28). For simplification, assume 1-V service even though the source voltage is 120 V. Resulting current values can then be multiplied by any given voltage to determine the corresponding current produced by the given voltage.

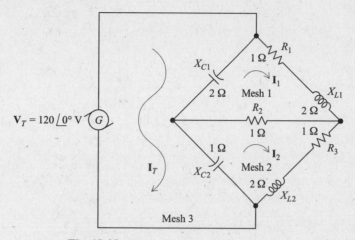

Fig. 18-28 Reactances in a 3-mesh network

Step 1. Find the impedance determinant Δ.

$$\Delta = \begin{vmatrix} \mathbf{Z}_{11} & -\mathbf{Z}_{12} & -\mathbf{Z}_{13} \\ -\mathbf{Z}_{21} & \mathbf{Z}_{22} & -\mathbf{Z}_{23} \\ -\mathbf{Z}_{31} & -\mathbf{Z}_{32} & \mathbf{Z}_{33} \end{vmatrix} \tag{18-19}$$

$$\mathbf{Z}_{11} = (R_1 + R_2) + jX_{L1} - jX_{C1} = 2 + j2 - j2 = 2\ \Omega$$

$$\mathbf{Z}_{12} = \mathbf{Z}_{21} = R_2 = 1\ \Omega$$

$$\mathbf{Z}_{13} = \mathbf{Z}_{31} = -jX_{C1} = -j2\ \Omega$$

$$\mathbf{Z}_{22} = (R_2 + R_3) + jX_{L2} - jX_{C2} = 2 + j2 - j1 = 2 + j1\ \Omega$$

$$\mathbf{Z}_{23} = \mathbf{Z}_{32} = -jX_{C2} = -j1\ \Omega$$

$$\mathbf{Z}_{33} = -jX_{C1} - jX_{C2} = -j1 - j2 = -j3\ \Omega$$

Substitute impedance values in Eq. *18-19*.

$$\Delta = \begin{vmatrix} 2 & -1 & j2 \\ -1 & 2+j1 & j1 \\ j2 & j1 & -j3 \end{vmatrix}$$

Expand the first column into second-order determinants.

$$\Delta = 2\begin{vmatrix} 2+j1 & j1 \\ j1 & -j3 \end{vmatrix} - (-1)\begin{vmatrix} -1 & j2 \\ j1 & -j3 \end{vmatrix} + j2\begin{vmatrix} -1 & j2 \\ 2+j1 & j1 \end{vmatrix}$$

$$= 2\big[(2+j1)(-j3) - (j1)(j1)\big] + \big[(-1)(-j3) - (j1)(j2)\big] + j2\big[-j - j2(2+j1)\big]$$

$$= (-j12 + 8) + (j3 + 2) + (10 + j4)$$

$$= 20 - j5$$

Step 2. Find N_{I_T} and then I_T.

Since I_T is the current in Mesh 3, we replace the 3rd column in Δ by the net voltages of each mesh to find N_{I_T}. Assume $V_3 = 1 \text{ V}$; $V_1 = V_2 = 0 \text{ V}$.

$$N_{I_3} = N_{I_T} = \begin{vmatrix} Z_{11} & -Z_{12} & V_1 \\ -Z_{21} & Z_{22} & V_2 \\ -Z_{31} & -Z_{32} & V_3 \end{vmatrix} \qquad \text{(from Eq. } 18\text{-}21\text{)}$$

$$= \begin{vmatrix} 2 & -1 & 0 \\ -1 & 2+j1 & 0 \\ j2 & j1 & 1 \end{vmatrix}$$

$$= 2 \begin{vmatrix} 2+j1 & 0 \\ j & 1 \end{vmatrix} - (-1) \begin{vmatrix} -1 & 0 \\ j1 & 1 \end{vmatrix} + j2 \begin{vmatrix} -1 & 0 \\ 2+j1 & 0 \end{vmatrix}$$

$$= 2(2+j1) + (-1) + 0$$

$$= 3 + j2$$

$$I_T = \frac{N_{I_T}}{\Delta}$$

$$= \frac{3+2j}{20-j5}$$

Rationalizing, $\qquad I_T = \dfrac{3+2j}{20-j5} \cdot \dfrac{20+j5}{20+j5} = \dfrac{50+j55}{425} = 0.118 + j0.129$

$$= 0.175 \underline{/47.6°} \text{ A}$$

At $V_T = 120 \text{ V}$, the current will be 120 times greater at the same phase angle, so

$$I_T = (120)(0.175 \underline{/47.6°}) = 21.0 \underline{/47.6°} \text{ A} \qquad Ans.$$

Step 3. Find the total impedance.

$$Z_T = \frac{V_T}{I_T} = \frac{120 \underline{/0°}}{21.0 \underline{/47.6°}} = 5.71 \underline{/-47.6°} \ \Omega \qquad Ans.$$

18.16 Show for parallel-connected networks that the total real power P_T and the total reactive power Q_T are the sum of the individual real power and the individual reactive power, respectively, in each branch.

Write $\qquad\qquad\qquad S = VI^*$ $\qquad\qquad\qquad\qquad\qquad\qquad\qquad$ (18-17)

$$S_T = VI_T^* = V\left(I_1^* + I_2^* + I_3^* + \cdots + I_n^*\right) \qquad \text{for } n\text{-branches}$$

$$= VI_1^* + VI_2^* + VI_3^* + \cdots + VI_n^*$$

Therefore $\qquad S_T = S_1 + S_2 + S_3 + \cdots + S_n$

But $\qquad\qquad S_T = P_T + jQ_T, \quad S_1 = P_1 + jQ_1, \quad S_2 = P_2 + jQ_2,$

$\qquad\qquad\qquad S_3 = P_3 + jQ_3, \quad S_n = P_n + jQ_n$

So $\qquad P_T + jQ_T = (P_1 + jQ_1) + (P_2 + jQ_2) + (P_3 + jQ_3) + \cdots + (P_n + jQ_n)$

from which $\qquad P_T = P_1 + P_2 + P_3 + \cdots + P_n \qquad Ans.$

$$Q_T = Q_1 + Q_2 + Q_3 + \cdots + Q_n \qquad Ans.$$

Supplementary Problems

18.17 Refer to Solved Problem 17.14. An applied voltage of $\mathbf{V} = 110\ \underline{/0°}$ V is applied to each series circuit pictured, (a)–(f). For each circuit find the current $\mathbf{I}$, voltage drops across each element, and draw the phasor diagram.

Ans.

(a) $\mathbf{I} = 22\ \underline{/0°}$ A

$\qquad$ $\mathbf{V}_R = 110\ \underline{/0°}$ V

Phasor diagrams

$\mathbf{V}_R = 110\ \underline{/0°}$ V

$\longrightarrow$ **V** reference

$\mathbf{I} = 22\ \underline{/0°}$ A

I and **V** in phase

(b) $\mathbf{I} = 11\ \underline{/-90°}$ A

$\qquad$ $\mathbf{V}_L = 110\ \underline{/0°}$ V

$\mathbf{V}_L = 110\ \underline{/0°}$ V

$\longrightarrow$ **V** reference

$-90°$

I lags **V** by 90°

$\mathbf{I} = 11\ \underline{/-90°}$ A

(c) $\mathbf{I} = 11\ \underline{/90°}$ A

$\qquad$ $\mathbf{V}_C = 110\ \underline{/0°}$ V

$\mathbf{I} = 11\ \underline{/90°}$ A

I leads **V** by 90°

$90°$

$\longrightarrow$ **V** reference

$\mathbf{V}_C = 110\ \underline{/0°}$ V

(d) $\mathbf{I} = 6.96\ \underline{/-71.6°}$ A

$\qquad$ $\mathbf{V}_R = 34.8\ \underline{/-71.6°}$ V

$\qquad$ $\mathbf{V}_L = 104.4\ \underline{/18.4°}$ V

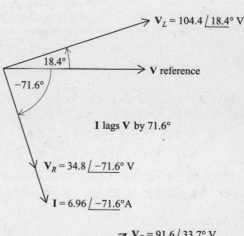

$\mathbf{V}_L = 104.4\ \underline{/18.4°}$ V

$18.4°$

$\longrightarrow$ **V** reference

$-71.6°$

I lags **V** by 71.6°

$\mathbf{V}_R = 34.8\ \underline{/-71.6°}$ V

$\mathbf{I} = 6.96\ \underline{/-71.6°}$ A

(e) $\mathbf{I} = 6.11\ \underline{/33.7°}$ A

$\qquad$ $\mathbf{V}_R = 91.6\ \underline{/33.7°}$ V

$\qquad$ $\mathbf{V}_C = 61.1\ \underline{/-56.3°}$ V

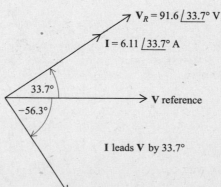

$\mathbf{V}_R = 91.6\ \underline{/33.7°}$ V

$\mathbf{I} = 6.11\ \underline{/33.7°}$ A

$33.7°$

$\longrightarrow$ **V** reference

$-56.3°$

I leads **V** by 33.7°

$\mathbf{V}_C = 61.1\ \underline{/-56.3°}$ V

(f) $\mathbf{I} = 25.9 \underline{/-45°}$ A

$\mathbf{V}_R = 77.7 \underline{/-45°}$ V

$\mathbf{V}_L = 207.2 \underline{/45°}$ V

$\mathbf{V}_C = 129.5 \underline{/-135°}$ V

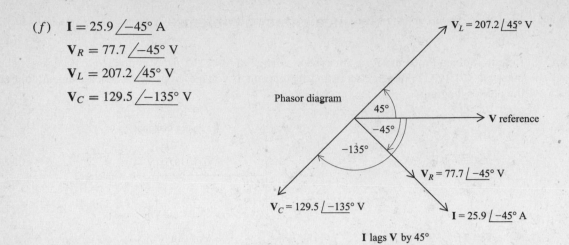

Phasor diagram

I lags **V** by 45°

18.18 120 V is applied to a series *RLC* circuit (Fig. 18-29). Find the total impedance, resultant current, and the voltage drops across each element in both polar and rectangular forms. Draw the phasor diagram. Check your solution by equating the given applied voltage to the sum of the voltage drops.

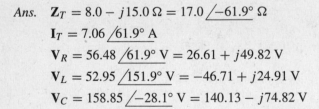

Fig. 18-29

Ans. $\mathbf{Z}_T = 8.0 - j15.0\ \Omega = 17.0 \underline{/-61.9°}\ \Omega$

$\mathbf{I}_T = 7.06 \underline{/61.9°}$ A

$\mathbf{V}_R = 56.48 \underline{/61.9°}$ V $= 26.61 + j49.82$ V

$\mathbf{V}_L = 52.95 \underline{/151.9°}$ V $= -46.71 + j24.91$ V

$\mathbf{V}_C = 158.85 \underline{/-28.1°}$ V $= 140.13 - j74.82$ V

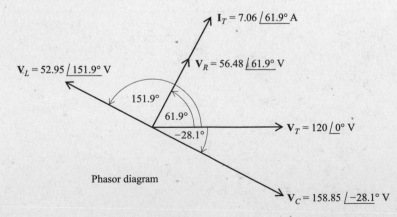

Phasor diagram

18.19   Find the branch currents I_1 and I_2 by the current division formula found in Solved Problem 18.3 (Fig. 18-30).

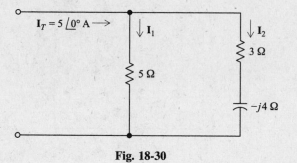

Fig. 18-30

Ans. $I_1 = 2.8 \angle{-26.6°}$ A
         $I_2 = 2.8 \angle{26.6°}$ A

18.20   If $I_T = 5 \angle{30°}$ A and $I_1 = 3 \angle{-30°}$ A, find Z_2 (Fig. 18-31).

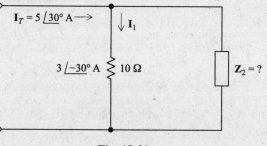

Fig. 18-31

Ans. $Z_2 = 0.79 - j6.84$ Ω
         Because the resistance is so small, the impedance Z_2 for practical purposes is a capacitor.

18.21   Find the equivalent impedance of the parallel circuit (Fig. 18-32).

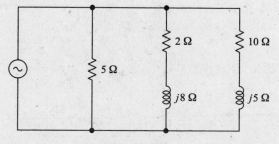

Fig. 18-32

Ans. $Z_T = 2.87 \angle{27°}$ Ω

18.22   In the series ac circuit (Fig. 18-33), find $\mathbf{Z}_T$ and $\mathbf{I}_T$. Is the circuit a lagging or leading one?

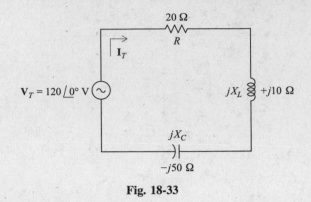

Fig. 18-33

Ans. $\mathbf{Z}_T = 20 - j40 = 44.7 \underline{/-63.4°}\ \Omega$

$\mathbf{I}_T = 2.68 \underline{/63.4°}\ \text{A}$

The circuit is leading because $X_C > X_L$, resulting in $\mathbf{I}_T$ leading $\mathbf{V}_T$ by 63.4°.

18.23   For the circuit in Fig. 18-33, find the phasor voltages across each element and verify your answers by showing that $\mathbf{V}_T$ equals the sum of the voltage drops.

Ans. $\mathbf{V}_R = 53.6 \underline{/63.4°} = 24 + j48\ \text{V}$

$\mathbf{V}_{X_L} = 26.8 \underline{/153.4°} = -24 + j12\ \text{V}$

$\mathbf{V}_{X_C} = 134 \underline{/-26.6°} = 120 - j60\ \text{V}$

18.24   For the parallel circuit (Fig. 18-34), find branch currents $\mathbf{I}_1$ and $\mathbf{I}_2$, and total current $\mathbf{I}_T$. Draw the current-phasor diagram with $\mathbf{V}_T$ as the reference.

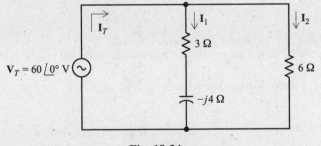

Fig. 18-34

Ans. $\mathbf{I}_1 = 12 \underline{/53.1°} = 7.2 + j9.6\ \text{A}$

$\mathbf{I}_2 = 10 \underline{/0°} = 10 + j10\ \text{A}$

$\mathbf{I}_T = 17.2 + j9.6 = 19.7 \underline{/29.2°}\ \text{A}$

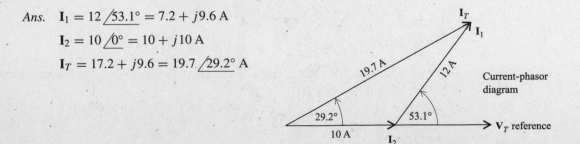

18.25   Figure 18-35 shows a coil with negligible resistance, a capacitor, and a resistor connected in parallel to a source voltage of 120 V. What will be the reading of ammeter A? Draw the current-phasor diagram.

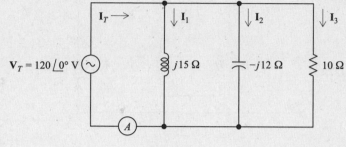

Fig. 18-35

Ans. $\mathbf{I}_T = 12 + j2 = 12.2 \underline{/9.46°}$ A. Ammeter will read effective value of 12.2 A.

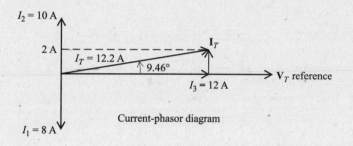

Current-phasor diagram

18.26   A voltmeter placed across the 5-Ω resistor (Fig. 18-36) reads 30 V. What does the ammeter read?

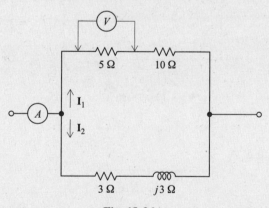

Fig. 18-36

Ans.   25.8 A
       $(\mathbf{I}_1 = 6 \underline{/0°}$ A, $\mathbf{I}_2 = 21.2 \underline{/-45°}$ A$)$

18.27   A voltmeter placed across the 5-Ω resistor (Fig. 18-37) indicates 45 V. What is the ammeter reading? What is the voltage reading between points A and B?

Ans.   18.0 A
       $(\mathbf{I}_1 = 9 \underline{/0°}$ A, $\mathbf{I}_2 = 9.7 \underline{/-31.3°}$ A$)$
       $V_{AB} = 25.2$ V

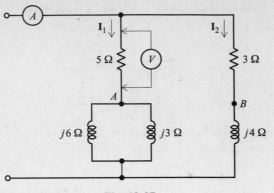

Fig. 18-37

18.28   For the circuit shown in Fig. 18-38 solve for

    (*a*)   equivalent series impedance $\mathbf{Z}_T$

    (*b*)   total current $\mathbf{I}_T$, and

    (*c*)   effective power drawn by the circuit.

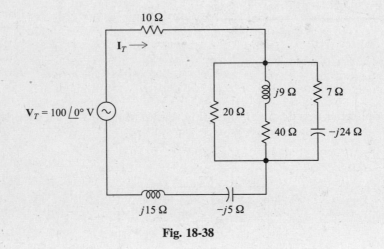

Fig. 18-38

Ans.   (*a*) $21.2\,\underline{/16.5^\circ}\ \Omega$

      (*b*) $4.7\,\underline{/-16.5^\circ}\ \text{A}$

      (*c*) $451\ \text{W}$

18.29   In the parallel circuit (Fig. 18-39), find (*a*) all the branch currents, (*b*) total current, (*c*) total impedance, and (*d*) power drawn.

Ans.   (*a*) $\mathbf{I}_1 = 10\,\underline{/0^\circ} = 10 + j0\ \text{A}$

             $\mathbf{I}_2 = 4\,\underline{/-73.7^\circ} = 1.12 - j3.84\ \text{A}$

             $\mathbf{I}_3 = 5\,\underline{/53.1^\circ} = 3 + j4\ \text{A}$

      (*b*) $\mathbf{I}_T = 14.1 + j0.16 = 14.1\,\underline{/1^\circ}\ \text{A}$

      (*c*) $\mathbf{Z}_T = 7.09\,\underline{/-1^\circ}\ \Omega$

      (*d*) $P = 1410\ \text{W}$

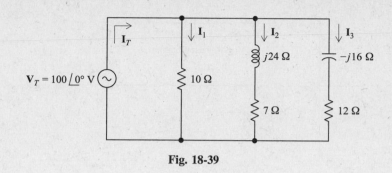

Fig. 18-39

18.30   In the series–parallel circuit (Fig. 18-40), (a) write the impedances of each branch and then find (b) the total impedance and (c) total current.

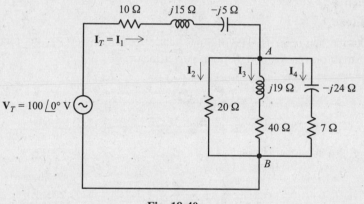

Fig. 18-40

Ans.   (a) $\mathbf{Z}_1 = 10 + j10 = 14.14 \; \underline{/45°} \; \Omega$

         $\mathbf{Z}_2 = 20 = 20 \; \underline{/0°} \; \Omega$

         $\mathbf{Z}_3 = 40 + j19 = 44.2 \; \underline{/25.4°} \; \Omega$

         $\mathbf{Z}_4 = 7 - j24 = 25 \; \underline{/-73.7°} \; \Omega$

    (b) $\mathbf{Z}_T = 20.9 + j6.16 = 21.8 \; \underline{/16.4°} \; \Omega$

    (c) $\mathbf{I}_T = 4.59 \; \underline{/-16.4°} \; \Omega$

18.31   For the circuit shown (Fig. 18-40), find the voltage drop across the parallel branch $\mathbf{V}_{AB}$. Then find the branch currents $\mathbf{I}_2, \mathbf{I}_3$, and $\mathbf{I}_4$. Finally, draw the current-phasor diagram with $\mathbf{V}_T$ as reference.

Ans. $\mathbf{V}_{AB} = 43 - j31 = 53.0 \; \underline{/-35.8°} \; \text{V}$

         $\mathbf{I}_2 = 2.65 \; \underline{/-35.8°} \; \text{A}$

         $\mathbf{I}_3 = 1.20 \; \underline{/-61.2°} \; \text{A}$

         $\mathbf{I}_4 = 2.12 \; \underline{/-37.9°} \; \text{A}$

Ans.

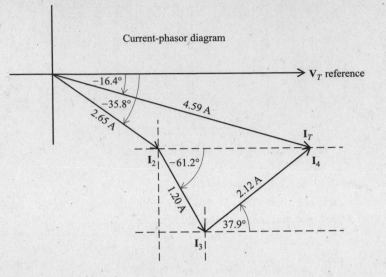

18.32 For the circuit shown (Fig. 18-41), find the total effective power P_T, the total reactive power Q_T, the total apparent power S_T, and the total power factor, PF_T.

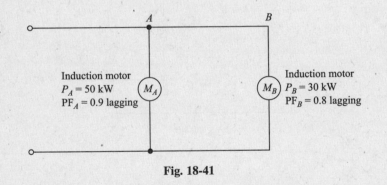

Fig. 18-41

Ans. $P_T = 80\,\text{kW}$

$Q_T = 46.7\,\text{kvars lagging}$

$S_T = 92.6\,\text{kVA}$

$PF_T = 0.863\,\text{lagging}$

18.33 A 3-kW lamp load is in parallel with a motor operating at a PF of 0.6 lagging and drawing 4 kW (Fig. 18-42). Find the power relations of the total circuit and the total current.

Ans. $\mathbf{S}_T = 7 + j5.33 = 8.80\,\underline{/37.3°}\,\text{kVA}$

from which $P_T = 7\,\text{kW}$

$Q_T = 5.33\,\text{kvars lagging}$

$S_T = 8.80\,\text{kVA}$

$PF_T = 0.795\,\text{lagging}$

$\mathbf{I}_T = 80\,\underline{/-37.3°}\,\text{A}$

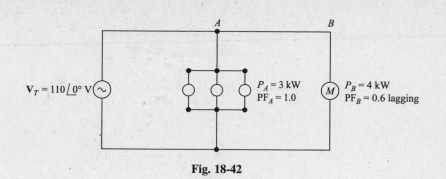

Fig. 18-42

Find the power relations of the following circuits.

18.34

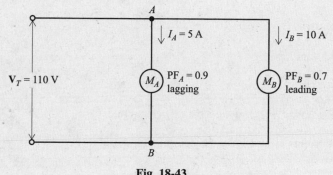

Fig. 18-43

Ans. $\mathbf{S}_T = 1265 - j547 = 1378 \, \underline{/-23.4°}$ VA

$P_T = 1265$ W

$Q_T = 547$ vars leading

$S_T = 1378$ VA

$\text{PF}_T = 0.918$ leading

18.35

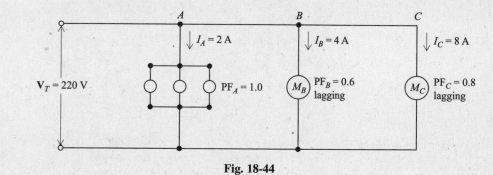

Fig. 18-44

Ans. $\mathbf{S}_T = 2376 + j1760 = 2957 \, \underline{/36.5°}$ VA

$P_T = 2376$ W

$Q_T = 1760$ VAR lagging

$S_T = 2957$ VA

$\text{PF}_T = 0.804$ lagging

18.36

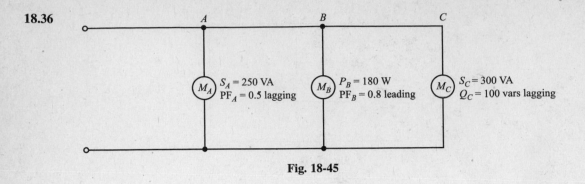

Fig. 18-45

Ans. $\mathbf{S}_T = 588 + j181 = 615 \underline{/17.1°}$ VA

$\mathbf{P}_T = 588$ W

$Q_T = 181$ vars lagging

$S_T = 615$ VA

$\text{PF}_T = 0.956$ lagging

18.37   Find the mesh currents $\mathbf{I}_1$ and $\mathbf{I}_2$ by the determinant method (Fig. 18-46). What is the voltage across the 20-Ω resistor?

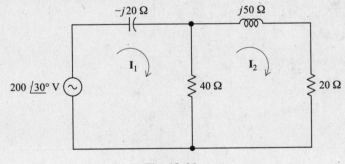

Fig. 18-46

Ans. $\mathbf{I}_1 = 7.93 \underline{/45.8°}$ A

$\mathbf{I}_2 = 4.06 \underline{/6.04°}$ A

$\mathbf{V}_{20\Omega} = 81.2 \underline{/6.04°}$ V

(*Hint*: Impedance determinant $\Delta = 1800 + j800$)

18.38   (*a*)   Find the current $\mathbf{I}_T$ by solving first for the total impedance $\mathbf{Z}_T$ (Fig. 18-47).
(*b*)   Confirm the $\mathbf{I}_T$ value by use of determinants.
Ans.   (*a*) $\mathbf{Z}_T = 43.1 \underline{/48.6°}\ \Omega$

$\mathbf{I}_T = 2.32 \underline{/-28.6°}$ A

(*b*)   (*Hint*: Impedance determinant $\Delta = 750 + j200$)

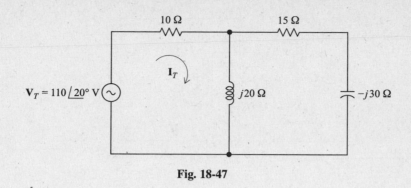

Fig. 18-47

18.39 Solve by determinants in a generator-bridge circuit (Fig. 18-48).

(a) impedance determinant Δ

(b) mesh currents I_1, I_2, I_3

(c) total impedance seen by the generator Z_T

(d) voltage drop across the 1-Ω resistor in the center branch

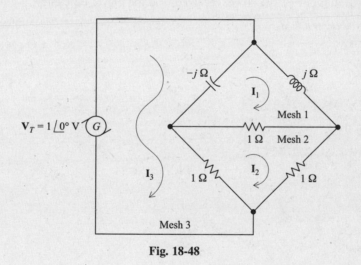

Fig. 18-48

Ans. (a) $\Delta = 4 \underline{/0°} = 4 + j0$

 (b) $I_1 = 0.791 \underline{/-71.65°} = 0.2490 - j0.7508 \text{ A}$

 $I_2 = 0.354 \underline{/-45°} = 0.2500 - j0.2500 \text{ A}$

 $I_3 = 0.500 \underline{/0°} = 0.5000 - j0 \text{ A}$

 (c) $Z_T = 2 \underline{/0°} = 2 + 0j \ \Omega$

 (d) $0 - j0.5 = 0.5 \underline{/-90°} \text{ V}$

18.40 How much capacitance must be provided by the capacitor bank (Fig. 18-49) to improve the power factor of the circuit to 0.95 lagging? What is the resulting decrease in apparent power?

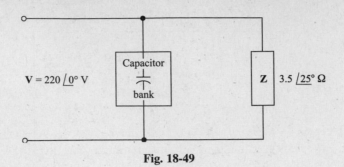

Fig. 18-49

Ans.   1725 vars leading. Decrease in apparent power 636 VA.

18.41   A load of $P = 1200\,\text{kW}$ with PF = 0.5 lagging is fed by a 5-kV generator. A capacitor is added in parallel to improve the PF to 0.85 lagging. Find the reduction in current drawn from the generator.

Ans.   With the same amount of real power, the current is reduced to 198 A (480 A − 282 A), representing a 41.3% current reduction.

18.42   A 24-V generator with impedance $600 + j150\,\Omega$ feeds a load impedance. Find the load impedance to obtain maximum power, the power received, and the power efficiency.

Ans. $\mathbf{Z}_L = 600 - j150\,\Omega$

        $P_L = 240\,\text{mW}$

        Efficiency = 50%

Chapter 19

Transformers

IDEAL TRANSFORMER CHARACTERISTICS

The basic transformer consists of two coils electrically insulated from each other and wound upon a common core (Fig. 19-1). Magnetic coupling is used to transfer electric energy from one coil to another. The coil which *receives* energy from an ac source is called the *primary*. The coil which *delivers* energy to an ac load is called the *secondary*. The core of transformers used at low frequencies is generally made of magnetic material, usually sheet steel. Cores of transformers used at higher frequencies are made of powdered iron and ceramics, or nonmagnetic materials. Some coils are simply wound on nonmagnetic hollow forms such as cardboard or plastic so that the core material is actually air.

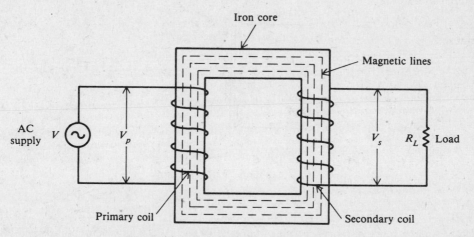

Fig. 19-1 Simple diagram of a transformer

If a transformer is assumed to be operating under an ideal or perfect condition, the transfer of energy from one voltage to another is accompanied by no losses.

Voltage Ratio

The voltage on the coils of a transformer is directly proportional to the number of turns on the coils. This relationship is expressed by the formula

$$\frac{V_p}{V_s} = \frac{N_p}{N_s} \qquad (19\text{-}1)$$

where V_p = voltage on primary coil, V
   V_s = voltage on secondary coil, V
   N_p = number of turns on primary coil
   N_s = number of turns on secondary coil

The ratio V_p/V_s is called the *voltage ratio* (VR). The ratio N_p/N_s is called the *turns ratio* (TR). By substituting these terms into Eq. (*19-1*), we obtain an equivalent formula

$$\text{VR} = \text{TR} \qquad (19\text{-}2)$$

A voltage ratio of $1:4$ (read as 1 to 4) means that for each volt on the transformer primary, there is 4 V on the secondary. When the secondary voltage is greater than the primary voltage, the transformer is called a *step-up* transformer. A voltage ratio of $4:1$ means that for each 4 V on the primary, there is only 1 V on the secondary. When the secondary voltage is less than the primary voltage, the transformer is called a *step-down* transformer.

Example 19.1   A filament transformer (Fig. 19-2) reduces the 120 V in the primary to 8 V on the secondary. If there are 150 turns on the primary and 10 turns on the secondary, find the voltage ratio and turns ratio.

$$VR = \frac{V_p}{V_s} = \frac{120}{8} = \frac{15}{1} = 15:1 \qquad Ans.$$

$$TR = \frac{N_p}{N_s} = \frac{150}{15} = \frac{15}{1} = 15:1 \qquad Ans.$$

Note that $VR = TR$ [Eq. (*19-2*)].

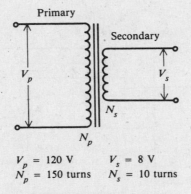

Primary

Secondary

V_p

V_s

N_s

N_p

$V_p = 120$ V $V_s = 8$ V
$N_p = 150$ turns $N_s = 10$ turns

Fig. 19-2   Filament transformer

Example 19.2   An iron-core transformer operating from a 120-V line has 500 turns in the primary and 100 turns in the secondary. Find the secondary voltage.

$$\frac{V_p}{V_s} = \frac{N_p}{N_s} \qquad\qquad (19\text{-}1)$$

Solve for V_s and substitute known values.

$$V_s = \frac{N_s}{N_p} V_p = \frac{100}{500} 120 = 24 \text{ V} \qquad Ans.$$

Example 19.3   A power transformer has a turns ratio of $1:5$. If the secondary coil has 1000 turns and the secondary voltage is 30 V, find the voltage ratio, the primary voltage, and the number of primary turns.

$$VR = TR \qquad\qquad (19\text{-}2)$$

$$= 1:5 \qquad Ans.$$

$$\frac{V_p}{V_s} = VR = 1:5 = \frac{1}{5}$$

$$V_p = \frac{1}{5} V_s = \frac{30}{5} = 6 \text{ V} \qquad Ans.$$

$$TR = \frac{N_p}{N_s} = \frac{1}{5}$$

$$N_p = \frac{1}{5} N_s = \frac{1000}{5} = 200 \text{ turns} \qquad Ans.$$

Current Ratio

The current in the coils of a transformer is inversely proportional to the voltage in the coils. This relationship is expressed by the equation

$$\frac{V_p}{V_s} = \frac{I_s}{I_p} \tag{19-3}$$

where I_p = current in primary coil, A
        I_s = current in secondary coil, A

From Eq. (19-1) we may substitute N_p/N_s for V_p/V_s, so we have

$$\frac{N_p}{N_s} = \frac{I_s}{I_p} \tag{19-4}$$

Example 19.4   Derive the current-ratio equation $V_p/V_s = I_s/I_p$.
For an ideal transformer, the power input to the primary is equal to the power output of the secondary. Thus, an ideal transformer is assumed to operate at an efficiency of 100 percent. Therefore,

$$\text{Power input} = \text{power output}$$

$$P_p = P_s$$

$$\text{Power input} = P_p = V_p I_p$$

$$\text{Power output} = P_s = V_s I_s$$

Substituting for P_p and P_s, $$V_p I_p = V_s I_s$$

from which $$\frac{V_p}{V_s} = \frac{I_s}{I_p} \qquad Ans.$$

Example 19.5   When the primary winding of an iron-core transformer is operated at 120 V, the current in the winding is 2 A. Find the current in the secondary winding load if the voltage is stepped up to 600 V.

$$\frac{V_p}{V_s} = \frac{I_s}{I_p} \tag{19-3}$$

Solve for I_s and substitute known values.

$$I_s = \frac{V_p}{V_s} I_p = \frac{120}{600} 2 = 0.4 \text{ A} \qquad Ans.$$

Example 19.6   A bell transformer with 240 turns on the primary and 30 turns on the secondary draws 0.3 A from a 120-V line. Find the secondary current.

$$\frac{N_p}{N_s} = \frac{I_s}{I_p} \tag{19-4}$$

Solve for I_s and substitute known values.

$$I_s = \frac{N_p}{N_s} I_p = \frac{240}{30}(0.3) = 2.4 \text{ A} \qquad Ans.$$

Efficiency

The efficiency of a transformer is equal to the ratio of the power output of the secondary winding to the power input of the primary winding. An ideal transformer is 100 percent efficient because it delivers all the energy it receives. Because of core and copper losses, the efficiency of even the best practical transformer is less than 100 percent. Expressed as an equation,

$$\text{Eff} = \frac{\text{power output}}{\text{power input}} = \frac{P_s}{P_p} \qquad\qquad (19\text{-}5)$$

where   Eff = efficiency
   P_s = power output from secondary, W
   P_p = power input to primary, W

Example 19.7   What is the efficiency of a transformer if it draws 900 W and delivers 600 W?

$$\text{Eff} = \frac{P_s}{P_p} \qquad\qquad (19\text{-}5)$$

$$= \frac{600}{900} = 0.667 = 66.7\% \qquad Ans.$$

Example 19.8   A transformer is 90 percent efficient. If it delivers 198 W from a 110-V line, find the power input and the primary current.

$$\text{Eff} = \frac{P_s}{P_p} \qquad\qquad (19\text{-}5)$$

Solve for power input P_p.

$$P_p = \frac{P_s}{\text{Eff}} = \frac{198}{0.90} = 220 \text{ W} \qquad Ans.$$

Write the power input formula.

$$P_p = V_p I_p$$

Solve for I_p.

$$I_p = \frac{P_p}{V_p} = \frac{220}{110} = 2 \text{ A} \qquad Ans.$$

Example 19.9   A transformer draws 160 W from a 120-V line and delivers 24 V at 5 A. Find its efficiency.

$$P_p = 160 \text{ W, given}$$
$$P_s = V_s I_s = 24(5) = 120 \text{ W}$$

Then

$$\text{Eff} = \frac{P_s}{P_p} = \frac{120}{160} = 0.75 = 75\% \qquad Ans.$$

TRANSFORMER RATINGS

Transformer capacity is rated in kilovoltamperes. Since power in an ac circuit depends on the power factor of the load and the current in the load, an output rating in kilowatts must specify the power factor.

Example 19.10   What is the kilowatt output of a 5-kVA 2400/120-V transformer serving loads with the following power factors: (*a*) 100 percent, (*b*) 80 percent, and (*c*) 40 percent? What is the rated output current of the transformer?

Power output:

(*a*) $P_s = \text{kVA} \times \text{PF} = 5(1.0) = 5 \text{ kW}$ *Ans.*

(*b*) $P_s = 5(0.8) = 4 \text{ kW}$ *Ans.*

(*c*) $P_s = 5(0.4) = 2 \text{ kW}$ *Ans.*

Current output:

$$P_s = I_s V_s$$

Solving for I_s,

$$I_s = \frac{P_s}{V_s} = \frac{5000}{120} = 41.7 \text{ A} \qquad Ans.$$

Since rated current is determined by the rated kilovoltamperage, the full-load current of 41.7 A is supplied by the transformer at the three different PFs even though the kilowatt output is different for each case.

IMPEDANCE RATIO

A maximum amount of power is transferred from one circuit to another when the impedances of the two circuits are equal or *matched*. If the two circuits have unequal impedances, a coupling transformer may be used as an impedance-matching device between the two circuits. By constructing the transformer's winding so that it has a definite turns ratio, the transformer can perform any impedance-matching function. The turns ratio establishes the proper relationship between the ratio of the primary and secondary winding impedances. This relationship is expressed by the equation

$$\left(\frac{N_p}{N_s}\right)^2 = \frac{Z_p}{Z_s} \qquad (19\text{-}6)$$

Taking the square root of both sides, we obtain

$$\frac{N_p}{N_s} = \sqrt{\frac{Z_p}{Z_s}} \qquad (19\text{-}7)$$

where N_p = number of turns on primary
   N_s = number of turns on secondary
   Z_p = impedance of primary, Ω
   Z_s = impedance of secondary, Ω

Example 19.11   Find the turns ratio of a transformer used to match a 14 400-Ω load to a 400-Ω load.

$$\frac{N_p}{N_s} = \sqrt{\frac{Z_p}{Z_s}} \qquad (19\text{-}7)$$

$$= \sqrt{\frac{14\,400}{400}} = \sqrt{36} = \frac{6}{1} = 6:1 \qquad Ans.$$

Example 19.12   Find the turns ratio of a transformer to match a 20-Ω load to a 72 000-Ω load. Use Eq. (*19-7*).

$$\frac{N_p}{N_s} = \sqrt{\frac{Z_p}{Z_s}} = \sqrt{\frac{20}{72\,000}} = \sqrt{\frac{1}{3600}} = \frac{1}{60} = 1:60 \qquad Ans.$$

Example 19.13 The secondary load of a step-down transformer with a turns ratio of $5:1$ is 900 Ω. Find the impedance of the primary.

$$\frac{Z_p}{Z_s} = \left(\frac{N_p}{N_s}\right)^2 \qquad (19\text{-}6)$$

Solve for Z_p and substitute given values.

$$Z_p = \left(\frac{N_p}{N_s}\right)^2 Z_s = \left(\frac{5}{1}\right)^2 (900) = 22\,500\,\Omega \qquad Ans.$$

AUTOTRANSFORMER

An autotransformer is a special type of power transformer. It consists of only one winding. By tapping, or connecting, at points along the length of the winding, different voltages may be obtained. The autotransformer (Fig. 19-3) has a single winding between terminals A and C. The winding is tapped and a wire brought out as terminal B. Winding AC is the primary while winding BC is the secondary. The simplicity of the autotransformer makes it economical and space-saving. However, it does not provide electrical isolation between primary and secondary circuits.

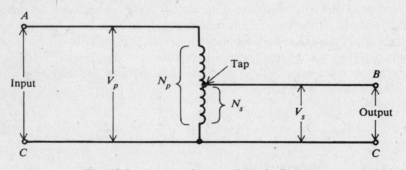

Fig. 19-3 Autotransformer schematic diagram

Example 19.14 An autotransformer having 200 turns is connected to a 120-V line (Fig. 19-3). To obtain a 24-V output, find the number of turns of the secondary and the turn number at which the transformer should be tapped, counting from terminal A.

$$\frac{V_p}{V_s} = \frac{N_p}{N_s} \qquad (19\text{-}1)$$

$$N_s = \frac{V_s}{V_p} N_p = \frac{24}{120} 200 = 40\,\text{turns} \qquad Ans.$$

Since the secondary turns include primary, the B tap should be where the turn number is 160 ($160 = 200 - 40$). If tap B is made movable, the autotransformer becomes a variable transformer. As the tap is moved downward toward C, the secondary voltage decreases.

TRANSFORMER LOSSES AND EFFICIENCY

Actual transformers have copper losses and core losses. Copper loss is the power lost in the primary and secondary windings due to the ohmic resistance of the windings. Copper loss in watts is obtained by the formula

$$\text{Copper loss} = I_p^2 R_p + I_s^2 R_s \qquad (19\text{-}8)$$

where I_p = primary current, A
I_s = secondary current, A
R_p = resistance of the primary winding, Ω
R_s = resistance of the secondary winding, Ω

Core loss is caused by two factors: hysteresis loss and eddy-current loss. Hysteresis loss is the energy lost by reversing the magnetic field in the core as the magnetizing alternating current rises and falls and reverses direction. Eddy-current loss is the result of induced currents circulating in the core material.

Copper loss in both windings may be measured by means of a wattmeter. The wattmeter is placed in the primary circuit of the transformer while the secondary is short-circuited. The voltage applied to the primary is then increased until the rated full-load current is flowing in the short-circuited secondary. At that point the wattmeter will read the total copper loss. Core loss may be determined also by a wattmeter in the primary circuit by applying the rated voltage to the primary with the secondary circuit open.

The efficiency of an actual transformer is expressed as follows:

$$\text{Eff} = \frac{\text{power output}}{\text{power input}} = \frac{P_s}{P_p} \qquad (19\text{-}5)$$

$$= \frac{\text{power output}}{\text{power output} + \text{copper loss} + \text{core loss}}$$

and
$$\text{Eff} = \frac{V_s I_s \times \text{PF}}{(V_s I_s \times \text{PF}) + \text{copper loss} + \text{core loss}} \qquad (19\text{-}9)$$

where PF = power factor of the load

Example 19.15 A 10 : 1 step-down 5-kVA transformer has a full-load secondary current rating of 50 A. A short-circuit test for copper loss at a full load gives a wattmeter reading of 100 W. If the resistance of the primary winding is 0.6 Ω, find the resistance of the secondary winding and the power loss in the secondary.
Use Eq. (19-8).

$$\text{Copper loss} = I_p^2 R_p + I_s^2 R_s = 100\,\text{W}$$

To find I_p at full load, write Eq. (19-4).

$$\frac{N_p}{N_s} = \frac{I_s}{I_p} \qquad (19\text{-}4)$$

from which
$$I_p = \frac{N_s}{N_p} I_s = \frac{1}{10} 50 = 5\,\text{A}$$

Solve for R_s from the copper-loss equation above.

$$I_s^2 R_s = 100 - I_p^2 R_p$$

$$R_s = \frac{100 - I_p^2 R_p}{I_s^2} = \frac{100 - 5^2(0.6)}{50^2} = 0.034\,\Omega \qquad Ans.$$

$$\text{Power loss in secondary} = I_s^2 R_s = 50^2(0.034) = 85\,\text{W} \qquad Ans.$$

or
$$\text{Power loss in secondary} = 100 - I_p^2 R_p = 100 - 5^2(0.6) = 85\,\text{W} \qquad Ans.$$

Example 19.16   An open-circuit test for core loss in the 5-kVA transformer of Example 19.15 gives a reading of 70 W. If the PF of the load is 85 percent, find the efficiency of the transformer at full load.

$$\text{Eff} = \frac{V_S I_S \times \text{PF}}{(V_S I_S \times \text{PF}) + \text{copper loss} + \text{core loss}} \qquad (19\text{-}9)$$

$$V_S I_S = \text{transformer rating} = 5\,\text{kVA} = 5000\,\text{VA}$$

$$\text{PF} = 0.85 \qquad \text{Copper loss} = 100\,\text{W} \qquad \text{Core loss} = 70\,\text{W}$$

Substitute known values and solve.

$$\text{Eff} = \frac{5000(0.85)}{5000(0.85) + 100 + 70} = \frac{4250}{4420} = 0.962 = 96.2\% \qquad \textit{Ans.}$$

NO-LOAD CONDITION

If the secondary winding of a transformer is left open-circuited (Fig. 19-4a), the primary current is very low and is referred to as the *no-load current*. The no-load current produces the magnetic flux and supplies the hysteresis and eddy-current losses in the core. Therefore, the no-load current I_E consists of two components: the magnetizing-current component I_M and the core-loss component I_H. The magnetizing current I_M lags the applied primary voltage V_p by 90°, while the core-loss component I_H is always in phase with V_p (Fig. 19-4b). Note also that the primary applied voltage V_p and the induced secondary voltage V_s are shown 180° out of phase with each other. Since in practice I_H is small in comparison with I_M, the magnetizing current I_M is very nearly equal to the total no-load current I_E. I_E is also called the *exciting current*.

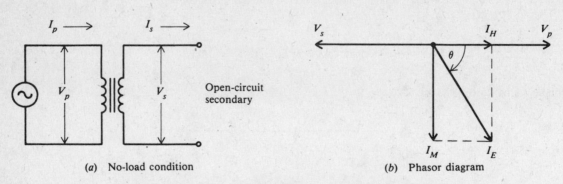

(a) No-load condition   (b) Phasor diagram

Fig. 19-4   Iron-core transformer with the secondary open-circuited

Example 19.17   When the secondary of a 120/240-V transformer is open, the primary current is 0.3 A at a PF of 20 percent. The transformer is rated at 4 kVA. Find (a) the full-load current I_p, (b) the no-load exciting current I_E, (c) the core-loss current I_H, and (d) the magnetizing current I_M. (e) Determine the percentages of each current with respect to full-load current. (f) Draw the phasor diagram.

(a)

$$\text{Full-load current} = \frac{\text{transformer kVA rating}}{\text{primary voltage}}$$

$$I_p = \frac{4000}{120} = 33.3\,\text{A} \qquad \textit{Ans.}$$

(b) The primary current measured at no load (secondary open) is the exciting current I_E. Thus,

$$I_E = 0.3\,\text{A} \qquad Ans.$$

(c) From Fig. 19-4b,

$$I_H = I_E \cos\theta = I_E \times \text{PF} = 0.3(0.2) = 0.06\,\text{A} \qquad Ans.$$

(d) From Fig. 19-4b,

$$I_M = I_E \sin\theta$$
$$\theta = \arccos 0.2 = 78.5°$$

Then $$I_M = 0.3\sin 78.5° = 0.3(0.980) = 0.294\,\text{A} \qquad Ans.$$

(e) Percent no-load primary current (exciting current) to full-load primary current:

$$\frac{0.3}{33.3} = 0.0090 = 0.90\% \qquad Ans.$$

Percent core-loss current to full-load current:

$$\frac{0.06}{33.3} = 0.0018 = 0.18\% \qquad Ans.$$

Percent magnetizing current to full-load current:

$$\frac{0.294}{33.3} = 0.0088 = 0.88\% \qquad Ans.$$

Notice that the magnetizing current (0.294 A) has nearly the same values as the no-load primary current (0.3 A).

(f) Phasor diagram: See Fig. 19-5.

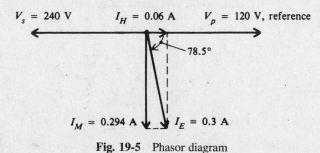

Fig. 19-5 Phasor diagram

COIL POLARITY

The symbol for a transformer gives no indication of the phase of the voltage across the secondary since the phase of that voltage actually depends on the direction of the windings around the core. To solve this problem, polarity dots are used to indicate the phase of primary and secondary signals. The voltages are either in phase (Fig. 19-6a) or 180° out of phase with respect to the primary voltage (Fig. 19-6b).

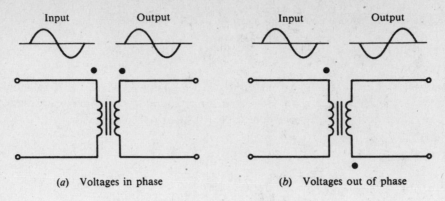

(a) Voltages in phase   (b) Voltages out of phase

Fig. 19-6 Polarity notation of transformer coils

Solved Problems

19.1 A power transformer is used to couple electric energy from a power-supply line to one or more components of the system. In one type of power transformer (Fig. 19-7), there are three separate secondary windings, each designed for a different voltage output. The primary of the transformer is connected to a 120-V source of supply and has 100 turns. Find the number of turns on each secondary.

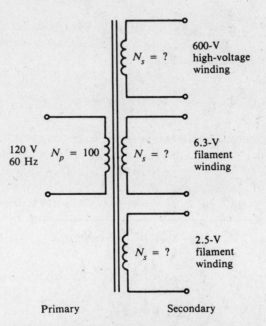

Fig. 19-7 Power transformer schematic diagram

Find N_s by using Eq. (*19-1*).

$$\frac{V_p}{V_s} = \frac{N_p}{N_s} \qquad \text{from which} \qquad N_s = \frac{V_s}{V_p} N_p$$

For the 600-V secondary: $N_s = \dfrac{600}{120} 100 = 500 \text{ turns} \qquad Ans.$

For the 6.3-V secondary: $N_s = \dfrac{6.3}{120}100 \approx 5$ turns   *Ans.*

For the 2.5-V secondary: $N_s = \dfrac{2.5}{120}100 \approx 2$ turns   *Ans.*

19.2   A transformer whose primary is connected to a 110-V source delivers 11 V. If the number of turns on the secondary is 20 turns, find the number of turns on the primary. How many extra turns must be added to the secondary if it must deliver 33 V?

Find N_p by using Eq. (*19-1*).

$$\frac{V_p}{V_s} = \frac{N_p}{N_s} \quad \text{from which} \quad N_p = \frac{V_p}{V_s}N_s = \frac{110}{11}(20) = 200 \text{ turns} \quad Ans.$$

For $V_s = 33$ V, $N_s = \dfrac{V_s}{V_p}N_p = \dfrac{33}{110}200 = 60$ turns

Hence 40 turns $(60 - 20)$ must be added.   *Ans.*

19.3   A step-down transformer with a turns ratio of 50 000 : 500 has its primary connected to a 20 000-V transmission line. If the secondary is connected to a 25-Ω load, find (*a*) the secondary voltage, (*b*) the secondary current, (*c*) the primary current, and (*d*) the power output.

$$\text{TR} = \frac{N_p}{N_s} = \frac{50\,000}{500} = \frac{100}{1}$$

(*a*)
$$\frac{N_p}{N_s} = \frac{V_p}{V_s} \qquad (19\text{-}1)$$

Then $V_s = \dfrac{N_s}{N_p}V_p = \dfrac{1}{100}(20\,000) = 200$ V   *Ans.*

(*b*) By Ohm's law,

$$I_s = \frac{V_s}{R_L} = \frac{200}{25} = 8 \text{ A} \qquad Ans.$$

(*c*)
$$\frac{V_p}{V_s} = \frac{I_s}{I_p} \qquad (19\text{-}3)$$

Then $I_p = \dfrac{200}{20\,000}8 = 0.08$ A   *Ans.*

(*d*) $P_s = V_s I_s = 200(8) = 1600$ W   *Ans.*

19.4   A 7 : 5 step-down transformer draws 2 A. Find the secondary current.

$$\text{TR} = \frac{N_p}{N_s} = \frac{7}{5}$$

$$\frac{N_p}{N_s} = \frac{I_s}{I_p} \qquad (19\text{-}4)$$

Then $I_s \dfrac{N_p}{N_s}I_p = \dfrac{7}{5}2 = 2.8$ A   *Ans.*

19.5   A transformer draws 2.5 A at 110 V and delivers 7.5 A at 24 V to a load with a PF of 100 percent. Find the efficiency of the transformer.

$$\text{Power in} = P_p = V_p I_p = 110(2.5) = 275 \text{ W}$$

$$\text{Power out} = P_s = V_s I_s = 24(7.5) = 180 \text{ W}$$

$$\text{Eff} = \frac{P_s}{P_p} \tag{19-5}$$

$$= \frac{180}{275} = 0.655 = 65.5\% \qquad Ans.$$

19.6   A transformer delivers 550 V at 80 mA at an efficiency of 90 percent. If the primary current is 0.8 A, find the power input in voltamperes and the primary voltage.

$$\text{Power out} = P_s = V_s I_s = 550(80 \times 10^{-3}) = 44 \text{ VA}$$

$$\text{Eff} = \frac{P_s}{P_p} \tag{19-5}$$

Then
$$\text{Power in} = P_p = \frac{P_s}{\text{Eff}} = \frac{44}{0.9} = 48.9 \text{ VA} \qquad Ans.$$

Since the PF of the load is not specified, power is expressed in voltamperes. Also

$$P_p = V_p I_p \qquad \text{so} \qquad V_p = \frac{P_p}{I_p} = \frac{48.9}{0.8} = 61.1 \text{ V} \qquad Ans.$$

19.7   The rating of a power-supply transformer that is to be operated from a 60-Hz 120-V power line may read as follows: 600 V CT (center tap) at 90 mA, 6.3 V at 3 A, 5 V at 2 A. Find the wattage rating of this transformer.

The wattage rating is the total power delivered at 100 percent PF. It is found by adding the power ratings of the individual secondary windings. The general formula to use is $P_s = V_s I_s$.

At 600 V tap: $P_s = 600(90 \times 10^{-3}) = 54 \text{ W}$
At 6.3 V tap: $P_s = 6.3(3)$ $= 18.9 \text{ W}$
At 5 V tap: $P_s = 5(2)$ $= \underline{10 \text{ W}}$

$$\text{Total power } P_T = 82.9 \text{ W} \qquad Ans.$$

19.8   The step-down autotransformer at a power factor of unity is designed to deliver 240 V to a load of 5 kW (Fig. 19-8). The autotransformer's primary winding is connected to a 600-V source. Find the current in (*a*) the load, (*b*) the primary winding, and (*c*) the secondary winding.

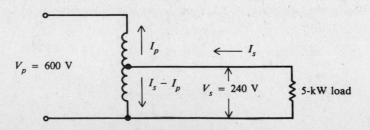

Fig. 19-8   Step-down autotransformer

(a)   Write $P_s = V_s I_s$ (I_s in this case is the load current). So

$$I_s = \frac{P_s}{V_s} = \frac{5000}{240} = 20.8 \text{ A} \qquad Ans.$$

(b)   At unity PF and 100 percent efficiency, $V_p I_p = V_s I_s$. So

$$I_p = \frac{V_s I_s}{V_p} = \frac{5000}{600} = 8.33 \text{ A} \qquad Ans.$$

(c)   The current in the secondary winding is $I_s - I_p$ by Kirchhoff's current law.

$$I_s - I_p = 20.8 - 8.3 = 12.5 \text{ A} \qquad Ans.$$

19.9   A 60:1 output transformer is used to match an output transistor to a 4-Ω voice coil. Find the impedance of the output circuit.

$$\frac{Z_p}{Z_s} = \left(\frac{N_p}{N_s}\right)^2 \qquad\qquad\qquad (19\text{-}6)$$

In this case the output transistor is in the primary circuit and the voice coil is in the secondary circuit.

$$Z_p = \left(\frac{N_p}{N_s}\right)^2 Z_s = \left(\frac{60}{1}\right)^2 (4) = 14\,400\ \Omega \qquad Ans.$$

19.10   A 1:10 step-up transformer is used to match a 500-Ω line to a circuit. Find the impedance of the circuit.

$$\frac{Z_p}{Z_s} = \left(\frac{N_p}{N_s}\right)^2 \qquad\qquad\qquad (19\text{-}6)$$

In this case the circuit is in the secondary.

$$Z_s = \left(\frac{N_s}{N_p}\right)^2 Z_p = \left(\frac{10}{1}\right)^2 (500) = 50\,000\ \Omega = 50\,\text{k}\Omega \qquad Ans.$$

19.11   A 240/720-V 5-kVA transformer undergoes a short-circuit test for copper loss. At the start of the test, the primary voltage is varied until the ammeter across the secondary indicates rated full-load secondary current. The measured resistance of the primary winding is 0.05 Ω and that of the secondary winding is 1.5 Ω. Calculate the total copper loss.

Step 1.   Calculate the copper loss in the secondary.

$$\text{Full-load secondary current } I_s = \frac{5000}{720} = 6.94 \text{ A}$$

So
$$I_s^2 R_s = (6.94)^2 (1.5) = 72.2 \text{ W}$$

Step 2. Calculate the copper loss in the primary.

$$\text{Full-load primary current } I_p = \frac{5000}{240} = 20.8 \text{ A}$$

So
$$I_p^2 R_p = (20.8)^2 (0.05) = 21.6 \text{ W}$$

Step 3. Calculate total copper loss. The total copper loss is the sum of the losses in both windings.

$$\text{Total copper loss} = I_p^2 R_p + I_s^2 R_s \qquad (19\text{-}8)$$
$$= 21.6 + 72.2 = 93.8 \text{ W} \qquad Ans.$$

The wattmeter in the primary circuit should read 93.8 W.

19.12 On an open-circuit test for core loss in the 5-kVA transformer of Problem 19.11, when the primary voltage is set at the rated voltage of 240 V, the wattmeter in the primary circuit indicates 80 W. If the power factor of the load is 0.8, find the efficiency of the transformer at full load.

Use the efficiency formula:

$$\text{Eff} = \frac{V_s I_s \times \text{PF}}{(V_s I_s \times \text{PF}) + \text{copper loss} + \text{core loss}} \qquad (19\text{-}9)$$

$$= \frac{5000(0.8)}{5000(0.8) + 93.8 + 80} = \frac{4000}{4174} = 0.958 = 95.8\% \qquad Ans.$$

19.13 When the secondary of a power transformer is open, the no-load current in the primary is 0.4 A. If the power factor of the input primary circuit is 0.10, find the exciting current I_E, the core-loss current I_H, and the magnetizing current I_M.

The exciting current is the same as the no-load primary current.

So
$$I_E = 0.4 \text{ A} \qquad Ans.$$

From the right-triangle relationships (see Fig. 19-9),

$$I_H = I_E \cos \theta = 0.4(0.10) = 0.04 \text{ A} \qquad Ans.$$

$$\text{PF} = \cos \theta = 0.10 \qquad \theta = \arccos 0.10 = 84.3°$$

Then
$$I_M = I_E \sin \theta = 0.4 \sin 84.3° = 0.4 \text{ A} \qquad Ans.$$

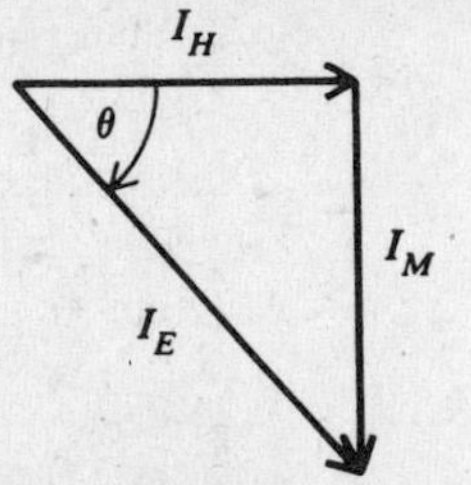

Fig. 19-9

19.14 The no-load current taken by a 110/220-V transformer is 0.7 A. The transformer is rated at 2.2 kVA. If the power factors of the primary and secondary circuits are equal, find the primary current when the secondary is supplying its rated 2.2 kVA to the load.

$$\text{Full-load secondary current } I_s = \frac{2200}{220} = 10 \text{ A}$$

Since the PFs for primary and secondary are equal at full load, the main component of load current in the primary is

$$I_p' = \frac{V_s}{V_p} I_s = \frac{220}{110} 10 = 2(10) = 20 \text{ A}$$

To I_p' we add directly the 0.7-A no-load current. So

$$I_p = 20 + 0.7 = 20.7 \text{ A} \qquad Ans.$$

Because the no-load components I_H and I_M of the primary current are much less than the load-current component I_p', the no-load current can be added arithmetically instead of vectorially to the total load-current.

19.15 Indicate the correct polarity dots for the secondary circuit (Fig. 19-10a).

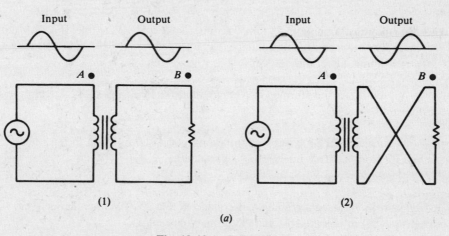

Fig. 19-10a Coil polarity

For diagram (1) (Fig. 19-10a), the voltage at point B with respect to ground has the same phase as the voltage at point A with respect to ground (Fig. 19-10b). For diagram (2) (Fig. 19-10a), the secondary windings are now reversed so that the output voltage at B is now 180° out of phase with the input voltage at A (Fig. 19-10c).

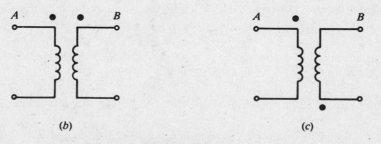

Fig. 19-10b, c

Supplementary Problems

19.16   A bell transformer reduces the voltage from 110 to 11 V. If there are 20 turns in the secondary, find the number of turns on the primary and the turns ratio.   *Ans.* $N_p = 200$ turns; TR $= 10:1$

19.17   Find the voltage at the spark plugs connected to the secondary of a coil with 60 turns on the primary and 36 000 turns on the secondary, if the primary is connected to a 12-V alternator.
Ans. $V_s = 7200$ V

19.18   A coil with a primary winding of 80 turns must supply 4800 V. If the primary is connected to an 8-V source, find the number of turns on the secondary.   *Ans.* $N_s = 48\,000$ turns

19.19   The 110-V primary of a power transformer has 220 turns. Three secondaries are to deliver (*a*) 600 V, (*b*) 35 V, and (*c*) 12.5 V. Find the number of turns needed on each secondary.
Ans.   (*a*) $N_s = 1200$ turns;   (*b*) $N_s = 70$ turns;   (*c*) $N_s = 25$ turns

19.20   The secondary coil of a transformer has 100 turns and the secondary voltage is 10 V. If the turns ratio is 18:1, find (*a*) the voltage ratio, (*b*) the primary voltage, and (*c*) the number of primary turns.
Ans.   (*a*) VR $= 18:1$;   (*b*) $V_p = 180$ V;   (*c*) $N_p = 1800$ turns

19.21   A step-down autotransformer with 55 turns is connected to a 110-V ac line. If a 28-V output is desired, find the secondary turns and the turn number to be tapped.
Ans. $N_s = 14$ turns; tap at turn 31

19.22   A 220/110-V step-down transformer in a stage-lighting circuit draws 12 A from the line. Find the current delivered.   *Ans.* $I_s = 24$ A

19.23   An ideal transformer with 2400 turns on the primary and 600 turns on the secondary draws 9.5 A at 100 percent PF from a 220-V line. Find I_s, V_s, and P_s.
Ans. $I_s = 38$ A; $V_s = 55$ V; $P_s = 2090$ W

19.24   A transformer with 96 percent efficiency is connected to a 2000-V line. If it delivers 10 000 VA, find the power input P_p in voltamperes and the primary current I_p.
Ans. $P_p = 10\,417$ VA; $I_p = 5.21$ A

19.25   A transformer with an efficiency of 85 percent delivers 650 V and 120 mA at 100 percent PF to a secondary load. The primary current is 0.6 A. Find the power input and the primary voltage.
Ans. $P_p = 91.8$ W; $V_p = 153$ V

19.26   The three secondary coils of a power-supply transformer deliver 84 mA at 300 V, 1.4 A at 12.6 V, and 1.9 A at 2.5 V. Find the power delivered to the secondary loads. Find also the efficiency if the transformer draws 55 W from a 110-V line. (Assume unity PF in both primary and secondary.)
Ans. $P_s = 47.9$ W; Eff $= 87.1\%$

19.27   Find the current rating of each winding of a 100-kVA 2400/120-V 60-Hz transformer.
Ans.   Primary winding, 41.7 A; secondary winding, 833.3 A

19.28   Find the turns ratio of a transformer used to match a 50-Ω load to a 450-Ω line.   *Ans.*   TR $= 3:1$

19.29   Find the turns ratio of a transformer used to match a 30-Ω load to a 48 000-Ω load.
Ans.   TR $= 1:40$

19.30 Find the turns ratio of the transformer needed to match a load of 4000 Ω to three 12-Ω speakers in parallel.   *Ans.*   TR = 31.6 : 1 ≈ 32 : 1

19.31 A 1:18 step-up output transformer is used to match a microphone with a grid circuit impedance of 35 kΩ. Find the impedance of the microphone.   *Ans.* $Z_p = 108\ \Omega$

19.32 A 6:1 step-down transformer matches an input load to a secondary load of 800 Ω. Find the impedance of the input.   *Ans.* $Z_p = 28.8\ k\Omega$

19.33 A step-up autotransformer requires 100 turns for its 120-V primary. To obtain an output of 300 V, find the number of turns that must be added to the primary.   *Ans.*   150 turns ($N_s = 250$ turns)

19.34 A load of 12 kW at 480 V and 100 percent PF is to be supplied by a step-down autotransformer whose high-voltage winding is connected to a 1200-V source. Find the current in (*a*) the load, (*b*) the primary winding, and (*c*) the secondary winding.
Ans.   (*a*) $I_s = 25\ A$;   (*b*) $I_p = 10\ A$;   (*c*) $I_s - I_p = 15\ A$

19.35 An autotransformer starter used to start an induction motor on a 440-V line applies 70 percent of line voltage to the motor during the starting period. If the motor current is 140 A at start-up, what is the current drawn from the line?   *Ans.*   98 A

19.36 A step-down 600/480-V autotransformer supplies a 10-kVA load. Find the primary and secondary line currents and the current in the winding common to both primary and secondary circuits.
Ans. $I_p = 16.7\ A$; $I_s = 20.8\ A$; $I_s - I_p = 4.1\ A$

19.37 A 5-kVA 480/120-V transformer is equipped with high-voltage taps so that it may be operated at 480, 456, or 432 V depending on the tap setting. Find the current in the high-voltage winding for each tap setting. The transformer supplies the rated kVA load at 120 V in each case.
Ans.   10.4 A at 480 V; 11.0 A at 456 V; 11.6 A at 432 V

19.38 A transformer with 800 turns in its primary and 160 turns in its secondary is rated 10 kVA at 480 V. Find (*a*) the VR, (*b*) the primary voltage, (*c*) the rated full-load secondary current, and (*d*) the rated full-load primary current, disregarding the no-load current.
Ans.   (*a*) 5:1;   (*b*) 2400 V;   (*c*) 20.8 A;   (*d*) 4.16 A

19.39 A 250-kVA 2400/480-V transformer has copper losses of 3760 W and core losses of 1060 W. What is the efficiency when the transformer is fully loaded at 0.8 PF?   *Ans.*   Eff = 97.6%

19.40 An open-circuit test for core loss in a 240/720-V 10-kVA transformer gives a reading of 60 W. The measured resistance of the low side winding is 0.03 Ω and that of the high side winding is 1.3 Ω. Find (*a*) the total copper loss and (*b*) the transformer efficiency when the power factor of the load is 0.85.   *Ans.*   (*a*) Total copper loss = 303 W;   (*b*) Eff = 95.9%

19.41 A short-circuit test for copper loss at full load gives a wattmeter reading of 175 W. The transformer undergoing the test is a 240/24-V step-down transformer that has a full-load secondary current rating of 60 A. If the resistance of the primary is 0.7 Ω, find the resistance of the secondary.
Ans. $R_s = 0.042\ \Omega$

19.42 On an open-circuit test for core loss, the transformer of Problem 19.41 takes 1.5 A from a 240-V ac source. The wattmeter reads 95 W. Determine (*a*) the copper loss at no-load condition and (*b*) the core loss.
Ans.   (*a*) 1.58 W;   (*b*) 93.4 W (In this case, the wattmeter reading of 95 W indicates core loss plus copper loss at no load.)

19.43   A 10-kVA 2400/240-V 60-Hz transformer has a primary winding resistance of 6 Ω and a secondary winding resistance of 0.06 Ω. The core loss is 60 W. Find (*a*) the full-load copper loss and (*b*) the efficiency of the transformer when it is fully loaded at 0.9 PF.   *Ans.*   (*a*) 208 W;   (*b*) Eff = 97.1%

19.44   If the transformer of Problem 19.43 had operated at 0.6 PF with the same kilovoltampere loading, what would be its efficiency?   *Ans.*   Eff = 95.7%

19.45   A 10-kVA 7200/120-V transformer has a resistance in the primary winding of 12 Ω and in the secondary winding of 0.0033 Ω. Find the copper loss (*a*) at full load, (*b*) at half load (5 kVA), and (*c*) at a load of 2 kVA.   *Ans.*   (*a*) 46.0 W;   (*b*) 11.5 W;   (*c*) 1.84 W

19.46   A 5-kVA 480/240-V transformer has its secondary open-circuited. Under this no-load condition, the primary current is 0.15 A at a PF of 0.6. Find (*a*) the full-load current I_p, (*b*) the core-loss component I_H, (*c*) the magnetizing current I_M, and (*d*) the percentage of each current with respect to full-load current; and (*e*) draw the phasor diagram.
   Ans.   (*a*) I_p = 10.4 A;   (*b*) I_H = 0.09 A;   (*c*) I_M = 0.12 A;   (*d*) percent exciting current = 1.44%;   percent core-loss current = 0.87%;   percent magnetizing current = 1.15%;   (*e*) see Fig. 19-11

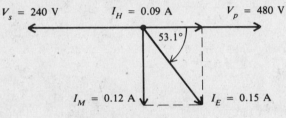

Fig. 19-11   Phasor diagram

19.47   If a transformer circuit has a polarity (Fig. 19-12) where the output is 180° out of phase with the input, show the correct polarity dots when the leads to the load are reversed.   *Ans.*   See Fig. 19-13

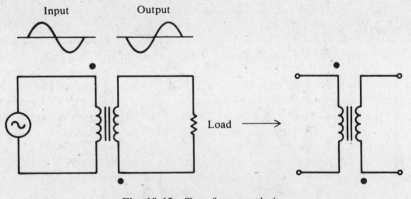

Fig. 19-12   Transformer polarity

19.48   A secondary center-tapped transformer is shown in Fig. 19-14. Indicate the correct output waveforms at points *A* and *B*.   *Ans.*   See Fig. 19-15

19.49   Two transformers can be connected together to obtain a higher voltage by connecting the primaries together in parallel and connecting their secondaries in series. If the secondaries are properly phased,

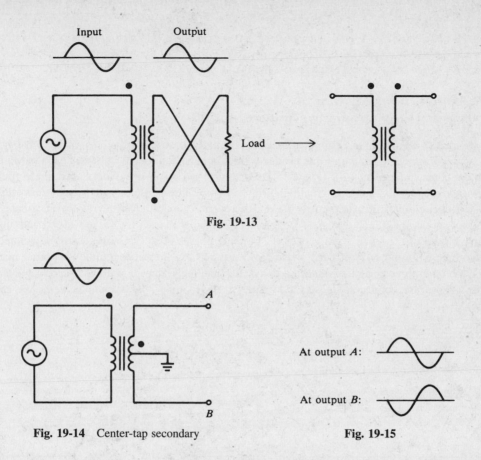

Fig. 19-13

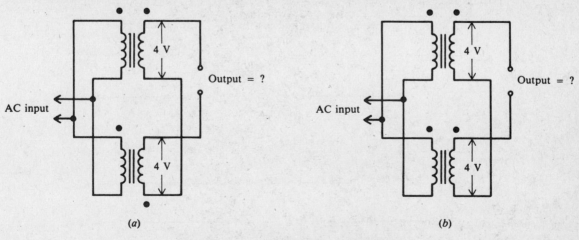

Fig. 19-14   Center-tap secondary   Fig. 19-15

the output will be the sum of the secondary voltages. If the output is the difference of the secondary voltages, the connection to one of the secondaries may be reversed, or one of the primary windings may be reversed. For a series connection of two transformers, each with a secondary output of 4 V (Fig. 19-16), find the output voltage.

Ans.   (*a*) Output = 8 V;   (*b*) output = 0 V (secondaries "bucking")

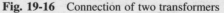

Fig. 19-16   Connection of two transformers

Chapter 20

Three-Phase Systems

CHARACTERISTICS OF THREE-PHASE SYSTEMS

A three-phase (3-ϕ) system is a combination of three single-phase (1-ϕ) systems. In a 3-ϕ balanced system, the power comes from an ac generator that produces three separate but equal voltages, each of which is out of phase with the other voltages by 120° (Fig. 20-1). Although 1-ϕ circuits are widely used in electrical systems, most generation and distribution of alternating current is 3-ϕ. Three-phase circuits require less weight of conductors than 1-ϕ circuits of the same power rating; they permit flexibility in the choice of voltages; and they can be used for single-phase loads. Also, 3-ϕ equipment is smaller in size, lighter in weight, and more efficient than 1-ϕ machinery of the same rated capacity. The three phases of a 3-ϕ system may be connected in two ways. If the three common ends of each phase are connected together at a common terminal marked N for neutral, and the other three ends are connected to the 3-ϕ line, the system is *wye*- or Y-connected (Fig. 20-2a). If the three phases are connected in series to form a closed loop, the system is *delta*- or Δ-connected (Fig. 20-2b).

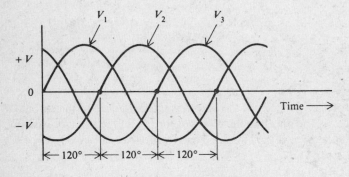
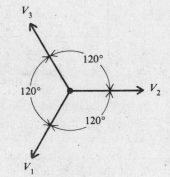
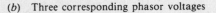

(a) Three sine wave voltages

(b) Three corresponding phasor voltages

Fig. 20-1   Three-phase alternating voltages with 120° between each phase

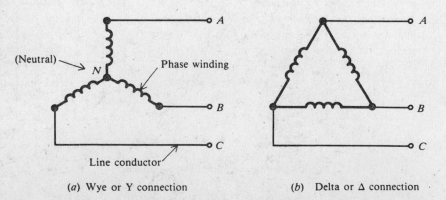

(a) Wye or Y connection

(b) Delta or Δ connection

Fig. 20-2   Connections for 3-ϕ ac power

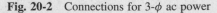

THREE-PHASE TRANSFORMER CONNECTIONS

Three-phase transformers may consist of three separate but identical 1-ϕ transformers or a single 3-ϕ unit containing three-phase windings. The transformer's windings (three in the primary and three in the secondary) may be connected to form a 3-ϕ bank in any one of four common ways (Fig. 20-3). Each primary winding is matched to the secondary winding drawn parallel to it.

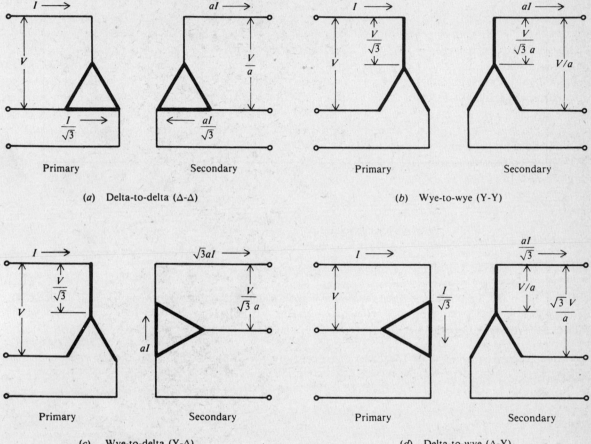

Fig. 20-3 Common 3-ϕ transformer connections. The transformer windings are indicated by the heavy lines. $a = N_1/N_2$

Shown are the voltages and currents in terms of the applied primary line-to-line voltage V and line current I, where $a = N_1/N_2$, the ratio of the number of primary to secondary turns. A *line voltage* is a voltage between two lines, while a *phase voltage* is a voltage across a transformer winding. A *line current* is a current in one of the lines, while a *phase current* is a current in the transformer winding. Voltage and current ratings of the individual transformers depend on the connections shown (Fig. 20-3) and are indicated in tabular form (Table 20-1) for convenience in calculations. Ideal transformers are assumed. The kilovoltampere rating of each transformer is one-third of the kilovoltampere rating of the bank, regardless of the transformer connections used.

Example 20.1 If the line voltage V is 2200 V to a 3-ϕ transformer bank, find the voltage across each transformer primary winding for all four types of transformer connection. Refer to Fig. 20-3 and Table 20-1.

Δ-Δ: Primary winding voltage $= V = 2200$ V   *Ans.*

Y-Y: Primary winding voltage $= \dfrac{V}{\sqrt{3}} = \dfrac{2200}{1.73} = 1270$ V   *Ans.*

Table 20-1 Voltage and Current Relationships for Common 3-ϕ Transformer Connections

Transformer Connection (Primary to Secondary)	Primary				Secondary			
	Line		Phase		Line		Phase	
	Voltage	Current	Voltage	Current	Voltage*	Current	Voltage	Current
Δ-Δ	V	I	V	$\dfrac{I}{\sqrt{3}}$	$\dfrac{V}{a}$	aI	$\dfrac{V}{a}$	$\dfrac{aI}{\sqrt{3}}$
Y-Y	V	I	$\dfrac{V}{\sqrt{3}}$	I	$\dfrac{V}{a}$	aI	$\dfrac{V}{\sqrt{3}a}$	aI
Y-Δ	V	I	$\dfrac{V}{\sqrt{3}}$	I	$\dfrac{V}{\sqrt{3}a}$	$\sqrt{3}aI$	$\dfrac{V}{\sqrt{3}a}$	aI
Δ-Y	V	I	V	$\dfrac{I}{\sqrt{3}}$	$\dfrac{\sqrt{3}V}{a}$	$\dfrac{aI}{\sqrt{3}}$	$\dfrac{V}{a}$	$\dfrac{aI}{\sqrt{3}}$

*$a = N_1/N_2$; $\sqrt{3} = 1.73$

Y-Δ: Primary winding voltage $= \dfrac{V}{\sqrt{3}} = \dfrac{2200}{1.73} = 1270$ V   *Ans.*

Δ-Y: Primary winding voltage $= V = 2200$ V   *Ans.*

It is clear that in any Δ connection the total voltage across any winding in the primary or secondary equals the line voltage in the primary or secondary, respectively, because each winding is directly across the line. The voltage in each winding will be out of phase by 120° with the voltages in the other windings.

Example 20.2   If the line current I is 20.8 A to a 3-ϕ transformer connection, find the current through each primary winding for all four transformer configurations. Refer to Fig. 20-3 and Table 20-1.

Δ-Δ: Primary winding current $= \dfrac{I}{\sqrt{3}} = \dfrac{20.8}{1.73} = 12$ A   *Ans.*

Y-Y: Primary winding current $= I = 20.8$ A   *Ans.*

Y-Δ: Primary winding current $= I = 20.8$ A   *Ans.*

Δ-Y: Primary winding current $= \dfrac{I}{\sqrt{3}} = \dfrac{20.8}{1.73} = 12$ A   *Ans.*

The current in each winding will be out of phase by 120° with the currents in the other windings.

Example 20.3   For each type of transformer connection, find the secondary line current and secondary phase current if the primary line current I is 10.4 A and the turns ratio is 2:1. Refer to Fig. 20-3 and Table 20-1.

Δ-Δ: Secondary line current $= aI = 2(10.4) = 20.8$ A   *Ans.*

                    Secondary phase current $= \dfrac{aI}{\sqrt{3}} = \dfrac{2(10.4)}{1.73} = 12$ A   *Ans.*

Y-Y: Secondary line current $= aI = 2(10.4) = 20.8$ A   *Ans.*

                    Secondary phase current $= aI = 2(10.4) = 20.8$ A   *Ans.*

Y-Δ: $\qquad\qquad$ Secondary line current $= \sqrt{3}aI = 1.73(2)(10.4) = 36$ A $\quad$ *Ans.*

$\qquad\qquad\qquad\qquad$ Secondary phase current $= aI = 2(10.4) = 20.8$ A $\quad$ *Ans.*

Y-Δ: $\qquad\qquad$ Secondary line current $= \dfrac{aI}{\sqrt{3}} = \dfrac{2(10.4)}{1.73} = 12$ A $\quad$ *Ans.*

$\qquad\qquad\qquad\qquad$ Secondary phase current $= \dfrac{aI}{\sqrt{3}} = \dfrac{2(10.4)}{1.73} = 12$ A $\quad$ *Ans.*

The current in each secondary line will be out of phase by $120°$ with the currents in the other secondary lines. Likewise, the current in each secondary winding will be out of phase by $120°$ with the currents in the other secondary windings.

POWER IN BALANCED THREE-PHASE LOADS

A balanced load has identical impedance in each secondary winding (Fig. 20-4). The impedance of each winding in the Δ load is shown equal to Z_Δ (Fig. 20-4a), and in the Y load equal to Z_Y (Fig. 20-4b). For either connection, the lines A, B, and C provide a three-phase system of voltages. The neutral point N in the Y-connection is the fourth conductor of the three-phase four-wire system.

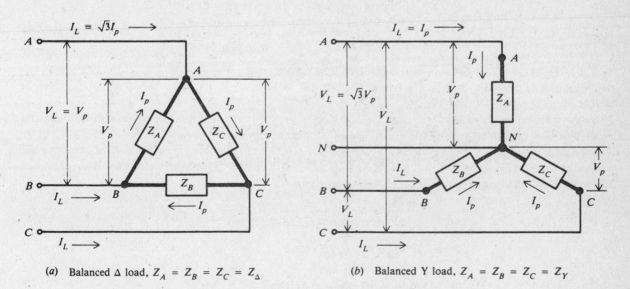

(a) Balanced Δ load, $Z_A = Z_B = Z_C = Z_\Delta$ $\qquad\qquad$ (b) Balanced Y load, $Z_A = Z_B = Z_C = Z_Y$

Fig. 20-4 Three-phase balanced load types

In a balanced Δ-connected load (Fig. 20-4a), as well as in the windings of a transformer, the line voltage V_L and the winding or phase voltages V_p are equal, and the line current I_L is $\sqrt{3}$ times the phase current I_p. That is,

Δ-load: $\qquad\qquad\qquad\qquad\qquad\qquad V_L = V_p \qquad\qquad\qquad\qquad\qquad\qquad$ (20-1)

$$I_L = \sqrt{3}I_p \qquad\qquad\qquad\qquad\qquad\qquad \text{(20-2)}$$

For a balanced Y-connected load (Fig. 20-4b), the line current I_L and the winding or phase current I_p are equal, the neutral current I_N is zero, and the line voltage V_L is $\sqrt{3}$ times the phase voltage V_p. That is,

Y load:
$$I_L = I_p \qquad \qquad (20\text{-}3)$$

$$I_N = 0 \qquad \qquad (20\text{-}4)$$

$$V_L = \sqrt{3}V_p \qquad \qquad (20\text{-}5)$$

(These relationships are also observed in Fig. 20-3 and Table 20-1.)

Since the phase impedance of balanced Y or Δ loads have equal currents, the phase power or power of one phase is one-third the total power. Phase power P_p is

$$P_p = V_p I_p \cos \theta \qquad \qquad (20\text{-}6)$$

and total power P_T is

$$P_T = 3V_p I_p \cos \theta \qquad \qquad (20\text{-}7)$$

Since $V_L = V_p$ [Eq. (20-1)] and $I_p = \sqrt{3}I_L/3$ from Eq. (20-2), for a balanced Δ load,

Δ load:
$$P_T = \sqrt{3}V_L I_L \cos \theta \qquad \qquad (20\text{-}8)$$

Since $I_L = I_p$ [Eq. (20-3)] and $V_p = \sqrt{3}V_L/3$ from Eq. (20-5), for balanced Y loads, substitution in Eq. (20-7) gives

Y load:
$$P_T = \sqrt{3}V_L I_L \cos \theta \qquad \qquad (20\text{-}8)$$

Thus the total-power formulas for Δ and Y loads are identical. θ is the phase angle between the voltage and current of the load impedance, so $\cos \theta$ is the power factor of the load.

The total apparent power S_T in voltamperes and the total reactive power Q_T in voltamperes reactive are related to total real power P_T in watts (Fig. 20-5). Therefore, a balanced three-phase load has the real power, apparent power, and reactive power given by the equations

$$P_T = \sqrt{3}V_L I_L \cos \theta \qquad \qquad (20\text{-}8)$$

$$S_T = \sqrt{3}V_L I_L \qquad \qquad (20\text{-}9)$$

$$Q_T = \sqrt{3}V_L I_L \sin \theta \qquad \qquad (20\text{-}10)$$

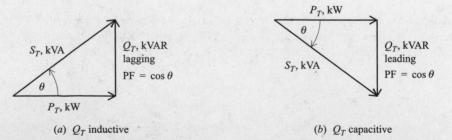

(a) Q_T inductive   (b) Q_T capacitive

Fig. 20-5 Power-triangle relationships for 3-ϕ circuit with net inductive or net capacitive loads

where P_T = total real power, W

        S_T = total apparent power, VA

        Q_T = total reactive power, VAR

        V_L = line voltage, V

        I_L = line current, A

        θ = load phase angle

        $\sqrt{3}$ = 1.73, a constant

Transformer ratings are specified generally in kilovoltamperes. The relationships of voltage, current, and power expressed in Eqs. *(20-1)–(20-10)* are applicable to all balanced 3-ϕ circuits.

Example 20.4   How much power is delivered by a balanced 3-ϕ system if each wire carries 20 A and the voltage between the wires is 220 V at a PF of unity?

    Using Eq. *(20-8)*,

$$P_T = \sqrt{3}\,V_L I_L \cos\theta = 1.73(220)(20)(1) = 7612 \text{ W} \qquad Ans.$$

Example 20.5   Each phase of a 3-ϕ Δ-connected generator supplies a full-load current of 100 A at a voltage of 240 V and at a PF of 0.6 lagging (Fig. 20-6). Find the *(a)* line voltage, *(b)* line current, *(c)* 3-ϕ power in kilovoltamperes, and *(d)* 3-ϕ power in kilowatts.

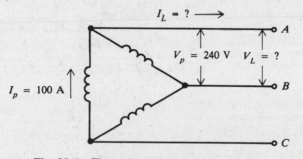

Fig. 20-6   Three-phase Δ-connected generator

(a)   Use Eq. *(20-1)*.

$$V_L = V_p = 240 \text{ V} \qquad Ans.$$

(b)   Use Eq. *(20-2)*.

$$I_L = 1.73 I_p = 1.73(100) = 173 \text{ A} \qquad Ans.$$

(c)   Use Eq. *(20-9)*.

$$S_T = \sqrt{3}\,V_L I_L = 1.73(240)(173) = 71\,800 \text{ VA} = 71.8 \text{ kVA} \qquad Ans.$$

(d)

$$P_T = S_T \cos\theta = 71.8(0.6) = 43.1 \text{ kW} \qquad Ans.$$

Example 20.6   Three resistances of 20 Ω each are Y-connected to a 240-V 3-ϕ line operating at a PF of unity (Fig. 20-7). Find the *(a)* current through each resistance, *(b)* line current, and *(c)* power taken by the three resistances.

(a) $V_L = \sqrt{3}\,V_p$ *(20-5)*

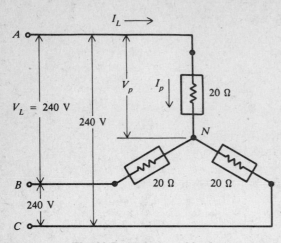

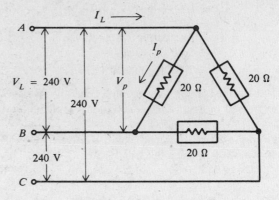

Fig. 20-7   Y-connected load   Fig. 20-8 Δ-connected load

$$V_p = \frac{V_L}{\sqrt{3}} = \frac{240}{1.73} = 138.7 \text{ V}$$

$$I_p = \frac{V_p}{Z_p} = \frac{V_p}{R_p} = \frac{138.7}{20} = 6.94 \text{ A} \qquad Ans.$$

(b)   Use Eq. (20-3).

$$I_L = I_p = 6.94 \text{ A} \qquad Ans.$$

(c)   Use Eq. (20-6).

$$P_T = 3P_p = 3V_p I_p \cos\theta = 3(138.7)(6.94)(1) = 2890 \text{ W} \qquad Ans.$$

Example 20.7   Repeat Example 20.6 if the three resistances are reconnected in delta (Fig. 20-8).

(a)   Use Eq. (20-1).

$$V_p = V_L = 240 \text{ V}$$

$$I_p = \frac{V_p}{Z_p} = \frac{V_p}{R_p} = \frac{240}{20} = 12 \text{ A} \qquad Ans.$$

(b)   Use Eq. (20-2).

$$I_L = \sqrt{3}I_p = 1.73(12) = 20.8 \text{ A} \qquad Ans.$$

(c)   Use Eq. (20-6).

$$P_T = 3V_p I_p \cos\theta = 3(240)(12)(1) = 8640 \text{ W} \qquad Ans.$$

Or use Eq. (20-8).

$$P_T = \sqrt{3}V_L I_L \cos\theta = 1.73(240)(20.8)(1) = 8640 \text{ W} \qquad Ans.$$

Example 20.8   A 3-ϕ Y-connected transformer secondary has a four-wire 208-V ABC system (Fig. 20-9). Thirty lamps, each rated at 120 V and 2 A, are to be connected across each phase. Show the connection of the lamp load if the load is

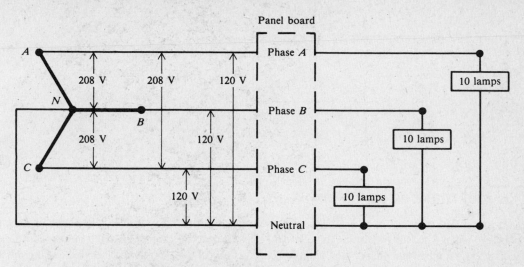

Fig. 20-9   Load connections for a balanced 3-ϕ circuit

to be balanced, and determine the power assumed by each phase and the power consumed by the system. (Assume lamps are resistive.)

The line voltage is given as $V_L = 208$ V. The line-to-neutral voltage or the phase voltage is V_p. Using Eq. (20-5),

$$V_L = \sqrt{3}V_p \quad \text{so} \quad V_p = \frac{V_L}{\sqrt{3}} = \frac{208}{1.73} = 120 \text{ V}$$

In order to have a balanced load, 30 lamps must be distributed equally across three 120-V phases. Thus, 10 lamps are connected across each 120-V phase (Fig. 20-9). The power per phase P_p is found by using Eq. (20-6):

$$P_p = V_p I_p \cos\theta = 120 \left(10 \text{ lamps} \times \frac{2 \text{ A}}{\text{lamp}} \right)(1) = 120(20)(1) = 2400 \text{ W} \qquad Ans.$$

and the total power is three times the phase power.

$$P_T = 3P_p = 3(2400) = 7200 \text{ W} \qquad Ans.$$

Example 20.9   A 3-ϕ three-wire system has a line current of 25 A and a line voltage of 1000 V. The power factor of the load is 86.6 percent lagging. Find (a) the real power delivered, (b) the reactive power, and (c) the apparent power, and (d) draw the power triangle.

(a)   Use Eq. (20-8).

$$P_T = \sqrt{3}V_L I_L \cos\theta = 1.73(1000)(25)(0.866) = 37\,500 \text{ W} = 37.5 \text{ kW} \qquad Ans.$$

(b)
$$\theta = \arccos 0.866 = 30°$$

$$\sin\theta = 0.5$$

$$Q_T = \sqrt{3}V_L I_L \sin\theta = 1.73(1000)(25)(0.5) = 21\,600 \text{ VAR} = 21.6 \text{ kVAR lagging} \qquad Ans.$$

(c)
$$S_T = \sqrt{3}V_L I_L = 1.73(1000)(25) = 43\,250 \text{ VA} = 43.3 \text{ kVA} \qquad Ans.$$

(d)   See Fig. 20-10.

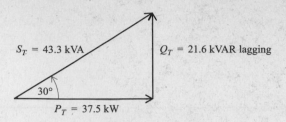

Fig. 20-10   Power triangle

UNBALANCED THREE-PHASE LOADS

A very important property of a 3-ϕ balanced system is that the phasor sum of the three line (or phase) voltages is zero and the phasor sum of the three line (or phase) currents is zero. When the three load impedances are not equal to each other, the phasor sum and the neutral current I_N are not zero, and we have an unbalanced load. An imbalance occurs when an open or short circuit appears at the load.

If a 3-ϕ system has an unbalanced power source and an unbalanced load, the methods for solving them are complex. We will consider an unbalanced load only with a balanced source.

Example 20.10   Consider a balanced 3-ϕ system (Fig. 20-11a) with a Y load. The line-to-line voltage is 173 V and the resistance in each branch is 10 Ω. Find the line current and neutral current under the following three load conditions: (a) balanced load, (b) open circuit in line A (Fig. 20-11b), and (c) short circuit in line A (Fig. 20-11c).

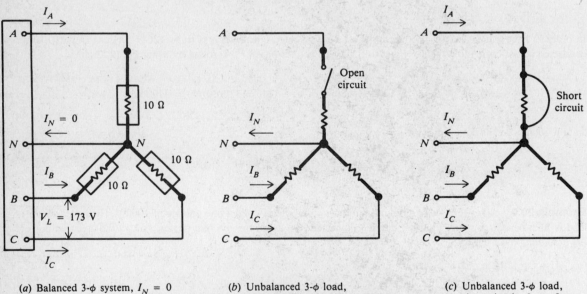

   (a) Balanced 3-ϕ system, $I_N = 0$ (b) Unbalanced 3-ϕ load,   (c) Unbalanced 3-ϕ load,
                                           open circuit, $I_N = I_B + I_C$ short circuit, $I_N = I_A$

Fig. 20-11   Three-phase unbalanced load conditions

(a)   Balanced 3-ϕ load (Fig. 20-11a):

$$V_p = \frac{V_L}{\sqrt{3}} = \frac{173}{1.73} = 100 \text{ V} \qquad\qquad I_p = \frac{V_p}{R_p} = \frac{100}{10} = 10 \text{ A}$$

$$I_L = I_p = 10 \text{ A} \qquad\qquad\qquad\qquad I_N = 0 \text{ A} \quad\quad Ans.$$

(b) Open Y circuit (Fig. 20-11b): Line voltage $V_L = 173$ V remains fixed. But current in lines B and C now becomes (two 10-Ω resistors are in series)

$$I_B = I_C = \frac{173}{2(10)} = 8.66 \text{ A} \qquad Ans.$$

which is less than the line current under balanced conditions.

$$I_N = I_B + I_C = 8.66 + 8.66 = 17.3 \text{ A} \qquad Ans.$$

(c) Short Y circuit (Fig. 20-11c): Line voltage $V_L = 173$ V remains fixed. But current in lines B and C now becomes

$$I_B = I_C = \frac{173}{10} = 17.3 \text{ A} \qquad Ans.$$

which is $\sqrt{3}$ times the value of the line current under balanced conditions. The current in line A is equal to the neutral line current, $I_A = I_N$. I_N is the phasor sum of I_B and I_C, which are 120° out of phase, so that

$$I_N = \sqrt{3}I_B = 1.73 \,(17.3) = 30 \text{ A} \qquad Ans.$$

which is three times its value under balanced conditions. Under a fault condition the neutral connector in the Y-connected load carries more current than the line or winding under a balanced load.

Solved Problems

20.1   Each phase of a Y-connected generator delivers a current of 30 A at a phase voltage of 254 V and a PF of 80 percent lagging (Fig. 20-12). What is the generator terminal voltage? What is the power developed in each phase? What is the 3-ϕ power developed?

$$V_L = \sqrt{3}V_p \qquad\qquad (20\text{-}5)$$

$$= 1.73(254) = 439 \text{ V} \qquad Ans.$$

$$P_p = V_p I_p \cos\theta \qquad\qquad (20\text{-}6)$$

$$= 254(30)(0.8) = 6096 \text{ W} \qquad Ans.$$

$$P_T = 3V_p I_p \cos\theta = 3P_p \qquad\qquad (20\text{-}7)$$

$$= 3(6096) = 18\,288 \text{ W} \qquad Ans.$$

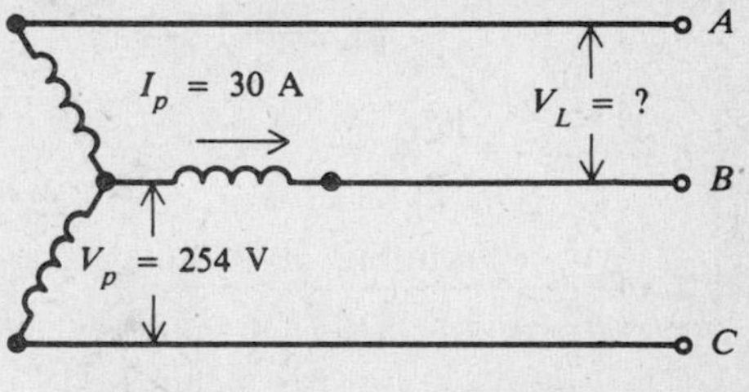

Fig. 20-12   Y-connected generator

20.2   Identify the three-phase transformer connections (Fig. 20-13a and b).

If you follow the wiring sequence for the primary winding connection and the secondary winding connection, you obtain the equivalent connections shown in Fig. 20-13c. So (1) is a Δ-Δ connection and (2) is a Y-Δ connection.

20.3   In a three-phase Y-Δ connection, each transformer has a voltage ratio of 4:1. If the primary line voltage is 660 V, find (a) the secondary line voltage, (b) the voltage across each primary winding, and (c) the voltage across each secondary winding.

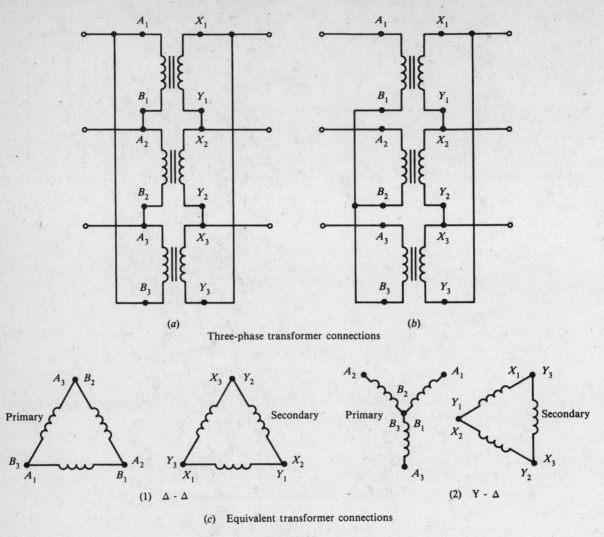

(a)   (b)

Three-phase transformer connections

(1) Δ - Δ   (2) Y - Δ

(c) Equivalent transformer connections

Fig. 20-13

From Fig. 20-3c, the Y-Δ connection is as shown in Fig. 20-14.

(a) Given are $V = 660$ V and VR $= 4 =$ TR $= a$. Then

$$\text{Secondary line voltage} = \frac{V}{\sqrt{3}a} = \frac{660}{1.73(4)} = 95.4 \text{ V} \qquad Ans.$$

(b) Primary winding voltage $= \dfrac{V}{\sqrt{3}} = \dfrac{660}{1.73} = 382$ V   *Ans.*

(c) Voltage across the secondary winding = voltage of secondary line = 95.4 V   *Ans.*

20.4 The secondary line voltage of a Δ-Y transformer bank is 411 V. The transformers have a turns ratio of 3 : 1. Find (a) the primary line voltage, (b) the current in each secondary winding or coil if the current in each secondary line is 60 A, and (c) the primary line current.

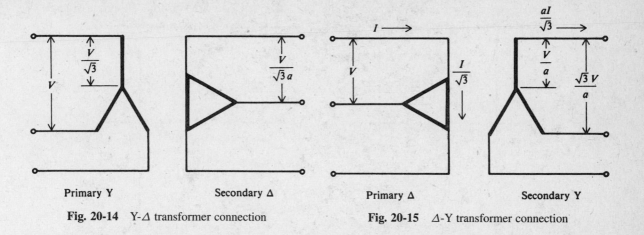

Primary Y **Secondary Δ**

Fig. 20-14 Y-Δ transformer connection

Primary Δ **Secondary Y**

Fig. 20-15 Δ-Y transformer connection

From Fig. 20-3d, the Δ-Y connection is as shown in Fig. 20-15.

(a) Secondary line voltage $= \dfrac{\sqrt{3}\,V}{a} = 411$ V   so

$$\text{Primary line voltage } V = \frac{411a}{\sqrt{3}} = \frac{411(3)}{1.73} = 713 \text{ V}   \textit{Ans.}$$

(b) When the secondaries are connected in Y, the current in each secondary coil equals the line current, or 60 A.

(c) Secondary line current $= \dfrac{aI}{\sqrt{3}} = 60$ A   so

$$\text{Primary line current } I = \frac{60\sqrt{3}}{a} = \frac{60(3)}{3} = 34.6 \text{ A}   \textit{Ans.}$$

20.5 What are the primary and secondary current ratings of a 500-kVA 3-ϕ transformer stepping down from a 480-V delta to 120/208-V wye (Fig. 20-16)? The 120/208-V wye four-wire transformer secondary is used in building electrical systems when both single-phase loads and three-phase motors are to be supplied.

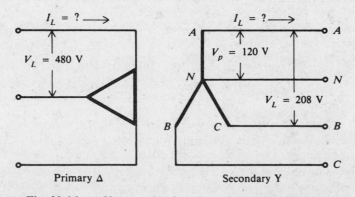

Primary Δ **Secondary Y**

Fig. 20-16 Δ-Y connection for a 3-ϕ step-down transformer

There are two secondary voltages in the Y transformer four-wire system: the phase voltage V_p of 120 V and the line voltage V_L of 208 V ($120 \times \sqrt{3} = 208$). On the primary side:

$$\text{Transformer rating} = \text{apparent power} = \sqrt{3}V_L I_L \qquad (20\text{-}9)$$

So

$$I_L = \frac{\text{transformer rating}}{\sqrt{3}V_L} = \frac{500\,000}{1.73(480)} = 602 \text{ A} \qquad Ans.$$

On the secondary side:

$$\text{Transformer rating} = \text{apparent power} = \sqrt{3}V_L I_L \qquad (20\text{-}9)$$

So

$$I_L = \frac{\text{transformer rating}}{\sqrt{3}V_L} = \frac{500\,000}{1.73(480)} = 1390 \text{ A} \qquad Ans.$$

20.6 Delta-connected secondaries are generally used for three-phase three-wire loads. Show how a Δ connection can be used to supply single-phase loads at two terminal voltages as well as three-phase loads.

A four-wire system is obtained by adding a fourth wire from the midpoint of one of the three Δ transformer windings (Fig. 20-17). The normal voltage is 240 V from phase to phase and 120 V from the neutral (or midphase) wire to each of the phasor conductors connected to the winding ends. With this system, single-phase loads are fed from 120-V terminals; and three-phase loads as well as single-phase loads are fed from the 240-V phase terminals A, B, and C.

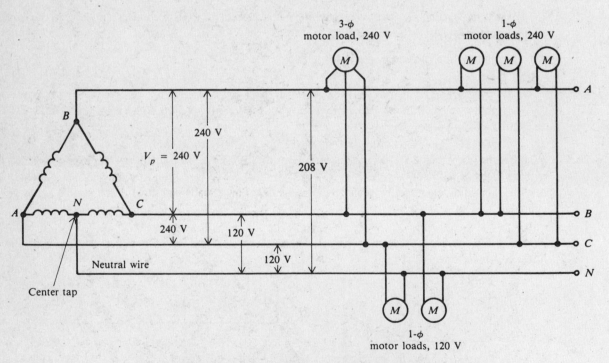

Fig. 20-17 Δ connections to four-wire line with neutral

20.7 A distribution transformer bank of three single-phase transformers is connected Δ-Y (Fig. 20-18a). The transformer turns ratio is 100:1. The secondaries of the bank supply power to a 208-V, 3-ϕ, four-wire system. The load in the system consists of a 72-kW 3-ϕ motor, PF = 1, terminal voltage 208 V; three 12-kW 1-ϕ lighting circuits, terminal voltage 120 V; and three 1-ϕ 10-kVA motors,

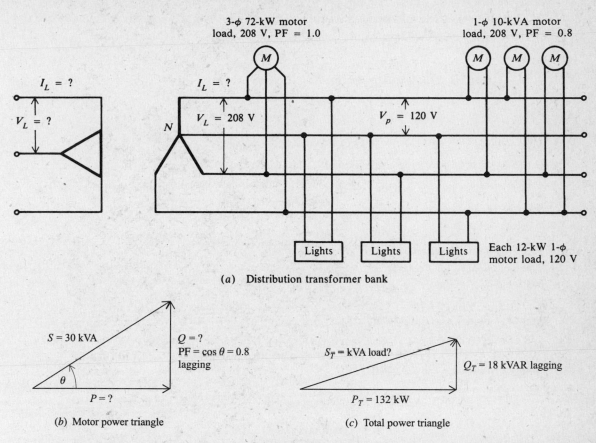

3-φ 72-kW motor
load, 208 V, PF = 1.0

1-φ 10-kVA motor
load, 208 V, PF = 0.8

$I_L = ?$

$V_L = ?$

$I_L = ?$

$V_L = 208$ V

$V_p = 120$ V

Each 12-kW 1-φ
motor load, 120 V

Lights   Lights   Lights

(a)   Distribution transformer bank

$S = 30$ kVA

$Q = ?$
PF = $\cos \theta = 0.8$
lagging

θ

$P = ?$

(b) Motor power triangle

S_T = kVA load?

$Q_T = 18$ kVAR lagging

$P_T = 132$ kW

(c) Total power triangle

Fig. 20-18

PF = 0.8 lagging, terminal voltage 208 V. Find (a) the total kilovoltampere load of the circuit; (b) the kilovoltampere rating of the transformer bank if ratings only of 100 kVA, 112.5 kVA, and 150 kVA are available; (c) the kilovoltampere rating of the individual transformers; (d) the secondary line current; (e) the primary line voltage; and (f) the primary line current.

Step 1.   Find the phase-to-neutral voltage V_p in the Y secondary.

$$V_p = \frac{V_L}{\sqrt{3}} = \frac{208}{\sqrt{3}} = 120 \text{ V}$$

Step 2.   Find the total kilovoltampere load and then the required kilovoltampere rating of the transformer bank.

Lighting load:	$P = 3 \times 12 \text{ kW} = 36 \text{ kW}$
PF = 1, motor load:	$P = 72 \text{ kW}$
PF = 0.8, motor load:	$S = 3 \times 10 \text{ kVA} = 30 \text{ kVA}$

In order to find the P and Q components of the motor load with PF = 0.8, we use relationships of the power triangle (Fig. 20-18b).

$$P = S \cos \theta = 30(0.8) = 24 \text{ kW} \qquad Q = S \sin \theta = 30(0.6) = 18 \text{ kVAR lagging}$$

The total kilowatt load is

$$P_T = 36 + 72 + 24 = 132 \text{ kW}$$

The total kilovoltampere-reactive load Q_T is

$$Q_T = 18 \, \text{kVAR}$$

So the total kilovoltampere load S_T (Fig. 20-18c) is

$$S_T = \sqrt{132^2 + 18^2}$$

(a) Kilovoltampere load $S_T = 133 \, \text{kVA}$ *Ans.*

(b) The next larger rating available is the 150-kVA transformer bank.   *Ans.*

(c) Therefore, this bank requires three 50-kVA single-phase transformers.   *Ans.*

Step 3. Find I_L in the secondary.

$$\text{Kilovoltampere load} = \sqrt{3} V_L I_L = 133 \, \text{kVA} \qquad (20\text{-}9)$$

(d) Then $$I_L = \frac{133(1000)}{\sqrt{3}V_L} = \frac{133\,000}{1.73(208)} = 370 \, \text{A} \quad \text{\textit{Ans.}}$$

Step 4. Find V_L in the primary.

(e)   Primary $V_L = (\text{TR})(\text{secondary } V_p) = 100(120) = 12\,000 \, \text{V}$ *Ans.*

Step 5. Find I_L in the primary.

$$\text{Kilovoltampere load} = \sqrt{3} V_L I_L = 133 \, \text{kVA}$$

(f) Then $$I_L = \frac{133(1000)}{\sqrt{3}V_L} = \frac{133\,000}{1.73(12\,000)} = 6.41 \, \text{A} \quad \text{\textit{Ans.}}$$

20.8 If the transformer bank of Problem 20.7 is rated at 150 kVA, find the full-load current on the high side and on the low side.

The low-side (secondary) voltage is given as 208 V.

$$\text{Kilovoltampere load} = \sqrt{3} V_L I_L = 150 \, \text{kVA} \qquad (20\text{-}9)$$

Low side: $$I_L = \frac{150(1000)}{\sqrt{3}V_L} = \frac{150\,000}{1.73(208)} = 417 \, \text{A} \quad \text{\textit{Ans.}}$$

The high-side (primary) voltage is the product of the turns ratio and the corresponding winding voltage on the low side (secondary), which is the phase voltage V_p. So

$$V_L = 100(120) = 12\,000 \, \text{V}$$

and, assuming ideal transformers without losses, on the high side

$$\text{Kilovoltampere load} = \sqrt{3} V_L I_L = 150 \, \text{kVA} \qquad (20\text{-}9)$$

High side: $$I_L = \frac{150(1000)}{\sqrt{3}V_L} = \frac{150\,000}{1.73(12\,000)} = 7.23 \, \text{A} \quad \text{\textit{Ans.}}$$

Since maximum currents are present under full-load conditions, the cables connected to the high- and low-voltage windings of the transformers would be sized according to these currents.

20.9   A Δ-connected load (Fig. 20-19) is drawing $600\,\mathrm{kW}$ from a 5000-V line at a lagging power factor. If each wire carries $75\,\mathrm{A}$, what is the power factor and angle of lag?

$$\mathrm{PF} = \cos\theta$$

$$P_T = \sqrt{3}V_L I_L \cos\theta \qquad\qquad (20\text{-}8)$$

Solve for $\cos\theta$:

$$\cos\theta = \frac{P_T}{\sqrt{3}V_L I_L}$$

Substitute known values:

$$\mathrm{PF} = \cos\theta = \frac{600(1000)}{1.73(5000)(75)} = 0.925 \text{ lagging} \qquad Ans.$$

Then $\theta = \arccos 0.925 = 22.4°$ \qquad Ans.

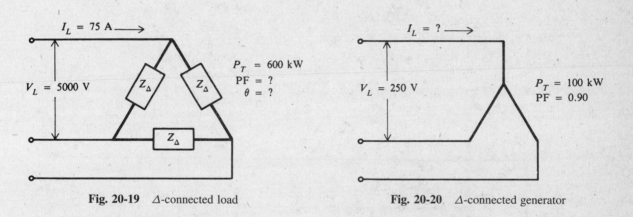

Fig. 20-19 Δ-connected load   **Fig. 20-20** Δ-connected generator

20.10   A Y-connected generator (Fig. 20-20) supplies $100\,\mathrm{kW}$ at a PF of 0.90. If the line voltage is $250\,\mathrm{V}$, find the line current.

$$\mathrm{PF} = \cos\theta = 0.90$$

$$P_T = \sqrt{3}V_L I_L \cos\theta \qquad\qquad (20\text{-}8)$$

Solve for I_L:

$$I_L = \frac{P_T}{\sqrt{3}V_L \cos\theta} = \frac{100(1000)}{1.73(250)(0.90)} = 257\,\mathrm{A} \qquad Ans.$$

20.11   Three loads, each having a resistance of $16\,\Omega$ and an inductive reactance of $12\,\Omega$, are Δ-connected to a 240-V 3-ϕ supply (Fig. 20-21). Find the (a) impedance per phase, (b) current per phase, (c) the 3-ϕ kilovoltamperage, (d) PF, and (e) the 3-ϕ power in kilowatts.

(a) $Z_p = \sqrt{R_p^2 + X_p^2} = \sqrt{16^2 + 12^2} = 20\,\Omega$ \qquad Ans.

$$\theta \text{ of load} = \arctan\frac{X_p}{R_p} = \arctan\frac{12}{16} = \arctan 0.75 = 36.9°$$

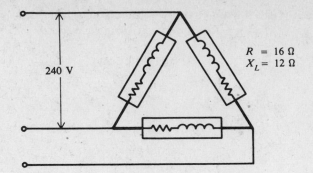

Fig. 20-21 Impedance in Δ-connected load

(b) $V_p = V_L = 240$ V

$$I_p = \frac{V_p}{Z_p} = \frac{240}{20} = 12 \text{ A} \qquad Ans.$$

(c) $S_T = 3V_p I_p = 3(240)(12) = 8640 \text{ VA} = 8.64 \text{ kVA} \qquad Ans.$

(d) $\text{PF} = \cos\theta = \cos 36.9° = 0.80 \text{ lagging} \qquad Ans.$

(e) $P_T = S_T \cos\theta = 8.64(0.8) = 6.91 \text{ kW} \qquad Ans.$

20.12 Find the indicated quantities (Fig. 20-22*a*), which show the connected loads for a balanced 3-ϕ Δ transformer bank.

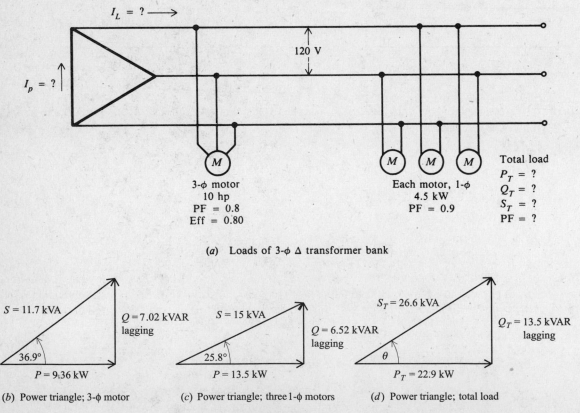

(*a*) Loads of 3-ϕ Δ transformer bank

(*b*) Power triangle; 3-ϕ motor   (*c*) Power triangle; three 1-ϕ motors   (*d*) Power triangle; total load

Fig. 20-22

Step 1. Find the power relationships for *each* load.

(a) 3-ϕ motor (Fig. 20-22b):

$$S = \text{motor load} = \frac{\text{hp} \times 746}{\text{PF} \times \text{Eff}} = \frac{10(746)}{0.8(0.8)} = 11\,656 = 11.7\,\text{kVA} \qquad (1\text{hp} = 746\,\text{W})$$

$$\theta = \arccos 0.8 = 36.9°$$

$$P = 11.7 \cos 36.9° = 11.7(0.8) = 9.36\,\text{kW}$$

$$Q = 11.7 \sin 36.9° = 11.7(0.6) = 7.02\,\text{kVAR lagging}$$

(b) Three 1-ϕ motors (Fig. 20-22c):

$$P = 13.5\,\text{kW} \quad \text{given } \theta = \arccos 0.9 = 25.8°$$

$$Q = 13.5 \tan 25.8° = 13.5(0.483) = 6.52\,\text{kVAR lagging}$$

$$S = \sqrt{P^2 + Q^2} = \sqrt{(13.5)^2 + (6.52)^2} = \sqrt{182 + 42.5} = \sqrt{224.5} = 15\,\text{kVA}$$

Step 2. Find the power relationships for the *total* load.

$$\text{Total kW load } P_T = 9.36 + 13.5 = 22.9\,\text{kW} \qquad Ans.$$

$$\text{Total kVAR load } Q_T = 7.02 + 6.52 = 13.5\,\text{kVAR lagging} \qquad Ans.$$

From the power-triangle relationships (Fig. 20-22d),

$$S_T = \sqrt{P_T^2 + Q_T^2} = \sqrt{(22.9)^2 + (13.5)^2} = \sqrt{524 + 182} = \sqrt{706} = 26.6\,\text{kVA} \qquad Ans.$$

$$\text{PF} = \cos\theta = \frac{P_T}{S_T} = \frac{22.9}{26.6} = 0.861\,\text{lagging} \qquad Ans.$$

Step 3. Find the current relationships.

$$S_T = \sqrt{3}\,V_L I_L \tag{20-9}$$

from which

$$I_L = \frac{S_T}{\sqrt{3}\,V_L} = \frac{26.5(1000)}{1.73(120)} = 128\,\text{A} \qquad Ans.$$

$$I_L = \sqrt{3}\,I_p \tag{20-2}$$

from which

$$I_p = \frac{I_L}{\sqrt{3}} = \frac{128}{1.73} = 74\,\text{A} \quad Ans.$$

20.13 On a 120/208-V four-wire Y system, two lamps are connected from phase A to neutral, five lamps from phase B to neutral, and four lamps from phase C to neutral (Fig. 20-23a). If all the lamps have the same rating and each draws 2 A, find the current in each phase and the neutral current.

Since each lamp draws 2 A,

$$I_A = 2 \text{ lamps} \times 2\,\text{A} = 4\,\text{A} \qquad Ans.$$

$$I_B = 5 \text{ lamps} \times 2\,\text{A} = 10\,\text{A} \qquad Ans.$$

$$I_C = 4 \text{ lamps} \times 2\,\text{A} = 8\,\text{A} \qquad Ans.$$

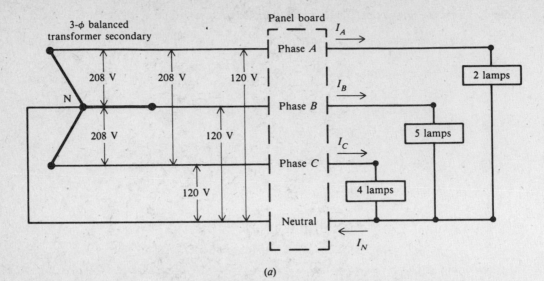

Fig. 20-23*a*   Unbalanced loading on four-wire, 3-φ Y circuit

Since the values of the load or phase currents are different, the load is unbalanced and current will flow in the neutral wire. The magnitude of I_N is the phasor sum of the three phase currents I_A, I_B, and I_C.

$$I_N = \sqrt{I_x^2 + I_y^2}$$

where I_x is the sum of the phasor currents along the x axis and I_y is the sum of the phasor currents along the y axis. We draw the phasor diagram of currents, calculate their component values along the x and y axes, add these components, and then solve for I_N.

Step 1.   Draw the current-phasor diagram (Fig. 20-23*b*).

We show I_A as the reference phasor and then show each load current 120° out of phase from the others.

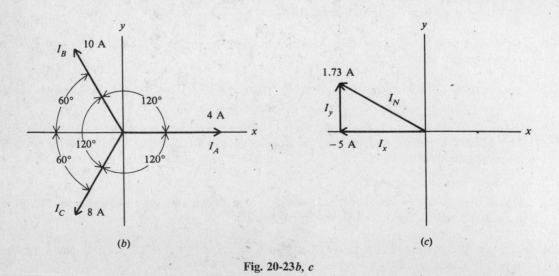

Fig. 20-23*b, c*

Step 2. Find the component values of the three load currents and add them. Refer to Fig. 20-23b.

Along the x axis: $I_A = 4$ A $I_B = -10 \cos 60° = -5$ A $I_C = -8 \cos 60° = -4$ A

Along the y axis: $I_A = 0$ A $I_B = 10 \sin 60° = 8.66$ A $I_C = -8 \sin 60° = -6.93$ A

Then $I_x = 4 - 5 - 4 = -5$ A $I_y = 0 + 8.66 - 6.93 = 1.73$ A

Step 3. Calculate I_N (Fig. 20-23c).

$$I_N = \sqrt{I_x^2 + I_y^2} = \sqrt{(-5)^2 + (1.73)^2} = 5.29 \text{ A} \qquad Ans.$$

Supplementary Problems

20.14 If the winding or phase voltage in the secondary is 120 V, find the secondary line voltage for Y and Δ connections.   *Ans.* Y-connection: $V_L = 208$ V; Δ-connection: $V_L = 120$ V

20.15 A Δ-connected generator supplies 100 V as line voltage and 25 A as line current. What is the voltage and current for each winding or phase?   *Ans.* $V_p = 100$ V; $I_p = 14.5$ A

20.16 A Y-connected generator supplies 40 A to each line and has a phase voltage of 50 V. Find the current through each phase and the line voltage.   *Ans.* $I_p = 40$ A; $V_L = 86.5$ V

20.17 In a Y-Δ transformer bank, each transformer has a voltage ratio of 3 : 1. If the primary line voltage is 625 V, find (a) the voltage across each primary winding, (b) the secondary line voltage, and (c) the voltage across each secondary winding.   *Ans.* (a) 361 V;   (b) 120 V;   (c) 120 V

20.18 The secondary line voltage of a Δ-Δ transformer bank is 405 V and the secondary line current is 35 A. If each transformer has a turns ratio of 5 : 1, find (a) the primary line voltage, (b) the secondary winding or phase current, (c) the primary line current, and (d) the primary winding or phase current.   *Ans.* (a) 2025 V;   (b) 20.2 A;   (c) 7 A;   (d) 4.05 A

20.19 Identify the three-phase transformer connections (Fig. 20-24a and b).   *Ans.* (a) Δ-Y connection;   (b) Y-Y connection

20.20 A 3-ϕ system with balanced load carries 30 A at 0.75 power factor. If the line voltage is 220 V, find the power delivered.   *Ans.* $P = 8.56$ kW

20.21 Find the kW and kVA drawn from a 3-ϕ generator when it is delivering 25 A at 240 V to a motor with 86 percent power factor.   *Ans.* $P = 8.93$ kW; $S = 10.4$ kVA

20.22 Find the full-load line and phase currents for both Δ- and Y-connected 3-ϕ alternators with ratings of 75 kVA at 480 V.   *Ans.* Δ-connected: $I_L = 90.3$ A; $I_p = 52.2$ A;   Y-connected: $I_L = 90.3$ A; $I_p = 90.3$ A

20.23 Of the four possible ways to connect transformers in three-phase systems, which one will give the highest secondary voltage for a given primary voltage?   *Ans.* Δ-Y

20.24 Find the line current in a balanced 3-ϕ system which delivers 36 kW at 0.95 power factor and 225 V.   *Ans.* $I_L = 97.4$ A

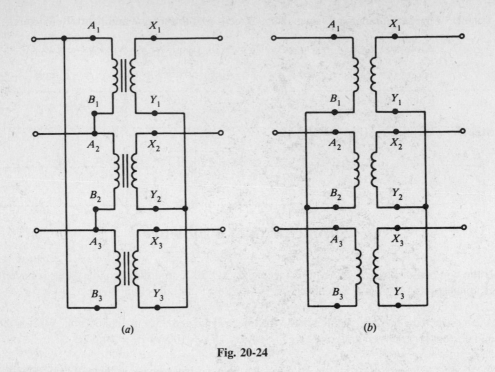

Fig. 20-24

20.25 A 3-ϕ system supplies 34.2 A at a line voltage of 208 V for a balanced load at 89 percent power factor. Find the kilovoltampere rating of the transformer.   *Ans.* $S = 12.3$ kVA

20.26 A balanced 3-ϕ line carries 14 A in each line at 1100 V. If the power delivered is 22 kW, find the power factor of the load.   *Ans.* · PF $= 0.826$ lagging

20.27 A three-phase system delivers a line current of 50 A at a line voltage of 220 V and 86.6 percent power factor. Find (*a*) the real power, (*b*) the reactive power, and (*c*) the apparent power.
Ans.   (*a*) $P = 16.5$ kW;   (*b*) $Q = 9.53$ kVAR;   (*c*) $S = 19.1$ kVA

20.28 How many 60-W 110-V lamps can be connected to a balanced 3-ϕ system if the line current is 28.4 A and the line voltage is 110 V on the load side?   *Ans.*   90 lamps

20.29 At full load, each of the three phases of a Y-connected generator delivers 150 A and 1329 V at a PF of 75 percent lagging. Find (*a*) the terminal voltage rating, (*b*) the kilovoltampere rating, and (*c*) the kilowatt rating.   *Ans.*   (*a*) 2300 V;   (*b*) 598 kVA;   (*c*) 448 kW

20.30 It is desired that a 10 000-kVA 3-ϕ 60-Hz generator have a rated terminal voltage of 13 800 V when it is Y-connected. (*a*) What must be the voltage rating of each phase? Find (*b*) the kilovoltampere rating per phase and (*c*) the rated line current.   *Ans.*   (*a*) 7977 V;   (*b*) 3333 kVA;   (*c*) 419 A

20.31 A 3-ϕ transformer bank is designed to supply a 240-V industrial distribution system. The connected balanced 3-ϕ loads are as follows: 50-kW motor at PF $= 0.80$, 40-kW motor at PF $= 0.85$, fluorescent lighting load of 10 kW at PF $= 0.90$, and an incandescent lighting load of 15 kW at PF $= 1.0$. Find the total real power, total apparent power, total reactive power, and total PF of the load.
Ans. $P_T = 115$ kW; $Q_T = 67.1$ kVAR lagging; $S_T = 133$ kVA;   PF $= 86.5\%$ lagging

20.32 Find the total line current drawn by the connected loads of Problem 20.31, assuming balanced conditions.   *Ans.* $I_L = 320$ A

20.33   For the distribution transformer bank (Fig. 20-25), find the total kilovoltampere load and the power factor of the load.
   Ans. $S_T = 170\,\text{kVA}$ ($P_T = 148\,\text{kW}$, $Q_T = 84.5\,\text{kVAR lagging}$); $\text{PF} = 0.868$ lagging

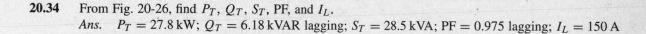
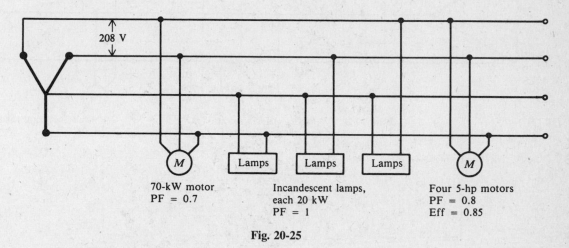

Fig. 20-25

20.34   From Fig. 20-26, find P_T, Q_T, S_T, PF, and I_L.
   Ans. $P_T = 27.8\,\text{kW}$; $Q_T = 6.18\,\text{kVAR lagging}$; $S_T = 28.5\,\text{kVA}$; $\text{PF} = 0.975$ lagging; $I_L = 150\,\text{A}$

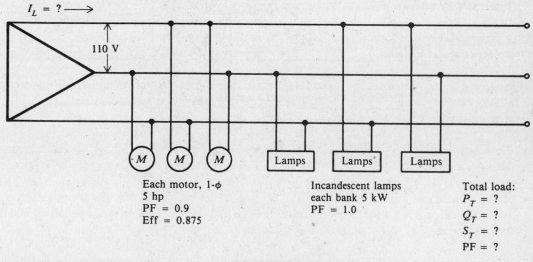

Fig. 20-26

20.35   From Fig. 20-27, find P_T, Q_T, S_T, PF, I_L, and I_p.
   Ans. $P_T = 20.1\,\text{kW}$; $Q_T = 15.9\,\text{kVAR lagging}$; $S_T = 25.6\,\text{kVA}$; $\text{PF} = 0.785$ lagging; $I_L = 67.3\,\text{A}$; $I_p = 38.9\,\text{A}$

20.36   From Fig. 20-28, find P_T, Q_T, S_T, PF, and I_L.
   Ans. $P_T = 13.1\,\text{kW}$; $Q_T = 6.24\,\text{kVAR lagging}$; $S_T = 14.5\,\text{kVA}$; $\text{PF} = 0.903$ lagging; $I_L = 43.9\,\text{A}$

20.37   A 6-hp 3-ϕ Δ-connected motor operates at a power factor of 0.85 with an efficiency of 80 percent. If the line voltage is 220 V, find the line current.   *Ans.* $I_L = 17.3\,\text{A}$

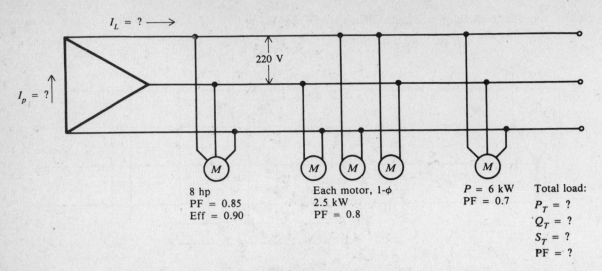

Fig. 20-27

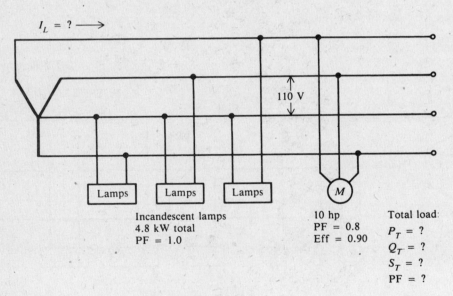

Fig. 20-28

20.38 A Y-connected load of three 10-Ω resistors is connected to a 208-V 3-ϕ supply. Find the (a) voltage supplied to each resistor, (b) line current, and (c) total power used.
Ans. (a) 120 V; (b) $I_L = 12$ A; (c) 4320 W

20.39 Three loads, each having a resistance of 16 Ω in series with an inductance of 12 Ω, are Y-connected to a 240-V 3-ϕ supply. Find the (a) impedance per phase, (b) current per phase, (c) 3-ϕ kilovoltamperage, (d) PF, and (e) the 3-ϕ power in kilowatts.
Ans. (a) $Z_p = 20$ Ω; (b) $I_p = 6.94$ A; (c) 2.89 kVA; (d) 0.80 lagging; (e) 2.31 kW

20.40 A Δ-connected load draws 28.8 kVA from a 3-ϕ supply. The line current is 69.2 A and the PF of the load is 80 percent lagging. Find the resistance and reactance of each phase (the kilovoltamperage of one phase is one-third of the total). *Ans.* 4.8 Ω resistance; 3.6 Ω reactance

20.41   Find the following indicated quantities:

	Three-Phase Connection	Line Voltage V_L, V	Line Current I_L, A	Phase Voltage V_p, V	Phase Current I_p, A
(a)	Δ	90	14	?	?
(b)	Y	?	?	50	10

Ans.

	Three-Phase Connection	V_L, V	I_L, A	V_p, V	I_p, A
(a)	Δ	....	....	90	8.09
(b)	Y	86.5	10	....	....

20.42   Find the following indicated quantities for a Δ-connected load:

	V_L, V	I_L, A	V_p, V	I_p, A	PF Lagging	P_T, kW
(a)	110	?	?	18	0.9	?
(b)	?	32	120	?	0.8	?
(c)	?	?	220	27	?	16

Ans.

	V_L, V	I_L, A	V_p, V	I_p, A	PF Lagging	P_T, kW
(a)	....	31.1	110	....	....	5.33
(b)	120	....	....	18.5	....	5.31
(c)	220	46.7	....	....	0.9	....

20.43   For the Δ-connected transformer secondary (Fig. 20-29), find the total line current.
Ans. $I_L = 75.1$ A

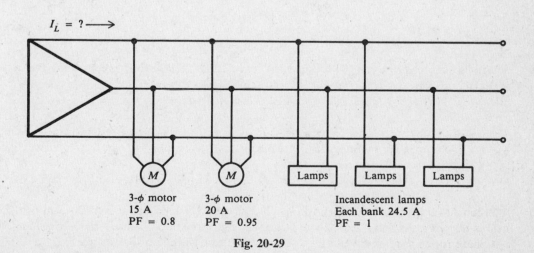

$I_L = ? \longrightarrow$

3-ϕ motor
15 A
PF = 0.8

3-ϕ motor
20 A
PF = 0.95

Incandescent lamps
Each bank 24.5 A
PF = 1

Fig. 20-29

20.44   For the Δ-connected loads (Fig. 20-30), find the total line current.   *Ans.* $I_L = 57.7$ A

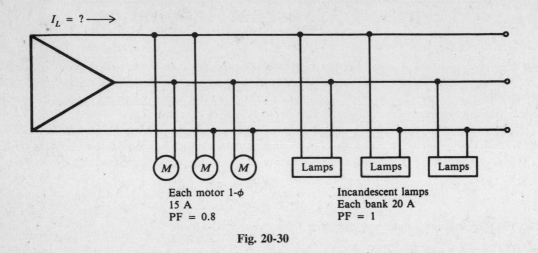

$I_L = ? \longrightarrow$

Each motor 1-ϕ
15 A
PF = 0.8

Incandescent lamps
Each bank 20 A
PF = 1

Fig. 20-30

20.45   Individual loads on separate feeders of a large power system are often unbalanced. The loads of a Y-connected generator, consisting of electric heaters, draw 150, 100, and 50 A. Find the neutral current.   *Ans.* $I_N = 86.6$ A

20.46   An unbalanced four-wire Y load has load currents of 3, 5, and 10 A. Calculate the value of the neutral current.   *Ans.* $I_N = 6.24$ A

Chapter 21

Series and Parallel Resonance

We have observed that in many circuits inductors and capacitors are connected in series or in parallel. Such circuits are often referred to as *RLC* circuits. One of the most important characteristics of an *RLC* circuit is that it can be made to respond most effectively to a single given frequency. When operated in this condition, the circuit is said to be in *resonance* with or *resonant* to the operating frequency.

A series or a parallel *RLC* circuit that is operated at resonance has certain properties that allow it to respond selectively to certain frequencies while rejecting others. A circuit operated to provide frequency selectivity is called a *tuned circuit*. Tuned circuits are used in impedance matching, bandpass filters, and oscillators.

SERIES RESONANCE

The *RLC* series circuit (Fig. 21-1a) has an impedance $Z = \sqrt{R^2 + (X_L - X_C)^2}$. The circuit is at resonance when the inductive reactance X_L is equal to the capacitive reactance X_C (Fig. 21-1b).

$$X_L = X_C$$

where $X_L = 2\pi f L$

$$X_C = \frac{1}{2\pi f C}$$

Then at resonance,

$$2\pi f L = \frac{1}{2\pi f C}$$

Solving for f,

$$f^2 = \frac{1}{(2\pi)^2 LC}$$

$$f = f_r = \frac{1}{2\pi\sqrt{LC}} = \frac{0.159}{\sqrt{LC}} \tag{21-1}$$

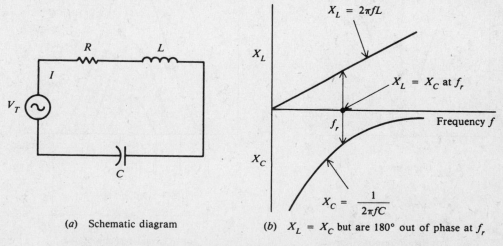

(a) Schematic diagram   (b) $X_L = X_C$ but are 180° out of phase at f_r

Fig. 21-1 Series resonance for *RLC* circuit at resonant frequency f_r

where f_r = resonant frequency, Hz
     L = inductance, H
     C = capacitance, F

For any LC product [Eq. (21-1)] there is only one resonant frequency. Thus, various combinations of L and C may be used to achieve resonance if the LC product remains the same. Equation (21-1) may be solved for L or C to find the inductance or capacitance needed to form a series resonant circuit at a given frequency.

$$L = \frac{1}{4\pi^2 f_r^2 C} = \frac{0.0254}{f_r^2 C} \qquad (21\text{-}2)$$

$$C = \frac{1}{4\pi^2 f_r^2 L} = \frac{0.0254}{f_r^2 L} \qquad (21\text{-}3)$$

Since $X_L = X_C$, $X_L - X_C = 0$ so that

$$Z = \sqrt{R^2 + (X_L - X_C)^2} = \sqrt{R^2} = R$$

Since the impedance at resonance Z equals the resistance R, the impedance is a minimum. With minimum impedance, the circuit has maximum current determined by $I = V/R$. The resonant circuit has a phase angle equal to 0° so that the power factor is unity.

At frequencies below the resonant frequency (Fig. 21-2a), X_C is greater than X_L so the circuit consists of resistance and capacitive reactance. However, at frequencies above the resonant frequency (Fig. 21-2a), X_L is greater than X_C so the circuit consists of resistance and inductive reactance. At resonance, maximum current is produced for different values of resistance (Fig. 21-2b). With a low resistance, the current increases sharply toward and decreases sharply from its maximum current as the circuit is tuned to and away from the resonant frequency. This condition where the curve is narrow at the resonant frequency provides good selectivity. With an increase of resistance, the curve broadens so that selectivity is less.

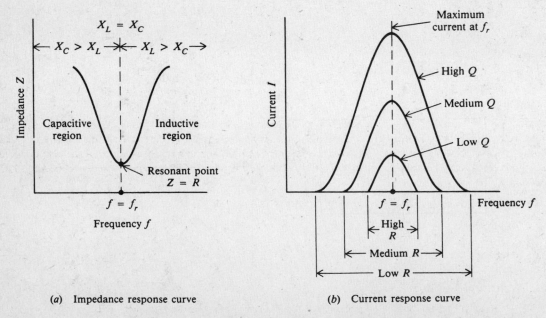

(a) Impedance response curve   (b) Current response curve

Fig. 21-2 Characteristics of series RLC circuit at resonance

Example 21.1 Find the resonant frequency of a simple radio receiver tuning circuit (Fig. 21-3) if the inductance is 200 μH and the capacitor is set at 200 pF.

$$L = 200\ \mu\text{H} = 200 \times 10^{-6}\ \text{H} = 2 \times 10^{-4}\ \text{H} \qquad C = 200\ \text{pF} = 200 \times 10^{-12}\ \text{F} = 2 \times 10^{-10}\ \text{F}$$

$$f_r = \frac{0.159}{\sqrt{LC}} \tag{21-1}$$

$$= \frac{0.159}{\sqrt{\left(2 \times 10^{-4}\right)\left(2 \times 10^{-10}\right)}} = 795\,000\ \text{Hz} = 795\ \text{kHz} \qquad Ans.$$

Notice that the voltage applied to this circuit is the voltage induced across the secondary winding, which is in series with the capacitor. Thus, while the circuit looks like a parallel circuit, it is a series circuit because of the series voltage input.

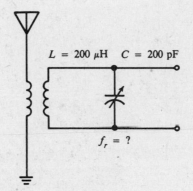

$$L = 200\ \mu\text{H} \quad C = 200\ \text{pF}$$
$$f_r = ?$$

Fig. 21-3 Radio receiver tuning circuit

Example 21.2 Find the value in picofarads of a tuning capacitor placed in series with a 400-μH inductance in order to provide resonance for a 500-kHz signal.

$$L = 400\ \mu\text{H} = 400 \times 10^{-6}\ \text{H} = 4 \times 10^{-4}\ \text{H} \qquad f_r = 500\ \text{kHz} = 500 \times 10^3\ \text{Hz} = 5 \times 10^5\ \text{Hz}$$

$$C = \frac{0.0254}{f_r^2 L} \tag{21-3}$$

$$= \frac{0.0254}{\left(5 \times 10^5\right)^2 \left(4 \times 10^{-4}\right)} = 2.54 \times 10^{-10}\ \text{F} = 254 \times 10^{-12}\ \text{F} = 254\ \text{pF} \qquad Ans.$$

Example 21.3 A voltage of 15-V ac is applied to a 150-μH coil connected in series with a 169-pF capacitor (Fig. 21-4). The total series resistance is 7.5 Ω, which includes the coil winding resistance, the resistance of the connecting leads, and the leakage resistance of the capacitor. This circuit is resonant at 1000 kHz. Find the magnitude of the current and of the voltage drops across each element. Describe the circuit when the frequency of the applied voltage is 800 kHz, 1000 kHz, and 1200 kHz.

Step 1.　Find the current at resonance.

At resonance, $$Z = R = 7.5\ \Omega$$

By Ohm's law, $$I = \frac{V_T}{R} = \frac{15}{7.5} = 2\ \text{A} \qquad Ans.$$

Step 2.　Find the voltage drops at resonance.

$$X_L = X_C = 2\pi f_r L = 6.28\left(10^6\right)\left(150 \times 10^{-6}\right) = 942\ \Omega$$

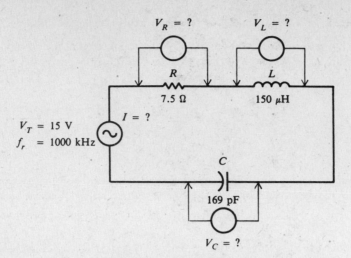

Fig. 21-4 Series *RLC* circuit

Then

$$V_R = IR = 2(7.5) = 15 \text{ V} \quad Ans.$$

$$V_L = IX_L = 2(942) = 1884 \text{ V} \quad Ans.$$

$$V_C = IX_C = 2(942) = 1884 \text{ V} \quad Ans.$$

Thus we see that the voltage drops across both the inductance and the capacitance are 125 times greater than the applied voltage. Since at resonance the high reactive voltages V_L and V_C cancel each other, they have no effect on the load. However, the insulation of the coil connectors and the dielectric of the capacitor must be designed to withstand a voltage about 125 times greater than the applied voltage.

Step 3. Describe the behavior of the circuit as a function of frequency. At frequency $f = 800$ kHz, which is *below* the resonant frequency:

$$X_L = 6.28fL = 6.28(8 \times 10^5)(150 \times 10^{-6}) = 754 \ \Omega$$

$$X_C = \frac{1}{6.28fC} = \frac{1}{6.28(8 \times 10^5)(169 \times 10^{-12})} = 1178 \ \Omega$$

Since $X_C > X_L$, the circuit is *capacitive* below the resonant frequency.   *Ans.*
At the resonant frequency $f_r = 1000$ kHz:

$$X_L = X_C = 942 \ \Omega$$

Therefore, at resonance the circuit is purely *resistive.*   *Ans.*
At a frequency $f = 1200$ kHz, which is *above* the resonant frequency:

$$X_L = 6.28fL = 6.28(12 \times 10^5)(150 \times 10^{-6}) = 1130 \ \Omega$$

$$X_C = \frac{1}{6.28fC} = \frac{1}{6.28(12 \times 10^5)(169 \times 10^{-12})} = 785 \ \Omega$$

Since $X_L > X_C$, the circuit is *inductive* above the resonant frequency.   *Ans.*

Refer to Fig. 21-2*a*, which confirms the nature of this circuit as a function of frequency.

Q OF SERIES CIRCUIT

The degree to which a series-tuned circuit is selective is proportional to the ratio of its inductive reactance to its resistance. This ratio X_L/R is known as the Q of the circuit and is expressed as follows:

$$Q = \frac{X_L}{R} \qquad (21\text{-}4)$$

where Q = quality factor or figure of merit
X_L = inductive reactance, Ω
R = resistance, Ω

The lower the reactance, the higher the value of Q. The higher the Q, the sharper and more selective is the resonant curve. Q has the same value if calculated with X_C instead of X_L since they are equal at resonance. $Q = 150$ is a high Q. Typical values are 50 to 250. Less than 10 is a low Q; more than 300 is a very high Q. See Fig. 21-2b.

The Q of the circuit is generally considered in terms of X_L since the coil has the series resistance of the circuit. In this case, the Q of the coil and the Q of the series resonant circuit are the same. If extra resistance is added, the Q of the circuit will be less than the Q of the coil. The highest possible Q for the circuit is the Q of the coil.

The Q of the resonant circuit can be considered a magnification factor that determines how much the voltage across L or C is increased by the resonant rise of current in a series circuit.

$$V_L = V_C = QV_T \qquad (21\text{-}5)$$

Example 21.4   Find the Q for the series resonant circuit and the rise of voltage across L or C due to resonance (Fig. 21-4).
Use Eq. (21-4) to find Q.

$$Q = \frac{X_L}{R} = \frac{942}{7.5} = 125.6 \qquad Ans.$$

Use Eq. (21-5) to find V_L or V_C.

$$V_L = V_C = QV_T = 125.6(15) = 1884 \text{ V} \qquad Ans.$$

This is the same value calculated for V_L or V_C at resonance in Example 21.3.

PARALLEL RESONANCE

Pure Parallel LC Circuit

In the pure LC parallel-tuned circuit (that is, one in which there is no resistance), the coil and capacitor are placed in parallel and the applied voltage V_T appears across these circuit components (Fig. 21-5a). In this parallel-tuned circuit, as in the series-tuned circuit, resonance occurs when the inductive reactance is equal to the capacitive reactance.

$$X_L = X_C$$

Because the applied voltage is common to both branches,

$$V_L = V_C$$

so that

$$\frac{V_L}{X_L} = \frac{V_C}{X_C}, \quad I_L = I_C$$

The current in the inductive branch I_L equals the current in the capacitive branch I_C. I_L lags the applied voltage V_T by 90°, while I_C leads the voltage by 90° (Fig. 21-5b). Since the phasor currents I_L and I_C are equal and out of phase by 180°, their vector sum is zero so that the total current I_T is zero. Under this condition, the impedance of the circuit at the resonant frequency must be infinite in value.

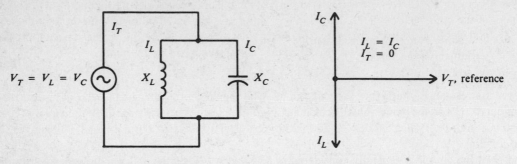

(a) Schematic diagram   (b) Current-phasor diagram at resonance

Fig. 21-5 Pure LC parallel circuit

The formula for the resonant frequency of a pure LC parallel-tuned circuit is the same as that for a series tuned circuit.

$$f_r = \frac{1}{2\pi\sqrt{LC}} = \frac{0.159}{\sqrt{LC}} \tag{21-1}$$

If the resonant frequency is known, then the inductance or capacitance for a parallel resonant LC circuit can be found by formulas.

$$L = \frac{0.0254}{f_r^2 C} \tag{21-2}$$

$$C = \frac{0.0254}{f_r^2 L} \tag{21-3}$$

Example 21.5 A parallel resonant circuit appears as an infinite impedance (or open circuit) at the resonant frequency. This makes it possible to reject or "trap" a wave of a definite frequency in antenna circuits. A 400-μH coil and a 25-pF capacitor are placed in parallel to form a "wave trap" in an antenna. Find the resonant frequency that the circuit will reject.
The resistance of the circuit is negligible.

$$L = 400\ \mu\text{H} = 4 \times 10^{-4}\ \text{H} \qquad C = 25\ \text{pF} = 25 \times 10^{-12}\ \text{F}$$

$$f_r = \frac{0.159}{\sqrt{LC}} \tag{21-1}$$

$$= \frac{0.159}{\sqrt{(4 \times 10^{-4})(25 \times 10^{-12})}} = 1590\ \text{kHz} \qquad Ans.$$

Example 21.6 The capacitance of a parallel resonant circuit used as a wave trap in an antenna circuit is 400 pF. Find the value of the parallel inductance in order to reject an 800-kHz wave.

$$f = 800\ \text{kHz} = 8 \times 10^{5}\ \text{Hz} \qquad C = 400\ \text{pF} = 400 \times 10^{-12}\ \text{F}$$

$$L = \frac{0.0254}{f_r^2 C} \tag{21-2}$$

$$= \frac{0.0254}{(8 \times 10^{5})(400 \times 10^{-12})} = 9.92 \times 10^{-5} = 99.2\ \mu\text{H} \qquad Ans.$$

Practical Parallel *LC* Circuit

In a practical *LC* parallel-tuned circuit (Fig. 21-6*a*), there is some resistance, most of which is due to the resistance of the inductor wire. The resonant frequency of a parallel circuit also is defined as that frequency at which the parallel circuit acts as a pure resistance. Therefore, the line current I_T must be in phase with the applied voltage V_T (unity power factor) (Fig. 21-6*b*). This means that the out-of-phase or quadrature component of the current through the inductive branch I_L must be equal to the current through the capacitive branch I_C; and the total line current I_T equals the in-phase component of the current through the inductive branch, or $I_T = I_R$ (Fig. 21-6*b*). Since the impedance is maximum, I_T is minimum.

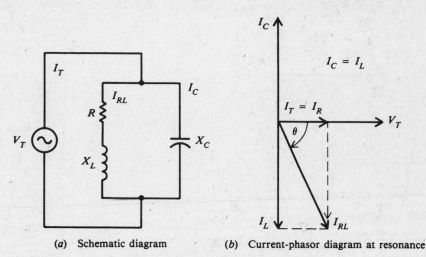

(*a*) Schematic diagram   (*b*) Current-phasor diagram at resonance

Fig. 21-6   Practical parallel *LC* circuit

The resonant frequency for the circuit (Fig. 21-6*a*) is

$$f_r = \frac{1}{2\pi} \sqrt{\frac{1}{LC} - \frac{R^2}{L^2}}$$

(*21-6*)

where f_r = resonant frequency, Hz
  L = inductance, H
  C = capacitance, F
  R = resistance, Ω

If the Q of the coil is high, say greater than 10, or the term $1/LC \gg R^2/L^2$, then, for practical purposes, the term R^2/L^2 can be disregarded. The result is that Eq. (*21-6*) becomes Eq. (*21-1*), the resonant-frequency formula for series resonance.

$$f_r = \frac{1}{2\pi\sqrt{LC}}$$

(*21-1*)

The total impedance at resonance of the practical *LC* parallel circuit is

$$Z_T = \frac{L}{RC}$$

(*21-7*)

In terms of quality factor Q, Z_T at resonance can also be found by

$$Z_T = X_L Q = 2\pi f_r L Q$$

(*21-8*)

or
$$Z_T = \frac{Q}{2\pi f_r C} \tag{21-9}$$

The impedance Z_T of a practical parallel circuit is maximum at the resonant frequency and decreases at frequencies below and above the resonant frequency (Fig. 21-7a). An increase in resistance decreases the impedance and causes the impedance to vary less "sharply" as the circuit is tuned over a band of frequencies below and above the resonant frequency (Fig. 21-7b). At frequencies below resonance, $X_C > X_L$ and $I_L > I_C$ so that the parallel-tuned circuit is inductive (Figs. 21-7a and c). At frequencies above resonance, the reverse condition is true, $X_L > X_C$ and $I_C > I_L$, so that now the circuit is capacitive (Figs. 21-7a and c). Since the impedance Z_T is maximum at parallel resonance, I_T is minimum (Fig. 21-7c).

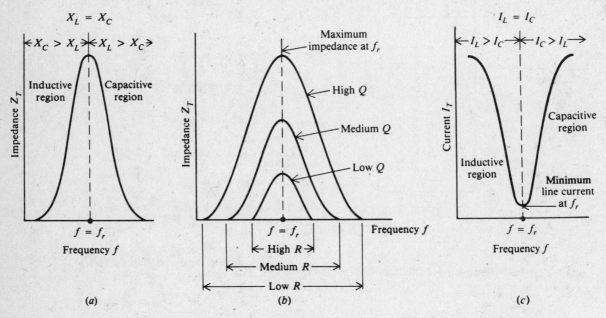

Fig. 21-7   Impedance and current response curves of practical parallel LC circuit at resonance

Q OF PARALLEL CIRCUIT

For a parallel resonant circuit in which R is very low compared with X_L,

$$Q = \frac{X_L}{R} \tag{21-4}$$

where R is the resistance of the coil in series with X_L (Fig. 21-6a). If the resistance of the source supply is very high and there is no other resistance branch shunting the tuned circuit, the Q of the parallel resonant circuit is the same as the Q of the coil.

Example 21.7   In the circuit shown (Fig. 21-8a) find (a) the resonant frequency of the circuit, (b) the impedance of the circuit, and (c) the total current at resonance; and (d) draw the phasor diagram.

(a)   Find f_r.

$$f_r = \frac{1}{2\pi}\sqrt{\frac{1}{LC} - \frac{R^2}{L^2}} \tag{21-6}$$

Compare $1/LC$ with R^2/L^2.

$$\frac{1}{LC} = \frac{1}{(203 \times 10^{-6})(500 \times 10^{-12})} = 9.85 \times 10^{12}$$

$$\frac{R^2}{L^2} = \frac{(6.7)^2}{(203 \times 10^{-6})^2} = \frac{44.9}{(4.12 \times 10^{-8})} = 10.9 \times 10^8$$

Because $9.85 \times 10^{12} \gg 10.9 \times 10^8$ or $1/LC \gg R^2/L^2$, the R^2/L^2 term can be disregarded and Eq. (21-1) can be used.

$$f_r = \frac{1}{2\pi}\sqrt{\frac{1}{LC}} = \frac{1}{6.28}\sqrt{9.85 \times 10^{12}} = 500\,000\ \text{Hz} = 500\ \text{kHz} \qquad Ans.$$

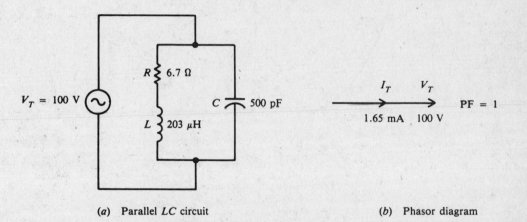

(a) Parallel LC circuit   (b) Phasor diagram

Fig. 21-8

(b) Find Z_T. Use Eq. (21-7).

$$Z_T = \frac{L}{RC} = \frac{203 \times 10^{-6}}{6.7(500 \times 10^{-12})} = 60\,600\ \Omega = 60.6\ \text{k}\Omega \qquad Ans.$$

(c) Find I_T by Ohm's law.

$$I_T = \frac{V_T}{Z_T} = \frac{100}{60.6 \times 10^3} = 1.65\ \text{mA} \qquad Ans.$$

(d) Draw the phasor diagram (see Fig. 21-8b).

Example 21.8 A coil with a Q of 71.6 is connected in parallel with a capacitor to produce resonance at 356 kHz. The impedance at resonance is found to be 64 kΩ. Find the value of the capacitor. Assume the Q of the resonant circuit is the same as the Q of the coil.

$$Z_T = \frac{Q}{2\pi f_r C} \tag{21-9}$$

from which
$$C = \frac{Q}{2\pi f_r Z_T} = \frac{71.6}{6.28(356 \times 10^3)(64 \times 10^3)} = 5 \times 10^{-10}\ \text{F} = 500\ \text{pF} \qquad Ans.$$

BANDWIDTH AND POWER OF RESONANT CIRCUIT

The width of the resonant band of frequencies centered around f_r is called the *bandwidth* of the tuned circuit. In Fig. 21-9a, the group of frequencies with a response equal to 70.7 percent or more of maximum is considered the bandwidth of the tuned circuit. For a series resonant circuit, the bandwidth is measured between the two edge frequencies f_1 and f_2, producing 70.7 percent of the maximum current at f_r (Fig. 21-9b). For a parallel resonant circuit, the bandwidth is measured between the two frequencies, allowing 70.7 percent of the maximum total impedance at f_r (Fig. 21-9c).

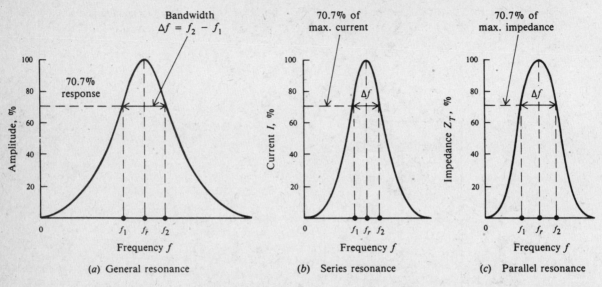

Fig. 21-9   Bandwidth of a tuned *LC* circuit

At each frequency f_1 and f_2, the net capacitive or net inductive reactance equals the resistance. Then Z_T of the series *RLC* resonant circuit is $\sqrt{2}$ or 1.4 times greater than R. The current then is $I/\sqrt{2} = 0.707\,I$. Since power is I^2R or V^2/R and $(0.707)^2 = 0.50$, the bandwidth at 70.7 percent response in current or voltage is also the bandwidth of *half-power* points.

Bandwidth (BW) in terms of Q is

$$\text{BW} = f_2 - f_1 = \Delta f = \frac{f_r}{Q} \qquad (21\text{-}10)$$

High Q means narrow bandwidth, whereas low Q yields greater bandwidth (Fig. 21-10).

Either f_1 or f_2 is separated from f_r by one-half of the total bandwidth (Fig. 21-9), so these edge frequencies can be calculated.

$$f_1 = f_r - \frac{\Delta f}{2} \qquad (21\text{-}11)$$

$$f_2 = f_r + \frac{\Delta f}{2} \qquad (21\text{-}12)$$

Example 21.9   An *LC* circuit resonant at 1000 kHz has a Q of 100. Find the total bandwidth Δf and the edge frequencies f_1 and f_2.

$$\Delta f = \frac{f_r}{Q} \qquad (21\text{-}10)$$

$$= \frac{1000 \times 10^3}{100} = 10 \times 10^3 = 10\,\text{kHz} \qquad Ans.$$

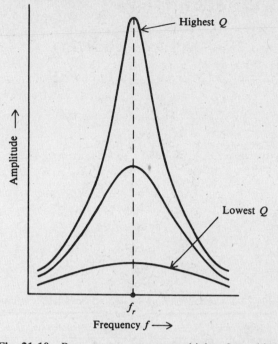

Fig. 21-10   Resonant response curves: higher Q provides sharper resonance, lower Q provides broader response

Since f_1 and f_2 have the same frequency spacing on each side of the resonant frequency, bandwidth can also be expressed as

$$\Delta f = \pm 5 \text{ kHz} \qquad Ans.$$

Use Eq. (*21-11*) to find f_1.

$$f_1 = f_r - \frac{\Delta f}{2} = 1000 \times 10^3 - \left(5 \times 10^3\right) = 995 \times 10^3 = 995 \text{ kHz} \qquad Ans.$$

Use Eq. (*21-12*) to find f_2.

$$f_2 = f_r + \frac{\Delta f}{2} = 1000 \times 10^3 + \left(5 \times 10^3\right) = 1005 \times 10^3 = 1005 \text{ kHz} \qquad Ans.$$

Example 21.10   For a series resonant circuit (Fig. 21-11), what is the real power consumed at resonance and at an edge frequency?

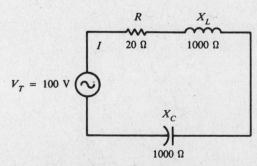

Fig. 21-11   Series resonant circuit

At resonance:
$$I = \frac{V_T}{R} = \frac{100}{20} = 5 \text{ A}$$

Power at resonance:
$$P_r = I^2 R = 5^2(20) = 500 \text{ W} \qquad Ans.$$

or
$$P_r = V_T I = 100(5) = 500 \text{ W} \qquad Ans.$$

Power at f_1 or f_2:
$$P_{f1} = \left(\frac{I}{\sqrt{2}}\right)^2 R = \frac{I^2 R}{2} = \frac{5^2(20)}{2} = 250 \text{ W} \qquad Ans.$$

or, since f_1 and f_2 are at the half-power points,

$$P_{f1} = P_{f2} = \frac{P_r}{2} = \frac{500}{2} = 250 \text{ W} \qquad Ans.$$

SUMMARY

Series and parallel resonance are compared in Table 21-1.

Table 21-1 Comparison of Series and Parallel Resonance

Series Resonance	Parallel Resonance ($Q > 10$)
$f_r = \dfrac{1}{2\pi\sqrt{LC}}$	$f_r = \dfrac{1}{2\pi\sqrt{LC}}$
I maximum at f_r with $\theta = 0°$	I_T minimum at f_r with $\theta = 0°$
Impedance Z minimum at f_r	Impedance Z_T maximum at f_r
$Q = \dfrac{X_L}{R}$	$Q = \dfrac{X_L}{R}$
Q rise in voltage $= QV_T$	Q rise in impedance $= QX_L$
Bandwidth $\Delta f = \dfrac{f_r}{Q}$	Bandwidth $\Delta f = \dfrac{f_r}{Q}$
Capacitive below f_r, but inductive above f_r	Inductive below f_r, but capacitive above f_r

Solved Problems

21.1 What value of inductance must be connected in series with a 300-pF capacitor in order that the circuit be resonant to a frequency of 500 kHz?

$$C = 300 \text{ pF} = 300 \times 10^{-12} \text{ F} = 3 \times 10^{-10} \text{ F} \qquad f = 500 \text{ kHz} = 500 \times 10^3 \text{ Hz} = 5 \times 10^5 \text{ Hz}$$

$$L = \frac{0.0254}{f_r^2 C} \tag{21-2}$$

$$= \frac{0.0254}{\left(5 \times 10^5\right)^2\left(3 \times 10^{-10}\right)} = 339 \text{ }\mu\text{H} \qquad Ans.$$

21.2   Prepare a table of various combinations of L and C required to produce resonance in a series RLC circuit at a frequency of 600 kHz.

Find the LC product for $f_r = 600$ kHz.

$$L = \frac{0.0254}{f_r^2 C} \qquad (21\text{-}2)$$

$$LC = \frac{0.0254}{f_r^2} = \frac{0.0254}{\left(6 \times 10^5\right)^2} = 7 \times 10^{-14}$$

If L is to be specified in microhenrys and C in microfarads, then

$$LC = \frac{7 \times 10^{-14}}{10^{-12}} = 0.07$$

Assume various values of L and solve for resulting values of C by Eq. (21-3). The result is the following table.

Ans.

Resonant Frequency, kHz	L, μH	C, μF	LC Product
600	100	0.0007	0.07
600	50	0.0014	0.07
600	10	0.007	0.07
600	1	0.07	0.07
600	0.2	0.35	0.07
600	0.05	1.4	0.07
600	0.005	14	0.07
600	0.0005	140	0.07

21.3   A 4-mH coil and a 50-pF capacitor form the secondary side of a transformer (Fig. 21-12). Find the resonant frequency.

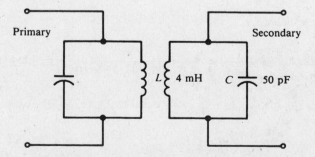

Fig. 21-12   Resonance in transformer secondary

$$L = 4\,\text{mH} = 4 \times 10^{-3}\,\text{H} \qquad C = 50\,\text{pF} = 50 \times 10^{-12}\,\text{F}$$

Using Eq. (*21-1*),

$$f_r = \frac{0.159}{\sqrt{LC}} = \frac{0.159}{\sqrt{(4 \times 10^{-3})(50 \times 10^{-12})}} = 356 \, \text{kHz} \qquad Ans.$$

21.4 A series circuit has $R = 40 \, \Omega$, $L = 0.5 \, \text{H}$, and $C = 0.2 \, \mu\text{F}$. Find (*a*) the impedance and Q of the circuit to a frequency of 400 Hz, (*b*) the capacitance that must be added in parallel with the 0.2 μF to produce resonance at this frequency, and (*c*) impedance and Q of the coil at resonance.

(*a*) Find Z at $f = 400 \, \text{Hz}$.

$$X_L = 6.28fL = 6.28(400)(0.5) = 1256 \, \Omega$$

$$X_C = \frac{1}{2\pi fC} = \frac{1}{6.28(400)(0.2 \times 10^{-6})} = 1990 \, \Omega$$

$$Z = \sqrt{R^2 + (X_L - X_C)^2} = \sqrt{40^2 + (1256 - 1990)^2} = \sqrt{40^2 + (-734)^2} = 735 \, \Omega \qquad Ans.$$

At nonresonance, $\qquad Q = \dfrac{X_L - X_C}{R} = \dfrac{734}{40} = 18.4 \qquad Ans.$

(*b*) Find the C that will produce resonance.

$$f = 400 \, \text{Hz} \qquad L = 0.5 \, \text{H}$$

$$C = \frac{0.0254}{f_r^2 L} \tag{21-3}$$

$$= \frac{0.0254}{400^2(0.5)} = 0.3175 \, \mu\text{F} \cong 0.318 \, \mu\text{F}$$

Since 0.318 μF is needed for resonance and we have only 0.2 μF, we must add 0.118 μF (0.318 μF – 0.2 μF) in parallel with 0.2 μF, so

$$C = 0.118 \, \mu\text{F} \qquad Ans.$$

(*c*) Find Z and Q at resonance.

$$Z = R = 40 \, \Omega \qquad Ans.$$

Find X_C at resonance.

$$X_C = \frac{1}{2\pi f_r C} = \frac{1}{6.28(400)(0.3175 \times 10^{-6})} = 1254 \, \Omega \cong 1256 \, \Omega$$

Due to rounding off, X_C is not exactly equal to X_L (1256 Ω). $X_L = X_C$ at resonance, so

$$Q = \frac{X_L}{R} = \frac{X_C}{R} = \frac{1256}{40} = 31.4 \qquad Ans.$$

21.5 An *RLC* series circuit with $R = 5 \, \Omega$, $C = 10 \, \mu\text{F}$, and a variable inductance L has an applied voltage $V_T = 110 \, \text{V}$ with a frequency of 60 Hz. L is adjusted until the voltage across the resistor is a maximum. Find the current and voltage across each element.

The maximum voltage across the resistor occurs at resonance. At resonance,

$$X_C = \frac{1}{2\pi f C} = \frac{1}{6.28(60)\left(10 \times 10^{-6}\right)} = 265\ \Omega$$

$$X_L = X_C = 265\ \Omega \quad \text{and} \quad Z = R = 5\ \Omega$$

Then
$$I = \frac{V_T}{Z} = \frac{110}{5} = 22\ \text{A} \qquad Ans.$$

and
$$V_R = IR = 22(5) = 110\ \text{V} \qquad Ans.$$

$$V_L = IX_L = 22(265) = 5830\ \text{V} \qquad Ans.$$

$$V_C = IX_C = 22(265) = 5830\ \text{V} \qquad Ans.$$

21.6   Find the minimum and maximum values of the capacitor needed to obtain resonance with a 300-μH coil to frequencies between 500 and 1500 kHz.

For a minimum C, f_r is equal to its maximum value of 1500 kHz.

$$C_{\min} = \frac{0.0254}{f_{r,\max}^2 L} = \frac{0.0254}{\left(1.5 \times 10^6\right)^2\left(300 \times 10^{-6}\right)} = 3.76 \times 10^{-11}\ \text{F} = 37.6\ \text{pF} \qquad Ans.$$

For a maximum C, f_r is equal to its minimum value of 500 kHz.

$$C_{\max} = \frac{0.0254}{f_{r,\min}^2 L} = \frac{0.0254}{\left(5 \times 10^5\right)^2\left(300 \times 10^{-6}\right)} = 3.39 \times 10^{-10}\ \text{F} = 339\ \text{pF} \qquad Ans.$$

21.7   A series RLC circuit with $R = 25\ \Omega$ and $L = 0.6$ H has a leading phase angle of $60°$ at a frequency of 40 Hz. Find the frequency at which this circuit will be resonant.

Step 1.   Draw the impedance triangle (Fig. 21-13) and solve for the net reactance X.

$$X = X_C - X_L = 25 \tan 60° = 25(1.73) = 43.3\ \Omega$$

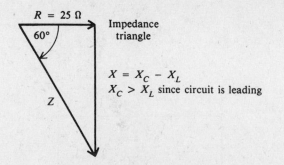

Fig. 21-13   Impedance triangle

Step 2.   Find X_L and then X_C.

$$X_L = 6.28 f L = 6.28(40)(0.6) = 151\ \Omega$$

$$X_C - X_L = 43.3 \quad \text{from Step 1}$$

Then
$$X_C = 43.3 + X_L = 43.3 + 151 = 194\ \Omega$$

Step 3. Find C.

$$C = \frac{1}{2\pi f X_C} = \frac{1}{6.28(40)(194)} = 20.5\ \mu\text{F}$$

Step 4. Knowing L and C, find f_r, using Eq. (*21-1*).

$$f_r = \frac{0.159}{\sqrt{LC}} = \frac{0.159}{\sqrt{0.6(20.5 \times 10^{-6})}} = 45.3\ \text{Hz} \qquad Ans.$$

Step 5. As a check, determine whether $X_L = X_C$ at $f_r = 45.3$ Hz.

$$X_L = 6.28 f_r L = 6.28(45.3)(0.6) = 171\ \Omega$$

$$X_C = \frac{1}{2\pi f_r C} = \frac{1}{6.28(45.3)(20.5 \times 10^{-6})} = 171\ \Omega$$

Therefore, $X_L = X_C = 171\ \Omega$.

21.8 A common application of resonant circuits is tuning a receiver to the carrier frequency of a desired radio station. The tuning is done by the air capacitor C, which can be varied with the plates completely in mesh (maximum capacitance) to out of mesh (minimum capacitance). Calculate the capacitance of the variable capacitor to tune for radio stations broadcasting at 500, 707, 1000, 1410, and 2000 kHz (Fig. 21-14).

$$C = \frac{0.0254}{f_r^2 L} \qquad\qquad (21\text{-}3)$$

At $f_r = 500$ kHz: $\quad C = \dfrac{0.0254}{(5 \times 10^5)^2(239 \times 10^{-6})} = 4.25 \times 10^{-10} = 425\ \text{pF} \qquad Ans.$

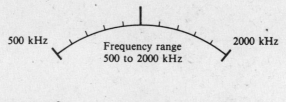

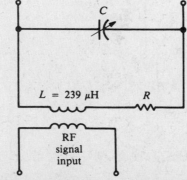

Fig. 21-14 Tuning *LC* circuit through the AM radio band of 500 to 2000 kHz

Since

$$C \propto \frac{1}{f_r^2} \quad C_n = C_o\left(\frac{f_{r,o}}{f_{r,n}}\right)^2$$

Let $C_o = 425$ pF, $f_{r,o} = 500$ kHz.

At $f_r = 707$ kHz: $\qquad C = 425\left(\frac{500}{707}\right)^2 = 213$ pF $\qquad$ *Ans.*

At $f_r = 1000$ kHz: $\qquad C = 425\left(\frac{500}{1000}\right)^2 = 106$ pF $\qquad$ *Ans.*

At $f_r = 1410$ kHz: $\qquad C = 425\left(\frac{500}{1410}\right)^2 = 53.4$ pF $\qquad$ *Ans.*

At $f_r = 2000$ kHz: $\qquad C = 425\left(\frac{500}{2000}\right)^2 = 26.6$ pF $\qquad$ *Ans.*

21.9 The Q of a coil in a series resonant circuit can be determined experimentally by measuring the Q rise in voltage across L or C and comparing this voltage with the generator or input voltage. As a formula

$$Q = \frac{V_{\text{out}}}{V_{\text{in}}} = \frac{V_L}{V_{\text{in}}} = \frac{V_C}{V_{\text{in}}}$$

This method is better than the X_L/R formula for determining Q because R is the ac resistance of the coil, which is not easily measured. A series circuit resonant at 400 kHz develops 100 mV across a 250-μH coil with a 2-mV input. Calculate the Q and ac resistance of the coil.

$$Q = \frac{V_{\text{out}}}{V_{\text{in}}} = \frac{100 \text{ mV}}{2 \text{ mV}} = 50 \qquad \textit{Ans.}$$

$$X_L = 2\pi f_r L = 6.28(4 \times 10^5)(250 \times 10^{-6}) = 628 \ \Omega$$

$$Q = \frac{X_L}{R} \tag{21-4}$$

$$R = \frac{X_L}{Q} = \frac{628}{50} = 12.6 \ \Omega \qquad \textit{Ans.}$$

21.10 A 0.1-mH coil and a 1200-pF capacitor are connected in parallel to form the primary of an IF transformer (Fig. 21-15). What is the resonant frequency?

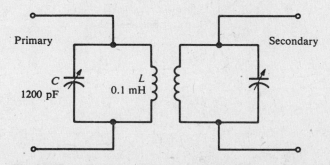

Fig. 21-15 *LC* tuning on the primary side of an IF transformer

Since this is a case of a pure LC parallel tuned circuit,

$$f_r = \frac{0.159}{\sqrt{LC}} \qquad (21\text{-}1)$$

$$L = 0.1 \,\text{mH} = 0.1 \times 10^{-3} \,\text{H} \qquad C = 1200 \,\text{pF} = 1200 \times 10^{-12} \,\text{F} = 1.2 \times 10^{-9} \,\text{F}$$

$$f_r = \frac{0.159}{\sqrt{(0.1 \times 10^{-3})(1.2 \times 10^{-9})}} = 459 \,\text{kHz} \qquad Ans.$$

21.11   The parallel tank circuit of an oscillator contains a coil of 320 μH. Find the value of the capacitance at the resonant frequency of 1 MHz. Resistance is negligible.

Use Eq. (21-3).

$$C = \frac{0.0254}{f_r^2 L} = \frac{0.0254}{(1 \times 10^6)^2(320 \times 10^{-6})} = 7.9 \times 10^{-11} \,\text{F} = 79 \,\text{pF} \qquad Ans.$$

21.12   For the practical LC parallel circuit (Fig. 21-16), the formula for the equivalent or total impedance Z_T is

$$Z_T = \frac{1}{R^2 + X^2}\sqrt{(RX_L X_C - RX_C X)^2 + (R^2 X_C + X_L X_C X)^2} \quad \text{where } X = X_L - X_C$$

Find the total impedance Z_T when $f = 1000$ kHz, $R = 4\,\Omega$, $L = 100\,\mu$H, and $C = 200$ pF. Round off Z_T to three significant figures.

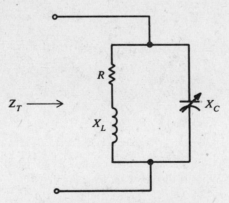

Fig. 21-16   Equivalent impedance of an
LC parallel circuit

Step 1.   Calculate X_L, X_C, and X.

$$X_L = 2\pi f L = 6.28(1 \times 10^6)(100 \times 10^{-6}) = 628 \,\Omega$$

$$X_C = \frac{1}{2\pi f C} = \frac{1}{6.28(1 \times 10^6)(200 \times 10^{-12})} = 796 \,\Omega$$

$$X = X_L - X_C = 628 - 796 = -168 \,\Omega$$

Step 2.   Substitute values of R, X, X_L, and X_C to calculate Z_T.

$$R^2 + X^2 = 4^2 + (-168)^2 = 2.82 \times 10^4$$

$$RX_L X_C - RX_C X = 4(628)(796) - 4(796)(-168) = 2.54 \times 10^6$$

$$R^2 X_C + X_L X_C X = 4^2(796) + 628(796)(-168) = -84 \times 10^6$$

$$Z_T = \frac{1}{2.82 \times 10^4} \sqrt{(2.54 \times 10^6)^2 + (-84 \times 10^6)^2} = \frac{84 \times 10^6}{2.82 \times 10^4} = 2979\,\Omega = 2980\,\Omega \qquad Ans.$$

In this case $X_C > X_L$ ($796\,\Omega > 628\,\Omega$) so that the circuit is inductive (more inductive current than capacitive current).

A simpler method for evaluating Z_T when $R \ll X_L$, $R \ll X_C$, and $R \ll X$ is by reducing the formula for Z_T. For these conditions, $X_L X_C X$ is the dominating term so that the other terms under the square root sign can be disregarded. The result is

$$Z_T \cong \frac{1}{X^2}\sqrt{(X_L X_C X)^2} \cong \frac{X_L X_C X}{X^2} \cong \frac{X_L X_C}{X}$$

Since the inequality conditions hold true for this circuit ($10\,\Omega \ll 628\,\Omega$, $10\,\Omega \ll 796\,\Omega$, and $10\,\Omega \ll 168\,\Omega$),

$$Z_T \cong \frac{628(796)}{168} \cong 2976\,\Omega \cong 2980, \text{ rounded off to three significant figures} \qquad Ans.$$

which equals the value previously calculated.

21.13   Find the resonant frequency and total impedance Z_T of the circuit (Fig. 21-16) when the component values remain the same as those in Problem 21.12, $R = 4\,\Omega$, $L = 100\,\mu\text{H}$, and $C = 200\,\text{pF}$.

Step 1.   Find f_r.

$$f_r = \frac{1}{2\pi}\sqrt{\frac{1}{LC} - \frac{R^2}{L^2}} \qquad\qquad (21\text{-}6)$$

Compare $1/LC$ with R^2/L^2.

$$\frac{1}{LC} = \frac{1}{(100 \times 10^{-6})(200 \times 10^{-12})} = 5 \times 10^{13}$$

$$\frac{R^2}{L^2} = \frac{4^2}{(100 \times 10^{-6})^2} = \frac{16}{10^{-8}} = 16 \times 10^8$$

Since $1/LC \gg R^2/L^2$, the R^2/L^2 term can be disregarded so f_r can be found by using Eq. (21-1).

$$f_r = \frac{1}{2\pi}\sqrt{\frac{1}{LC}} = \frac{1}{6.28}\sqrt{5 \times 10^{13}} = 1.13 \times 10^6\,\text{Hz} = 1130\,\text{kHz} \qquad Ans.$$

Step 2.   Find Z_T at resonance, using Eq. (21-7).

$$Z_T = \frac{L}{RC} = \frac{100 \times 10^{-6}}{4(200 \times 10^{-12})} = 125\,\text{k}\Omega \qquad Ans.$$

21.14   Find the total impedance of Z_T of the circuit (Fig. 21-16) when $f = 1300$ kHz and the component values remain the same as those in Problem 21.12, $R = 4\ \Omega$, $L = 100\ \mu$H, and $C = 200$ pF. Round off Z_T to three significant figures.

Step 1.   Calculate X_L, X_C, and X.

$$X_L = 2\pi fL = 6.28(1.3 \times 10^6)(100 \times 10^{-6}) = 816\ \Omega$$

$$X_C = \frac{1}{2\pi fC} = \frac{1}{6.28(1.3 \times 10^6)(200 \times 10^{-12})} = 612\ \Omega$$

$$X = X_L - X_C = 816 - 612 = 204\ \Omega$$

Step 2.   Substitute in the formula for Z_T given in Problem 21.12.

$$R^2 + X^2 = 4^2 + 204^2 = 4.16 \times 10^4$$

$$RX_LX_C - RX_CX = 4(816)(612) - 4(612)(204) = 1.5 \times 10^6$$

$$R^2X_C + X_LX_CX = 4^2(612) + 816(612)(204) = 1.02 \times 10^8$$

$$Z_T = \frac{1}{4.16 \times 10^4}\sqrt{(1.5 \times 10^6)^2 + (1.02 \times 10^8)^2} = \frac{1.02 \times 10^8}{4.16 \times 10^4} = 2452\ \Omega = 2450\ \Omega \qquad Ans.$$

In this case $X_L > X_C$ (816 Ω > 612 Ω) so that the circuit is capacitive (more capacitive current than inductive current).

A simpler method is to use the formula developed in Problem 21.12.

$$Z_T \cong \frac{X_LX_C}{X} \cong \frac{816(612)}{204} \cong 2448\ \Omega$$

$$\cong 2450\ \Omega, \text{ rounded off to three significant figures}$$

which equals the value calculated by the exact formula.

In summary, the accompanying table shows the nature of a practical LC parallel circuit below the resonant frequency, at the resonant frequency, and above the resonant frequency.

Problem Number	Frequency	Z_T	Nature of LC Parallel Circuit
21.12	1000 kHz: Below resonant frequency	2980 Ω	Inductive ($X_C > X_L$)
21.13	1130 kHz: At resonant frequency	125 kΩ	Resistive ($X_C = X_L$), maximum impedance, minimum current
21.14	1300 kHz: Above resonant frequency	2450 Ω	Capacitive ($X_L > X_C$)

21.15   A parallel LC circuit tuned to 200 kHz with a 350-μH coil has a measured impedance Z_T of 19 800 Ω. Calculate Q.

$$X_L = 2\pi f X_L = 6.28 (2 \times 10^5)(350 \times 10^{-6}) = 440\ \Omega$$

$$Z_T = X_L Q \qquad\qquad (21\text{-}8)$$

$$Q = \frac{Z_T}{X_L} = \frac{19\,800}{440} = 45 \qquad Ans.$$

Supplementary Problems

21.16   Find the resonant frequency of a series circuit if the inductance is 300 μH and the capacitance is 0.005 μF.   *Ans.* $f_r = 130$ kHz

21.17   Find the resonant frequency of a transmitting antenna that has a capacitance of 300 pF, a resistance of 40 Ω, and an inductance of 300 μH.   *Ans.* $f_r = 530$ kHz

21.18   Find the resonant frequency of a series resonant section of a bandpass filter when $L = 350\ \mu$H and $C = 20$ pF.   *Ans.* $f_r = 1500$ kHz

21.19   What is the resonant frequency of a series circuit that consists of an inductance of 500 μH and a capacitance of 400 pF?   *Ans.* $f_r = 356$ kHz

21.20   How much capacitance is needed to obtain resonance at 1500 kHz with an inductance of 45 μH?   *Ans.* $C = 251$ pF

21.21   Find the value of inductance that will produce resonance to 50 Hz if it is placed in series with a 20-μF capacitor.   *Ans.* $L = 0.508\ \mu$H

21.22   What is the capacitance of an antenna circuit whose inductance is 50 μH if it is resonant to 1200 kHz?   *Ans.* $C = 353$ pF

21.23   What is the inductance of a series resonant circuit with a 300-pF capacitor at a frequency of 1000 kHz?   *Ans.* $L = 84.7\ \mu$H

21.24   A voltage of 100-V ac at a frequency of 10 kHz is impressed across a series circuit that consists of a 220-pF capacitor and an 800-mH coil with an internal resistance of 125 Ω. Find (*a*) the current in the circuit, (*b*) the voltage drops across the capacitor and the coil, (*c*) the power dissipated by the circuit, and (*d*) the Q of the coil.
Ans.   (*a*) $I = 4.52$ mA ($Z = 22\,100\ \Omega$);   (*b*) $V_C = 327$ V; $V_L = 227$ V;   (*c*) $P = 2.55$ mW;
     (*d*) $Q = 402$

21.25   Find (*a*) the resonant frequency of the circuit in Problem 21.24. As a series resonant circuit, find (*b*) the current in the circuit, (*c*) the voltage drops across the capacitor and the coil, (*d*) the power dissipated by the resonant circuit, (*e*) the Q of the coil, (*f*) the total bandwidth and edge frequencies, and (*g*) the power dissipated by the circuit operating at the edge frequency.
Ans.   (*a*) $f_r = 12.0$ kHz;   (*b*) $I = 0.8$ A;   (*c*) $V_C = 48\,230$ V; $V_L = 48\,230$ V;   (*d*) $P = 80$ W;
(*e*) $Q = 482$;   (*f*) $\Delta f = 25$ Hz; $f_1 = 11\,988$ Hz; $f_2 = 12\,012$ Hz. Rounding off has distorted these answers. If f_r were rounded to unit hertz, $f_r = 11\,985$ Hz (five significant figures). Then $f_1 = 11\,973$ Hz and $f_2 = 11\,997$ Hz.   (*g*) $P = 40$ W

21.26 What type and value of "pure reactance" must be added to the circuit in Problem 21.24 to make it resonant at 10 kHz?   *Ans.* $L = 355$ mH added in series

21.27 A series circuit has a resistance of 30 Ω, an inductance of 0.382 H, and a capacitance of 0.2 μF. Find (a) the impedance of the circuit to a frequency of 500 Hz, (b) the Q of the circuit, (c) the capacitance that must be added in parallel with the 0.2-μF capacitor to produce resonance at this frequency, (d) the impedance of the circuit at resonance, (e) the Q of the resonant circuit, and (f) the bandwidth.
     Ans.   (a) $Z = 391$ Ω;   (b) $Q = 13$;   (c) $C = 0.065\ \mu$F (Total $C = 0.265\ \mu$F at resonance);
            (d) $Z = 30$ Ω;   (e) $Q = 40$;   (f) $\Delta f = 12.5$ Hz

21.28 A tuning capacitor is continuously variable between 20 and 350 pF. Find (a) the inductance that must be connected in series with it to produce a lowest resonant frequency of 550 kHz, and then (b) the highest resonant frequency.   *Ans.*   (a) $L = 0.239$ mH;   (b) $f_r = 2300$ kHz

21.29 A 0.1-H inductance, a 1-μF capacitor, and a 5-Ω resistor are connected in series across a supply voltage of 50 V at a frequency of 503 Hz. (a) Is the circuit resonant? (b) Find the impedance of the circuit, and (c) find the amount of voltage across each component.
     Ans.   (a) Yes ($X_C = X_L = 316$ Ω);   (b) $Z = 5$ Ω;   (c) $V_R = 50$ V; $V_L = 3160$ V; $V_C = 3160$ V

21.30 In a series RLC circuit, $R = 10$ Ω, $L = 20$ mH, and $C = 1.26\ \mu$F. Fill in the indicated quantities in the accompanying table, and plot the impedance as a function of frequency.

f	X_L	X_C	Z	Nature*	Q
800 Hz	?	?	?	?	?
900 Hz	?	?	?	?	?
1 kHz	?	?	?	?	?
1.1 kHz	?	?	?	?	?
1.2 kHz	?	?	?	?	?

*Inductive, capacitive, or resistive (resonant)

Ans.

f, Hz	X_L, Ω	X_C, Ω	Z, Ω	Nature	Q
800	101	158	57.9	Capacitive	5.7
900	113	140	28.8	Capacitive	2.7
1000	126	126	10	Resonant, resistive	12.6
1100	138	115	25.1	Inductive	2.4
1200	151	105	47.1	Inductive	4.6

See Fig. 21-17 for plot of impedence as a function of frequency.

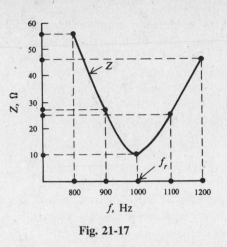

Fig. 21-17

21.31   A series RLC circuit with $R = 250\ \Omega$ and $L = 0.6$ H has a lagging phase angle of 30° at a frequency of 60 Hz. Find the frequency at which this circuit will be resonant.   *Ans.* $f_r = 36.1$ Hz

21.32   A series RLC circuit with $R = 10\ \Omega$ and $C = 10\ \mu$F has a leading phase angle of 45° at a frequency of 500 Hz. Find the frequency at which this circuit will be resonant.   *Ans.* $f_r = 604$ Hz

21.33   A series resonant circuit produces 240 mV across the coil with a 2-mV input. (*a*) What is the Q of the coil? (*b*) Find R if the coil is 5 mH and f_r is 300 kHz. (*c*) How much C is needed for this f_r? (*d*) What are the bandwidth and edge frequencies?
Ans.   (*a*) $Q = 120$;   (*b*) $R = 78.5\ \Omega$;   (*c*) $C = 56.4$ pF;   (*d*) $\Delta f = 2.5$ kHz; $f_1 = 298.75$ kHz; $f_2 = 301.25$ kHz

21.34   For $f_r = 450$ kHz and $Q = 50$, determine the bandwidth Δf and the edge frequencies f_1 and f_2.
Ans. $\Delta f = 9$ kHz; $f_1 = 445.5$ kHz; $f_2 = 454.5$ kHz

21.35   Find the lowest and highest values of C needed with 0.1-μH coil to tune through the commercial FM broadcast band of 88–108 MHz.
Ans. $C_{min} = 21.7$ pF; $C_{max} = 32.7$ pF

21.36   Find the resonant frequency of a filter made of a 150-μH coil and 40-pF capacitor in parallel.
Ans. $f_r = 2053$ kHz

21.37   A 0.001-μF capacitor and a coil are connected in parallel to form the primary of an IF transformer. Find the inductance of the coil so that the circuit is resonant to a frequency of 460 kHz.
Ans. $L = 120\ \mu$H

21.38   A 16-μH coil and a 50-pF capacitor are connected in parallel (Fig. 21-18). If the effective resistance of the coil is 25 Ω, find the resonant frequency, the Q of the coil at resonance, the bandwidth, and the edge frequencies.
Ans. $f_r = 5620$ kHz; $Q = 22.6$; $\Delta f = 249$ kHz; $f_1 = 5495.5$ kHz; $f_2 = 5744.5$ kHz

21.39   An inductor is connected in parallel with a 200-pF capacitor so that the circuit is resonant to 113 kHz. A circuit magnification meter indicates that the Q of the inductor is 800. Find (*a*) the value of the inductance, (*b*) the effective resistance of the inductor, and (*c*) the impedance of the circuit at resonance.   *Ans.*   (*a*) $L = 9.91$ mH;   (*b*) $R = 8.79\ \Omega$;   (*c*) $Z_T = 5.64$ MΩ

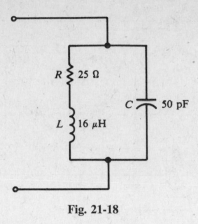

Fig. 21-18

21.40 An inductor with a measured Q of 100 resonates with a capacitor at 7500 kHz with an impedance of 65.9 kΩ. Find the value of the inductance and the capacitance of the test capacitor.
    Ans. $L = 14\ \mu$H; $C = 32.2$ pF

21.41 An inductor of 0.1 mH with a Q of 90 is connected in parallel with a 253-pF capacitor. Find (*a*) the resonant frequency and bandwidth, (*b*) the impedance of the circuit at resonance, and (*c*) the effective resistance of the inductor.
    Ans.   (*a*) $f_r = 1000$ kHz; $\Delta f = 11.1$ kHz;   (*b*) $Z_T = 56.5$ kΩ;   (*c*) $R = 7\ \Omega$

21.42 Find the resonant frequency of the parallel circuit (Fig. 21-19). (Note that this is a very low Q circuit.)
    Ans. $f_r = 159$ Hz [use Eq. (*21-6*)]

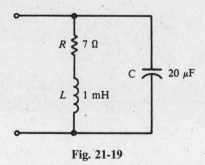

Fig. 21-19

21.43 Fill in the indicated quantities for the low-Q circuit (Fig. 21-19) for the various given frequencies. Round off Z_T to three significant figures. (See Solved Problem 21.12.) What is the Q of the circuit at resonance?

	f, Hz	Z_T, Ω	Nature
(*a*)	127	?	?
(*b*)	159	?	?
(*c*)	191	?	?

Ans.

	f, Hz	Z_T, Ω	Nature
(a)	127	7.09	20% below f_r, essentially resistive
(b)	159	7.14	At f_r, purely resistive
(c)	191	7.21	20% above f_r, essentially resistive

$Q = 7.14$

21.44 Find the resonant frequency of the parallel LC circuit (Fig. 21-20). (Note that this is a high-Q circuit.)
Ans. $f_r = 500$ kHz

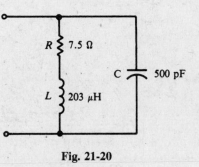

Fig. 21-20

21.45 Fill in the indicated quantities for the high-Q circuit (Fig. 21-20) for the given frequencies. Round off Z_T to three significant figures. (See Solved Problems 21.12–21.14.)

	f, kHz	Z_T, Ω	Nature	Q
(a)	400	?	?	?
(b)	500	?	?	?
(c)	600	?	?	?

Ans.

	f, kHz	Z_T, Ω	Nature	Q
(a)	400	1420	20% below f_r, highly inductive	38.0
(b)	500	54.1 k	At f_r, high resistance	84.9
(c)	600	1730	20% above f_r, highly capacitive	31.3

21.46 In a parallel resonant circuit with $X_L = 500\ \Omega$ and $Q = 50$, calculate Z_T.   *Ans.* $Z_T = 25$ kΩ

Chapter 22

Waveforms and Time Constants

RL SERIES CIRCUIT WAVEFORMS

When a switch S is closed in a circuit having only resistance R (Fig. 22-1a), the readings on the ammeter A and the voltmeter V both rise quickly to a value and remain steady at that value (Fig. 22-1b and c). If the resistance R and the current I are known, the voltage V can be found by Ohm's law,

$$V = IR \qquad (22\text{-}1)$$

When switch S is opened, the current and voltage drop quickly to zero.

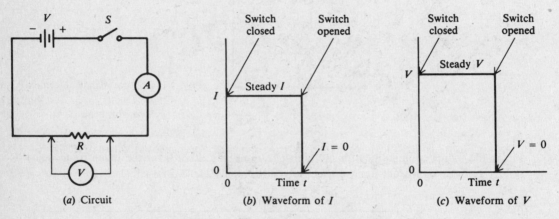

Fig. 22-1 Response of circuit with R only

When a coil of wire is added in series with a resistor, we have an *RL* series circuit (Fig. 22-2a), where V is the dc voltage, i is the instantaneous current, v_R is the instantaneous voltage across the resistor, and v_L is the instantaneous voltage across the coil. When the switch S is closed, the current begins to rise rapidly from zero until it reaches a steady value I determined by the resistance R of the circuit and the dc voltage V of the source (Fig. 22-2b). The shape that the current curve takes is exponential. When the switch is opened, the voltage source is removed and the current decays to zero at an exponential rate (Fig. 22-2b). With regard to voltage waveforms across R and L, when the switch is closed, the full dc voltage V appears across the coil

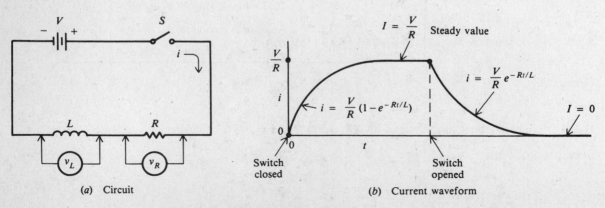

Fig. 22-2 Response of circuit with R and L in series

524

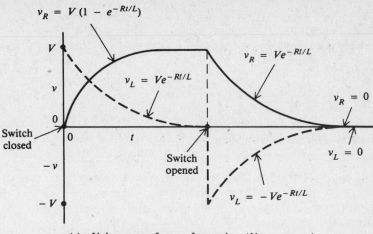

(c)   Voltage waveforms of v_R and v_L ($V = v_R + v_L$)

Fig. 22-2   (cont.)

and v_L decays to zero at an exponential rate; and v_R begins at zero and rises exponentially to a steady value of V (Fig. 22-2c). When the switch is opened, the net series voltage across R and L, $v_R + v_L$, must equal zero. Because v_R cannot change quickly, v_L drops to $-V$ and decays to 0 V (Fig. 22-2c).

The term *exponential* is used to describe the current and voltage waveforms because the formulas for calculating i, v_R, and v_L when time t is given contain an exponent of e (constant equal to 2.718).

The specific formulas to describe the RL waveforms follow:

RL Series Charging Formulas When Switch Is *Closed*

Current (Fig. 22-2b)
$$i = \frac{V}{R}(1 - e^{-Rt/L}) \qquad\qquad (22\text{-}2)$$

$$I = \frac{V}{R} \qquad \text{when } t \text{ is large} \qquad\qquad (22\text{-}3)$$

Voltage (Fig. 22-2c)
$$V = v_R + v_L \qquad\qquad (22\text{-}4)$$

$$v_R = V(1 - e^{-Rt/L}) \qquad\qquad (22\text{-}5)$$

$$v_L = Ve^{-Rt/L} \qquad\qquad (22\text{-}6)$$

RL Series Discharging Formulas When Switch Is *Opened*

Current (Fig. 22-2b)
$$i = \frac{V}{R}e^{-Rt/L} \qquad\qquad (22\text{-}7)$$

$$I = 0 \quad \text{when } t \text{ is large} \qquad\qquad (22\text{-}8)$$

Voltage (Fig. 22-2c)
$$0 = v_R + v_L \qquad\qquad (22\text{-}9)$$

$$v_R = Ve^{-Rt/L} \qquad\qquad (22\text{-}10)$$

$$v_L = -Ve^{-Rt/L} \qquad\qquad (22\text{-}11)$$

where i = instantaneous current, A
        V = applied dc voltage, V
        R = resistance of the circuit, Ω
        L = inductance of the circuit, H
        t = time, s
        e = natural log base, constant equal to 2.718
        I = final or steady-state value of current, A
      v_R = instantaneous voltage across the resistor, V
      v_L = instantaneous voltage across the coil, V

Example 22.1   Find the value of $e^{-Rt/L}$ for the following values of R, L, and t.

	R, Ω	L, H	t, s
(a)	15	15	0
(b)	15	15	1
(c)	30	15	1
(d)	15	30	1
(e)	15	15	3
(f)	30	20	1

The simplest way to evaluate exponentials is by use of an electronic calculator. First calculate $-Rt/L$ and then $e^{-Rt/L}$. Round off numerical answers to three significant digits.

(a) $-\dfrac{Rt}{L} = \dfrac{-15(0)}{15} = 0$

$e^{-Rt/L} = e^0 = 1$ *Ans.*

Any quantity raised to a zero exponent equals 1.

(b) $-\dfrac{Rt}{L} = \dfrac{-15(1)}{15} = -1$

$e^{-1} = 0.368$ *Ans.*

(c) $-\dfrac{Rt}{L} = \dfrac{-30(1)}{15} = -2$

$e^{-2} = 0.135$ *Ans.*

(d) $-\dfrac{Rt}{L} = \dfrac{-15(1)}{30} = -0.5$

$e^{-0.5} = 0.607$ *Ans.*

(e) $-\dfrac{Rt}{L} = \dfrac{-15(3)}{15} = -3$

$e^{-3} = 0.050$ *Ans.*

(f) $-\dfrac{Rt}{L} = \dfrac{-30(1)}{20} = -1.5$

$e^{-1.5} = 0.223$ *Ans.*

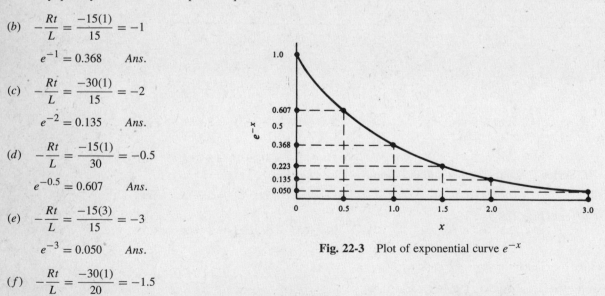

Fig. 22-3   Plot of exponential curve e^{-x}

If we let $x = (R/L)t$ and plot these values, we obtain the curve e^{-x} (Fig. 22-3).

Example 22.2   A 20-H coil and a 20-Ω resistor are connected in series across a 120-V dc source (Fig. 22-4a). (a) What is the current 1, 2, 3, 4, and 5 s after the circuit is closed? (b) What is the initial current at the instant the circuit is closed? (c) What is the voltage across the coil and across the resistor at 1 s after the circuit is closed? (d) Plot the curve of current versus time from $t = 0$ to $t = 5$ s.

(a)   **Step 1.**   Write the formula for series RL current when the switch is closed.

$$i = \frac{V}{R}(1 - e^{-Rt/L}) \tag{22-2}$$

Step 2.   Find the value of $e^{-Rt/L}$ for $t = 1, 2, 3, 4,$ and 5 s.

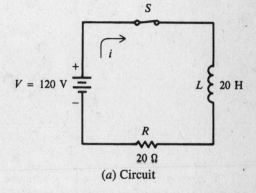

For $t = 1$ s: $-\dfrac{Rt}{L} = \dfrac{-20(1)}{20} = -1$

$e^{-1} = 0.368$ *Ans.*

For $t = 2$ s: $-\dfrac{Rt}{L} = \dfrac{-20(2)}{20} = -2$

$e^{-2} = 0.135$ *Ans.*

For $t = 3$ s: $-\dfrac{Rt}{L} = \dfrac{-20(3)}{20} = -3$

$e^{-3} = 0.050$ *Ans.*

Fig. 22-4   Series RL circuit with switch closed

For $t = 4$ s: $-\dfrac{Rt}{L} = \dfrac{-20(4)}{20} = -4$

$e^{-4} = 0.018$ *Ans.*

For $t = 5$ s: $-\dfrac{Rt}{L} = \dfrac{-20(5)}{20} = -5$

$e^{-5} = 0.007$ *Ans.*

Step 3.   Find i by substituting values for $e^{-Rt/L}$, V, and R.

$$i = \frac{V}{R}(1 - e^{-Rt/L}) \tag{22-2}$$

For $t = 1$ s: $i = \dfrac{120}{20}(1 - 0.368) = 6(0.632) = 3.79$ A   *Ans.*

For $t = 2$ s: $i = \dfrac{120}{20}(1 - 0.135) = 6(0.865) = 5.19$ A   *Ans.*

For $t = 3$ s: $i = \dfrac{120}{20}(1 - 0.050) = 6(0.950) = 5.70$ A   *Ans.*

For $t = 4$ s: $i = \dfrac{120}{20}(1 - 0.018) = 6(0.982) = 5.89$ A   *Ans.*

For $t = 5$ s: $i = \dfrac{120}{20}(1 - 0.007) = 6(0.993) = 5.96$ A   *Ans.*

(b)   The initial current is that current when $t = 0$ s. So

$$-\frac{Rt}{L} = \frac{-20(0)}{20} = 0 \quad \text{and} \quad e^0 = 1$$

Then $i = \dfrac{120}{20}(1 - 1) = 6(0) = 0$ A   *Ans.*

(c) Write Eqs. (22-5) and (22-6) and substitute proper values when $t = 1$ s.

$$V_R = V\left(1 - e^{-Rt/L}\right) = 120(1 - 0.368) = 120(0.632) = 75.8 \quad Ans.$$

$$v_L = Ve^{-Rt/L} = 120(0.368) = 44.2 \text{ V} \quad Ans.$$

As a check, substitute values for v_R and v_L into Eq. (22-4).

$$V = v_R + v_L$$
$$120 = 75.8 + 44.2$$
$$120 \text{ V} = 120 \text{ V}$$

(d) Plot of current curve is shown in Fig. 22-4b. Note that at $t = 5$ s, the current has almost reached its steady-state value of 6 A.

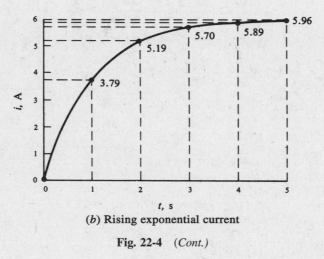

(b) Rising exponential current

Fig. 22-4   (Cont.)

Example 22.3 In Example 22.2, the switch is opened after having been closed for 7 s. What is the value of current 2 s after the switch is opened?

For practical purposes, at time $t = 7$ s, the value of current has reached 6 A, its steady-state value. See Fig. 22-4b. So write Eq. (22-7) and substitute values for $t = 2$ s.

$$i = \frac{V}{R}e^{-Rt/L} \quad \text{or} \quad i = Ie^{-Rt/L}$$

$$-\frac{Rt}{L} = -\frac{20(2)}{20} = -2$$

$$e^{-2} = 0.135$$

$$i = \frac{120}{20}(0.135) = 6(0.135) = 0.81 \text{ A} \quad Ans.$$

RL TIME CONSTANTS

Any value of t can be inserted in the formula for current [Eq. (22-2)], and a corresponding value of the current i can be found. Theoretically, the current never reaches the steady-state value of V/R on rise, zero, or decay. However, from a practical standpoint, the ammeter does show steady current values, a finite number of seconds after the circuit is closed or opened.

A convenient unit of time measurement is the time constant T, which is equal to the ratio L/R in an RL circuit.

$$t = T = \frac{L}{R} \qquad (22\text{-}12)$$

We say that

$$\text{One time constant} = 1T = \frac{L}{R}$$

$$\text{Two time constants} = 2T = \frac{2L}{R}$$

$$\text{Three time constants} = 3T = \frac{3L}{R}$$

and so on.

By substituting one time constant, $t = 1T = L/R$, in Eq. $(22\text{-}2)$,

$$i = \frac{V}{R}(1 - e^{-(R/L)t}) \qquad (22\text{-}2)$$

we obtain

$$i = \frac{V}{R}(1 - e^{-(R/L)(L/R)}) \quad \text{or} \quad i = \frac{V}{R}(1 - e^{-1})$$

Since

$$e^{-1} = 0.368$$

then

$$i = \frac{V}{R}(1 - 0.368) = (0.632)\frac{V}{R}$$

Therefore, one time constant is the time in seconds required for the current in an RL circuit to rise to 0.632 times its steady-state value V/R, or at $t = L/R$ the current is 63.2 percent of its final value.

Table 22-1 shows the value of the quantities $e^{-Rt/L}$ and $1 - e^{-Rt/L}$ for the number of time constants ranging from 0 to 6.

Table 22-1 Time Constant Factors

Number of Time Constants	$e^{-Rt/L}$	$1 - e^{-Rt/L}$
0	1.000	0.000
0.5	0.607	0.393
1.0	0.368	0.632
1.5	0.223	0.777
2.0	0.135	0.865
2.5	0.082	0.918
3.0	0.050	0.950
3.5	0.030	0.970
4.0	0.018	0.982
4.5	0.011	0.989
5.0	0.007	0.993
5.5	0.004	0.996
6.0	0.002	0.998

The table shows that the current reaches more than 99 percent (0.993) of its steady-state value in five time constants. Since the rise during the sixth time constant is only 0.005 (0.998 − 0.993), the current is considered, from a practical standpoint, to have reached its steady-state value at the end of five time constants.

Example 22.4   A circuit has total resistance of 10 Ω and an inductance of 50 H. In how many seconds will the current reach its steady-state value?

Write the formula for a time constant [Eq. (22-12)] and substitute $L = 50$ H and $R = 10$ Ω.

$$T = L/R = 50/10 = 5 \text{ s, one time constant}$$

The total time to reach the steady-state value is equal to five time constants.

$$\text{Total time} = 5T = 5(5) = 25 \text{ s} \qquad Ans.$$

Example 22.5   Draw a plot of a rising current versus time constants in an RL series circuit.

Refer to Table 22-1 for values of $1 - e^{-Rt/L}$ or $1 - e^{-t/T}$ and the corresponding time constants. The plot is shown in Fig. 22-5. Compare the similarity of this plot with that in Fig. 22-4b of Example 22.2d, where $V/R = 6$ A and one time constant equals $L/R = 20/20 = 1$ s. In that example, therefore, the horizontal scale of time is the same as that of time constants for this example. To illustrate, at $t = 4$ s, $i = 5.89$ A (Fig. 22-4b); in this example, $t = 4 = T$, so $i = 0.982(V/R) = 0.982(6) = 5.89$ A.

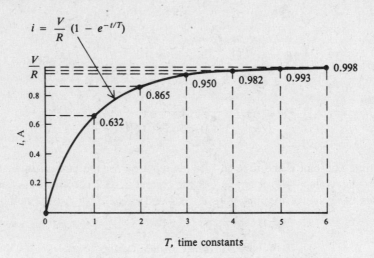

Fig. 22-5   Plot of rising current in RL series circuit

RC SERIES CIRCUIT WAVEFORMS

When a capacitor is added in series with a resistor, we have the RC series circuit (Fig. 22-6a), where V is the dc voltage, i is the instantaneous current, v_R is the instantaneous voltage across the resistor, and v_C is the instantaneous voltage across the capacitor.

When switch S_1 is closed and switch S_2 is opened, the current rises instantly to its maximum value of V/R and then decays exponentially until it reaches zero, its steady-state or final value (Fig. 22-6b). Also, the full voltage V of the battery is across the resistor, $V = V_R$. As the current decreases, the voltage across the resistance v_R decreases and the voltage across the capacitance v_C rises so that the sum of the voltages v_R and v_C is equal to V; v_R reduces to zero and the capacitor charges to $v_C = V$ (Fig. 22-6c).

After steady-state conditions are reached ($i = 0$, $v_R = 0$, $v_C = V$), switch S_1 is opened and switch S_2 is closed. Current now flows out of the capacitor in the direction opposite to the direction it flowed while the capacitor was being charged. The current drops instantly to $-V/R$ and gradually decays to zero (Fig. 22-6b).

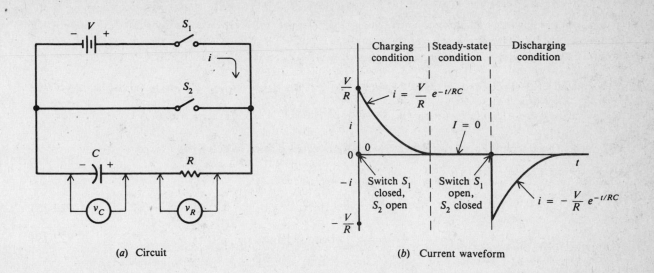

(a) Circuit

(b) Current waveform

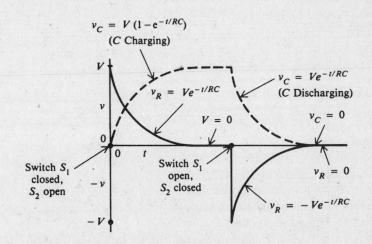

(c) Voltage waveforms of v_R and v_C ($V = v_R + v_C$)

Fig. 22-6 Response of circuit with R and C in series

Also, the voltage V across the capacitor cannot change instantly, so voltage V appears across the resistance with opposite polarity, that of $-V$ (Fig. 22-6c). Then the two voltages v_R and v_C decay exponentially to zero.

The specific formulas to describe the RC waveforms follow:

RC Series Charging Formulas When Switch S_1 Is *Closed* and Switch S_2 Is *Opened*

Current (Fig. 22-6b)

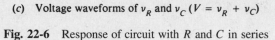

$$i = \frac{V}{R}e^{-t/RC} \qquad (22\text{-}13)$$

$$I = 0 \quad \text{when } t \text{ is large} \qquad (22\text{-}14)$$

Voltage (Fig. 22-6c)

$$V = v_R + v_C \tag{22-15}$$

$$v_R = Ve^{-t/RC} \tag{22-16}$$

$$v_C = V\left(1 - e^{-t/RC}\right) \tag{22-17}$$

RC Series Discharging Formulas When Switch S_1 Is *Opened* and Switch S_2 Is *Closed*

Current (Fig. 22-6b)

$$i = -\frac{V}{R}e^{-t/RC} \tag{22-18}$$

$$I = 0 \quad \text{when } t \text{ is large} \tag{22-19}$$

Voltage (Fig. 22-6c)

$$0 = v_R + v_C \tag{22-20}$$

$$v_R = -Ve^{-t/RC} \tag{22-21}$$

$$v_C = Ve^{-t/RC} \tag{22-22}$$

where $i =$ instantaneous current, A
        $V =$ applied dc voltage, V
        $R =$ resistance of the circuit, Ω
        $C =$ capacitance of the circuit, F
        $t =$ time, s
        $e =$ natural log base, constant equal to 2.718
        $I =$ final or steady-state value of current, A
        $v_R =$ instantaneous voltage across the resistor, V
        $v_C =$ instantaneous voltage across the capacitor, V

Example 22.6   Find the value of $e^{-t/RC}$ for the following values of R, C, and t.

	R	C	t
(a)	2 Ω	0.5 F	1 s
(b)	1 MΩ	10 μF	5 s
(c)	1 kΩ	500 μF	2 s
(d)	200 Ω	1000 μF	0.3 s

The simplest way to find values of exponential expressions is to use an electronic calculator.

(a) $-\dfrac{t}{RC} = \dfrac{1}{2\,(0.5)} = -1$

$e^{-t/RC} = e^{-1} = 0.368$ *Ans.*

(b) $-\dfrac{t}{RC} = -\dfrac{5}{\left(1 \times 10^6\right)\left(10 \times 10^{-6}\right)} = -\dfrac{5}{10} = -0.5$

Remember to convert R into ohms and C into farads before you calculate.

$e^{-0.5} = 0.607$ *Ans.*

(c) $-\dfrac{t}{RC} = -\dfrac{2}{\left(1 \times 10^3\right)\left(500 \times 10^{-6}\right)} = -4$

$e^{-4} = 0.018$ Ans.

(d) $-\dfrac{t}{RC} = -\dfrac{0.3}{\left(2 \times 10^2\right)\left(10^3 \times 10^{-6}\right)} = -\dfrac{0.3}{0.2} = -1.5$

$e^{-1.5} = 0.223$ Ans.

Example 22.7   For an RC series circuit (Fig. 22-7), find (a) the current 1 s after the circuit is closed and (b) the voltage across the resistance and the capacitor at that instant of time.

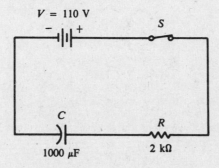

Fig. 22-7   Series RC circuit with switch closed

(a)   **Step 1.**   Write the formula for series RC current when the switch is closed.

$$i = \dfrac{V}{R} e^{-t/RC} \qquad (22\text{-}13)$$

Step 2.   Find the value of $e^{-t/RC}$ for $t = 1$ s.

$$-\dfrac{t}{RC} = -\dfrac{1}{\left(2 \times 10^3\right)\left(10^3 \times 10^{-6}\right)} = -\dfrac{1}{2} = -0.5 \qquad e^{-0.5} = 0.607$$

Step 3.   Find i by substituting values in Eq. (22-13).

$$i = \dfrac{110}{2000}(0.607) = 0.0334\ \text{A} = 33.4\ \text{mA} \qquad Ans.$$

(b)   Write the formulas for voltage when the switch is closed and then substitute the proper values.

$$v_R = V e^{-t/RC} \qquad (22\text{-}16)$$

$$= 110\,(0.607) = 66.8\ \text{V} \qquad Ans.$$

$$v_C = V\left(1 - e^{-t/RC}\right) \qquad (22\text{-}17)$$

$$= 110(1 - 0.607) = 110(0.393) = 43.2\ \text{V} \qquad Ans.$$

Check by using Eq. (22-15).

$$V = v_R + v_C$$

$$110 = 66.8 + 43.2$$

$$110\ \text{V} = 110\ \text{V}$$

Example 22.8   After the circuit (Fig. 22-7) has reached steady state when the capacitor has charged to 110 V, the switch is opened. Find now (*a*) the current 1 s after the circuit is opened and (*b*) the voltages across the resistance and the capacitor at that time.

(*a*)   **Step 1.**   Write the formula for series *RC* current when the switch is opened.

$$i = -\frac{V}{R}e^{-t/RC} \qquad (22\text{-}18)$$

Step 2.   Find $e^{-t/RC}$.

$$e^{-t/RC} = e^{-0.5} = 0.607$$

Step 3.   Find *i* by substituting values in Eq. (*22-18*).

$$i = -\frac{110}{2000}(0.607) = -0.0334\,\text{A} = -33.4\,\text{mA} \qquad Ans.$$

(*b*)   Write the formulas for voltage when the switch is opened and then substitute proper values.

$$v_R = -Ve^{-t/RC} \qquad (22\text{-}21)$$

$$= -110(0.607) = -66.8\,\text{V} \qquad Ans.$$

$$v_C = Ve^{-t/RC} \qquad (22\text{-}22)$$

$$= 110(0.607) = 66.8\,\text{V} \qquad Ans.$$

Check:
$$0 = v_R + v_C$$

$$0 = -66.8 + 66.8$$

$$0\,\text{V} = 0\,\text{V}$$

RC TIME CONSTANTS

The time constant *T* for a capacitive circuit is

$$T = RC \qquad (22\text{-}23)$$

We say that

$$\text{One time constant} = 1T = RC$$
$$\text{Two time constants} = 2T = 2RC$$
$$\text{Three time constants} = 3T = 3RC$$

and so on.

The time constant of a capacitive circuit is usually very short because the capacitance of a circuit may be only a few microfarads or even picofarads.

Voltage and current values in capacitive circuits are assumed to reach their steady-state or final values after five time constants, just as in inductive circuits. Table 22-1 applies also to capacitive circuits if $e^{-Rt/L}$ is replaced by $e^{-t/RC}$.

CALCULATION FOR TIME t

The time required for a specific voltage decay can be calculated by transposing Eq. (22-21) or Eq. (22-22).

$$v = Ve^{-t/RC}$$

where V is the higher voltage at the start and v (v_R or v_C) is the lower voltage at the finish.

$$e^{-t/RC} = \frac{v}{V}$$

Take the natural logarithm (ln) of each side.

$$-\frac{t}{RC}\ln e = \ln \frac{v}{V}$$

$$\ln e = 1$$

So
$$-\frac{t}{RC} = \ln \frac{v}{V} = \ln v - \ln V$$

$$\frac{t}{RC} = -\ln v + \ln V = \ln \frac{V}{v}$$

$$t = RC \ln \frac{V}{v} \qquad\qquad (22\text{-}24)$$

Example 22.9   A dc circuit of source voltage V has a 10-MΩ resistance in series with a 1-μF capacitor. What is the time constant of the circuit and how long will current flow in the circuit when the RC combination is short-circuited?
    Write the formula for time constant in a capacitive circuit, Eq. (22-23).

$$T = RC = (10 \times 10^6)(1 \times 10^{-6}) = 10\,\text{s} \qquad Ans.$$

Since the steady-state current is assumed to be zero after five time constants, the time that current will flow is

$$5T = 5(10) = 50\,\text{s} \qquad Ans.$$

Example 22.10   Find the resistance necessary in an RC series circuit if the circuit has a capacitance of 10 μF and a time constant of 1 s is desired.
    Rewrite Eq. (22-23) and solve for R.

$$R = \frac{T}{C} = \frac{1}{10 \times 10^{-6}} = 10^5 = 100\,\text{k}\Omega \qquad Ans.$$

Example 22.11   One time constant for a series RC circuit is also defined as the time in seconds required for the capacitor to charge to 63.2 percent of its final value. Show that this statement is true.
    Write the capacitance voltage equation v_C when the capacitor is being charged.

$$v_C = V\left(1 - e^{-t/RC}\right) \qquad\qquad (22\text{-}17)$$

One time constant $T = RC$. Substitute RC for t.

$$v_C = V\left(1 - e^{-RC/RC}\right) = V\left(1 - e^{-1}\right) = V\,(1 - 0.368) = 0.632\,\text{V}$$

So v_C is 63.2 percent of V in one time constant.

Example 22.12   If an RC circuit has a time constant of 1 s, how long will it take for v_R to drop from 100 to 50 V? Write the time equation for voltage decay, Eq. (22-24).

$$t = RC \ln \frac{V}{v} = 1 \left(\ln \frac{100}{50} \right) = \ln 2 = 0.693 \text{ s} \qquad \textit{Ans.}$$

Solved Problems

22.1   A series circuit contains a resistance of 20 Ω and an inductance of 10 H connected across a voltage source of 110 V. (*a*) What is the current 1 s after the circuit is closed? (*b*) What are v_R and v_C at this time?

Some time after the current reaches a steady value, the switch is opened. When the series circuit is open, the RL circuit opposes the decay of current toward the steady-state value of zero. (*c*) What is the current 2 s after the circuit is opened? (*d*) What are v_R and v_C at this time?

(*a*)   **Step 1.**   Write the formula for charging or rising current when the switch is closed.

$$i = \frac{V}{R}\left(1 - e^{-Rt/L}\right) \tag{22-2}$$

Step 2.   Find the value of $e^{-Rt/L}$ for $t = 1$ s.

$$-\frac{Rt}{L} = -\frac{20(1)}{10} = -2 \qquad e^{-2} = 0.135$$

Step 3.   Substitute values for $e^{-Rt/L}$, V, and R in Eq. (22-2).

$$i = \frac{110}{20}(1 - 0.135) = 5.5(0.865) = 4.76 \text{ A} \qquad \textit{Ans.}$$

(*b*)   Write the formulas for v_R and v_L after the switch is closed and solve them when $t = 1$ s.

$$v_R = V(1 - e^{-Rt/L}) \tag{22-5}$$

$$= 110(0.865) = 95.2 \text{ V} \qquad \textit{Ans.}$$

Also $v_R = iR = 4.76(20) = 95.2 \text{ V}$.

$$v_L = V e^{-Rt/L}$$

$$= 110(0.135) = 14.8 \text{ V} \qquad \textit{Ans.}$$

Check: $V = v_R + v_L \tag{22-4}$

$$110 = 95.2 + 14.8$$

$$110 \text{ V} = 110 \text{ V}$$

(*c*)   The steady-state value of current is

$$I = \frac{V}{R} = \frac{110}{20} = 5.5 \text{ A}$$

Write the formula for discharging or decaying current when the switch is opened and substitute values with $t = 2$ s.

$$i = \frac{V}{R}e^{-Rt/L} \tag{22-7}$$

$$-\frac{Rt}{L} = -\frac{20(2)}{10} = -4$$

$$e^{-4} = 0.018$$

$$i = 5.5(0.018) = 0.10 \text{ A} \qquad Ans.$$

(d)   Write the formulas for v_R and v_L after the switch is opened and solve them when $t = 2$ s.

$$v_R = Ve^{-Rt/L} \tag{22-10}$$

$$= 110(0.018) = 1.98 \text{ V} \qquad Ans.$$

$$v_L = -Ve^{-Rt/L} \tag{22-11}$$

$$= -110(0.018) = -1.98 \text{ V} \qquad Ans.$$

Check:

$$0 = v_R + v_L \tag{22-9}$$

$$0 = 1.98 - 1.98$$

$$0 \text{ V} = 0 \text{ V}$$

22.2   An RL series circuit has the following values: $V = 75$ V dc, $R = 50 \ \Omega$, and $L = 15$ H. (a) Find the time constant. (b) How long after the circuit is energized will the current reach a steady value? (c) What is the steady-state value of current?

(a)   Write the formula for the time constant, Eq. (22-12), and substitute given values.

$$T = \frac{L}{R} = \frac{15}{50} = 0.3 \text{ s} \qquad Ans.$$

(b)   A steady-state value of current is reached after five time constants.

$$5T = 5(0.3) = 1.5 \text{ s} \qquad Ans.$$

(c) $I = \dfrac{V}{R} = \dfrac{75}{50} = 1.5$ A $Ans.$

22.3   In the circuit shown (Fig. 22-8), find (a) the total inductance and (b) the time constant. Assume that a current of 10 A is flowing when the switch S is opened. Find (c) the current 2 s later.

(a)   Reduce the series–parallel combination of inductances to its equivalent.

Parallel:

$$L_a = \frac{L_1 L_2}{L_1 + L_2} = \frac{5(1)}{5+1} = \frac{5}{6} = 0.83 \text{ H}$$

Parallel:

$$L_b = \frac{L_3 L_4}{L_3 + L_4} = \frac{4(4)}{4+4} = \frac{16}{8} = 2.0 \text{ H}$$

Then series: $L_T = L_a + L_b = 0.83 + 2.0 = 2.83$ H $Ans.$

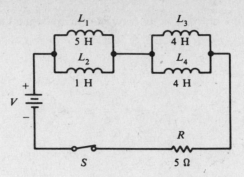

Fig. 22-8   Circuit with branch inductances in
series with R

(b)   The time constant of the circuit is the ratio of total inductance to the total resistance of the
circuit.

$$T = \frac{L_T}{R} = \frac{2.83}{5} = 0.57 \text{ s} \quad Ans.$$

(c)   Write the formula for decaying current. The V/R value is 10 A.

$$i = \frac{V}{R} e^{-Rt/L} \tag{22-7}$$

$$-\frac{Rt}{L_T} = -\frac{5(2)}{2.83} = -3.5$$

$$e^{-3.5} = 0.030$$

Then $$i = 10(0.030) = 0.30 \text{ A} \quad Ans.$$

22.4   An RL circuit (Fig. 22-9) is used to generate a high voltage to light a neon bulb, which requires 90 V
for ionization at which time it glows. When the circuit is open, the high resistance R_2 produces a low
time constant L/R so that the current drops toward zero much faster than the rise of current when the
switch is closed. The result is a high value of self-induced voltage across a coil when the RL circuit is
open. This voltage can be greater than the applied voltage. (a) Find the time constants of the circuit
when the switch is opened and when the switch is closed. (b) When the switch is closed, find the
voltage across the neon bulb. Is it sufficient to start ionization? (c) When the switch is open, what is
the voltage across the neon bulb, and is it great enough to light it?

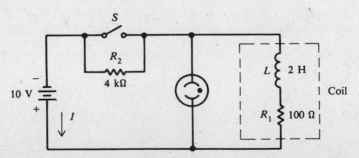

Fig. 22-9 RL circuit produces a high voltage when switch is opened

(a) The time constant is the ratio of total inductance to the total resistance.

Switch open:

$$T = \frac{L}{R_1 + R_2} = \frac{2}{4000 + 100} = \frac{2}{4100} = 4.9 \times 10^{-4} = 0.49 \text{ ms} \qquad \textit{Ans.}$$

Switch closed (R_2 short-circuited):

$$T = \frac{L}{R_1} = \frac{2}{100} = 2 \times 10^{-2} = 20 \text{ ms} \qquad \textit{Ans.}$$

When the switch is open, there is a much shorter time constant for current decay. The current decays practically to zero after five time constants, or 2.5 ms.

(b) When the switch is closed, there is 10 V across the neon bulb, which is far less than the 90 V needed to ionize it to glow.

(c) When the switch is closed, R_2 is short-circuited so that the 100-Ω R_1 is the only resistance. The steady-state current $I = V/R_1 = 10/100 = 0.10$ A. When the switch is opened, the rapid drop in current results in a magnetic field collapsing at a fast rate, inducing a high voltage across L. The energy stored in the magnetic field maintains I at 0.10 A for an instant before the current decays. With 0.10 A in the 4-kΩ R_2, its potential difference is 0.10(4000) = 400 V. This 400-V pulse is sufficient to light the bulb.

22.5 For an *RC* series circuit (Fig. 22-10), find (*a*) the time constant of the circuit, (*b*) v_C and v_R one time constant after the switch is closed and at 5 s, (*c*) v_C and v_R one time constant after discharge starts, assuming the capacitor is fully charged to 10 V.

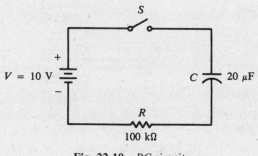

Fig. 22-10   *RC* circuit

(a) Write Eq. (*22-23*) for the time constant in an *RC* series circuit and substitute values for *R* and *C*.

$$T = RC = (100 \times 10^3)(20 \times 10^{-6}) = 2 \text{ s} \qquad \textit{Ans.}$$

(b) Write the formula for charging voltage, substitute values, and solve for v_C and v_R.

$$v_C = V(1 - e^{-t/RC}) \qquad (22\text{-}17)$$

When $t = T = 2$ s:

$$v_C = 10(1 - e^{-2/2}) = 10(1 - e^{-1}) = 10(1 - 0.368) = 10(0.632) = 6.32 \text{ V} \qquad \textit{Ans.}$$

$$V = v_R + v_C \qquad (22\text{-}15)$$

$$v_R = V - v_C = 10 - 6.32 = 3.68 \text{ V} \qquad \textit{Ans.}$$

When $t = 5$ s:

$$v_C = 10(1 - e^{-5/2}) = 10(1 - e^{-2.5}) = 10(1 - 0.082) = 10(0.918) = 9.18 \text{ V} \qquad Ans.$$

$$v_R = V - v_C = 10 - 9.18 = 0.82 \text{ V} \qquad Ans.$$

(c) Write the formula for discharging voltage, substitute values, and solve for v_C and v_R.

$$v_C = Ve^{-t/RC} \tag{22-22}$$

When $t = T = 2$ s:

$$v_C = Ve^{-2/2} = Ve^{-1} = 10(0.368) = 3.68 \text{ V} \qquad Ans.$$

$$0 = v_R + v_C \tag{22-20}$$

$$v_R = -v_C = -3.68 \text{ V} \qquad Ans.$$

22.6 For the circuit shown (Fig. 22-11), a 40-μF capacitor is added across the 20-μF capacitor in the circuit of Fig. 22-10. Find the time constant of this circuit and the voltage across this circuit 3 s after the start of charge.

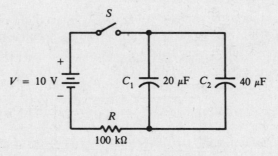

Fig. 22-11   Circuit with C branch in series with R

Find the total capacitance for two capacitors in parallel.

Parallel:
$$C_T = C_1 + C_2 = 20 + 40 = 60 \ \mu\text{F}$$

Then
$$T = RC_T = (100 \times 10^3)(60 \times 10^{-6}) = 6 \text{ s} \qquad Ans.$$

Write the formula for charging voltage and substitute values to find v_C.

$$v_C = V(1 - e^{-t/RC}) \tag{22-17}$$

$$= 10(1 - e^{-3/6}) = 10(1 - e^{-0.5}) = 10(1 - 0.607) = 10(0.393) = 3.93 \text{ V} \qquad Ans.$$

22.7 A simple switching circuit for producing a sawtooth wave is shown (Fig. 22-12a). The switch S is closed and then opened very quickly so that the capacitor is not fully charged but only charged to the linear portion of its charging exponential curve (Fig. 22-12b). (The linear portion is the straightest portion at the start of the charge cycle.) The switch is opened and closed at specific intervals to produce a sawtooth voltage wave across the capacitor. Find the magnitude of the v_C curve when the switching time interval is one-fifth the time constant of the circuit. What is the time interval?

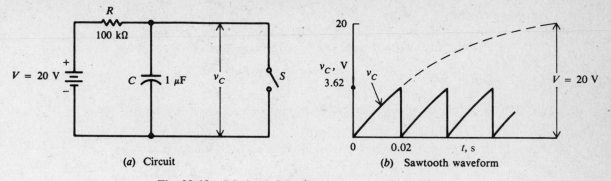

Fig. 22-12   *RC* circuit for generating a sawtooth waveform

Write the voltage formula for charging the capacitor.

$$v_C = V(1 - e^{-t/RC}) \qquad (22\text{-}17)$$

If $t = \dfrac{1}{5}RC$, then

$$e^{-t/RC} = e^{-(1/5)RC/RC} = e^{-1/5} = e^{-0.2} = 0.819$$

Then $v_C = 20(1 - 0.819) = 20(0.181) = 3.62 \text{ V} \qquad Ans.$

Time interval $t = \dfrac{1}{5}RC = \dfrac{(100 \times 10^3)(1 \times 10^{-6})}{5} = 0.02 \text{ s} \qquad Ans.$

22.8   An *RC* series circuit with a rectangular pulse input is shown (Fig. 22-13). A rectangular pulse is a special case of a constant dc source because the voltage is a constant *V* when the pulse is on and a constant zero when the pulse is off. If a single pulse is applied to the circuit, find the voltage across the resistance at the time of 0.5, 1, and 2 ms. The pulse has a maximum voltage of 10 V and lasts for 1 ms.

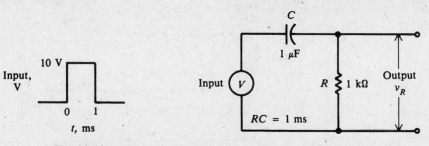

Fig. 22-13   *RC* circuit with rectangular pulse of applied voltage

Find the time constant of the circuit.

$$T = RC = (1 \times 10^3)(1 \times 10^{-6}) = 1 \times 10^{-3} = 1 \text{ ms}$$

Write the formula for v_R when the circuit is charging.

$$v_R = Ve^{-t/RC} \qquad (22\text{-}16)$$

When $t = 0.5$ ms,

$$v_R = 10e^{-0.5/1} = 10e^{-0.5} = 10(0.607) = 6.07 \text{ V} \qquad Ans.$$

When $t = 1$ ms,

$$v_R = 10e^{-1/1} = 10e^{-1} = 10(0.368) = 3.68 \text{ V} \qquad Ans.$$

At $t = 1$ ms, the pulse is turned off. At that instant the source voltage is zero, so

$$v_R + v_C = 0$$
$$v_R = -v_C$$

Since the voltage across the capacitor v_C cannot change instantly, the voltage across the resistor becomes $-v_C$. Now at an instant *before* $t = 1$ ms,

$$v_R + v_C = 10$$
$$v_C = 10 - v_R = 10 - 3.68 = 6.32 \text{ V}$$

Therefore, at $t = 1$ ms, $v_R = -v_C = -6.32$ V and then v_R decays to zero. So the formula for v_R at $t = 1$ ms is:

$$v_R = -6.32e^{-t/RC}$$

Then in the next 1 ms or when $t = 2$ ms measured from the origin,

$$v_R = -6.32e^{-1} = -6.32(0.368) = -2.33 \text{ V} \qquad Ans.$$

The plot for v_R is as shown in Fig. 22-14.

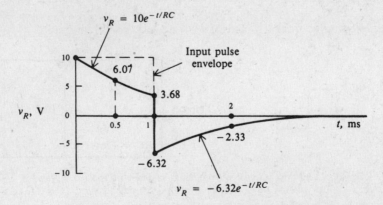

Fig. 22-14   Charge and discharge of an RC circuit

22.9   An RC series circuit with a square-wave input is shown (Fig. 22-15). The input is a periodic train of pulses with an amplitude of 10 V and a width of 1 ms, with each pulse generated every 2 ms. Plot the output voltage curve across the resistor.

$$T = RC = (1 \times 10^3)(0.1 \times 10^{-6}) = 0.1 \times 10^{-3} = 0.1 \text{ ms}$$

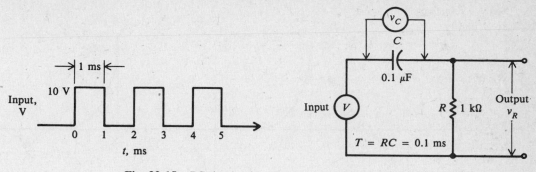

Fig. 22-15 *RC* circuit with square wave for input voltage

Find some values of v_R within the 0 to 1 ms that the pulse is on, say at $T = 0.1$ ms, $2T = 0.2$ ms, and $3T = 0.3$ ms. First, write the formula for v_R when the circuit is charging.

$$v_R = Ve^{-t/RC} \qquad (22\text{-}16)$$

When $t = T = 0.1$ ms, $v_R = 10e^{-1} = 10(0.368) = 3.68$ V

When $t = 2T = 0.2$ ms, $v_R = 10e^{-2} = 10(0.135) = 1.35$ V

When $t = 3T = 0.3$ ms, $v_R = 10e^{-3} = 10(0.05) = 0.5$ V

At $t = 5T$ or 0.5 ms, v_R will reach its steady-state value of 0 V.

When the pulse is turned off between 1 and 2 ms, v_R will instantly drop to -10 V because the capacitor charged to 10 V requires a finite time to discharge. Then v_R will gradually rise to 0 V.

The formula for v_R when the circuit is discharging is

$$v_R = -Ve^{-t/RC} \qquad (22\text{-}21)$$

where t represents time from 1 to 2 ms. So when $t = 1.1$ ms from origin,

$$v_R = -3.68 \text{ V}$$

When $t = 1.2$ ms, $v_R = -1.35$ V

When $t = 1.3$ ms, $v_R = -0.5$ V

When $t = 1.5$ ms, $v_R = 0$ V

The symmetry of the output curve is shown in the plot (Fig. 22-16). Because the circuit has changed the waveform of input pulses to peaks, it is called an *RC peaker*. It is called also a *differentiating circuit* because v_R can change instantaneously. We obtain a peak output when the circuit time constant is small compared with the half-period of the input waveform. In this case $T = 0.1$ ms compared with the half-period of 1 ms, so the ratio of the time constant to the half-period is 1 to 10.

In summary, to explain a short *RC* time circuit, a 10-V input is applied for 1 ms (10 time constants of the *RC* circuit), allowing *C* to become completely charged and v_R to be 0 V (Figs. 22-15 and 22-16). After *C* is charged, v_C remains at 10 V with v_R at 0 V ($V = v_C + v_R$). After 1 ms, the total voltage *V* drops to zero, *C* discharges completely in five time constants, and v_C and v_R remain at 0 V while there is no applied voltage. On the next cycle, *C* charges and discharges

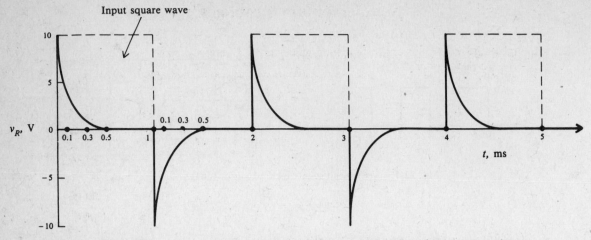

Fig. 22-16 Charge and discharge of an *RC* circuit with a short time constant

completely again. The interval between pulses when the input voltage is 0 V acts as a "short" to the *RC* circuit.

22.10 For the circuit shown (Fig. 22-17), (*a*) find the time constant. (*b*) If at time $t = 0$, the voltage across the capacitor C_1 is $v_1 = 10$ V and the voltage across the capacitor C_2 is $v_2 = 20$ V with the polarity shown, find the current when $t = 0.26$ s.

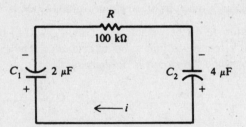

Fig. 22-17 *RC* circuit during discharge

(*a*) First, find the total capacitance for two capacitors in series with a single discharge path.

$$C_T = \frac{C_1 C_2}{C_1 + C_2} = \frac{2(4)}{2 + 4} = 1.33 \; \mu\text{F}$$

The time constant is

$$T = RC_T = \left(100 \times 10^3\right)\left(1.33 \times 10^{-6}\right) = 0.133 \; \text{s} \quad Ans.$$

(*b*) The net voltage V is $v_2 - v_1 = 20 - 10 = 10$ V so C_1 is being charged in the direction of i as shown.

$$i = \frac{V}{R}e^{-t/RC} \quad \text{(Here } V \text{ is the net voltage at } t = 0\text{)}$$

$$= \frac{10}{100 \times 10^3}e^{-0.26/0.133} = 10^{-4}e^{-1.95} = (0.1 \times 10^{-3})(0.142) = 0.0142 \; \text{mA} \quad Ans.$$

Supplementary Problems

22.11   Find the value of $e^{-Rt/L}$ for the following values of R, L, and t.

	R, Ω	L, H	t, s
(a)	10	10	1
(b)	5	10	1
(c)	5	10	2
(d)	10	5	1
(e)	10	5	2

Ans. (a) 0.368; (b) 0.607; (c) 0.368; (d) 0.135; (e) 0.018

22.12   Calculate the time constants of the following inductive circuits:

	L	R
(a)	20 H	400 Ω
(b)	20 μH	500 kΩ
(c)	50 mH	50 Ω
(d)	40 μH	5 Ω
(e)	20 mH	100 Ω

Ans. (a) 50 ms; (b) 40 μs; (c) 1 ms; (d) 8 μs; (e) 0.2 ms

22.13   In the circuit shown (Fig. 22-18) find (a) the current 1 s after the switch is closed, (b) the time constant, and (c) the steady-state or final value of the current.
Ans. (a) $i = 1.11$ A; (b) $T = 4$ s; (c) $I = 5$ A

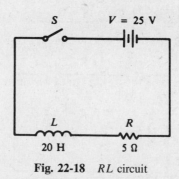

Fig. 22-18   *RL* circuit

22.14   The circuit (Fig. 22-18) is closed for 20 s. Then the switch is opened so that current decays through L and R. Calculate i, v_R, and v_L after 2 s.
Ans. $i = 3.03$ A; $v_R = 15.2$ V; $v_L = -15.2$ V

22.15   A current of 20 A is flowing in an *RL* series circuit the instant before the switch is opened. If $R = 10\ \Omega$ and $L = 10$ H, find the current after 3 s has elapsed. Also find v_R and v_L at that time.
Ans. $i = 0.996$ A; $v_R = 9.96$ V; $v_L = -9.96$ V

22.16 A 5-Ω resistor and a 10-H inductor are connected in series across a 60-V dc line. Find the time constant, steady-state current, and time to reach the steady-state current value.
Ans. $T = 2$ s; $I = 12$ A; $5T = 10$ s

22.17 A series circuit has the following values: $V = 80$ V, $R = 20\ \Omega$, and $L = 10$ H. (*a*) What is the time constant? (*b*) What is the current 1 s after the circuit is closed? (*c*) What are v_R and v_L at that time? (*d*) How long after the circuit is closed will the current reach a constant value?
Ans.   (*a*) $T = 0.5$ s;   (*b*) $i = 3.46$ A;   (*c*) $v_R = 69.2$ V; $v_L = 10.8$ V;   (*d*) $t = 2.5$ s

22.18 A circuit contains three inductances in series with a resistance across a 120-V source. If each inductor is 8 H and the resistor is 12 Ω, find (*a*) the time constant. Find the value of the current (*b*) 1 s, (*c*) 2 s, and (*d*) 3 s after the circuit is closed.
Ans.   (*a*) $T = 2$ s;   (*b*) $i = 3.94$ A;   (*c*) $i = 6.32$ A;   (*d*) $i = 7.77$ A

22.19 A circuit (Fig. 22.19) contains two inductances in parallel, a series inductance, and a series resistance across a 110-V source. Find the time constant and the current 0.45 s after the circuit is closed.
Ans. $T = 0.895$ s ($L_T = 13.43$ H); $i = 2.90$ A

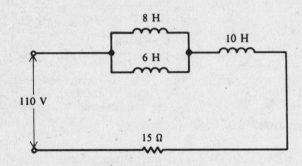

Fig. 22-19 Combination *RL* circuit

22.20 An *RL* series circuit is shown (Fig. 22-20*a*).

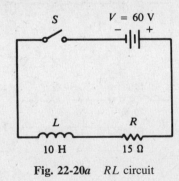

Fig. 22-20a *RL* circuit

(*a*) Plot the rise of current after the switch is closed when time is equal to 0.5, 1, 2, and 3 s.

(*b*) After the current reaches its final value, the switch is opened. Plot the decay of current when time is equal to 0.5, 1, 2, and 3 s after the switch is closed.

(*c*) Plot the voltages across the resistance and inductance when the switch is closed and opened for the same time values at 0.5, 1, 2, and 3 s.   *Ans.* See Fig. 22-20*b*, parts 1, 2, 3, and 4.

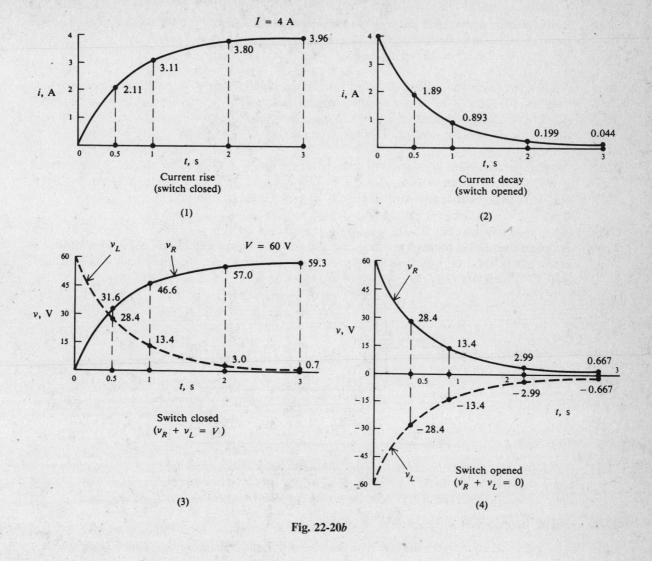

Fig. 22-20b

22.21 Find the value of $e^{-t/RC}$ for the following values of R, C, and t.

	R	C	t
(a)	1 MΩ	1 μF	1 s
(b)	1 MΩ	0.5 μF	1 s
(c)	100 kΩ	1 μF	0.1 s
(d)	100 kΩ	10 μF	0.5 s
(e)	10 kΩ	2 μF	10 ms

Ans. (a) 0.368; (b) 0.135; (c) 0.368; (d) 0.607; (e) 0.607

22.22 Calculate the time constants of the following capacitive circuits:

	R	C
(a)	1 MΩ	0.001 μF
(b)	1 MΩ	1 μF
(c)	250 kΩ	0.05 μF
(d)	110 kΩ	100 pF
(e)	5 kΩ	300 pF

Ans.   (a) 1 ms;   (b) 1 ms;   (c) 12.5 ms;   (d) 1 μs;   (e) 1.5 μs

22.23 In the circuit shown (Fig. 22-21), what are (a) the current 1 s after the circuit is closed, and (b) v_R and v_C at that instant of time?   *Ans.*   (a) $i = 36$ mA;   (b) $v_R = 72.8$ V;   (c) $v_C = 47.2$ V

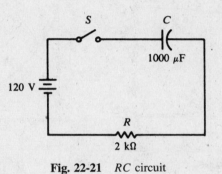

Fig. 22-21   *RC* circuit

22.24 After the circuit (Fig. 22-21) has reached its steady-state condition where the capacitor has been charged to 120 V, the switch is opened. Find (a) the time required for the circuit to have reached steady state, (b) the current 1.5 s after the circuit is opened, and (c) v_R and v_C at that time. *Ans.*   (a) $t = 10$ s;   (b) $i = -28$ mA;   (c) $v_R = -56.7$ V; $v_C = 56.7$ V

22.25 In a series *RC* circuit with the capacitor *charging*, fill in the missing values. Time t is measured at the instant the switch is closed.

	V, V	R	$C, \mu F$	T	t	I	i	v_R	v_C
(a)	115	100 kΩ	1	?	0.1 s	?	?	?	?
(b)	220	1 kΩ	?	1 ms	2 ms	?	?	?	?
(c)	110	?	10	10 s	5 s	?	?	?	?
(d)	115	10 kΩ	100	?	6 s	?	?	?	?

Ans.

	V, V	R	$C, \mu F$	T	t	I	i	v_R, V	v_C, V
(a)	....	....	....	0.1 s	....	1.15 mA	0.42 mA	42.3	72.7
(b)	....	....	1	....	....	0.22 A	29.8 mA	29.8	190.2
(c)	....	1 MΩ	....	....	....	0.11 mA	0.067 mA	66.8	43.2
(d)	....	....	....	1 s	....	11.5 mA	0 A	0	115

22.26   In a series *RC* circuit with the capacitor *discharging*, fill in the same missing values as in the table accompanying Problem 22.25. Time *t* is measured at the instant the switch is opened. The circuit constants of *V*, *R*, and *C* are the same as those in Problem 22.25.

Ans.

	V, V	R	C, μF	T	t	I	i	v_R, V	v_C, V
(a)	....	....	....	0.1 s	....	−1.15 mA	− 0.42 mA	−42.3	42.3
(b)	....	....	1	....	....	−0.22 A	−29.8 mA	−29.8	29.8
(c)	....	1 MΩ	....	....	....	−0.11 mA	− 0.067 mA	−66.8	66.8
(d)	....	....	....	1 s	....	−11.5 mA	0 A	0	0

22.27   Find the time constant of the circuits shown (Fig. 22-22).

   Ans.   (a) $T = 1\,\text{s}$

          (b) $T = 8.33\,\mu\text{s}\ (R_T = 8.33\,\Omega)$

          (c) $T = 0.2\,\text{ms}\ (C_T = 0.2\,\mu\text{F})$

          (d) $T = 90\,\mu\text{s}\ (C_T = 0.09\,\mu\text{F})$

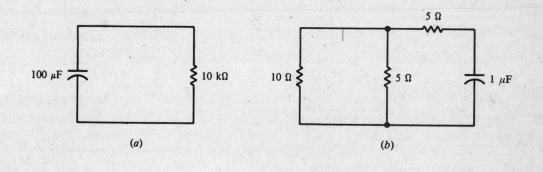

(a)   (b)

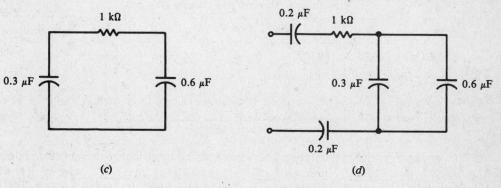

(c)   (d)

Fig. 22-22   *RC* circuit configurations

22.28   The *RC* circuit (Fig. 22-23*a*) has its switch closed at time $t = 0$. Plot curves for *i*, v_R, and v_C for $t = 0, 2, 6, 10,$ and 14 ms during the charting cycle.   *Ans.*   See Fig. 22-23*b*.

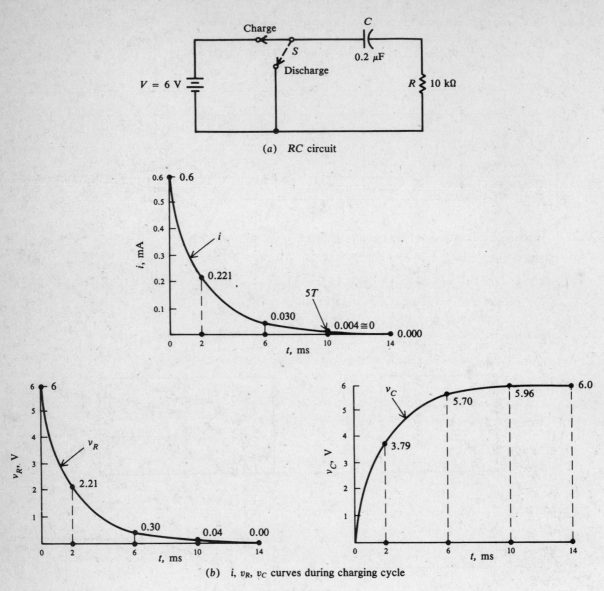

(a)   RC circuit

(b)   i, v_R, v_C curves during charging cycle

Fig. 22-23a, b

22.29   After the circuit (Fig. 22-23a) has reached steady-state values, the switch is opened. Plot curves for i, v_R, and v_C for t = 0, 2, 6, 10, and 14 ms during this discharging cycle.   *Ans.*   See Fig. 22-23c.

22.30   The dc source of 6 V in the circuit (Fig. 22-23a) is replaced by a single pulse of 6 V amplitude lasting 2 ms (Fig. 22-24a). Plot the voltage across the 10-kΩ resistance for t = 0, 2, 4, 6, 10, and 14 ms after the switch is closed. (Round off answers to two significant figures.)   *Ans.*   See Fig. 22-24b.

22.31   The source voltage for the circuit (Fig. 22-24a) is now replaced by a train of repetitive pulses (Fig. 22-25a). Each pulse has a 6-V amplitude and a pulse width of 2 ms. The period of the pulse train is 4 ms. Draw the plot of v_R for t = 0, 2, 4, 6, 8, 10, 12, and 14 ms. (Round off answers to two significant figures.)   *Ans.*   See Fig. 22-25b.

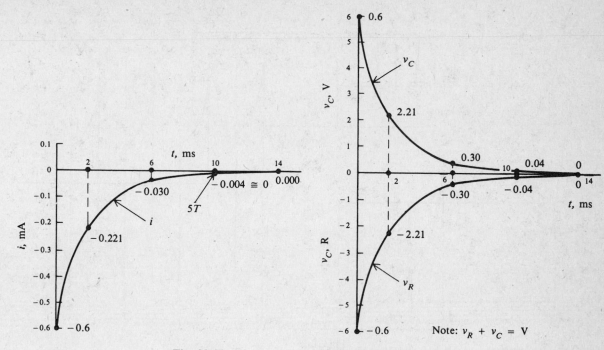

Fig. 22-23c i, v_R, v_C curves during discharging cycle

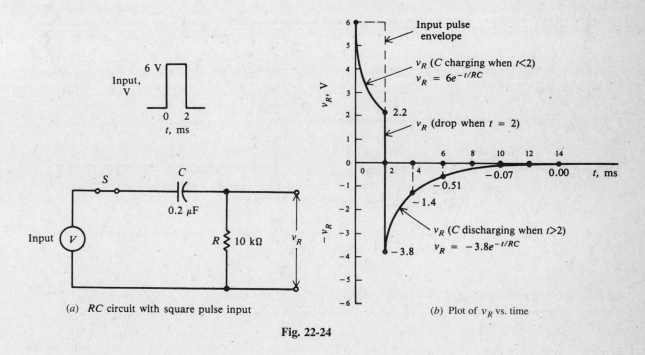

(a) *RC* circuit with square pulse input

(b) Plot of v_R vs. time

Fig. 22-24

22.32 For the circuit shown (Fig. 22-26), find (*a*) the time constant, and (*b*) the current at $t = 10$ ms when the voltage across C_1 is 15 V and the voltage across C_2 is 25 V, with the polarity as indicated at $t = 0$.
Ans. (*a*) $T = 5$ ms; (*b*) $i = 0.541$ mA

22.33 If the polarity of capacitor C_2 is reversed (Fig. 22-26), find the current at $t = 10$ ms.
Ans. $i = -0.135$ mA

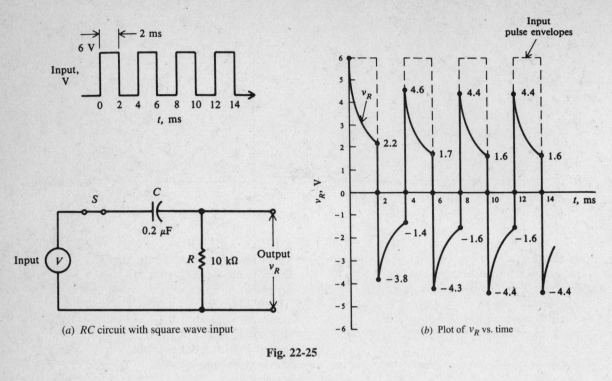

(a) RC circuit with square wave input   (b) Plot of v_R vs. time

Fig. 22-25

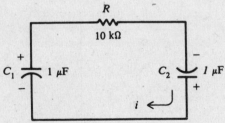

Fig. 22-26 RC discharging circuit

22.34   A 0.05-μF capacitor is charged to 264 V. It discharges through a 40-kΩ resistor. How much time is required for v_C to discharge down to 66 V?   *Ans.* $t = 2.77$ ms

22.35   A 100-V source is in series with a 2-MΩ resistor and a 2-μF capacitor. How much time is required for v_C to charge to 63.2 V?   *Ans.* $t = 4$ s $(1T)$

Index

The letter *p* following a page number refers to a problem.